Digital
Transmission
Systems

Second Edition

DAVID R. SMITH

Chief, Transmission Systems Development Division
Defense Information Systems Agency

Associate Professorial Lecturer
Department of Electrical Engineering and Computer Science
George Washington University

Visiting Associate Professor
Department of Electrical and Computer Engineering
George Mason University

KLUWER ACADEMIC PUBLISHERS
BOSTON/DORDRECHT/LONDON

Distributors for North, Central and South America:
Kluwer Academic Publishers
101 Philip Drive
Assinippi Park
Norwell, Massachusetts 02061 USA
Telephone (781) 871-6600
Fax (781) 871-6528
E-Mail <kluwer@wkap.com>

Distributors for all other countries:
Kluwer Academic Publishers Group
Distribution Centre
Post Office Box 322
3300 AH Dordrecht, THE NETHERLANDS
Telephone 31 78 6392 392
Fax 31 78 6546 474
E-Mail <orderdept@wkap.nl>

 Electronic Services <http://www.wkap.nl>

Library of Congress Cataloging-in-Publication

Smith, David R. (David Russell)
  Digital transmission systems, 2/c/David R. Smith
          p.   cm.
  Includes bibliographical references and index.
  ISBN 0-442-00917-8
  1. Digital communications. I. Title.
  TK5103.S65   1992   92-23962
  621.382--dc20   CIP

Copyright © 1993 by Van Nostrand Reinhold
Eighth printing by Kluwer Academic Publishers 1999

Printed on acid-free paper.

Printed in the United States of America

To my wife, Carolyn, and our children
Christy, Stacy, Allison, and Andrew.

Preface

In the development of telecommunication networks throughout the world, digital transmission has now replaced analog transmission as the predominant choice for new transmission facilities. This trend began in the early 1960s when the American Telephone and Telegraph Company first introduced pulse code modulation as a means of increasing capacity in their cable plant. Since that time, digital transmission applications have grown dramatically, notably in the United States, Canada, Japan, and Western Europe. With the rapidity of digital transmission developments and implementation, however, there has been a surprising lack of textbooks written on the subject.

This book grew out of my work, research, and teaching in digital transmission systems. My objective is to provide an overview of the subject. To accomplish this end, theory has been blended with practice in order to illustrate how one applies theoretical principles to actual design and implementation. The book has abundant design examples and references to actual systems. These examples have been drawn from common carriers, manufacturers, and my own experience. Considerable effort has been made to include up-to-date standards, such as those published by the CCITT and CCIR, and to interpret their recommendations in the context of present-day digital transmission systems.

The intended audience of this book includes engineers involved in development and design, system operators involved in testing, operations, and maintenance, managers involved in system engineering and system planning, and instructors of digital communications courses. For engineers, managers, and operators, the book should prove to be a valuable reference because of its practical approach and the insight it provides into state-of-the-art digital transmission systems. In the classroom the book will allow the student to relate theory and practice and to study material not covered by other textbooks. Theoretical treatments have been kept to a minimum by summarizing derivations or simply stating the final result. Moreover, by consulting the extensive list of references given with each chapter the interested reader can explore the theory behind a particular subject in greater detail. Even so, to gain from this book the reader should have a fundamental knowledge of communication systems and probability theory.

This second edition is an expanded and updated version of the first edition. The tremendous growth in fiber optic communication systems has led to a new, separate chapter on that topic. Coverage of digital coding and modulation has been expanded, to include spread spectrum and trellis coded modulation. The development of digital transmission networks is now given ample treatment. Finally, problems have been added to improve the book's utility as an academic textbook. The organization of this book's second edition follows a natural sequence of topics:

- Chapter 1:   Historical background and perspective on worldwide digital transmission systems

- Chapter 2:   System design including services to be provided, parameters used to characterize performance, and means of allocating performance as a function of the service, media, and transmission equipment

- Chapter 3:   Analog-to-digital conversion techniques, starting with the most commonly used technique, pulse code modulation, and ending with specialized voice coders

- Chapter 4:   Time-division multiplexing using both asynchronous and synchronous techniques; digital multiplex hierarchies

- Chapter 5:   Baseband transmission including coding, filtering, equalization, scrambling, and spread spectrum techniques

- Chapter 6:   Digital modulation, starting with binary (two-level) forms and generalizing to M-ary and coded forms, showing comparisons of error performance, bandwidth, and implementation complexity

- Chapter 7:   Digital cable systems for twisted-pair and coaxial metallic cable, showing design and performance

- Chapter 8:   Fiber optic transmission systems, including sources, detectors, fibers, system design, and optical hierarchies

- Chapter 9: Digital radio systems including propagation and interference effects, radio equipment design, and link calculations

- Chapter 10: Digital network timing and synchronization, describing fundamentals of time and frequency and means of network synchronization

- Chapter 11: Testing, monitoring, and control techniques used to measure and maintain system performance and manage the network

- Chapter 12: Digital transmission over the telephone network, from voice-channel modems for low data rates to wideband modems for high data rates

- Chapter 13: Assessment of digital transmission networks and services with emphasis on the emerging integrated services digital network (ISDN)

The material presented in Chapters 3 through 8 has been used for graduate courses in digital communications at both George Washington University and George Mason University. The contents of the entire book form the basis for a short course on digital transmission, offered by George Washington University.

David R. Smith

Acknowledgments

I first wish to acknowledge the support of my colleagues at the Defense Information Systems Agency, especially K. R. Belford, J. J. Cormack, W. A. Cybrowski, B. S. McAlpine, T. L. McCrickard, J. L. Osterholz, M. J. Prisutti, D. O. Savoye, D. O. Schultz, P. S. Selvaggi, S. Soonachan, and H. A. Stover.

My students at George Washington University and George Mason University used early versions of chapters of this book, both first and second editions, and made many helpful contributions. Participants in numerous short courses offered encouragement and suggested improvements. I am indebted to all those who have taken my courses and at the same time helped me with the writing of this book.

I would like to pay special thanks to Dr. Heinz Schreiber of Grumman Aerospace Corporation and the State University of New York, who carefully reviewed the entire manuscript of the first edition and provided many valuable comments.

In writing a book of this size, I have made considerable use of the open literature, especially the *IEEE Transactions on Communications,* the *IEEE Communications Magazine* and proceedings of various technical conferences. In describing standards applicable to digital transmission systems, I have extracted from the reports and recommendations of the CCIR and CCITT.*

Finally, I would like to express my thanks to Linda Thomas who diligently typed the original and revised versions of the manuscript for the first edition.

---

*The reproduction in this book of material taken from the publications of the International Telecommunications Union (ITU), Place des Nations, 1211 Geneva 20, Switzerland, has been authorized by the ITU. Complete volumes from which I have extracted figures and tables can be obtained from the ITU.

Contents

1

Introduction to Digital Transmission Systems

OBJECTIVES

- Describes the development of digital communications from the ancient Greeks to the global telecommunications networks of today
- Discusses the present-day use of digital transmission in the United States and abroad
- Explains national and international standards that have been established for digital transmission
- Describes the advantages of digital over analog transmission
- Defines the fundamental components of a digital transmission system

1.1 Historical Background

From the biblical passage:

> But let your communication
> be yea, yea, nay, nay.
> Matthew 5:37

one could argue that digital (actually, binary with redundancy!) transmission was divined two thousand years ago. In fact, early written history tells us that the Greeks used a form of digital transmission based on an array of torches in 300 B.C. and the Roman armies made use of semaphore signaling. The modern history of digital transmission, however, begins with the invention of the telegraph. Of the sixty or more different methods that had been proposed for telegraphy by 1850, the electromagnetic telegraph of Samuel Morse emerged as the most universally accepted one [1]. The Morse

1

telegraph consisted of a battery, a sending key, a receiver in the form of an electromagnet, and the connecting wire. Pressure on the key activated the armature of the electromagnet, which produced a clicking sound. Although the original intention was to make a paper recording of the received signals, operators found that they could read the code faster by listening to the clicks. Morse's original code was based on a list of words in which individual numbers were assigned to all words likely to be used. Each number was represented by a succession of electrical pulses—for example, a sequence of 10 pulses signified the tenth word in the code. This cumbersome code was replaced by the now familiar dots and dashes with the assistance of Alfred Vail. Morse and Vail constructed a variable-length code in which the code length was matched to the frequency of occurrence of letters in newspaper English. Vail estimated letter probabilities by counting the numbers of letters in type boxes at the office of a local newspaper. This idea of assigning the shortest codes to the letters most frequently used was an early application of information theory.

Multiplexing techniques were also introduced in early telegraphy to allow simultaneous transmission of multiple signals over the same line. Time-division multiplexing was used in which a rotary switch connected the various signals to the transmission line. A basic problem then (and now) was the synchronizing of the transmitting and receiving commutators. A practical solution introduced by Baudot in 1874 involved the insertion of a fixed synchronizing signal once during each revolution of the switch. Baudot also introduced a five-unit code in place of the Morse type. Up to six operators could work on a single line; each operator had a keyboard with five keys. For each character to be transmitted the operator would depress those keys corresponding to marks in the code. The keys remained locked in place until connection was made with the rotary switch and the signals were transmitted down the line. The keys were then unlocked and an audible signal given to the operator to allow resetting of the keys for the next character transmission. Practical limitation for this multiplexing scheme was about 90 bits per second (b/s).

The telephone was born in 1876 when Alexander Graham Bell transmitted the first sentence to his assistant: "Mr. Watson, come here, I want you." The first telephones were primarily a transmitter or receiver, but not both. Thus initial use was limited to providing news and music, much in the fashion of radio in the twentieth century. Even when refinements provided the telephone with both transmitting and receiving capabilities, there was the problem of educating customers about its use. Read one Bell circular in 1877: "Conversation can be easily carried on after slight practice and with occasional repetition of a word or sentence." But by 1880 the telephone had begun to dominate the development of communication networks, and digital transmission in the form of telegraphy had to fit with the characteristics of analog transmission channels. Today telephony remains dominant in the world's communication systems, but data, video, and facsimile requirements are growing more rapidly than telephony. As for Bell, he continued research in speech after the telephone became an economic success in the early twentieth century. But in his study there was a conspicuous absence: no telephones. Bell did not like interruptions [2].

Alexander Graham Bell is also credited as the inventor of modern lightwave communications. In 1880, Bell demonstrated the transmission of speech over a beam of light using a device he called the "photophone." One version of this invention used sunlight as a carrier, reflected from a voice-modulated thin mirror and received by a light-sensitive selenium detector. Although the photophone was Bell's proudest invention, the practical application of lightwave communication was made possible only after the invention of reliable light sources (such as the laser, invented in 1958) and the use of low-loss glass fiber. In 1966, Kao first proposed the use of low-loss glass fibers as a practical optical waveguide. In the early 1970s, Corning developed and introduced practical fiber cables having a loss of only a few decibels per kilometer. Field demonstrations of digital telephony transmission over fiber optic cable began in the mid-1970s. Since that time, the use of fiber optics for communications has grown rapidly, notably in the United States, Canada, Europe, and Japan.

Although many researchers contributed to the introduction of "wireless" transmission, Marconi is usually designated as the inventor of radio. He began his experiments in Italy in 1895. In 1899 he sent a wireless message across the English Channel, and in 1901 across the Atlantic Ocean between St. John's, Newfoundland, and Cornwall, England. Marconi's radiotelegraph produced code signals by on-off keying similar to that used for land telegraphy. Pressure on the key closed an electrical circuit, which caused a spark to jump an air gap between two metal balls separated by a fraction of an inch. The spark gave off radio waves that went through the antenna and over the air to the receiver. Still, radio transmission had limited use until the invention and perfection of the vacuum tube. Used for both amplification and detection, the vacuum tube allowed longer transmission distances and better reception. With the introduction of modulation, radio-telephony was possible and became the dominant mode of radio transmission by the 1920s. Digital radio for telephony was still a long way off, however, and did not become available until the 1970s.

With the advent of power amplifiers, limitations in basic noise levels were discovered and, in 1928, Nyquist described noise theory in a classic paper [3]. At the same time, the relationship between bandwidth and sampling rate was being investigated, and in 1928 Nyquist published his famous account of digital transmission theory [4]. Nyquist showed mathematically that a minimum sampling rate of $2W$ samples per second is required to reconstruct a band-limited analog signal of bandwidth W. Twenty years later, Shannon linked noise theory with digital transmission theory, which led to his famous expression for the information capacity C of a channel with noise [5]:

$$C = W \log_2 (1 + S/N) \text{ bits/second} \tag{1.1}$$

where W is the bandwidth, S is the signal power, and N is the noise power. This expression implies that data transmission at any rate is possible for an arbitrary bandwidth provided an adequate S/N can be achieved, the actual data rate R is less than C, and a sufficiently involved encoder/decoder is used.

Digital transmission of analog signals was made practical with the invention of pulse code modulation (PCM) by A. Reeves in 1936–1937. Reeves first discovered pulse-position modulation and then recognized the noise immunity possible by quantizing the amplitude and expressing the result in binary form. The first fully operational PCM system did not emerge until ten years later in the United States [6]. PCM systems in the late 1940s were large, unreliable, and inefficient in power consumption. The key invention responsible for the success of PCM was of course the transistor, which was invented in 1948 and in practical use by the early 1950s. Initial applications of PCM systems involved twisted-pair cable, whose characteristics limited analog multichannel carrier operation but were acceptable for PCM. By the mid-1960s, PCM at 64 kb/s had become a universal standard for digital transmission of telephony and was used with terrestrial radio, satellites, and metallic cable. Because of the coding inefficiency of 64 kb/s for voice, various alternative techniques were introduced to reduce the coding rate. In the mid-1980s, 32-kb/s adaptive differential PCM became an international digital telephony standard that has since rivaled 64-kb/s PCM in popularity.

The history of satellite communications began with Echo, a passive satellite experiment conducted in 1960. Telstar, launched in 1962 for use in the Bell System, was the first satellite to use active repeaters, or transponders. These early satellites were placed in low orbits that required complex earth stations to track the motion of the satellite. Beginning with the Early Bird (INTELSAT I) satellite in 1965, communication satellite orbits have since been geostationary—that is, stationary above a fixed point on the earth's surface. Using geostationary satellites, complete earth coverage is possible with only three or four satellites. The international telecommunications satellite consortium known as INTELSAT was organized in 1964 and by 1970 had established a global telecommunications network. Present-day satellites contain a number of transponders, each providing a separate service or leased separately to a customer.

In 1962, the Bell System introduced the T1 cable carrier system in the United States [7]. This carrier system has a capacity of 24 PCM voice channels and a transmission rate of 1.544 megabits per second (Mb/s). This 24-channel PCM system is now the standard PCM multiplexer in North America and Japan. By 1982 the Bell System had over 100 million circuit miles of T1 carrier systems. These systems, operating over twisted-pair or fiber optic cable, already dominate the U.S. metropolitan networks. The prime application of PCM over twisted-pair cable was to take advantage of the local telephone companies' investment in buried copper cable. Frequency-division multiplex (FDM) equipment was expensive and did not provide economic advantage over the cost of copper pairs for individual voice channels over short distances between telephone offices (typically less than 10 mi). PCM carrier equipment reduced hardware costs so that T1 carrier became more economical than individual copper pairs for short distances. The ability to regenerate digital signals also allowed better performance than analog transmission. Digital ratio was used to connect metropolitan islands of T carrier and to interconnect digital switches. The first standard fiber optic system was introduced by the Bell System in 1980. It operated at the T3 standard rate 44.736 Mb/s and provided 672 voice channels on a fiber pair.

Although the United States may have been the early leader in developing digital transmission technology, other countries, notably Japan, Canada, and several European nations, were also exploiting digital transmission. From the mid-1960s to early 1970s, T1 carrier systems were introduced in many national systems and led to considerable growth in digital transmission throughout the world. In Japan, Nippon Telegraph and Telephone (NTT) developed and applied higher-order PCM multiplex for coaxial cable, digital radio, and fiber optics systems, beginning in 1965 [8]. Although the United States, Japan, and Canada [9] based their digital hierarchy on the original 24-channel, 1.544 Mb/s PCM standard, European countries developed another standard based on a 30-channel grouping and a transmission rate of 2.048 Mb/s. This second PCM standard now forms the basis for the European digital hierarchy. Italy was the first European nation to use PCM, beginning in 1965 with the 24-channel standard. Since 1973, only 30-channel systems have been installed—principally by the Societa Italiana per l'Esercizio Telefonico (SIP), which is the major Italian operating company. In 1979, some 25 percent of Italian carrier systems were digital; by 1993 about 85 percent of installed carrier channels were digital [10]. In the United Kingdom, the British Post Office (BPO) introduced 24-channel PCM in the late 1960s, but since 1978 only 30-channel systems have been installed [11]. France was also an early leader in the use of PCM transmission facilities in conjunction with digital switching.

1.2   Present-Day Digital Transmission

Within the United States, the major carriers have all converted their networks to digital facilities. In 1988, US Sprint was the first carrier to claim an all-digital network in which all transmission facilities are fiber optic. Both AT&T and MCI completed digitization in 1992, which was accomplished by accelerating the replacement of aging analog equipment with new digital microwave or fiber optic facilities. As recently as 1987, AT&T had projected that its public switched network would not be completely digitized until 2010, a goal actually realized almost twenty years early [12]. Today these three carriers offer a variety of digital services including T3 (44.736 Mb/s), T1, fractional T1 (either $56 \times N$ or $64 \times N$ kb/s up to 1.544 Mb/s), and digital data (for example, 2.4, 4.8, 9.6, 19.2, and 56 kb/s). A number of specialized private line and bypass carriers have also invested heavily in digital transmission facilities. For example, the Williams Telecommunications (Will-Tel) Group has an extensive private line network consisting of digital microwave and fiber optic facilities, which provide digital services from digitized voice to T3. Will-Tel is a member of the National Telecommunications Network (NTN), which is a consortium of several digital network providers whose regional networks are interconnected to provide national coverage. The transmission medium of choice by U.S. long-distance carriers has clearly been fiber optics, with over 100,000 route miles in place by 1993. Both AT&T and MCI use digital radio for spurs off their fiber optic backbones or to provide regional coverage where fiber optic systems are not yet available. Regional Bell Operating Companies (RBOCs) have been slower to complete digitization plans, because of the predominance of analog local loops. With much of the long-haul plant already converted to fiber optic transmission, most of the growth in fiber optic cable installation in the coming decade will be accomplished by the RBOCs in the local loop to replace existing copper twisted-pair cable.

Other countries have had similar conversions to an all-digital network over the last decade. Japan has an extensive digital microwave and fiber optic network, while Canada and several European countries—notably the United Kingdom, France, Germany, Italy, and Spain—have also invested in digital transmission facilities. The NTT network in Japan is currently based on the Japanese digital hierarchy, with digital radio and fiber optic systems both providing rates up to 400 Mb/s. With the international development of a new synchronous digital hierarchy, NTT has developed new transmission facilities based on this new hierarchy, including digital radio operating at rates of 52, 156, and 312 Mb/s as well as fiber optic systems operating at rates of 52, 156, 622, and 2488 Mb/s [13]. NTT also has an objective of connecting all subscribers with optical fiber loops by the beginning of the twenty-first century [14], while providing a variety of narrowband and broadband services via the concept of an Integrated Services Digital Network (ISDN). Telecom Canada, a consortium of Canadian telecommunications companies, has spearheaded the installation of a transcontinental fiber optic system, making it one of the world's longest fiber routes, spanning more than 4,000 miles coast-to-coast. Deregulation, privatization, and competition in the European telecommunications industry, particularly in the European Community (EC) nations, has led to modernization and the simultaneous digitization of telecommunications. The existing British Telecom (BT) network is a mixture of coaxial cable, microwave radio, and fiber optic links. A major modernization program by BT now provides fiber optic connectivity between main switching nodes and will eventually introduce optical fiber into the local loop [15]. Although the U.K. is the biggest European user of fiber optic transmission, France Telecom has been installing fiber optic transmission in both its long-haul and local networks to bring "fiber to the home." The existing French network is based on 140-Mb/s coaxial cable and digital radio, and on 565-Mb/s coaxial and optical fiber cable. However, this long-haul network is being upgraded with new optical fiber to allow introduction of the Synchronous Digital Hierarchy (SDH) [16]. In the Italian digital network, the existing infrastructure is based on twisted-pair cable, coaxial cable, fiber optic cable, and digital radio. However, evolution of the network is dominated on all levels (intercity, urban, and local) by massive deployment of fiber optics, which will allow the introduction of SDH [17]. The next generation networks in the European Community will introduce not only the SDH [16, 17, 18, 19] but also ISDN [20] and intelligent networks (IN) [21, 22].

Submarine fiber optic cable installations have dramatically increased the capacity and performance of international networks. Several transatlantic and transpacific cables are in place, and more are planned (see Tables 1.1 and 1.2). Other submarine cable systems are planned or in existence in the Mediterranean region, the Caribbean region, and the British Isles region. The first transatlantic fiber optic cable, known as the Trans-Atlantic Telecommunications (TAT)-8, was commissioned in 1988; it has a total capacity of 565 Mb/s, consists of two operational fiber pairs plus one spare pair, and interconnects the U.S., U.K., and France. The first transpacific fiber optic cable, known as the Hawaii-4/Trans-Pacific Cable-3 (HAW-4/TPC-3), was commissioned in 1989; it uses the same technology as TAT-8, carries a total of 565 Mb/s, and interconnects the U.S. West Coast, Hawaii, Guam, and Japan. These cables are currently backed up by satellite channels, but as

TABLE 1.1   Characteristics of Transatlantic Fiber Optic Cables

Name	Capacity	Operational Date	Cable Station Locations
TAT-8	560 Mb/s	1988	U.S., U.K., France
TAT-9	1.12 Gb/s	1992	U.S., U.K., France, Spain, Canada
TAT-10	1.12 Gb/s	1992	U.S., Germany, Netherlands
TAT-11	1.12 Gb/s	1993	U.S., U.K., France
TAT-12	2.4/4.8 Gb/s	1995	U.S., U.K.
TAT-13	2.4/4.8 Gb/s	1995	U.S., France
PTAT-1	1.26 Gb/s	1989	U.S., U.K., Bermuda, Ireland
PTAT-2	1.26 Gb/s	1992	U.S., U.K.

additional submarine cables are installed, redundancy will be provided by other submarine cable systems rather than by satellite. Transatlantic fiber optic systems are listed in Table 1.1, and Table 1.2 gives characteristics of cable systems for the Pacific region. With the deployment of these planned systems, the resulting capacities will allow broadband services to be realized on international connections. Of special significance are the TAT-12/13 and TPC-5/6 systems, because of their planned use of the latest generation of fiber optics technology; that is, SDH standards at the terminating locations and optical rather than electronic amplifiers in the repeaters.

The tremendous growth in fiber optic systems has reduced costs to the customer for digital services, and also forced lower prices from competing services such as those provided by satellite carriers. The cost and perfor-

TABLE 1.2   Characteristics of Major Submarine Fiber Optic Cables in the Pacific Region

Name	Capacity	Operational Date	Cable Station Locations
HAW-4/TPC-3	560 Mb/s	1989	California, Hawaii, Guam, Japan
NPC	1.26 Gb/s	1990	Oregon, Alaska, Japan
TPC-4	1.12 Gb/s	1992	California, Japan, Canada
HAW-5	1.12 Gb/s	1993	California and Hawaii
TPC-5/6	2.4/4.8 Gb/s	1995/6	U.S. West Coast, Guam, Hawaii, and Japan
G-P-T	280 Mb/s	1989	Guam, Philippines, Taiwan
H-J-K	280 Mb/s	1990	Hong Kong, Japan, Korea
TASMAN-2	1.12 Gb/s	1991	Australia, New Zealand
PacRim (East)	1.12 Gb/s	1993	Hawaii, New Zealand
PacRim (West)	1.12 Gb/s	1994	Australia, Guam
KJG	1.12 Gb/s	1995	Korea, Japan, Guam

mance advantages of fiber optic transmission have led to a greater number of carriers and businesses who select fiber over satellite transmission for both private line and switched services. Of course, satellite systems have certain advantages over fiber optic systems, including the ability to deploy earth stations adjacent to carrier facilities or customer premises, the ability to access remote areas or developing countries where international fiber connectivity is not available, the ability to serve the mobile user, and the capability to broadcast services such as audio or video programs. Satellite services in the United States and increasingly abroad are based on an *open skies* policy, in which Very Small Aperture Terminals (VSATs) as used by businesses do not require regulatory approval for receive-only operation and require only frequency-coordination approval for two-way operation. The various U.S. domestic and international satellite carriers provide switched and private line services using both analog and digital transmission techniques. One of the most popular commercial services used with VSATs and based on digital transmission is time-division multiple access (TDMA), in which several users time-share the same carrier frequency by transmitting data in bursts. U.S. domestic satellites, about 40 altogether, are geostationary and operate in both the C-band (4 and 6 GHz) and in the Ku-band (11/12 and 14 GHz). All C-band satellites have 24 transponders, each with 36-MHz bandwidth, while Ku-band satellites vary in transponder design. INTELSAT is the primary international satellite carrier, with about 20 satellites in geostationary orbits operating in both the C-band and Ku-band. Satellite connectivity to mobile users (mainly ships) can be provided by INMARSAT (International Maritime Satellite), which operates 9 satellites using C-band and L-band (1.5/1.6 GHz). Other mobile satellite systems, both low earth orbit and geostationary earth orbit, have been proposed. For example, IRIDIUM would establish a network of 77 satellites, with 11 satellites in each of seven coplanar low earth orbits, to provide a satellite-based cellular system similar to those now used on earth [23].

Services provided by today's digital transmission systems are still dominated by voice, accounting for about 85 percent of the U.S. domestic public network. Data services make up another 10 percent, and the remainder consists of video and specialized services. Data and facsimile are the fastest growing services, where at the current rate of growth of 15 to 20 percent per year, data and facsimile transmission will account for half the AT&T domestic network by the mid-1990s [12]. This rapid growth in data communications has given rise to the integration of voice and data in digital networks, where once the two services were segregated. High-speed data and digital video services have also led to the development of prototype *gigabit networks* within the U.S. government, research, and education communities [24]. Despite these advances in data communication networks, the analog interface to the user remains dominant. Voice-band modems have improved in performance and capacity, with 19.2 kb/s modems now available. Advances in source coding have also improved the bandwidth efficiency of terrestrial radio and satellite systems in accommodating voice, video, and other analog services. Digital video coding has become an important subject with the growing popularity of video teleconferencing and the introduction of high-definition television (HDTV). Although digital compression standards already exist for video teleconferencing, there are several competing techniques for HDTV, where a standard is to be selected in the United States in 1993. Another fast-growing service is cellular telephone, where TDMA combined with low-data-rate voice coding is a popular choice of technology. In the coming decade, these cellular services will be expanded to provide *personal communication services (PCS)*, which will allow the user to have one telephone for home, business, and mobile use. Once again, digital transmission techniques are being touted for PCS, with its advantages in digital radio technology, multiple access to the same medium, and voice coding for compression.

1.3   Digital Transmission Standards

Standards play an important role in establishing compatibility among communications facilities and equipment that are designed by different manufacturers and used by the many communications carriers throughout the world. These standards are developed by various national and international groups. The organization that has the greatest effect on digital transmission standards is the International Telecommunications Union (ITU), which was founded in 1865 and in 1987 had 160 member countries. There are three main organizations within the ITU: the International Frequency Registration Board, which registers and standardizes radio frequency assignments; the International Radio Consultative Committee (CCIR), which deals with standards for radio communications; and the International Telegraph and Telephone Consultative Committee (CCITT). Both the CCIR and CCITT are divided into a number of study groups that make recommendations on various aspects of telephony, telegraphy, and radio communications. These study groups report their results at the CCITT and CCIR plenary assemblies held every three or four years. Here recommendations are formally adopted and work is organized for the next study period. Table 1.3 indicates the study groups that are mostly concerned with digital transmission. Examples of CCITT and CCIR standards will be given throughout this book.

Another important source of data transmission standards is the International Organization for Standardization (ISO). The ISO was organized in

TABLE 1.3   CCITT and CCIR Study Groups for Digital Transmission Topics

Designation	Title	Recommendations and Reports
CCITT Study Group VII	Data Communication Networks	CCITT 1989 Vol. VIII.2 through VIII.8 (X Series Rec.)
CCITT Study Group XV	Transmission Systems	CCITT 1989 Vol. III.1 through III.5 (G Series Rec.) CCITT 1989 Vol. III.6 (H, J Series Rec.)
CCITT Study Group XVII	Data Communication Over the Telephone Network	CCITT 1989 Vol. VIII.1 (V Series Rec.)
CCITT Study Group XVIII	Digital Networks including ISDN	CCITT 1989 Vol. III.7 through III.9 (I Series Rec.)
CCIR Study Group 4	Fixed Satellite Service	CCIR 1990 Vol. IV—Pt. 1
CCIR Study Group 5	Propagation in Nonionized Media	CCIR 1990 Vol. V
CCIR Study Group 7	Standard Frequencies and Time Signals	CCIR 1990 Vol. VII
CCIR Study Group 9	Fixed Service Using Radio-Relay Systems	CCIR 1990 Vol. IX—Pt. 1
CCIR Study Group 11	Broadcasting Service (Television)	CCIR 1990 Vol. XI—Pt. 1
Joint CCIR/CCITT Study Group CMTT	Transmission of Sound Broadcasting and Television Signals Over Long Distances	CCIR 1990 Vol. XII

1926 with the objective of developing and publishing international standards. Currently, 89 countries are represented in the ISO; the American National Standards Institute (ANSI) is the U.S. member. The ISO is divided into more than 150 technical committees. The technical committee responsible for data transmission standards is TC-97, Information Processing Systems. Within ANSI, X3S3 is the technical committee on data communications. Another organization playing a significant role in data transmission standards in the United States is the Telecommunications (formerly Electronic) Industries Association (TIA), established in 1924 to assist U.S. industry in legislation, regulation, and standards. Examples of ISO and TIA standards will be given in data transmission sections of the book. The interrelationship of these various standards organizations is shown in Figure 1.1.

Prior to the divestiture of the Bell System in 1984, standards for telecommunications in the United States were provided by various AT&T publications. In the aftermath of divestiture, an organization

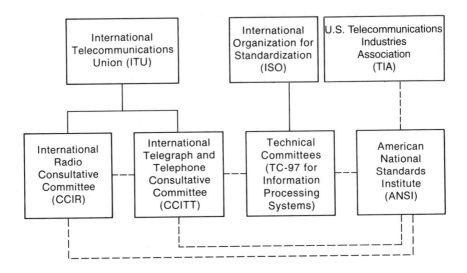

FIGURE 1.1   Relationships Among International and U.S. National Standards
Organizations

known as the Exchange Carriers Standards Association (ECSA) was created to assume the functions formerly provided by AT&T. The ECSA organization is accredited by ANSI, and ECSA standards are published as ANSI standards. The focal point within ECSA for telecommunication standards is the Committee T1 on Telecommunications, whose membership is drawn from North American carriers, manufacturers, users, and government organizations. The Committee T1 has many international liaisons, for example, with the CCITT, CCIR, and national standards organizations. The Committee T1 is responsible for the development of standards and reports dealing with those functions—such as switching, signaling, transmission, performance, operation, and maintenance—necessary to ensure proper interconnection and interoperability of telecommunication systems. The organization of Committee T1 as of 1993 is shown in Table 1.4; six technical subcommittees cover the various technology areas.

The U.S. government has a wide variety of agencies involved in aspects of telecommunications standards. The Federal Communications Commission (FCC) authorizes certain carrier services, prescribes technical and operational criteria, and certifies equipment for compliance with certain technical standards. Within the Department of Commerce, the National Telecommunications and Information Administration (NTIA) and the

TABLE 1.4   ECSA Committee T1 Technical Subcommittees and Working Groups

Technical Subcommittees	Technical Working Groups
T1A1: Specialized Subjects	Specialized Video and Audio Services
	Specialized Voice and Data Processing
	Environmental Standards for Exchange
	and Interexchange Networks
T1E1: Network Interfaces	Analog Access
	Wideband Access
	Connectors and Wiring Arrangements
	Digital Subscriber Line Access
T1M1: Internetwork Operations,	Internetwork Planning and Engineering
Administration, Maintenance, and	Internetwork Operations
Provisioning (OAM&P)	Testing and Operations Systems and Equipment
	OAM&P Architecture, Interfaces, and
	Protocols
T1P1: Systems Engineering, Standards	Program Management
Planning, and Program Management	Wireless Access
	Network and Services
T1S1: Services Architecture and	Architecture and Services
Provisioning	Switching and Signaling Protocols
	Common Channel Signaling
	Broadband ISDN
T1X1: Digital Hierarchy and	Synchronization Interfaces
Synchronization	Metallic Hierarchical Interfaces
	Optical Hierarchical Interfaces
	Tributary Analysis Interfaces

National Institute for Standards and Technology (NIST) help promote cooperative development of technology and standards within the United States and between the U.S. and foreign nations. Within the Department of Defense, the Defense Information Systems Agency (DISA) is responsible for the development and publication of military standards (MIL-STD) for use by tactical and strategic communication systems; and the National Communications System (NCS) is responsible for federal standards (FED-STD) for use by the civil agencies. These U.S. military and federal standards generally adopt national and international standards unless unique military requirements dictate otherwise.

Other regional or national standards organizations play a significant role in the development of international standards. The European Conference of Postal and Telecommunications Administrations, more commonly known as CEPT, coordinates postal and telecommunications policies of European governments and promulgates recommendations. The European Telecommunications Standards Institute (ETSI) was established by CEPT in the late 1980s to accelerate the development of standards in the European Community (EC). The ETSI, much like the ECSA Committee T1 for North America, provides for direct participation at the European level by all parties interested in standards. Both CEPT and ETSI have liaisons with the CCITT and many other standards organizations. In the Pacific area, both Japan and Korea have established national standards organizations, which have liaisons with other national and international standards organizations. Foreign military organizations follow procedures similar to those of the United States. Within NATO, the United States, Canada, and European member nations have formed standardization agreements (STANAGS) as a means of ensuring compatible military communication systems among these various nations.

The worldwide evolution of digital transmission networks has led to three digital hierarchies in use today. These are referred to as North American, Japanese, and European (CEPT) and have been recognized by international standards organizations, most importantly the CCITT. The transmission rate structure of the three hierarchies is shown in Table 1.5. A more detailed comparison is deferred until Chapter 4. A new standard has emerged even more recently that will replace or supplement these three different digital hierarchies with a single international hierarchy. In the United States this hierarchy has been documented in a series of ANSI standards and is referred to as the Synchronous Optical Network (SONET); internationally, this same hierarchy has been recommended by the CCITT and is called the Synchronous Digital Hierarchy (SDH). A more detailed description is deferred until Chapter 8.

1.4   Advantages of Digital Transmission

Since most of today's transmission systems are now digital, while most of the services provided (voice, video, and facsimile) are analog, the obvious question is "why digital?" To answer this question, we consider the various advantages of digital versus analog transmission.

TABLE 1.5   Transmission Rates for Digital Hierarchy of North America, Japan, and Europe

Multiplex Level	North America (Mb/s)	Japan (Mb/s)	Europe (Mb/s)
1	1.544	1.544	2.048
2	6.312	6.312	8.448
3	44.736	32.064	34.368
4	274.176	97.728	139.264
5		397.200	

- Digital performance is superior. The use of regenerative repeaters eliminates the noise accumulation of analog systems. Performance is nearly independent of the number of repeaters, system length, or network topology.
- Sources of digital data, such as teletypewriters or computers, can be more efficiently accommodated. Analog systems use individual modems (modulator-demodulators), one per data source, making inefficient use of bandwidth.
- With the advent of space and computer technologies, advances in digital logic and use of medium to large-scale integration have led to superior reliability and maintainability of digital transmission equipment.
- The cost of digital transmission equipment has effectively decreased over the last thirty years. The cost and poor reliability of vacuum tube and discrete transistor technology made digital transmission expensive in its early years. Today PCM and digital multiplex equipment is generally less costly than frequency-division multiplex (FDM) equipment.
- Digital radio systems are less susceptible to interference than are analog radio systems. Time-division multiplex signals are tolerant of single-frequency interference, since that interference is spread over all channels. Frequency-division multiplex, however, is susceptible to the single-frequency interferer since the interference is translated directly to a few baseband channels. Digital radio systems commonly employ scramblers to smooth the power spectrum and use modulation schemes that suppress the carrier; hence the overall radio system's performance is immune to single-frequency interference.
- The use of digital transmission greatly facilitates the protection of communications against eavesdropping. The simple addition of **cryptographic equipment,** a scrambling device based on digital logic, provides privacy to the users. In military applications, digital transmission has led to extensive use of **bulk encryption,** a process in which multiple signals are encrypted by a single cryptographic apparatus, and **secure voice,** in which voice communications are encrypted by cryptographic equipment. Encryption has always been desirable for military communications, but it is becoming more important in other communication systems for protection of electronic banking, industrial secrets, and the like.
- The proliferation of fiber optics has provided low-cost, high-speed digital transmission for national and international telecommunications. Digital transmission is a natural choice for fiber optic communications, since analog modulation of an optical fiber is more difficult and more expensive. The bandwidth and performance advantages of fiber optics make it the medium of choice and reinforce the selection of digital transmission.

1.5   A Simplified Digital Transmission System

To introduce the terminology for the remaining chapters in this book, consider the block diagram of a digital transmission system shown in Figure 1.2. The fundamental parts of the system are the *transmitter,* the *medium* over which the information is transmitted, and the *receiver,* which estimates the original information from the received signal. At the transmitter the information source may take several forms—for example, a computer, teletypewriter, telemetry system, voice signal, or video signal. The source *encoder* formats data signals—for example, by creating alphanumeric characters for computer output—or digitizes analog signals such as voice and video. The *modulator* then interfaces the digital signal with the transmission medium by varying certain parameters (such as frequency, phase, or amplitude) of a particular square wave or sinusoidal wave called the *carrier.* A transmission medium such as cable or radio provides the *channel,* a single path that connects the transmitter and receiver. Transmitted signals are corrupted by noise, interference, or distortion inherent in the transmission medium. The receiver must then demodulate and decode the degraded signal to make an estimate of the original information.

The block diagram of Figure 1.2 represents a simplified digital transmission system. The direction of transmission is indicated as one-way, or *simplex,* between two points. Transmission systems usually operate *duplex,* in which simultaneous two-way information is passed point to point. The electrical path between the two endpoints, called a *circuit,* may be dedicated for exclusive use by two users or may be switched to provide temporary connection of the two users. Further, a circuit may be composed of several types of transmission media, or channels. A long-haul circuit, for example, may require a mix of satellite, radio, and cable media. Transmission systems may be devised for single circuits or for multiple cir-

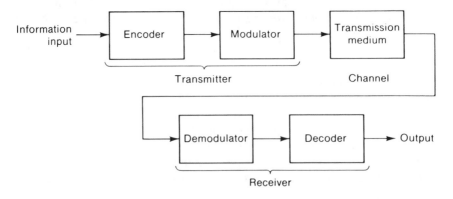

FIGURE 1.2   Block Diagram of Digital Transmission System

cuits through the use of multiplexing techniques. Transmission systems may also interconnect many users to provide a *network* for communication. As the size of a network grows, it becomes impractical to provide dedicated transmission channels for each circuit. Hence, most networks employ switches that provide a temporary circuit at the request of one of the connected users.

The theme of this book is transmission systems used in digital communication systems. Emphasis is placed on long-haul, multichannel, terrestrial transmission systems, although the theory and practices presented here are also applicable to single-channel transmission systems such as mobile radio and subscriber loops and to satellite communication systems. Inevitably, discussion of certain digital communication topics had to be curtailed or omitted to keep the book to a manageable size. A separate, thorough treatment of error correction codes has not been attempted, although their use with transmission systems is included in several places (for example, in discussing performance monitoring and coded modulation). Switching systems have been discussed only to the extent that they influence transmission system design and performance. The growth in data communications has led to the development of many technologies, such as packet switching, which are not given thorough treatment here. Chapter 13 does attempt to show how digital transmission networks are evolving toward the integration of all services—voice, data, and video—and what technologies are being used in such networks. Moreover, certain topics relevant to military communications are not covered in detail, including the use of cryptography for privacy. For further information on these related topics, the reader is asked to make use of the references [25, 26, 27, 28].

1.6   Summary

The use of digital transmission dates back to early written history, but the modern history of digital transmission barely stretches over the last hundred years. The major events in the development of digital transmission, summarized in Figure 1.3, have culminated in worldwide applications, led by the United States, Canada, Japan, and several European nations. Today these developments in digital transmission are facilitated by standards and hierarchies established by the ITU, ISO, and various national standards organizations. Advantages of digital transmission—performance, bandwidth efficiency for data, reliability, maintainability, equipment cost, ease of encryption, and worldwide deployment of fiber optics—have led to universal acceptance of digital transmission for telecommunication systems. The remaining chapters of this book describe the theory and practices of digital transmission systems and explain in more detail the terminology, standards, and examples given in this first chapter.

1850—Invention of telegraph by Morse

1874—Invention of time-division multiplexing by Baudot

1876—Invention of telephone by Bell

1880—Invention of photophone by Bell

1899—Invention of radio by Marconi

1928—Development of sampling theory by Nyquist

1936—Invention of pulse code modulation by Reeves

1948—Development of channel capacity theory by Shannon
    Invention of transistor by Bell Laboratories

1962—First 1.544-Mb/s T1 cable carrier system by Bell System
    Telstar, first communications satellite by Bell System

1965—Early Bird, first geostationary communications satellite by
    INTELSAT

1966—Low-loss optical fiber proposed by Kao

1980—AT&T introduces fiber optic transmission at T3 data rate

1984—Divestiture of the Bell System increases competition and accel-
    erates introduction of digital transmission in the United States

1985—Introduction of 32-kb/s adaptive differential PCM doubles
    voice-channel capacity of digital transmission

1988—First transatlantic fiber optic cable (TAT-8) installed

1989—First transpacific fiber optic cable (HAW-4/TPC-3) installed

1992—Conversion of major U.S. networks to digital transmission
    completed

1990s—Deployment of digital transmission worldwide in support of the
    Synchronous Digital Hierarchy and Integrated Services Digital
    Network

FIGURE 1.3   Significant Events in the Modern History of Digital Transmission

References
1. W. R. Bennett and J. R. Davey, *Data Transmission* (New York: McGraw-Hill, 1965).
2. J. Brooks, *Telephone: The First Hundred Years* (New York: Harper & Row, 1976).

3. H. Nyquist, "Thermal Agitation of Electricity in Conductors," *Phys. Rev.* 32(1928):110–113.
4. H. Nyquist, "Certain Topics in Telegraph Transmission Theory," *Trans. AIEE* 47(1928):617–644.
5. C. E. Shannon, "A Mathematical Theory of Communication," *Bell System Technical Journal* 27(July and October 1948):379–423 and 623–656.
6. H. S. Black and J. O. Edson, "Pulse Code Modulation," *Trans. AIEE* 66(1947):895–899.
7. K. E. Fultz and D. B. Penick, "The T1 Carrier System," *Bell System Technical Journal* 44(September 1965):1405–1451.
8. T. Mirrami, T. Murakami, and T. Ichikawa, "An Overview of the Digital Transmission Network in Japan," *International Communications Conference,* 1978, pp. 11.1.1–11.1.5.
9. J. A. Harvey and J. R. Barry, "Evolution and Exploitation of Bell Canada's Integrated Digital Network," *International Communications Conference,* 1981, pp. 17.4.1–17.4.6.
10. M. R. Aaron, "Digital Communications—The Silent (R)evolution," *IEEE Comm. Mag.,* January 1979, pp. 16–26.
11. J. F. Boag, "The End of the First Pulse Code Modulation Era in the U.K.," *Post Office Elec. Eng. J.* 71(April 1978):2.
12. D. A. Lee, "Digitization of the AT&T Worldwide Intelligent Network," *AT&T Technology* 5(1990):14–19.
13. H. Kasai, T. Murase, and H. Ueda, "Synchronous Digital Transmission Systems Based on CCITT SDH Standard," *IEEE Comm. Mag.,* August 1990, pp. 50–59.
14. T. Uenoya and others, "Operation, Administration and Maintenance Systems of the Optical Fiber Loop," *1990 Global Telecommunications Conference,* pp. 802.5.1–802.5.5.
15. S. Hornung, R. Wood, and P. Keeble, "Single-Mode Optical Fibre Networks to the Home," *1990 International Conference on Communications,* pp. 341.5.1–341.5.9.
16. G. Bars, J. Legras, and X. Maitre, "Introduction of New Technologies in the French Transmission Networks," *IEEE Comm. Mag.,* August 1990, pp. 39–43.
17. U. Mazzei and others, "Evolution of the Italian Telecommunication Network Towards SDH," *IEEE Comm. Mag.,* August 1990, pp. 44–49.
18. S. Whitt, I. Hawker, and J. Callaghan, "The Role of SONET-Based Networks in British Telecom," *1990 International Conference on Communications,* pp. 321.4.1–321.4.5.
19. H. Klinger and others, "A 2.4 Gbit/s Synchronous Optical Fiber Transmission System," *1990 International Conference on Communications,* pp. 305.1.1–305.1.10.
20. R. Liebscher, "ISDN Deployment in Europe," *1990 Global Telecommunications Conference,* pp. 705A.3.1–705A.3.5.
21. P. Collet and R. Kung, "The Intelligent Network in France," *IEEE Comm. Mag.,* February 1992, pp. 82–89.
22. K. Schulz, H. Kaufer, B. Heilig, "The Deutsche Bundespost Telekom Implements IN," *IEEE Comm. Mag.,* February 1992, pp. 90–96.

23. J. L. Grubb, "The Traveller's Dream Come True," *IEEE Comm. Mag.,* November 1991, pp. 48–51.

24. D. Fisher, "Getting up to Speed on Stage-Three NREN: Testbed Lessons on Architectures and Applications," *1991 Global Telecommunications Conference,* paper 28.1.

25. J. J. Spilker, *Digital Communications by Satellite* (Englewood Cliffs, NJ: Prentice-Hall, 1977).

26. W. W. Peterson and E. J. Weldon, Jr. *Error Correcting Codes* (Cambridge, MA: MIT Press, 1972).

27. W. Stallings, *Data and Computer Communications,* 3rd ed. (New York: Macmillan, 1991).

28. D. J. Torrieri, *Principles of Secure Communication Systems* (Dedham, MA: Artech House, 1985).

2

Principles of System Design

OBJECTIVES

- Describes how performance objectives are established for a digital transmission system
- Discusses the three main services provided by a digital communications system: voice, data, and video
- Explains how the hypothetical reference circuit is used to determine overall performance requirements
- Discusses the two main criteria for system performance: availability and quality

2.1 General Plan

The first step in transmission system design is to establish performance objectives. This process begins by stating the overall performance from circuit end to circuit end. The requirements of the user depend on the service being provided. For human communications such as voice and video, satisfactory performance is largely a subjective judgment; for data users, performance can be better quantified through measures such as throughput, efficiency, and error rate. For any type of service, the measure used to indicate performance must be based on statistics, since performance will vary with time due to changes in equipment configuration and operating conditions. Once these end-to-end objectives have been stated, the allocation of objectives to individual subsystems and equipment may take place.

The parameters used to characterize performance are based on the service and the type of transmission media and equipment. Here we will use two basic parameters for all cases: availability and quality. Circuit availability indicates the probability or percentage of time that a circuit is usable. During periods of circuit availability, other quality-related parameters such as error rate express the soundness of the circuit. These quality parameters are selected according to the service and transmission system. In many cases, however, performance objectives are based on telephony service because of its dominance in today's communication systems. Other services provided in the same transmission network may demand greater performance calling for special procedures.

The allocation process can be done for each combination of transmission equipment unique to a particular circuit, but this approach is unnecessarily redundant and clearly tedious. Instead, hypothetical reference circuits are used to represent the majority of practical cases and to reduce the system designer's job to the allocation of performance for just this single circuit. Once this hypothetical circuit has been formulated, the overall (end-to-end) performance objective can be allocated to the individual subsystems and equipment. Numerous factors must be considered in constructing the hypothetical reference circuit, such as type of service, mix of transmission media, mix of analog and digital transmission, and multiplex hierarchy. Thus, because of diverse requirements in the type of service or transmission system, more than one reference circuit may be required to establish performance objectives for all cases. Figure 2.1 summarizes the general process for establishing performance allocations in a digital transmission system.

2.2   Transmission Services

The main services provided by a digital communications system are voice, data, and video. Each of these services is described in the following sections from the standpoint of how digital transmission systems accommodate the service. Emerging digital services are described in Chapter 13.

2.2.1   Voice

Voice signals are the major service provided by today's communication systems. For digital transmission applications, the voice signal must first be converted to a digital representation by the process of analog-to-digital (A/D) conversion. The principles of A/D conversion are illustrated in Figure 2.2. The input voice signal is first band-limited by a low-pass filter and then periodically sampled to generate a time-sampled, band-limited signal. These samples are then made discrete in amplitude by the process of quantization in which the continuous sample value is replaced by the closest discrete value from a set of quantization levels. The coder formats the quantized signal for transmission or multiplexing with other digital signals. **Pulse code**

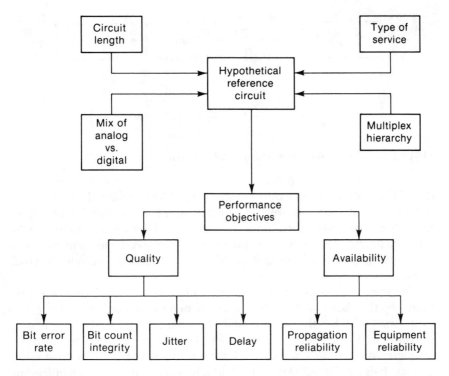

FIGURE 2.1   Process for Performance Allocation of a Digital Transmission System

modulation (PCM) represents each quantization level with a binary code. Thus if there are 256 quantization levels, an 8-bit PCM code is required for each sample. Rather than code the absolute value of each voice sample as is done in PCM, the difference or delta between samples can be coded in an attempt to eliminate the redundancy in speech and, in the process, lower the transmission bit rate. As shown in the dashed lines of Figure 2.2, the quantized sample can be fed back through a prediction circuit and subtracted from the next sample to generate a difference signal. This form of A/D conversion is known as **differential PCM** or **delta modulation.** A detailed description of these and other A/D techniques is given in Chapter 3.

The universal standard for voice digitization is PCM at a transmission rate of 64 kb/s. This rate results from a sampling rate of 8000 samples per second and 8 bits of quantization per sample. For multichannel digital transmission of voice, digitized signals are time-division multiplexed. The resulting composite transmission bit rate depends on the coding rate per voice signal, the number of voice signals to be multiplexed, and the overhead required for multiplex synchronization. In the North American stan-

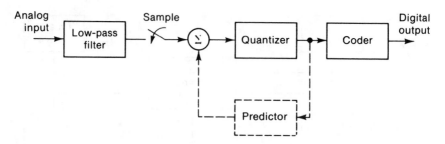

FIGURE 2.2   Block Diagram of Analog-to-Digital Converter

dard PCM multiplex, for example, there are 24 voice channels, each coded at 64 kb/s, plus 8 kb/s of overhead for synchronization, with a total transmission rate of 1.544 Mb/s. In the European CEPT standard, there are 30 voice channels, each at 64 kb/s, plus two 64 kb/s overhead channels used for signaling and synchronization, with a total transmission rate of 2.048 Mb/s.

The transmission quality of digitized voice channels is determined primarily by the choice of A/D technique, the number of tandem A/D conversions traversed by the circuit, and the characteristics of transmission impairment. For a choice of 64-kb/s PCM as the voice coder, the transfer characteristics of the voice channel are essentially the same as the analog channel being replaced. A/D conversion, however, introduces **quantization noise,** caused by the error of approximation in the quantizing process. With successive A/D conversions in the circuit, quantization noise will accumulate, although certain encoding techniques such as 64-kb/s PCM provide tolerable performance for a large number of A/D conversions. Regarding the effects of transmission impairment on voice quality, certain bit errors and timing slips will cause audible "clicks" and jitter will cause distortion in the speech signal recovered from 64-kb/s PCM [1]. Effects of a 10^{-4} error rate are considered tolerable and a 10^{-6} error rate negligible for 64-kb/s PCM. Depending on the technique employed, low-bit-rate voice (LBRV) coders may be more susceptible to degradation from greater quantization noise, multiple A/D conversions, and transmission impairment (Chapter 3).

2.2.2 Data

Data communication requirements are a fast-growing service, due largely to the emergence of networks that connect computer-based systems to geographically dispersed users. Most current data services are provided by modems operating over analog channels within a frequency-division multiplex (FDM) hierarchy as described in Chapter 12. Rates up to 19200 b/s can be operated over a single voice channel. Higher data rate modems require proportionally larger FDM bandwidths; for example, 56-kb/s modems have been developed for a 48-kHz bandwidth and 1.544-Mb/s modems for a 480-kHz bandwidth. Digital transmission facilities can be designed to handle data requirements more efficiently. However, early implementations of PCM for digital telephony did not include direct data interfaces. Thus modems were still required to provide the necessary conversion for interface of data with a 64-kb/s PCM voice channel. PCM channels can provide satisfactory performance for modems at rates up to 19.2 kb/s (see Chapter 3 for details), but at the expense of wasted bandwidth. To be more efficient, data channels can be directly multiplexed with PCM channels. In this case, a single 64-kb/s PCM channel might be replaced with, say, six channels of 9.6-kb/s data. Submultiplexing of 64-kb/s channels is now common in digital transmission facilities, as evidenced by the number of low-speed TDM equipment available today.

To understand the organization of emerging data transmission networks, it is necessary to define some specialized terminology and review existing interface standards. Figure 2.3 illustrates the basic interface between the **data terminal equipment** (DTE) and **data circuit terminating equipment** (DCE) of a data transmission network. The DTE is the user device, typically a computer, while the DCE is the communication equipment, say modem or multiplex, that connects the user to the network. The DTE/DCE interface is characterized by three types of signals—data, timing, and control—and by several levels of organization necessary for network applications.

Timing Signals
The use of the timing signal is determined by the type of DTE. **Asynchronous** terminals transmit data characters independently, separated by start-stop pulses. The use of unique start and stop pulses allows the receiving terminal to identify the exact beginning and end of each character. The stop pulse is made longer than the data bits to compensate for differences in

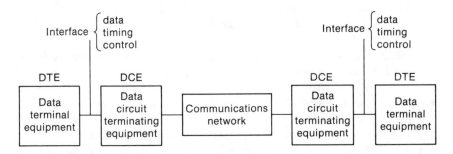

FIGURE 2.3   Data Transmission Network Interfaces

clock frequency at the transmitter and receiver. With this arrangement, no timing line is required between the DTE and DCE.

With **synchronous** transmission, characters are transmitted continuously in large blocks without separation by start or stop pulses. The transmitting and receiving terminals must be operating at the same speed if the receiver is to know which bit is first in a block of data. Likewise, the speed of the DTE must be matched to the communication line. The various possible arrangements of the timing lines for synchronous transmission are shown in Figure 2.4. With **codirectional timing,** the DTE supplies the timing line that synchronizes the DCE frequency source (oscillator) to the DTE. Conversely, with **contradirectional timing,** the DTE is slaved to the timing of the DCE. A third configuration, **buffered timing,** is often used when the communication line interfaces with several different terminals as in the case of time-division multiplexing. Here buffers allow each DTE to operate from its own timing source by compensating for small differences in data rate between each DTE and the DCE. These buffers are typically built into the DTE or DCE and are sized to permit slip-free operation for some minimum period of time. (See Chapter 10 for calculation of the buffer slip rate.) Receive timing in all cases is derived from the incoming received signal, which then synchronizes the receive terminal to the far-end transmitter. In **loop timing,** transmit timing of a terminal is locked to its receive timing, a technique that synchronizes two ends of a communications link.

Interface Standards
The various levels of interface can be best described by examining the standards that exist for data transmission. Only a brief summary of the topic is given here; a detailed account is beyond the scope of this book and has been given ample treatment elsewhere [2]. The basic architecture that has emerged identifies seven interface layers (Figure 2.5):

- Layer 1: the physical, electrical, functional, and procedural layer used to establish, maintain, and disconnect the physical link between the DTE and the DCE
- Layer 2: the data link control layer for interchange of data between the DTE and the network in order to provide synchronization control and error detection/correction functions
- Layer 3: the network control layer that defines the formatting of messages or packets and the control procedures for establishing end-to-end connections and transferring user data through the network
- Layer 4: the transport layer that provides error recovery and flow control for end-to-end transfer of data

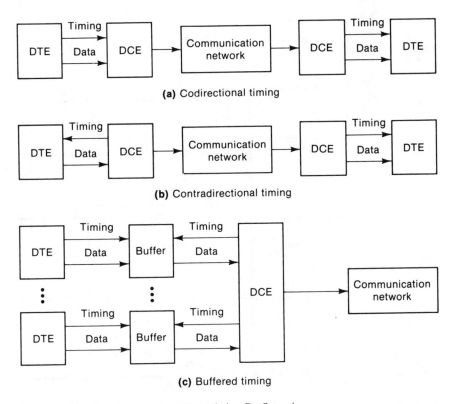

(a) Codirectional timing

(b) Contradirectional timing

(c) Buffered timing

FIGURE 2.4   Synchronous Data Transmission Configurations

- Layer 5: the session layer used to establish and manage connections between applications
- Layer 6: the presentation layer that allows applications to be independent of the data representation
- Layer 7: the application layer for interface of the user to the data network

The architecture described above and shown in Figure 2.5 is known as the *open systems interconnection* (*OSI*) model. The OSI model was developed by the ISO as a framework for the development of standards in data and computer communications. The development of these standards has been the responsibility of several organizations. Layer 1 standards have been developed by the ISO and CCITT, and in the United States by the ANSI and EIA. At layer 2, the ISO has had responsibility internationally and ANSI in the United States. Higher layer standards have been developed by the ISO, CCITT, and ANSI. These layers are not independent of

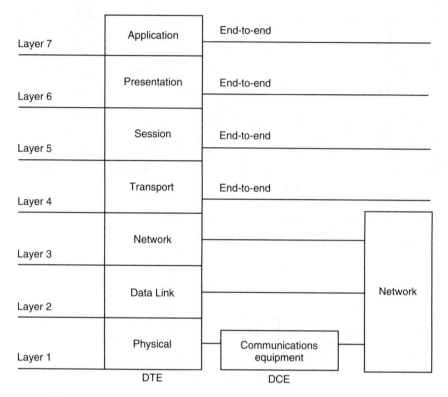

FIGURE 2.5   Architecture of Open Systems Interconnection Model

each other. Interface at a particular layer requires that all subordinate layers also be specified.

The interface of digital transmission systems with data transmission networks can be characterized primarily by the physical interface level. The standards for this layer 1 interface have undergone two generations of development, which are summarized in Table 2.1. The first generation is based on the EIA "Recommended Standard" RS-232E and its international counterpart, CCITT V.24 for functional characteristics and V.28 for electrical characteristics. These standards were limited to data transmission over telephone networks using modems at rates up to 20 kb/s. The newer, second-generation standards apply both to modems and to the new public data networks offering circuit-switching, packet-switching, and leased-circuit services. These new standards are the EIA RS-422 for balanced circuits (using a differential generator with both outputs aboveground) and RS-423 for unbalanced circuits (using a single-ended generator with one side grounded) and their international equivalents, CCITT X.27 (V.11) and

TABLE 2.1 Layer 1 Interface Standards for Data Transmission

Type	First Generation		Second Generation	
	Standard	Characteristics	Standard	Characteristics
Electrical	EIA RS-232E CCITT V.28	Unbalanced interface 0–20 kb/s data rate 50 ft maximum distance	EIA RS-422 CCITT X.27 (V.11)	Balanced interface 0–10 Mb/s data rate Interoperable with RS-423 and X.26 (V.10) Distance limitation: 4000 ft at 100 kb/s, 40 ft at 10 Mb/s
			EIA RS-423 CCITT X.26 (V.10)	Unbalanced interface 0–300 kb/s data rate Interoperable with RS-232E, V.28, RS-422, and X.27 (V.11) Distance limitation: 4000 ft at 3 kb/s, 40 ft at 300 kb/s
Functional	EIA RS-232C CCITT V.24	Definitions of data, timing, and control signals 25-pin connector	EIA RS-530* CCITT V.24	Definitions of data, timing, and control signals 25-pin connector
Mechanical	EIA RS-232C ISO 2110		EIA RS-530* ISO 2110	interoperable with RS-232E

*RS-530 replaces RS-449, an earlier standard based on a 37-pin connector.

29

X.26 (V.10). The CCITT X and V standards are equivalent; the X version is for public data network interfaces and the V version for modems operating in a telephone network. Standard RS-530 specifies the mechanical connector design and functional pin assignments for implementation of RS-422 and RS-423. RS-449 was an earlier second-generation standard for functional and mechanical interfaces, also applicable to RS-422 and RS-423, but because its 37-pin connector never gained popularity, RS-530 has been adopted and is gradually replacing RS-449. Salient characteristics of these layer 1 standards are listed in Table 2.1. Note the interoperability specification for second-generation standards, which allows continued use of older equipment in modern data networks.

Performance
Data transmission quality is measured by response time and block error rate. Digital transmission impairments that affect data quality include delay, slips, and bit errors. Most data users employ some form of error detection/ correction such as automatic request for repetition (ARQ) or forward error correcting (FEC) codes to augment the performance provided by a transmission channel. FEC techniques are based on the use of overhead bits added at the transmitter to locate and correct erroneous bits at the receiver. ARQ systems detect errors at the receiver and, after a request for repeat, automatically retransmit the original data. These error-correcting schemes act on blocks of data. Hence, the data user is concerned with block error rate, where the length of a block is determined either by the basic unit of traffic, such as an alphanumeric character, or by standard ARQ or FEC coding schemes.

2.2.3   Video
Digital coding has been applied to a variety of video signals, from high-definition television (HDTV) to video conferencing. The coding technique and bit rate depend, of course, on the application and the performance requirements. Coding techniques used for voice, such as pulse code modulation, are commonly used for video signals when the necessary bandwidth is available. For example, 8-bit PCM applied to a 6-MHz television signal would result in a bit rate of 96 Mb/s. Video compression schemes are even more popular, however, largely because of the need for digital video to be as bandwidth-efficient as analog video transmission. Differential PCM as shown by the dashed lines in Figure 2.2 is a common choice for coding, but transform and subband coding provide even greater compression (see Chapter 3). These waveform coding techniques are often used in combination with video compression schemes that take advantage of the redundancy in the signals and the perceptions of human vision. Compression in

space, known as *intraframe compression,* uses the correlation in a still picture to remove redundancy. Compression in time, known as *interframe compression,* exploits the correlation in time found in motion pictures. The human eye is much more perceptive of resolution in the luminance (brightness) signal than in the chrominance (color) signal. Consequently, the luminance signal is typically sampled at a higher rate and assigned more quantization bits than the chrominance signals. The eye is also more sensitive to lower-frequency components in the video signal than to its high-frequency components. Thus, coding techniques that operate on the signal in a transform domain will code low-frequency coefficients with more bits than high-frequency coefficients. Since video compression will distort moving pictures, some form of *motion compensation* is commonly used. Motion compensation can be performed on either the luminance or chrominance signals or both. Motion estimation is used to predict motion from a sequence of frames, or alternatively, from a sequence of points or regions within the frames.

Television
Encoding of color television falls into two separate categories. One operates directly on the composite television signal and the other divides the signal into three components that are separately encoded. Because studio cameras provide output in the form of these three components, component coding is more practical to achieve an all-digital television system. The three components of color television signals can be represented in two basic forms: the R-G-B, which stands for the three primary colors red, green, and blue; and the Y-U-V, where Y is the luminance signal and U and V are color difference signals. The format of the National Television System Committee (NTSC) color television signal, used in the United States, is based on a luminance component and two chrominance components. Likewise, the other two standards for color television systems, PAL and SECAM, used principally in Western Europe and Eastern Europe, respectively, are based on three separate components. Various coding schemes have been applied to digitize these existing television systems [3]. International standards have been the responsibility of a joint CCITT/CCIR committee, the Television Transmission Joint Committee (CMTT), which deals with transmission of sound and color television signals for long-haul transmission. The CMTT in Recommendations 721 and 723 and Reports 1235 and 1234 has provided general specifications of coding systems with bit rates of 32, 34, 45, 68, and 140 Mb/s [4,5]. These bit rates are based on standard rates from either the European, North American, or Japanese digital hierarchy. However, the main form of transmission for conventional television has remained analog to allow compatibility with existing television sets and bandwidth allocations.

The introduction of **high-definition television (HDTV)**, however, has led to the development of several digital coding schemes for applications to the studio camera, transmission lines, and subscribers. HDTV requires a much wider bandwidth than conventional television because of increased luminance definition, a larger aspect ratio, which means a wider screen, and greater sampling rate applied to the luminance and two chrominance signals. Using 8-bit PCM without any form of video compression, a bit rate of 1 Gb/s would be needed. Although such a bit rate might be used for studio camera applications, digital transmission of HDTV must be much more efficient.

In the United States, all-digital systems employing video compression are in the forefront of HDTV technology. The FCC has mandated that HDTV be capable of simulcast so that NTSC signals can be broadcast terrestrially on one channel and HDTV signals can be carried on a second channel, without interference of one signal type with the other. The restriction of HDTV signals to the existing 6-MHz bandwidth channel used for analog television broadcasting has led to highly efficient compression and error correction algorithms that typically operate at a transmitted bit rate of approximately 20 Mb/s. Similar HDTV coding schemes have been developed in the United States for satellite broadcasting and cable television using bit rates in the range of 15 to 30 Mb/s. Europe and Japan have focused on the use of satellite broadcasting for HDTV, but mostly using analog transmission. The use of optical fiber and SONET transmission for HDTV has also been proposed, which would allow larger bandwidths and simpler coders, with resulting transmission rates in the range of 100 Mb/s to 1 Gb/s [6,7]. CCIR Rep. 1092-1 gives examples of HDTV digital transmission for optical fiber and satellite channels, with bit rates between 100 and 400 Mb/s [5].

Video Teleconferencing
Video teleconferencing permits conferees at two or more facilities to have real-time, interactive, visual contact. Bandwidths used for broadcast television are too costly and unnecessary for teleconferencing. Typically, the picture to be transmitted consists of graphics, slides, or conferees whose movement is very limited. Due to this limited motion, bandwidth compression techniques can be applied while maintaining good video resolution. Video teleconferencing services now commercially available use digital encoding, video compression, and digital transmission to achieve lower costs.

Standard bit rates for teleconferencing services have included 6.3 Mb/s [8], 3 Mb/s [9], 1.544 Mb/s [10], and 2.048 Mb/s [11]. Standards for video teleconferencing were initially based on video broadcasting standards such as those established by the U.S. NTSC. However, the CCITT has published

Rec. H.261, which prescribes video teleconferencing bit rates of $p \times 64$ kb/s, for p equals 1 to 32. This standard is based on a common set of nonproprietary algorithms that use a combination of compression techniques, including transform coding, interframe prediction, variable-length coding, motion compensation, and error correction [12]. Commercially acceptable picture quality at 64 kb/s is the ultimate objective, which would provide transmission compatibility with dial-up ISDN channels.

Performance
Digital transmission impairments that affect video performance include jitter, slips, and bit errors. Depending on the choice of A/D conversion method, another impairment is the distortion caused by the sampling and filtering process, amplitude quantization, and bit rate reduction coding. Error performance requirements for broadcast-quality television are usually more stringent than those for voice and data because of the disturbing effects of loss of TV synchronization. Permissible error rates are in the range between 1×10^{-4} and 1×10^{-9}, depending on the choice of A/D conversion and the use of forward error correction techniques [5].

2.3 Hypothetical Reference Circuits

The task of assigning performance allocations to a digital transmission system is greatly facilitated by use of a hypothetical reference circuit composed of interconnected segments that are representative of the media and equipment to be used. Reference circuits are a standard tool used by communication system designers, including the CCITT, CCIR, and AT&T, to provide the basis for development of:

- Overall user-to-user performance requirements for satisfactory service
- Performance specifications of typical segments of the overall circuit that when connected will meet the overall performance objectives

The hypothetical reference circuit should represent a majority of all circuits but is not necessarily the worst-case design. The resulting design should provide satisfactory service to a majority of users. The numerous factors to be considered in constructing a reference circuit are discussed in the following paragraphs.

Path Length
The dominant factor in formulating the reference circuit is the path length. International circuits may span the globe and use several media types, including terrestrial radio, satellite, fiber optic cable, and metallic cable. National reference circuits are considerably shorter in length and use pri-

marily cable and terrestrial radio. A national reference circuit may be connected to an international reference circuit—as in the case of the longest reference circuit described by the CCITT, shown in Figure 2.6a. This circuit has a 25,000-km international connection with two 1250-km national systems. Separate reference circuits have also been developed by the CCITT and CCIR for specific transmission media including coaxial cable (CCITT Rec. G.332), line-of-sight radio relay (CCIR Rec. 556-1), tropospheric scatter radio relay (CCIR Rec. 396-1), and communication satellite systems (CCIR Rec. 521-2).

Multiplex Hierarchy
Another factor in structuring the reference circuit is the number of multiplex stages and their location in the network. The reference circuit should represent the tandem multiplex and demultiplex functions that occur from one end of the circuit to the other. If the service is telephony, the circuit may also undergo a number of tandem analog-to-digital conversions. The reference circuit should show the relative composition of multiplex and A/D functions at appropriate locations. Typically, higher-level multiplexers appear more often than lower-level multiplexers. The A/D conversion required for digital telephony would occur at the lowest level in the hierarchy and at analog switching or patching locations. The CCIR's hypothetical reference circuit for digital radio-relay systems shown in Figure 2.6b includes nine sets of digital multiplex equipment at the hierarchical levels recommended by the CCITT.

Analog vs. Digital
Since most existing transmission systems are analog, reference circuits in the past have been based on the use of analog equipment—for example, a choice of frequency-division multiplex for telephony. As analog transmission networks incorporate and evolve toward digital facilities, the reference circuit must show a representative mix of analog and digital transmission facilities. Interfaces of digital and analog equipment also must be reflected in the choice of reference circuit. An example of such a hybrid system is the use of digital multiplexing for telephony but with circuit switching done on an analog basis, requiring that A/D conversion equipment (such as PCM) appear at each switching center. Such hybrid reference circuits are important in planning the transition strategy for exchange of analog with digital transmission. The only all-digital reference circuits that have been developed by the CCITT or CCIR are shown in Figures 2.6 and 2.7. These circuits are limited principally to telephony. Reference circuits for other services and other media using digital transmission are under development by both organizations.

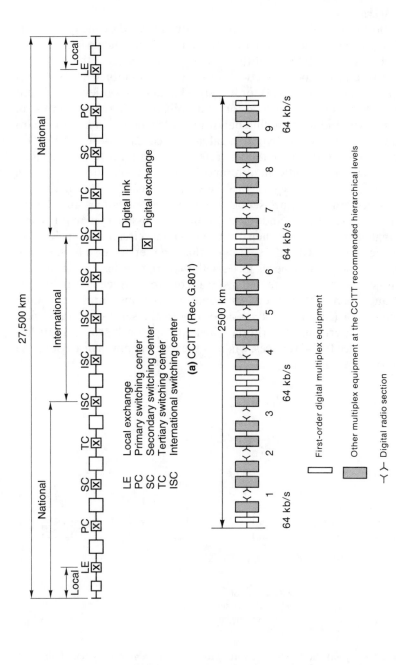

FIGURE 2.6 Hypothetical Reference Circuits for 64-kb/s Digital Transmission [13, 14] (Courtesy CCITT and CCIR)

(a) CCITT (Rec. G.801)

(b) CCIR (Rec. 556-1)

LE Local exchange
PC Primary switching center
SC Secondary switching center
TC Tertiary switching center
ISC International switching center

☐ Digital link
☒ Digital exchange

☐ First-order digital multiplex equipment
▨ Other multiplex equipment at the CCITT recommended hierarchical levels
⌒⟩⌒ Digital radio section

27,500 km

National

International

National

Local

Local

2500 km

64 kb/s   64 kb/s   64 kb/s   64 kb/s   64 kb/s

1 2 3 4 5 6 7 8 9

Type of Service

The composition of the reference circuit may be affected by the type of service being provided. The hypothetical reference circuits shown in Figure 2.6 are based on telephony. Should the service be data instead, the multiplex hierarchy would likely be affected—showing either additional levels of multiplex (below the 64-kb/s level) for low-speed data or fewer levels of multiplex for high-speed data. Television transmission generally requires an entire terrestrial radio or satellite bandwidth allocation, eliminating multiplex equipment from the hypothetical reference circuit—as shown, for example, by the CCIR's hypothetical reference circuit for television (Rec. 567-3) [4]. The most significant difference in these various services, as seen by the designer of the hypothetical reference circuit, is the method of describing performance. Each service—data, voice, and video—may have a different set of parameters and values. If the transmission system is to carry only a single type of service, then a single reference circuit and single set of performance criteria will suffice. If more than one service is to be supplied, however, the system designer is faced with two approaches: (1) Set performance objectives to satisfy the largest class of users, which is usually telephony, and force other users to adopt special measures for better performance, or (2) set objectives according to the service with the most stringent set of performance criteria. The latter approach provides satisfactory performance for all services, whereas the former approach accounts for the majority of users.

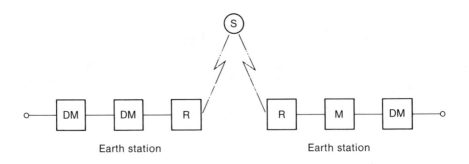

S: Space station
DM: Digital multiplex equipment
M: Modem equipment
R: IF/RF equipment

FIGURE 2.7   Hypothetical Reference Circuit for Digital Transmission Systems Using Satellite Communications (CCIR Rec. 521-2) [15] (Courtesy CCIR)

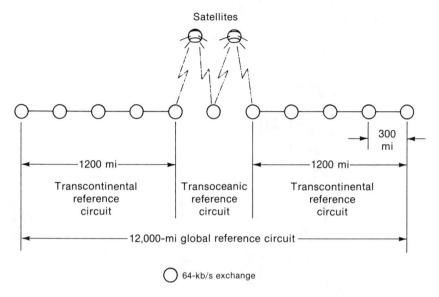

FIGURE 2.8   Example of Global Reference Circuit

Example 2.1

Figure 2.8 shows a typical global reference circuit spanning two continents and one ocean. The transoceanic circuit consists of two satellite hops, although one or both of these hops could well be transoceanic cable. Each transcontinental circuit is 1200 mi in length and is composed of four terrestrial reference circuits, which are each 300 mi in length. Figure 2.9 shows the terrestrial reference circuit to be composed of line-of-sight (LOS) digital radio along the backbone and digital cable, both metallic and fiber optic, at the tails. The repeater sites provide regeneration of the digital signal but do not include multiplexers. The sites labeled M2 utilize second-level multiplexers to interconnect 1.544-Mb/s digital signals. Finally, the sites at the end of the terrestrial reference circuit will use first and second levels of multiplex equipment to derive individual 64-kb/s voice and data channels.

2.4   Performance Objectives

Performance objectives for a digital transmission system may be stated with two characteristics: availability and quality. **Availability** is defined as the probability or fraction of time that circuit continuity is maintained and usable service is provided. The quality of transmission, then, is considered only when

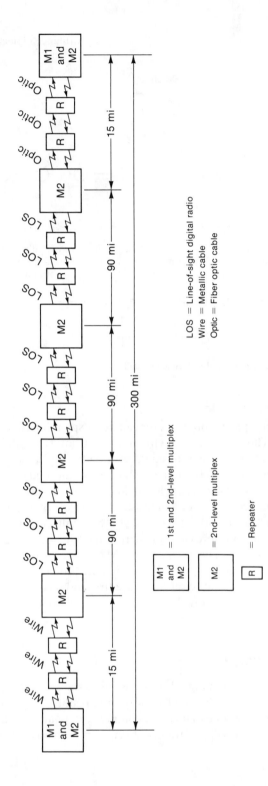

FIGURE 2.9 Example of Terrestrial Reference Circuit

LOS = Line-of-sight digital radio
Wire = Metallic cable
Optic = Fiber optic cable

| M1 and M2 | = 1st and 2nd-level multiplex |

| M2 | = 2nd-level multiplex |

| R | = Repeater |

38

the circuit is available. For digital transmission, quality parameters commonly used include bit error rate, timing slips, jitter, and delay. These parameters apply to all types of services, although parametric values may differ and additional parameters may be called out that are related to service. Here we will consider availability and the quality parameters just cited.

Before proceeding, a distinction must be made between design objectives discussed here and operation and maintenance (O&M) standards of performance. In operation, the performance of a service may deteriorate for various reasons: aging, excessive environmental conditions, human error, and so forth. Therefore design objectives should provide a suitable margin to assure that the service in operation meets O&M performance standards. In addition, the commissioning of real circuits requires another set of performance values to be met. The actual makeup of a real circuit will vary from the hypothetical reference circuit so that allowance must be made for the unique configuration of the actual system. The values selected for commissioning represent the minimum acceptable performance during installation and acceptance testing. Commissioning tests should be conducted during periods of normal operating conditions since time does not permit tests to be conducted over all operating conditions. When performance degrades significantly below the commissioning values, maintenance actions are required. The O&M performance standards represent limits beyond which the service is either degraded or unacceptable. When these limits are exceeded, maintenance actions are required to restore the circuit performance or to remove the circuit from service.

2.4.1 Availability Objectives

Periods of time when a circuit is not usable, or **unavailable,** may be categorized as long term or short term. This distinction is made because of the relative impact of long-term and short-term outages on the user. A voice user subjected to a long-term outage (say, greater than a few seconds in length) will be automatically disconnected from the multiplexer or switch or will impatiently hang up his instrument. A data user subjected to a long-term outage will have messages corrupted, since the outage length will exceed the ability of coding or requested retransmission to provide error correcting. Short-term outages (less than a few seconds) will be a disturbance to voice users but probably will not cause termination of the connection. Likewise, short-term outages in the data user's channel may be compensated by error correction or retransmission techniques. Multipath fading in radio transmission, resynchronization in time-division multiplexers, and automatic switching actions for restoral in redundant equipment are all sources of short-term outages. Long-term outages consist primarily of catastrophic failure of equipment such as radio, antennas, or power supply, although certain propagation conditions such as precipitation attenuation and power fading also produce long-term outages.

There are two basic approaches to allocating availability: (1) Lump all outages together, long and short term, or (2) deal only with long-term outages. The second approach places short-term outages in the category of circuit quality and not circuit availability. In either case, an end-to-end availability is levied that takes on typical values of 0.9999 to 0.99. Further suballocation is then made to individual segments and finally equipment within the hypothetical reference circuit.

Following the definition given by the CCITT and CCIR (shown later in Tables 2.3 and 2.4), we will consider a circuit to be unavailable when the service is degraded (for example, bit error rate is worse than a specified threshold) for periods exceeding n seconds. Typically the values of n that have been used or proposed are in the range of a few seconds to a minute. Thus this definition tends to exclude short-term outages and includes only long-term outages caused principally by equipment failure.

Definitions

To allocate performance, the system designer must understand and apply the science of statistics as used in the prediction and measurement of equipment availability. The following definitions are essential to this understanding.

Failure refers to any departure from specified performance, whether catastrophic or gradual degradation. For prediction and system design purposes, the types of failure considered are limited to those that can be modeled statistically, which excludes such instances as natural disasters or human error. **Failure rate** (λ) refers to the average rate at which failures can be expected to occur throughout the useful life of the equipment. Equipment failure rates can be calculated by using component failure rate data or obtained from laboratory and field tests.

Reliability, $R(t)$, refers to the probability that equipment will perform without failure for a period of time t given by

$$R(t) = e^{-\lambda t} \tag{2.1}$$

Studies of failure rates in electronic equipment have shown that the reliability of components decays exponentially with time. It is known, however, that electronic equipment may contain parts subject to early failure, or "infant mortality." It is also known that the rate of failures increases as the device nears the end of its life. During the useful life of the equipment, however, the rate of failure tends to remain constant.

Mean time between failures (MTBF) refers to the ratio of total operating time divided by the number of failures in the same period:

$$\text{MTBF} = \int_0^\infty t\left(\lambda e^{-\lambda t}\right) dt = \frac{1}{\lambda} \qquad (2.2)$$

Mean time to repair (MTTR) is defined as the average time to repair failed equipment, including fault isolation, equipment replacement or repair, and test time, but excluding administrative time for travel and locating spare modules or equipment. **Mean time to service restoral (MTSR)** is the average time to restore service, including repair and administrative time.

Availability (A) is the probability that equipment will be operable at any given point in time or the fraction of time that the equipment performs the required function over a stated period of time. It is also defined as the ratio of uptime to total time:

$$A = \frac{\text{MTBF}}{\text{MTBF} + \text{MTSR}} \qquad (2.3)$$

Unavailability (U) is the complement of availability:

$$\begin{aligned} U &= 1 - A \\ &= \frac{\text{MTSR}}{\text{MTBF} + \text{MTSR}} \end{aligned} \qquad (2.4)$$

For large MTBF and small MTSR (usually the case)

$$U \approx \frac{\text{MTSR}}{\text{MTBF}} \qquad (2.5)$$

Outage (O) is the condition whereby the user is deprived of service due to failure within the communication system. For redundant equipment, a single failed unit should not cause an outage but should cause switchover to a backup unit. **Mean time between outages (MTBO)** refers to the ratio of total operating time divided by the number of outages in the same period.

Redundant Equipment
In equipment designed with redundancy, an outage can occur either from simultaneous failure of all redundant units or from a failure to properly rec-

ognize and switch out a failed unit. If the equipment consists of n identical units, each with the same MTBF, for which at least r must be operable for the system to be operable, then the first type of outage can be characterized by

$$\text{MTBO}_1 \simeq \frac{\text{MTBF}(\text{MTBF}/\text{MTTR})^{n-r}}{n(n-1)!/(r-1)!(n-r)!}, \text{ for MTTR} \ll \text{MTBF} \qquad (2.6)$$

where the sensing of a failure and resulting switchover is itself assumed to occur without failure. Expression (2.6) thus accounts for the situation where the backup units all fail within the MTTR for the first failed unit. As an example, consider two identical units, one used on-line and the other used as a redundant standby. For this case, $n = 2$, $r = 1$, and

$$\text{MTBO}_1 \simeq \frac{\text{MTBF}^2}{2(\text{MTTR})}, \text{ for MTTR} \ll \text{MTBF} \qquad (2.7)$$

The monitoring and control necessary for redundancy switching has some probability P_s of successfully detecting failure in the on-line unit and completing the switching action to remove the failed unit. This second type of outage for redundant equipment has an MTBO given by

$$\text{MTBO}_2 = \frac{\text{MTBF}}{1 - P_S} \qquad (2.8)$$

The total MTBO of the redundant equipment is then

$$\frac{1}{\text{MTBO}} = \frac{1}{\text{MTBO}_1} + \frac{1}{\text{MTBO}_2} \qquad (2.9)$$

With double redundancy, for example, the total MTBO is found from (2.7) and (2.8):

$$\text{MTBO} = \frac{\text{MTBF}^2}{2(\text{MTTR}) + \text{MTBF}(1 - P_S)} \qquad (2.10)$$

Availability for redundant equipment can be expressed by using the basic definition given in (2.3) with MTBO used in place of MTBF:

$$A = \frac{\text{MTBO}}{\text{MTBO} + \text{MTSR}} \qquad (2.11)$$

whereas the unavailability can be expressed from (2.5) as

$$U \approx \frac{MTSR}{MTBO} \tag{2.12}$$

Provision for redundancy in equipment design allows uninterrupted service when a single unit fails. For redundancy to be effective, the system designer must be able to guarantee that:

1. Failures in the active (on-line) unit are recognized and successfully switched out. This requires a sophisticated monitoring and control capability (see Chapter 11). Both on-line and off-line units should be continuously monitored and compared with some performance factor such as error rate.
2. A prescribed MTBO can be met with the design proposed. This requires a failure analysis to demonstrate that a certain percentage of all failures will be recognized and removed by automatic switchover.
3. Failure sensing and switching times are short enough to be classified as a short-term outage, so that they are not counted as a contribution to system unavailability.

Example 2.2

A doubly redundant system has the following characteristics:

$$MTBF = 3200 \, hr$$
$$MTTR = MTSR = 15 \, min$$
$$P_s = 0.97$$

Find the total MTBO and unavailability. What effect will $P_s = 0.9$ have on the system?

Solution   From (2.10) we have, for a doubly redundant system,

$$MTBO = \frac{(3200)^2}{2(0.25) + 3200(1 - 0.97)}$$

$$= 106,114 \, hr$$

The unavailability is given by (2.12) as

$$U = \frac{0.25}{106,114} = 2.4 \times 10^{-6}$$

For $P_s = 0.9$, the corresponding MTBO and U are

$$\text{MTBO} = 31,950\,\text{hr}$$
$$U = 7.8 \times 10^{-6}$$

System Availability

To calculate the availability of a system composed of interconnected equipment, one needs only to apply the basic laws of probability to the definitions and equations given here. Failures can be assumed to be independent from one piece of equipment to another. Availability expressions are given here for typical system configurations.

Series Combination.   For a series combination (Figure 2.10) the system availability is the product of individual availabilities, and system failure rate is the sum of individual failure rates:

$$A_N = \prod_{i=1}^{N} A_i \tag{2.13}$$

$$\lambda_N = \sum_{i=1}^{N} \lambda_i \tag{2.14}$$

The unavailability is then

$$U_N = 1 - \prod_{i=1}^{N} A_i \tag{2.15}$$

It can be shown that for small outages system unavailability is determined by summing the outages of individual equipment. Therefore

$$A_N \approx 1 - \sum_{i=1}^{N} U_i \tag{2.16}$$

Parallel Combination.   If only one piece of equipment must be available for the whole system to be available, then for the parallel configuration of Figure 2.11

$$A_M = 1 - \prod_{i=1}^{M} U_i \tag{2.17}$$

where the product of individual unavailabilities yields the probability that all M units are unavailable.

Series-Parallel Combination.   Using (2.13) and (2.17), the availability for a series-parallel combination (Figure 2.12) can be written as

$$A_{NM} = \prod_{i=1}^{N}\left(1 - \prod_{j=1}^{M} U_{ij}\right) \qquad (2.18)$$

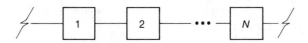

FIGURE 2.10   Series Combination

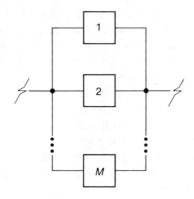

FIGURE 2.11   Parallel Combination

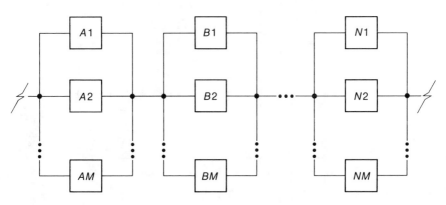

FIGURE 2.12   Series-Parallel Combination

Parallel-Series Combination.   Using (2.15) and (2.17), the availability for a parallel-series combination (Figure 2.13) can be written as

$$A_{MN} = 1 - \prod_{i=1}^{M}\left(1 - \prod_{j=1}^{N}A_{ij}\right)$$

(2.19)

Effect of Maintenance and Logistics Practices

Maintenance and logistics practices have a strong impact on system availability because of potential outage extension beyond the MTTR. This outage extension is influenced by:

- *The sparing of modules and equipment:* Without a complete supply of replacement parts, failed equipment may go unrepaired for a time much in excess of the MTTR. Logistics support must maintain a readily available supply of spare parts to avoid jeopardizing system availability.
- *Travel time:* Outages at unmanned sites are largely determined by the time required for a maintenance team to arrive on site. Unavailability of spares at the unmanned site is another potential cause of outage. Moreover, failures at such sites must be recognized and isolated by a fault reporting system before any repair action can be taken.
- *Periodic inspection and test:* Although digital transmission equipment generally provides extensive self-diagnostics, periodic inspection and testing can prevent later outages. Suppose, for example, the off-line unit of a redundant piece of equipment fails without the system operator's knowledge. Periodic manual switching between redundant units would allow recognition and repair of this failure, thus preventing later outages.

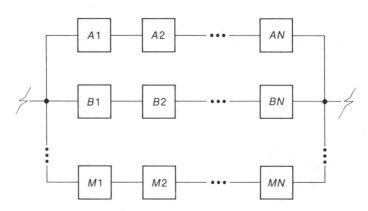

FIGURE 2.13   Parallel-Series Combination

Example 2.3

Based on an end-to-end availability objective of 0.99 for the global reference circuit of Example 2.1, the following unavailability allocations are proposed:

Transcontinental 1	0.004
Satellite 1	0.001
Satellite 2	0.001
Transcontinental 2	0.004
Total unavailability	0.01

Since the two transcontinental circuits are each composed of four identical 300-mi terrestrial reference circuits in tandem, each terrestrial reference circuit has an unavailability allocation of 0.001. To proceed with further suballocation some assumptions must be stated about the terrestrial reference circuit of Example 2.1 and the unavailability allocations shown in Table 2.2:

1. The equipment's MTBO is based on current design practices in digital communications. Redundancy with automatic switchover is employed in radio and cable systems and with higher-level multiplex (level 2 and above) due to the number of circuits that are affected when this equipment fails.
2. The MTSR values shown in Table 2.2 are ½ hr at an attended location and 3 hrs at an unattended site. The MTSR values for radio, cable, and

TABLE 2.2   Unavailability Allocation for Hypothetical Reference Circuit Shown in Figure 2.9

Equipment	Quantity	MTSR (hr)	MTBO (hr)	Unavailability ($\times 10^4$)
Digital LOS radio	18	2.16	125,000	3.12
Digital fiber optic cable system	6	2.16	125,000	1.04
Digital metallic cable system	6	2.16	50,000	2.60
Second-level multiplex				
Common equipment	10	0.5	200,000	0.25
Channel equipment	10	0.5	220,000	0.27
First-level multiplex				
Common equipment	2	0.5	7,000	1.43
Channel equipment	2	0.5	170,000	0.06
Station power	16	2.06	500,000	0.66
Total				9.43

station power equipment are averages and result from the ratio of attended locations to unattended locations.
3. All repeater sites are unattended. Sites with multiplex equipment are all attended.

Based on these assumptions, the unavailability contribution from each type of equipment was calculated using Equation (2.12). Unavailabilities were then summed according to the quantity of each piece of equipment and entered in the last column of Table 2.2. The total unavailability is then simply the sum of that column's entries. The resulting value of 0.000943 for this example is less than 0.001, thus meeting the objective for unavailability of this terrestrial reference circuit.

2.4.2   Error Performance

The basic quality measure for a digital transmission system is its error performance. Transmission errors occur with varying statistical characteristics, depending on the media. On digital satellite links, in the absence of forward error correcting, errors tend to occur at random, suggesting that average bit error rate (BER) can be satisfactorily used for error allocation in satellite transmission. For cable and terrestrial radio media, the error performance is characterized by long periods with no or few errors interspersed with short periods of high error rate. For such media, error allocation is best done by the percentage of time spent below a specified error threshold. The choice of the error parameter, be it average BER or percentage of time below threshold, is also determined by the service's requirements (voice, data, or video). One service may be tolerant of randomly distributed errors but intolerant of error bursts whereas another service may perform better under the opposite conditions. Finally, the system designer should also be motivated to select error parameters that are easy to allocate and apply to link design, easy to verify by measurement, and easy for the user to understand and apply to the performance criteria.

Error Parameters
The probability of error in a transmitted bit is a statistical property. The measurement or prediction of errors can be expressed in various statistical ways, but four main parameters have been traditionally used:

- *Bit error rate (or ratio) (BER)*: ratio of errored bits to the total transmitted bits in some measurement interval. The *residual bit error rate (RBER)* or *background bit error rate (BBER)* can be used to measure BER with outage periods not included in the measurement. The RBER is particularly useful for radio systems as a way to measure BER in the absence of fading.

- *Error-free seconds (EFS) or error seconds (ES)*: percentage or probability of one-second measurement intervals that are error free (EFS) or in error (ES)
- *Percentage of time (T_1) that the BER does not exceed a given threshold value:* percentage of specified measurement intervals (say, 1 min) that do not exceed a given BER threshold (say, 10^{-6}). Examples include the *severely errored second (SES)*, which is based on seconds with a BER threshold of 10^{-3}, and the *degraded minute (DM)*, which is based on minutes with a BER threshold of 10^{-6}.
- *Error-free blocks (EFB)*: percentage or probability of data blocks that are error free

These parameters all attempt to characterize the same performance characteristic—bit errors—but yield different results for the same link and measurement interval. Before examining specific applications of each parameter, we will establish the relationships among them. First, for statistically independent errors with an average probability of error $p = $ BER, a binomial distribution of errors can be assumed, which leads to the relationship

$$\% \, EFS = 100(1 - p)^R \tag{2.20}$$

for a bit rate R. This probability may also be expressed in terms of the Poisson distribution when p is small and the number of transmitted bits is large. Then the probability of observing x errors in n transmitted bits is given by

$$P(x) = \frac{e^{-\mu}\mu^x}{x!} \text{ for } p \ll 1, n \gg 1 \tag{2.21}$$

where $\mu = np$.

The probability of no errors is given by

$$P(0) = e^{-\mu} \tag{2.22}$$

For a time period of 1 s, the Poisson distribution may be related to percentage of error-free seconds by

$$\% \, EFS = 100e^{-pR} \tag{2.23}$$

This relationship is shown in Figure 2.14 for a rate $R = 64$ kb/s. As an example, a mean probability of error of 1.3×10^{-6} is required to meet an objective of 92 percent error-free seconds for a 64-kb/s channel rate.

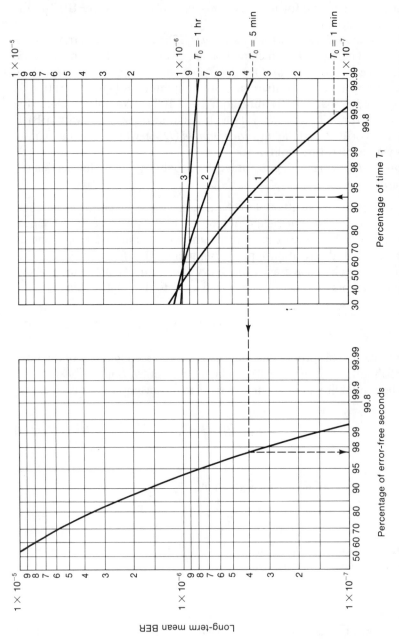

FIGURE 2.14 Relationship Between Percentage of Error-Free Seconds and Long-Term Mean BER for 64 kb/s [16] (Reprinted by Permission of Bell-Northern Research)

FIGURE 2.15 Relationship Between Percentage of Time T_1 and Long-Term Mean BER for Mean BER of 1×10^{-6} of Various Averaging Periods T_0 for 64 kb/s [16] (Reprinted by Permission of Bell-Northern Research)

The percentage of time T_1 that the BER does not exceed a given threshold value may be related to long-term average error rate, p, by use of the Poisson distribution:

$$\% \ T_1 \ (\text{BER} \leq \text{threshold}) = 100P \ (x \leq T_0 Rp)$$

$$= 100 \sum_{x=0}^{T_0 Rp} \frac{(T_0 Rp)^x}{x!} \ e^{-(T_0 Rp)} \tag{2.24}$$

where
x = number of errors in time T_0
R = bit rate
T_0 = measurement interval
$T_0 Rp$ = error threshold (integer value)

To take an example, assume the BER threshold to be 1×10^{-6} averaged over 1-min intervals for a 64-kb/s bit rate. The error threshold to meet this BER averaged over 1-min intervals is $T_0 Rp = (60)(64 \times 10^3)(10^{-6}) = 3.84$; therefore the 64-kb/s connection must have three or fewer errors, where

$$P \ (x \leq 3) = \sum_{x=0}^{3} \frac{(60 \ Rp)^x}{x!} \ e^{-(60 \ R \ p)} \tag{2.25}$$

This relationship is plotted as curve 1 in Figure 2.15. To illustrate the use of Figure 2.15, consider the following two examples:

1. If 90 percent of 1-min intervals must have a BER better than 10^{-6}, a long-term mean BER of $\leq 4.5 \times 10^{-7}$ would be required.
2. A long-term mean BER of 10^{-6} would yield only 46 percent of the 1-min intervals meeting the 10^{-6} BER threshold.

Note by comparison of Figure 2.15 with Figure 2.14 that the percentage of time spent below BER threshold can be related to the percentage of EFS through a common parameter: long-term average BER.

For statistically independent errors with average probability of error p, the probability of block error is given by the binomial distribution

$$P \ (\text{block error}) = \sum_{k=1}^{n} \binom{n}{k} p^k \ (1 - p)^{n-k} = 1 - (1 - p)^n \tag{2.26}$$

where n = block length

$$\binom{n}{k} = \text{binomial coefficient} = \frac{n!}{k!\,(n-k)!}$$

Conversely, the probability of error-free blocks (EFB) is simply $(1 - p)^n$. Expression (2.26) may be approximated for two cases that are commonly assumed:

$$P\,(\text{block error}) \approx 1 - e^{-np} \qquad \text{for } p \ll 1, n \gg 1 \tag{2.27a}$$

$$P\,(\text{block error}) \approx np \qquad \text{for } np \ll 1 \tag{2.27b}$$

The relationship of block error probability to bit error probability and error-second probability is illustrated in Figure 2.16 and summarized here:

1. For a block of length 1, the bit error probability is equal to the block error probability.
2. For a block length equal to the data rate, the error-second probability is equal to the block error probability.

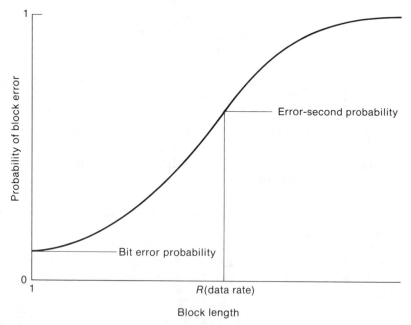

FIGURE 2.16   Probability of Block Error vs. Block Length

3. As the block length increases, the probability of block error also increases, approaching 1 asymptotically.

Error Models
The error parameters and their interrelationships have been described here under the assumption that errors occur at random. Most transmission systems experience errors in bursts or clusters, however, which invalidates the use of the Poisson (random) error model. To characterize this clustering effect, various models for channels with memory have been developed, among them the Neyman Type A contagious distribution [17, 18, 19]. This distribution is a compound Poisson model in which error clusters have a Poisson distribution and errors within a cluster also have a Poisson distribution. Each distribution can be described by its mean:

M_1 = mean number of clusters per sample size of transmitted bits
M_2 = mean number of errors per cluster

The probability of a sample containing exactly r errors is given by

$$P(r) = \frac{M_2^r}{r!} e^{-M_1} \sum_{j=0}^{\infty} \frac{z^j j^r}{j!} \qquad (2.28)$$

where $z = M_1 e^{-M_2}$. The probability of a sample containing no errors is given by

$$P(0) = \exp\left\{-M_1 \left(1 - e^{-M_2}\right)\right\} \qquad (2.29)$$

The mean of the distribution is $M_1 M_2$; this is equal to np, where n is the number of bits in the sample and p is the long-term mean bit error probability. The variance of the distribution is $M_1 M_2 (1 + M_2)$. For a given error distribution, defined in terms of M_1 and M_2, it is possible to calculate the percentage of:

- Error-free seconds (Figure 2.17)
- Minutes during which the error rate is better than a specified threshold, say 10^{-5} or 10^{-6} (Figure 2.18)

Figure 2.17 indicates that the random error model represents worst case results for percentage of error-free seconds. As the errors become more clustered, the performance expressed in terms of EFS is significantly better than would be observed for a random error channel. Figure 2.18a also indicates that the random error model represents worst case results for percentage of minutes below a 10^{-6} BER. For the same measurement interval of one minute but a BER threshold of 10^{-5}, Figure 2.18b reveals that error clustering up to a mean value of errors per cluster (M_2) of about 50 actually reduces the percentage of acceptable minutes for a given long-term mean error rate. For M_2 greater than 50, the percentage of acceptable minutes increases monotonically. The difference in the curves of Figure 2.18a versus Figure 2.18b is explained by the maximum acceptable number of errors per minute, which is 38 for a 10^{-5} BER threshold but only three for a 10^{-6} BER threshold.

The Neyman Type A distribution is an example of a *descriptive* error model that attempts to characterize channels by use of statistics that describe the general behavior of the channel. The Neyman model has been effectively applied to channels that have relatively few error-causing mechanisms, such as cable [19], but has not been effectively used with more complex channels such as digital radio. Another approach is based on *generative* error models that match the channel by iteratively expanding the "size" of the model to fit the error data. Such generative models are usually based on a finite-state Markov chain. Early Markov chain models such as Gilbert's

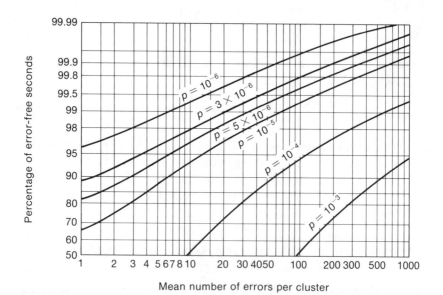

FIGURE 2.17   The Variation with Degree of Clustering of the Percentage of Seconds of 64-kb/s Transmission That Are Error-Free for Different Values of Long-Term Mean Error Ratio (p) [18] (Reprinted by Permission of IT&T)

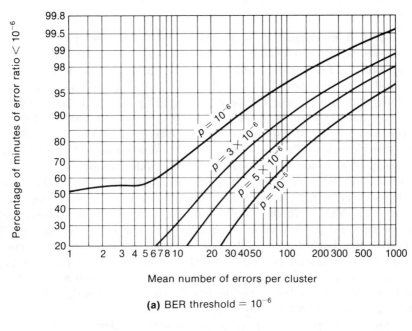

(a) BER threshold $= 10^{-6}$

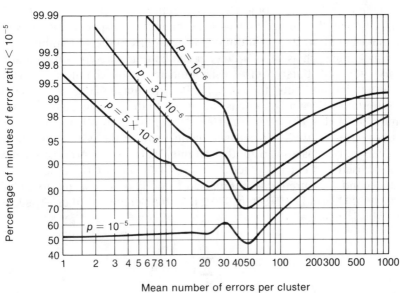

(b) BER threshold $= 10^{-5}$

FIGURE 2.18   The Variation with Degree of Clustering of the Percentage of Minutes of 64-kb/s Transmission Having an Error Ratio Less Than Threshold for Different Values of Long-Term Mean Error Ratio (p) [18] (Reprinted by Permission of IT&T)

used just two states, labeled good and bad [20]. Subsequent applications of the Markov chain model used a larger number of states to better fit the error statistics. Fritchman introduced an N-state Markov chain partitioned into N-1 error-free states and a single error state [21]. His model requires $2N$ independent parameters that are used in fitting a sum of exponentials to the error data. Markov chain models have also been used to describe error performance for tandem links; one strategy uses a concatenation process to combine individual link models into an equivalent end-to-end model [22]. Thus it is possible to use these Markov chain models to allocate error performance for end-to-end connections for error parameters such as bit error rate, block error rate, error-second rate, and error-threshold rate. The drawback to Markov chain models is that the number of fitting parameters can become large for complex channels. Another mathematical model called the Bond distribution has been used to describe bursty error channels using just two parameters, one for burstiness and the other for the background error rate [23]. Measurement data of errored seconds and severely errored seconds have shown that the Bond distribution provides a better fit than a sum of exponentials [24].

Error performance measurements are usually conducted at the system bit rate, but performance objectives such as those prescribed by the CCITT and CCIR are often based on a 64-kb/s channel. Any specification or measurement based on BER will directly translate from the system bit rate to any other rate such as 64 kb/s, since it can be reasonably assumed that bit errors, even when bursty, will be distributed among the channels comprising the system bit rate. Thus, for example, the percentage of degraded minutes measured at the system bit rate is equivalent to the percentage of degraded minutes at 64 kb/s. However, for errored seconds or any other error parameter not based on the BER, measurements made at the system bit rate can not be directly translated to another bit rate. Methods of extrapolation have been devised and even adopted by the CCITT and CCIR, especially for error-second measurements normalized to 64 kb/s. According to CCITT Rec. G.821 [13] and CCIR Rep. 930-2 [25], the percentage of error seconds at 64 kb/s is given by

$$\% \, \text{ES}_{64} = \frac{1}{j} \sum_{i=1}^{i=j} \left(\frac{n}{N} \right)_i \times 100\% \tag{2.30}$$

where

(1) n is the number of errors in the ith second at the measurement bit rate;
(2) N is the system bit rate divided by 64 kb/s;

(3) j is the integer number of one-second periods (excluding unavailable time) that comprises the total measurement period;

(4) the ratio $(n/N)_i$ for the ith second is n/N if $0 < n < N$, or 1 if $n \geq N$, and represents the probability of an errored second at 64 kb/s for the ith second.

The relationship given by (2.30) assumes that errors at the system bit rate are uniformly distributed among the constituent 64-kb/s channels. In practice, most transmission channels exhibit bursty errors and therefore the performance at 64 kb/s will be generally better than that estimated by Equation (2.30). Field measurements indicate that a better extrapolation is given by use of a Poisson distribution [26], in which case

$$\% \, ES_{64} = (1 - e^{-n/N}) \times 100\% \tag{2.31}$$

where n and N are as defined above. Other mathematical models have been developed that relate errored seconds at 64 kb/s to error parameters used on other standard rates such as 1.544 Mb/s and 44.736 Mb/s [27]. Another possible approach reported in CCIR Rep. 930-2 is based on selective sampling of the errors collected at the system bit rate, whereby a secondary error signal is created that corresponds to the bit rate of interest, say 64 kb/s [25].

Service Requirements
As noted earlier, the choice of error parameter is determined largely by the effects of errors on the user, which are different for various services—voice, data, and video. For the voice user, a suitable parameter is the percentage of time in which the BER does not exceed a specified threshold. Based on informal listening tests, subjective evaluations of disturbances in 64-kb/s PCM speech indicate that the total outage time is the most important parameter, not the frequency or duration of outages [1].

For typical data applications, EFS or EFB is better for evaluating the effect of errors. Data are transmitted in block or message form, often accompanied by some form of parity to allow a check on data integrity at the receiver. For this application the number of errors is unimportant, since a single error requires retransmission of the whole block or message. If an error-correcting code is employed in a data application, the effects of single errors are minimized. However, error correction coding can only cope with error bursts of finite length. Hence the data user employing error-correcting codes will be interested in the probability of error bursts exceeding a certain length. In video applications, with the use of differential coding and various image compression techniques, much of the redundancy in video is usually removed. As the transmission rate is reduced, the effects of errors become more pronounced. A single error may persist through many frames. Hence the choice of error characterization by the video user is similar to that of the data user and is influenced by whether or not error-correcting codes are employed.

For transmission systems providing more than one service, say voice, video, and data, the choice of appropriate error parameters for system design requires further comparison. Since voice and video signals contain much redundancy and because of the tolerance to errors by the human eye and ear, the data user often has the greatest requirement for error-free performance. Yet the use of error-correcting codes or retransmissions upon error makes a data system more tolerant of error. Further, with the use of bandwidth compression schemes, voice and video systems may become the most demanding service with regard to error performance.

International Standards

CCITT Rec. G.821 has defined error performance objectives using three parameters—errored seconds, severely errored seconds, and degraded minutes—reflecting the requirements of both voice and data services. In practice, it is expected that all parts of the objective should be met simultaneously. These performance objectives are stated for a 64-kb/s circuit switched connection operating over the CCITT international hypothetical reference circuit identified in Figure 2.6a. The error objectives apply to the period when the connection is available. A connection is considered unavailable when the BER in each second is worse than 1×10^{-3} for a period of ten or more consecutive seconds; a new period of availability begins with the first second of ten consecutive seconds each of which has a BER better than 10^{-3}. The one-minute intervals used in measuring degraded minutes are derived by removing unavailable time and also severely errored seconds. Table 2.3 summarizes these CCITT error performance objectives, and Figure 2.19 provides a methodology for measurement of G.821 error parameters. Each of the three error parameters is further allocated in G.821 to a local-, medium-, and high-grade section of the G.801 hypothetical reference circuit, as follows:

- 15 percent to each local-grade portion
- 15 percent to each medium-grade (for instance, national) portion
- 40 percent to the high-grade (for instance, international) portion

For digital radio systems, the CCIR has defined its error performance objectives using three BER thresholds and measurement averaging times. These objectives are based on a 64-kb/s connection operating over the 2500-km hypothetical reference circuit identified in Figure 2.6b. The error

TABLE 2.3   CCITT Performance Objectives for Digital Transmission (Rec. G.821) [13]

Reference Channel
   64-kb/s channel traversing a 27,500-km hypothetical reference connection
User Types
   1. Voice
   2. Data
Availability
   A connection is considered to be unavailable when the BER in each second is worse than
   1×10^{-3} for periods of 10 seconds or more.
Error Performance*

BER in 1 Min	Percentage of Available Minutes	BER in 1 Sec	Percentage of Available Seconds
Worse than 10^{-6}	<10% (% DM)	Worse than 10^{-3}	<0.2% (% SES)
Better than 10^{-6}	≥90%	0	≥92% (% EFS)

*Measurement time for error rate is unspecified. A period of 1 month is suggested.

performance objectives apply only when the circuit is available. The CCIR error and availability performance objectives are summarized in Table 2.4. For each of the three error rate thresholds, performance is stated as a percentage of time in which the error rate should not be exceeded. The availability objective is considered provisional with an allowed range of 99.5 to 99.9 percent. It should be noted that this availability figure accounts only for radio-related outages and does not include multiplex equipment.

These CCIR error objectives are based on a proration of the error objectives contained in CCITT Rec. G.821, from the 27,500-km CCITT reference circuit to the 2500-km CCIR reference circuit. Thus, Rec. G.821 forms the overall performance objective for the network, and the corresponding CCIR objectives are intended to meet G.821. For real radio links (as opposed to the hypothetical reference circuit), the CCIR provides additional guidance in Recs. 634-1 and 695 [14] that further prorates the objectives given in Table 2.4. For links of length L, where L is between 280 km and 2500 km, the objectives for bit error rate and availability shown in Table 2.4 are prorated by the factor $L/2500$, meaning that the allowed percentages for the three error parameters and for unavailability are reduced by the factor $L/2500$, as given by:

$$\% \, DM \leq 0.4 \, (L/2500) \, \% \tag{2.32}$$

$$\% \, SES \leq 0.054 \, (L/2500) \, \% \tag{2.33}$$

$$\% \, ES \leq 0.32 \, (L/2500) \, \% \tag{2.34}$$

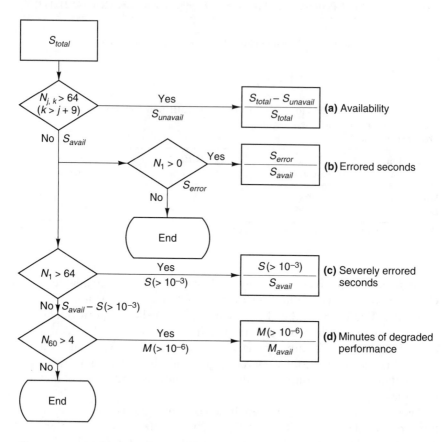

S_{total}:   total measured seconds: one month
N_j: number of bit errors in j second intervals
$N_{j,\,k}$:   number of bit errors in each second interval between jth second and kth second
$S(> 10^{-3})$:   total time during which the BER exceeds 10^{-3} in each second interval (s)
$S_{unavail}$:   unavailable time (s)
S_{avail}:   available time (s)
S_{error}:   total time where the number of errors is one or more in each second interval (s)
M_{avail}:   available time (min) $= \dfrac{S_{total} - S_{unavail}}{60}$
$M(> 10^{-6})$:   total time that the BER exceeds 10^{-6}, measured in blocks of 60 consecutive 1 s intervals derived by excluding any 1 s intervals during which the BER exceeds 10^{-3} (min)

FIGURE 2.19   Measurement Algorithm for CCITT and CCIR Performance Parameters [14] (Courtesy CCIR)

and

$$\% A \geq 100 - (0.3 \times L/2500) \, \% \qquad (2.35)$$

In addition, CCIR Rec. 634-1 introduces a requirement for real radio links to meet a residual bit error rate (RBER) of

$$RBER \leq 5 \times 10^{-9} \, (L/2500). \qquad (2.36)$$

The provisional method of measurement of RBER calls for taking 15-min BER measurements over a one-month period, discarding 50 percent of the 15-min intervals containing the worst BER measurements, and using the remaining 15-min intervals to calculate the RBER.

For transmission of 64-kb/s channels via satellite, the CCIR has defined error and availability performance using an approach similar to digital radio objectives. Two different satellite services are distinguished by the CCIR for a 64-kb/s channel, one for PCM telephony and the other for ISDN. Both services are defined by the same hypothetical reference circuit shown in Figure 2.7, and both are specified to have the same availability, but the two differ in error performance, with the ISDN channel having the more stringent requirement. As summarized in Table 2.5, for each of the two types of

TABLE 2.4   CCIR Performance Objectives for Digital Radio Systems [14, 25]

Reference Channel
   64-kb/s channel traversing a 2500-km hypothetical reference path (Rec. 556-1)
Bit Error Rate (Rec. 594-2)
   $\geq 10^{-6}$:   not more than 0.4% of any month averaged over 1-min intervals (degraded minutes).
   $\geq 10^{-3}$:   not more than 0.054% of any month averaged over 1-s intervals (severely errored seconds).
   >0:   not more than 0.32% of any month for 1-s intervals (errored seconds).
Availability (CCIR Rec. 557-2)
   The objective for a hypothetical reference digital path is 99.7% with a possible range of 99.5 to 99.9%.
Unavailability (CCIR Rec. 557-2)
   The concept of unavailability of a hypothetical reference digital path should be as follows. In at least one direction of transmission, one or both of the two following conditions occur for at least 10 consecutive seconds:
   1. The digital signal is interrupted (alignment or timing is lost).
   2. The bit error rate in each second is greater than 10^{-3}.
Under Study
   Burst errors [Rep. 930-2]
   Outages <10 s [Rec. 557-2, note 8]

TABLE 2.5   CCIR Performance Objectives for Digital Transmission on Fixed Satellite Systems [15]

Reference Channel
    64-kb/s channel traversing hypothetical reference path (Rec. 521-2).
Bit Error Rate for PCM Channel (CCIR Rec. 522-3)
    $\geq 10^{-6}$:   not more than 20% of any month averaged over 10-min intervals
    $\geq 10^{-4}$:   not more than 0.3% of any month averaged over 1-min intervals
    $\geq 10^{-3}$:   not more than 0.05% of any month averaged over 1-s intervals
Bit Error Rate for ISDN Channel (CCIR Rec. 614-1)
    $\geq 10^{-6}$:   fewer than 2% of the 1-min intervals over any month
    $\geq 10^{-3}$:   fewer than 0.03% of the 1-s intervals over any month
    > 0:   fewer than 1.6% of the 1-s intervals over any month
Availability (CCIR Rec. 579-1)
    The provisional objective is $\geq 99.8\%$.
Unavailability (CCIR Rec. 579-1)
    The reference channel is considered unavailable if one or both of the two following
    conditions occur for 10 or more consecutive seconds:
    1. The bit error rate in each second exceeds 10^{-3}.
    2. The digital signal is interrupted (alignment or timing is lost).

64-kb/s channels, there are three bit error rate thresholds prescribed along with measurement intervals and allowed percentages of time in which the BER is not to be exceeded. These error objectives include effects of interference, atmospheric absorption, and rain, but exclude unavailable periods. The error objectives set forth for the ISDN channel are further intended to meet CCITT G.821 error objectives, as evidenced by the use of degraded minutes, severely errored seconds, and errored seconds. Specifically, for the SES objective, CCITT G.821 allocates 15 percent out of the total of 0.2 percent allowed, or 0.03 percent, to the satellite reference circuit; for the degraded minute objective, CCITT G.821 allocates 20 percent out of the total of 10 percent allowed, or 2 percent, to the satellite reference circuit; and for the errored second objective, CCITT G.821 allocates 20 percent out of a total of 8 percent allowed, or 1.6 percent, to the satellite reference circuit. CCIR Rec. 614-1 [15] and Rep. 997-1 [28] also prescribe long-term, average bit error rate performance objectives in which 10^{-7}, 10^{-6}, and 10^{-3} BER thresholds are not to be exceeded for more than 10 percent, 2 percent, and 0.03 percent, respectively, for any month. These average BER objectives are also intended to meet CCITT Rec. G.821. Areas under study by the CCIR include burst error characterization and short-term (< 10 s) propagation outages for both digital radio and satellite systems.

CCITT Rec. M.550 deals with error performance limits for bringing into service and maintaining digital circuits [29]. The purpose of these performance limits is to achieve the performance objectives given in CCITT Rec.

G.821. An aging margin is included to distinguish performance objectives for the hypothetical reference circuit and limits for bringing a circuit into service. Two limits are given for bringing circuits into service, one for unconditional acceptance and the other for conditional acceptance. Two maintenance limits are defined, one for a degraded state and the other for an unacceptable state. Test intervals are also defined as three or four days for bringing into service and from 15 minutes to 24 hours for maintenance. A methodology based on linear scaling is described for the allocation of performance objectives derived from Recs. G.801 and G.821 down to smaller-length circuits. The bringing-into-service and maintenance allocations are given as the allowed number of ES, SES, and DM for a 64-kb/s circuit. Translation of these performance limits to other bit rates can be accomplished by use of Equation (2.30). Values for performance limits are given in Rec. M.550 for various length circuits, but these are listed as examples only and require further study.

Error Allocation
Once the end-to-end objectives have been established for the hypothetical reference circuit, the allocation of error performance to individual segments and links can be accomplished. Several methods have been used. CCITT Rec. G.821, for example, allocates degraded minutes as 0.00016 percent per kilometer and errored seconds as 0.000128 percent per kilometer for the 25,000-km international segment of the hypothetical reference circuit. This results in an end-to-end objective for the international segment of 4 percent degraded minutes and 3.2 percent errored seconds. Such an allocation procedure is practical only when the hypothetical reference circuit is homogeneous—that is, where each kilometer has the same expected performance. When the hypothetical reference circuit consists of various media and equipment producing different error characteristics, error objectives are best allocated to reference segments and reference links.

To prescribe link performance, a model of error accumulation is needed for n tandem links. If p is the probability of an individual bit being in error on a given link, then assuming statistical independence of errors from link to link, the probability of error per bit for n links will be [30]

$$P_n(e) = P[\text{error in } n \text{ links}] = \frac{1}{2}\left[1 - (1 - 2p)^n\right] \qquad (2.37)$$

For small enough p, the probability of multiple errors on the same bit is negligible. Then (2.37) simplifies to a sum of individual link error probabilities or

$$P_n(e) \approx np \qquad (np \ll 1) \qquad (2.38)$$

These results can be extended to any of the error parameters discussed earlier: bit error rate, error-free seconds, percentage of time below error threshold, and block error rate. For example, if the percentage EFS and percentage T_1 are known for a particular type of link, then for n such links the percentage of error-free seconds is

$$\% \text{ EFS}_n \approx 100 - n(100 - \% \text{ EFS}) \quad (n)(100 - \% \text{ EFS}) \ll 1 \qquad (2.39)$$

and the percentage of time where the BER does not exceed threshold is

$$\% \, T_{1n} \approx 100 - n(100 - \% \, T_1) \quad (n)(100 - \% \, T_1) \ll 1 \qquad (2.40)$$

Again note that Expressions (2.38) to (2.40) assume that multiple errors in the same bit, second, or measurement interval (T_0) do not occur from link to link. For large probability of error or for a large number of links, however, this assumption becomes invalid and forces the system designer to use an exact expression as in (2.37).

This model is easily extended to combinations of links with different error characteristics. Designating link types by numbers $(1, 2, \ldots, m)$, the probability of error expression for a hypothetical reference circuit (hrc) is

$$P[\text{error in hrc}] = \frac{1}{2} \left[1 - \prod_{i=1}^{m} (1 - 2p_i)^{n_i} \right]$$

$$\approx \sum_{i=1}^{m} n_i p_i \quad (n_i p_i \ll 1 \text{ for all } i) \qquad (2.41)$$

This result is directly applicable to other error parameters. For example, if $\% \text{ EFS}_1, \% \text{ EFS}_2, \ldots, \% \text{ EFS}_m$ are the percentages of error-free seconds for each link type, then the percentage of error-free seconds for the hypothetical reference circuit is

$$\% \text{ EFS}[\text{hrc}] \approx 100 - \sum_{i=1}^{m} n_i (100 - \% \text{ EFS}_i) \qquad (2.42)$$

$$(n_i)(100 - \% \text{ EFS}_i) \ll 1 \text{ for all } i$$

and similarly

$$\% \, T_1[\text{hrc}] \approx 100 - \sum_{i=1}^{m} n_i (100 - \% \, T_{1i}) \qquad (2.43)$$

$$(n_i)(100 - \% \, T_{1i}) \ll 1 \text{ for all } i$$

Example 2.4 ———————————————————————————————————————

For the global reference circuit of Example 2.1, let

$$\% \, ES_L = \%[\text{ error seconds for LOS radio reference link}]$$
$$\% \, ES_F = \%[\text{ error seconds for fiber optic cable reference link}]$$
$$\% \, ES_M = \%[\text{ error seconds for metallic cable reference link}]$$
$$\% \, ES_S = \%[\text{ error seconds for satellite reference link}]$$

Assume the relative error performance of these different links to be given by

$$\% \, ES_S = (10)(\% \, ES_M) = (100)(\% \, ES_L) = (100)(\% \, ES_F)$$

(This relationship is somewhat arbitrary; the actual case would be determined by link error statistics, perhaps based on experimental data.) That is, the satellite link is allocated 10 times the error-second percentage of metallic cable and 100 times that of line-of-sight radio and fiber optic cable reference links. Further assume an end-to-end objective of 99 percent EFS. From Example 2.1, the numbers of each medium contained in the hypothetical reference circuit are

$$n_S = \text{number of satellite links} = 2$$
$$n_M = \text{number of metallic cable links} = 24$$
$$n_F = \text{number of fiber optic cable links} = 24$$
$$n_L = \text{number of LOS radio links} = 72$$

From Equation (2.42) the following link allocations then result for percentage of EFS:

Fiber optic cable reference link:	99.998%
LOS radio reference link:	99.998%
Metallic cable reference link:	99.98%
Satellite reference link:	99.8%

2.4.3   Bit Count Integrity Performance

Bit count integrity is defined as the preservation of the precise number of bits (or characters or frames) that are originated in a message or unit of time. Given two enumerable bits in a transmitted bit stream that are separated by n bits, bit count integrity is maintained when the same two bits are separated by n intervals in the received bit stream. Losses of bit count integrity (BCI) cause a short-term outage that may be extended if equipment resynchronization (such as multiplex reframe) is required.

When a digital signal is shifted (or slipped) by some number of bits, the resulting loss of BCI is known as a **slip.** Slips may be controlled or uncontrolled. If controlled, the slip is limited to repetition or deletion of a controlled number of bits, characters, or frames at a controlled instant, which then limits the effect on the user. As an example, for 64-kb/s PCM, the slippage can be controlled to 8-bit characters (octets) in order to maintain character synchronization. For multichannel PCM, the slippage must be confined to whole frames in order to avoid loss of frame alignment. The effect of such controlled slips on PCM speech is very slight, but it can be significant for other signal types such as secure (encrypted) voice, digital data, and high-speed data via modems on telephone channels. Table 2.6 shows the performance impact of one slip on several types of services. Table 2.7 gives the controlled octet slip rate objectives from CCITT Rec. G.822 for the 27,500-km international connection (as shown in Figure 2.6). These objectives apply to both telephone and nontelephone services at 64 kb/s. Table 2.8 is the CCITT allocation of these slip rate objectives to the local, national, and international portions of the hypothetical reference circuit.

Uncontrolled slips or high error rates can cause loss of frame alignment in multiplexing equipment. For these cases, the outage caused by loss of bit count integrity is extended by the time required to regain frame synchronization. The effect on the user is much more evident than in the case of a controlled slip. In most PCM multiplexers, for example, speech outputs are suppressed during periods of frame loss. Frame resynchronization times are typically several milliseconds for PCM multiplex, thus affecting on the order of 100 frames. The CCITT has included a specification on the percentage of short-term outages for BER greater than 10^{-3} as a safeguard against excessive short interruptions such as loss of frame alignment. A 1-s measurement time was selected to distinguish between an unacceptable performance (≥ 1 s) and a period of unavailability (≥ 10 s), both of which have the same BER threshold but different measurement periods. As shown in Table 2.3, the specification for severely errored seconds is that less than 0.2 percent of available seconds should have a BER worse than 10^{-3} for the 64-kb/s hypothetical reference circuit [13].

After determining the end-to-end objectives for bit count integrity, allocations are established for the segments and links of the hypothetical reference circuit. First we must develop the model for loss of BCI in n tandem links. Since this parameter is probabilistic, we can associate a probability q that each received bit (or character or frame) leads to a loss of BCI. Assuming statistical independence of the parameter q from link to link, then for n links each with the same q, per bit

$$P_n(\text{LBCI}) = P[\text{loss of BCI in } n \text{ links}] - \frac{1}{2}\left[1 - (1 - 2q)^n\right] \qquad (2.44)$$

TABLE 2.6   Performance Impact of One Slip [31] (Courtesy ANSI)

Service	Potential Impact
Encrypted Text	Encryption key must be resent
Video	Freeze frame for several seconds.
	Loud pop on audio
Digital Data	Deletion or repetition of data.
	Possible reframe.
Facsimile	Deletion of 4–8 scan lines.
	Drop call.
Voice Band Data	Transmission errors for 0.01 to 2 sec.
	Drop call.
Voice	Possible click.

TABLE 2.7   CCITT Performance Objectives for Slip Rate on 64-kb/s International Connections (Rec. G.822) [13]

Performance Classification	Mean Slip Rate Thresholds	Measurement Averaging Period	Percentage of Total Time
Unacceptable	>30 slips in 1 hr	1 yr	<0.1%
Degraded	>5 slips in 24 hrs and ≤30 slips in 1 hr	1 yr	<1.0%
Acceptable	≤5 slips in 24 hrs	1 yr	≥98.9%

Notes:
1. These objectives are primarily addressed to unencrypted transmission.
2. Rec. G.822 notes that further study is required to confirm that these values are compatible with other objectives such as the error performance objectives.

TABLE 2.8   CCITT Allocation of Controlled Slip Performance Objectives (Rec. G.822) [13]

Portion of Hypothetical Reference Circuit from Figure 2.6a	Allocated Proportion of Each Objective in Table 2.7	Objectives as Proportion of Total Time	
		Degraded	Unacceptable
International Portion	8.0%	0.08%	0.008%
Each National Portion	6.0%	0.06%	0.006%
Each Local Portion	40.0%	0.4%	0.04%

For small enough q and n, Equation (2.44) reduces to

$$P_n\,(\text{LBCI}) \approx nq \qquad (nq \ll 1) \tag{2.45}$$

This model is easily extended to combinations of links with different BCI characteristics. For link types designated $(1,2,\ldots,m)$ in a hypothetical reference circuit,

$$P_{\text{hrc}}\,(\text{LBCI}) = \frac{1}{2}\left[1 - \prod_{i=1}^{m}(1-2q_i)^{n_i}\right] \tag{2.46}$$

which simplifies to

$$P_{\text{hrc}}\,(\text{LBCI}) \approx \sum_{i=1}^{m} n_i q_i \qquad (n_i q_i \ll 1 \text{ for all } i) \tag{2.47}$$

With the assumption of statistically independent losses of BCI, the number of losses of BCI has a binomial distribution. Therefore, the expected number of losses of BCI per unit time is

$$E[\,\text{LBCI}\,] = qR \tag{2.48}$$

where R represents bit, character, or frame rate, as determined by the definition of q. The mean time to loss of BCI, $\overline{X}$, is then the reciprocal of (2.48), or

$$\overline{X} = \frac{1}{qR} \tag{2.49}$$

The rate of BCI losses given in Equation (2.48) and mean time to loss of BCI given in (2.49) are the two parameters commonly used in allocating BCI performance. When they are applied to the hypothetical reference circuit, we obtain

$$E[\,\text{LBCI}_{\text{hrc}}\,] = \frac{R}{2}\left[1 - \prod_{i=1}^{m}(1-2q_i)^{n_i}\right]$$
$$\approx R\sum_{i=1}^{m} n_i q_i \qquad (n_i q_i \ll 1 \text{ for all } i) \tag{2.50}$$

and

$$\overline{X}_{\text{hrc}} = \frac{1}{E[\,\text{LBCI}_{\text{hrc}}\,]} \tag{2.51}$$

A more convenient form of (2.51) is given by

$$\frac{1}{\overline{X}_{hrc}} \approx R \sum_{i=1}^{m} n_i q_i = \sum_{i=1}^{m} \frac{n_i}{\overline{X}_i} \qquad (n_i q_i \ll 1 \text{ for all } i) \qquad (2.52)$$

where $\overline{X}_i = (q_i R)^{-1}$.

From link allocations of BCI, performance requirements can be determined for those functions that can cause loss of BCI, such as:

- Bit synchronization in a timing recovery loop
- Buffer lengths and clock accuracy required for network synchronization
- Frame and pulse stuffing synchronization in a digital multiplexer
- Protection switching in redundant equipment

Example 2.5 _____

For the global reference circuit of Example 2.1, let

$\overline{X}_L$ = mean time to loss of BCI on LOS radio reference link
$\overline{X}_F$ = mean time to loss of BCI on fiber optic cable reference link
$\overline{X}_M$ = mean time to loss of BCI on metallic cable reference link
$\overline{X}_S$ = mean time to loss of BCI on satellite reference link

Assume the relative BCI performance of these different links to be given by

$$\overline{X}_L = \overline{X}_F = 10\overline{X}_M = 100\overline{X}_S$$

That is, the line-of-sight radio and fiber optic cable links are allocated a mean time to loss of BCI that is 10 times that of metallic cable and 100 times that of satellite reference links. Further assume an end-to-end objective of 1 hr mean time to loss of BCI. From Example 2.1, the numbers of each medium contained in the hypothetical reference circuit are: $n_S = 2$, $n_M = 24$, $n_F = 24$, and $n_L = 72$. The following link allocations of mean time to loss of BCI result from application of Equation (2.52):

Fiber optic cable reference link:	536	hr
LOS radio reference link:	536	hr
Metallic cable reference link:	53.6	hr
Satellite reference link:	5.36	hr

_____

2.4.4 Jitter Performance

Jitter is defined as a short-term variation of the sampling instant from its intended position in time or phase. Longer-term variation of the sampling instant is sometimes called **wander** or **drift**. Jitter causes transmission impairment in three ways:

1. Displacement of the ideal sampling instant leads to a reduction in the noise margin and a degradation in system error rate performance.
2. Slips in timing recovery circuits occur when excessive jitter causes a loss of phase lock or cycle slip.
3. Irregular spacing of the decoded samples of digitally encoded analog signals introduces distortion in the recovered analog signal.

Wander causes buffers used for frame or clock alignment to fill or empty, resulting in a repetition or deletion (slip) of a frame or data bit.

Contributions to *jitter* arise primarily from two sources: regenerative repeaters and digital multiplexers. These two sources of jitter are briefly discussed here and described in more detail in Chapters 4 and 7. Regenerative repeaters must derive a train of regularly spaced clock pulses in order to properly sample the digital signal. Jitter is introduced in the clock recovery process due to the following:

- Mistuning of a resonant circuit causes accumulation of phase error, especially during the absence of data transitions when the circuit drifts toward its natural frequency.
- Misalignment of the threshold detector from ideal (zero amplitude) causes variation in the clock triggering time for any amplitude variation of the signal.
- Imperfect equalization results in **intersymbol interference** in which the pulses may be skewed, shifting the signal from the ideal transition time and causing a phase shift in the clock triggering time.
- Certain repetitive data patterns produce a characteristic phase shift in the recovered clock. A change in pattern can cause a relative change in clock timing resulting in a form of timing jitter.

Jitter associated with digital multiplexing operation stems from the process of buffering input signals prior to multiplexing and the inverse process for demultiplexing. The insertion of overhead bits for synchronization causes phase differences between the input message traffic and the composite transmitted bit stream. Jitter is created when deriving the timing signal for the message traffic at the receiving demultiplexer. When pulse stuffing is used in a multiplexer, an additional form of jitter called **stuffing jitter** can arise. This jitter results from the periodic insertion of stuff bits in the multiplexer and their removal at the demultiplexer.

Contributions to *wander* arise primarily from two sources: oscillator instability and propagation delay variation. These two effects are important considerations in the design of network synchronization and are described in detail in Chapter 10. The magnitude of wander determines the buffer sizes required at each node in a network.

The specification and measurement of jitter is based on three standard parameters:

1. *Tolerable input jitter,* defined as the *maximum jitter,* which an equipment input can tolerate without causing bit errors.
2. *Output jitter,* defined either as the *intrinsic jitter,* which an equipment generates in the absence of input jitter, or as the *network jitter,* which a network generates at hierarchical interfaces.
3. *Jitter transfer characteristic,* defined as the *ratio of output jitter to input jitter.*

Each of these parameters is specified by peak-to-peak jitter amplitude versus frequency, usually in the form of a mask. Measurement of these parameters is discussed in Chapter 11. The CCITT specifies each of these jitter parameters, at hierarchical levels, in the G.700 series recommendations for individual equipments, in the G.800 series recommendations for digital networks, and in the G.900 series recommendations for digital sections. Examples of these CCITT recommendations are shown in Figures 2.20, 2.21, and 2.22. Figure 2.20 specifies the maximum permissible output jitter in a digital network, as given by CCITT Rec. G.824 for digital rates within the North American and Japanese digital hierarchies. Network limits on output jitter will in turn place limits on the amount of intrinsic jitter than can be allowed in individual equipment, since jitter will accumulate in a network of cascaded equipment. These limits on network jitter are also intended to be compatible with the tolerable input jitter for equipment input ports, as specified by CCITT Rec. G.824 and shown here in Figure 2.21 for North American and Japanese hierarchical rates. Figure 2.21 thus provides the envelope of lower limits on the maximum tolerable jitter and ensures that the equipment can accommodate network output jitter up to the levels given in Figure 2.20. An example of a jitter transfer characteristic is given in Figure 2.22 for North American hierarchical rates. The jitter transfer characteristic determines the gain or attenuation of jitter by a particular equipment and controls the accumulation of jitter in a network of cascaded equipment.

The allocation of jitter performance in a reference circuit requires knowledge of the jitter characteristics of the components in the circuit and a model of jitter accumulation from component to component. Jitter accumulation depends on the structure of the reference path and the sources of jitter considered. Certain jitter sources are systematic and lead to accumulation in a predetermined manner. Sources of jitter associated with repeatered cable systems tend to be systematic and therefore accumulate with a known characteristic, as discussed in Chapter 7. Multiplexers tend to contribute nonsystematic jitter and work as jitter reducers. Therefore, a reference circuit that has frequent multiplex points tends to have less jitter accumulation than the case for no multiplex points. Without the jitter reducing effect of demultiplexers, means of reducing jitter (called *dejitterizers*) may be necessary to meet reference circuit performance objectives.

2.4.5 Delay

Absolute delay in transmission affects any interactive service, such as voice and certain types of data circuits. Delay in voice transmission imparts an unnatural quality to speech, while data circuits using some form of automatic response are affected if the transmission delay exceeds the allowed

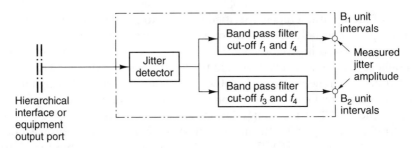

(a) Measurement arrangement for output jitter

	Network limit (UI peak-to-peak)		Band-pass filter having a lower cut-off frequency f_1 or f_3 and a minimum upper cut-off frequency f_4		
Digital rate (kb/s)	B_1	B_2	f_1 (Hz)	f_3 (kHz)	f_4 (kHz)
1,544	5.0	0.1 (Note)	10	8	40
6,312	3.0	0.1 (Note)	10	3	60
32,064	2.0	0.1 (Note)	10	8	400
44,736	5.0	0.1	10	30	400
97,728	1.0	0.05	10	240	1000

UI = unit interval.
Note: This value still under study.

(b) Maximum permissible output jitter at hierarchical rates

FIGURE 2.20   Limits on Network Jitter Output, Based on CCITT Rec. G.824 [13] (Courtesy CCITT)

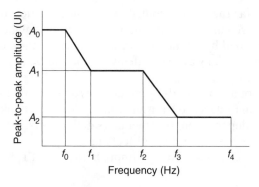

(a) Mask of tolerable jitter

Bit rates (kb/s)	Jitter amplitude (peak-to-peak)			Frequency				
	A_0 (μs)	A_1 (UI)	A_2 (UI)	f_0 (Hz)	f_1 (Hz)	f_2 (Hz)	f_3 (kHz)	f_4 (kHz)
1,544	18 (Note)	5.0	0.1 (Note)	1.2×10^{-5}	10	120	6	40
6,312	18 (Note)	5.0	0.1	1.2×10^{-5}	10	50	2.5	60
32,064	18 (Note)	2.0	0.1	1.2×10^{-5}	10	400	8	400
44,736	18 (Note)	5.0	0.1 (Note)	1.2×10^{-5}	10	600	30	400
97,728	18 (Note)	2.0	0.1	1.2×10^{-5}	10	12,000	240	1000

Note: This value still under study.

(b) Values for the mask of tolerable jitter

FIGURE 2.21   Tolerance of Equipment Input to Jitter, Based on CCITT Rec. G.824 [13] (Courtesy CCITT)

response time. Sources of transmission delay are propagation time and equipment processing. Propagation delays are a function of distance and are independent of the bit rate. Equipment delays are due to buffering—as used in digital processing (multiplexing; error correction coding; packet, message, or circuit switching) and in network synchronization—and are inversely proportional to bit rate. The total delay is therefore a complex function of the geographical, media, and equipment configuration. Delay in a satellite link is by far the largest contributor to propagation time, however; delays in terrestrial links usually can be considered negligible. Excessive delay can be avoided by eliminating or minimizing the use of satellite links or by using commercially available delay compensation units designed for interactive data circuits.

The specification of allowed end-to-end delay is determined by user tolerance. The allocation of delay to transmission segments and links of a reference circuit then depends on the media (satellite versus terrestrial), multiplex hierarchy, use of error control techniques, type of network synchronization, and type of switching. Although the CCITT has not yet

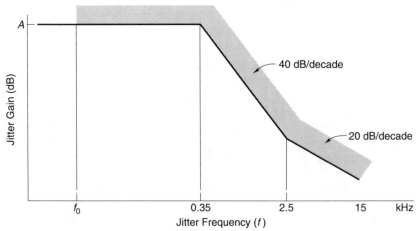

Note: The frequency of f_0 should be as low as possible taking into account the limitations of measuring equipment.

(a) Jitter transfer characteristic

Bit Rate (kb/s)	Jitter Gain A(dB)	CCITT Recommendation
1544	0.5	G.743
6312	0.1	G.752

(b) Values of jitter transfer characteristic

FIGURE 2.22   Jitter Transfer Characteristic, Based on CCITT Recommendations [32] (Courtesy CCITT)

addressed delay allocations for the hypothetical reference circuits, delay in the AT&T network has been specified to have a one-way maximum of 60 ms for T1 service [33] and 10 ms for T3 service [34].

2.5 Summary

Digital transmission system planning is based on the following steps:

1. Identify services and their performance requirements.
2. Formulate a hypothetical reference circuit.
3. Specify performance objectives for the end-to-end circuit.
4. Allocate performance objectives to individual segments, links, and equipment.

The performance requirements of various services are a complex function involving human perception for voice and video and transmission efficiency and response time for data. Once these requirements are established, the system designer can assign performance objectives to end-to-end service. Hypothetical reference circuits are a valuable tool used by the system designer in allocating performance to representative portions of the circuit. These circuits can be constructed with different media and equipment in order to reflect an appropriate combination for the transmission system being designed.

The performance parameters used in communication systems can be divided into two areas: availability and quality. Availability is defined as the probability or percentage of time that a connection is usable. A typical criterion for digital transmission is to define a circuit as unavailable if the bit error rate exceeds a specified threshold for some stated period of time. This definition distinguishes short-term interruptions that affect quality from long-term interruptions that are periods of unusable time. For periods of available service, various parameters are used to indicate the quality of service. Although these parameters and their value depend on the type of service, four quality parameters play a dominant role in digital transmission: error rate, bit count integrity, jitter, and delay.

Error rate can be stated in several different ways, such as average bit error rate (BER), percentage of error-free seconds, percentage of time the BER does not exceed a given threshold value, and percentage of error-free blocks. Each of these methods has application to certain types of service or media, although different error performance readings may result under the same test conditions, depending on the distribution of bit errors. Standard error performance criteria and allocation techniques are proposed for various media and services.

Bit count integrity (BCI) is defined as the preservation of the precise number of bits that are originated in a message or unit of time. Losses of

BCI are due to loss of bit synchronization, buffer slips, loss of multiplex synchronization, and protection switching in redundant equipment. The result of BCI losses is a short interruption due to deletion or repetition of a bit, character, or frame, and in some cases this interruption is extended due to the time required to resynchronize equipment. The effect on a voice user is minimal, but data and secure (encrypted) voice users suffer more serious effects due to retransmission and resynchronization. BCI performance is most commonly expressed as the rate of BCI losses or the time between losses of BCI, either of which can be stated for end-to-end connections and individual links in a reference circuit.

Jitter is defined as a short-term variation of the sampling instant from its intended position. The effects of jitter include degradation in error rate performance, slips in timing recovery circuits, and distortion in analog signals recovered from digital (PCM) representations. The principal sources of jitter are regenerative repeaters and digital multiplexers. Long-term variation of the sampling instant, called wander or drift, is caused by oscillator instability and propagation delay variations. The specification of jitter should include allowed output jitter and tolerable input jitter. From knowledge of the jitter transfer function for components of a reference circuit, a jitter specification can be developed for the end-to-end connection and for individual equipment.

Absolute delays in transmission affect all interactive service, but primarily the data user. Since most interactive data transmission systems require automatic responses within a specified time, excessive delays will affect such data users. Delay is due to equipment processing and propagation time. The most significant contributor is the delay inherent in satellite transmission. Allocation of delay performance thus depends on the media and equipment of the reference circuit.

References
1. J. Gruber, R. Vickers, and D. Cuddy, "Impact of Errors and Slips on Voice Service," *1982 International Conference on Communications,* pp. 2D.5.1–2D.5.7.
2. W. Stallings, *Data and Computer Communications,* 3rd ed. (New York: Macmillan, 1991).
3. J. O. Limb, C. B. Rubinstein, and J. E. Thompson, "Digital Coding of Color Video Signals—A Review," *IEEE Trans. on Comm.,* vol. COM-25, no. 11, November 1977, pp. 1349–1385.
4. Recommendations of the CCIR, 1990, Vol. XII, *Television and Sound Transmission (CMTT)* (Geneva: ITU, 1990).
5. Reports of the CCIR, 1990, Annex to Vol. XII, *Television and Sound Transmission (CMTT)* (Geneva: ITU, 1990).
6. K. Sawada, "Coding and Transmission of High-Definition Television—Technology Trends in Japan," *1990 International Conference on Communications,* pp. 320.3.1–320.3.7.

7. A. Jalali and others, "622 Mbps SONET-Like Digital Coding, Multiplexing, and Transmission of Advanced Television Signals on Single-Mode Optical Fiber," *1990 International Conference on Communications*, pp. 325B.1.1–325B.1.6.

8. K. Takikawa, "Simplified 6.3 Mbit/s Codec for Video Conferencing," *IEEE Trans. on Comm.*, vol. COM-29, no. 12, December 1981, pp. 1877–1882.

9. H. S. London and D. B. Menist, "A Description of the AT&T Video Teleconferencing System," *1981 National Telecommunications Conference*, pp. F5.4.1–F5.4.5.

10. T. Ishiguro and K. Iinuma, "Television Bandwidth Compression by Motion-Compensated Interframe Coding," *IEEE Comm. Mag.*, November 1982, pp. 24–30.

11. J. E. Thompson, "European Collaboration on Picture Coding Research for 2 Mbit/s Transmission," *IEEE Trans. on Comm.*, vol. COM-29, no. 12, December 1981, pp. 2003–2004.

12. CCITT Blue Book, vol. III.6, *Line Transmission of Non-Telephone Signals* (Geneva: ITU, 1989).

13. CCITT Blue Book, vol. III.5, *Digital Networks, Digital Sections, and Digital Line Systems* (Geneva: ITU, 1989).

14. Recommendations of the CCIR, 1990, Vol. IX—Part 1, *Fixed Service Using Radio-Relay Systems* (Geneva: ITU, 1990).

15. Recommendations of the CCIR, 1990, Vol. IV—Part 1, *Fixed-Satellite Service* (Geneva: ITU, 1990).

16. Bell Northern Research, *Considerations on the Relationship Between Mean Bit Error Ratio, Averaging Periods, Percentage of Time and Percent Error-Free Seconds*, COM XVIII, no. 1-E (Geneva: CCITT, 1981), pp. 73–75.

17. J. Neyman, "On a New Class of Contagious Distribution, Applicable in Entomology and Bacteriology," *Ann. Math. Statist.* 10(35)(1939).

18. International Telephone and Telegraph Corp., *Error Performance Objectives for Integrated Services Digital Network (ISDN)*, COM XVIII, no. 1-E (Geneva: CCITT, 1981), pp. 89–95.

19. D. Becham and others, "Testing Neyman's Model for Error Performance of 2 and 140 Mb/s Line Sections," *1984 International Conference on Communications*, pp. 1362–1365.

20. E. N. Gilbert, "Capacity of a Burst-Noise Channel," *Bell System Technical Journal* 39 (September 1960): 1253–1265.

21. B. D. Fritchman, "A Binary Channel Characterization Using Partitioned Markov Chains," *IEEE Trans. on Information Theory*, vol. IT-13, April 1967, pp. 221–227.

22. S. Dravida, M. J. Master, and C. H. Morton, "A Method to Analyze Performance of Digital Connections," *IEEE Trans. on Comm.*, vol. COM-36, no. 6, March 1988, pp. 298–305.

23. D. J. Bond, "A Theoretical Study of Burst Noise," *British Telecom Technology Journal* 5 (October 1987): (51–60).

24. "Error Frequency Distribution Model," AT&T submission at the CCITT Study Group IV Meeting on Q16/IV, 6–10 November 1989, Montreal.

25. Reports of the CCIR, 1990, Annex to Vol. IX—Part 1, *Fixed Service Using Radio-Relay Systems* (Geneva: ITU, 1990).

26. E. Damosso and others, "Experimental Results on Digital Radio-Relay Quality Compared with the New CCIR Objectives," *European Conference on Radio-Relay Systems,* November 1986, pp. 368–375.
27. S. D. Akers and others, "Uses of In-Service Monitoring Information to Estimate Customer Service Quality," *1989 Global Communications Conference,* pp. 1856–1860.
28. Reports of the CCIR, 1990, Annex to Vol. IV—Part 1, *Fixed-Satellite Service* (Geneva: ITU, 1990).
29. CCITT Blue Book, vol. IV.1, *General Maintenance Principles: Maintenance of International Transmission Systems and Telephone Circuits* (Geneva: ITU, 1989).
30. H. D. Goldman and R. C. Sommer, "An Analysis of Cascaded Binary Communication Links," *IRE Trans. on Comm. Systems,* vol. CS-10, no. 3, September 1962, pp. 291–299.
31. Draft ANSI PN-2198, "Private Digital Network Synchronization," July 1990.
32. CCITT Blue Book, vol. III.4, *General Aspects of Digital Transmission Systems; Terminal Equipments* (Geneva: ITU, 1989).
33. AT&T Technical Report TR 62411, "ACCUNET T1.5 Service Description and Interface Specification," December 1990.
34. AT&T Technical Report TR 54014, "ACCUNET T45 Service Description and Interface Specifications," June 1987.

Problems

2.1 Derive Equation (2.2) from Equation (2.1).

2.2 Consider a system with two elements in series, E_1 and E_2 that have failure rates λ_1 and λ_2. (a) Find the probability that the system has failed in some time t. (b) Find the probability that E_1 fails before E_2.

2.3 Consider a system consisting of n identical elements, each with the same MTBF and MTTR. Derive an expression for the probability of exactly j of the n elements being available at a randomly selected instant.

2.4 Consider a system of n digital radio hops in tandem, each hop having a probability of error $P(e)$ equal to p.
 (a) Derive an exact expression for the total (end-to-end) probability of error for $n = 2$.
 (b) Show for n hops that the total $P(e)$ is given by Equation (2.37).
 (c) Expand the expression given in 2.4 (b) to show that $P(e) \approx np$ for an n-hop system with $np \ll 1$.

2.5 Using Equations (2.32) through (2.35):
 (a) Plot the percentage of availability, degraded minutes, severely errored seconds, and errored seconds versus distance L for L up to 2500 km.

(b) Plot the number of seconds or minutes of availability, degraded minutes, severely errored seconds, and errored seconds versus distance L for L up to 2500 km.

(c) For a random error channel, which error parameter is the most stringent to meet?

2.6 As shown in Equations (2.30) and (2.31), there are two ways of extrapolating 64-kb/s error-second performance from measured data at the system bit rate. Using these two equations, create two plots of the probability of an errored second at 64 kb/s versus n/N ($0 \leq n/N \leq 2$), where n is the number of errors in a given second at the system bit rate and N is the system bit rate divided by 64 kb/s. Comment on the differences between the two plots.

2.7 Consider the following digital transmission system consisting of line-of-sight (LOS), fiber optic (FIBER), and satellite (SAT) hops connected in tandem, with an associated allocation of bit error rate (BER) and availability (A) for a 64-kb/s circuit on each hop as shown:

$$(\text{----})\ (\text{----})\ (\text{----})\ (\text{----})\ (\text{----})\ (\text{----})\ (\text{----})$$

| LOS | LOS | FIBER | SAT | FIBER | LOS | LOS |

	A	BER
SAT	0.999	1×10^{-7}
FIBER	0.9999	1×10^{-8}
LOS	0.99995	5×10^{-9}

(a) What is the total (end-to-end) availability and BER?

(b) If the MTTR = 1 hr for each link, what is the MTBF for each LOS, FIBER, and SAT hop, assuming that unavailability is due only to equipment failures?

(c) Based on the given BER allocations, and assuming random error distribution, what is the: (1) total (end-to-end) percent of error-free seconds for a data rate of 64 kb/s? (2) total (end-to-end) probability of block error for a block length of a 100 bits?

2.8 A 50-km digital section of a 64-kb/s reference circuit is to have an allocation of 2 percent of the total error performance for the end-to-end hypothetical reference circuit of G.821. Using a linear scaling approach as specified in CCITT Rec. M.550, determine the number of errored seconds, severely errored seconds, and degraded minutes allowed:

(a) In 24 hours for performance objectives.

(b) In 15 minutes for the unacceptable threshold of the maintenance limits. Assume that the unacceptable performance limit is ten times worse than the reference performance objective.

(c) In 24 hours for performance objectives but for a 44.736-Mb/s system bit rate.

(d) In 15 minutes for the unacceptable threshold of the maintenance limits, but for a 44.736-Mb/s system bit rate. Assume that the unacceptable performance limit is ten times worse than the reference performance objective.

2.9 A given system consists of N identical elements connected in series. If A is the availability of each element and A_N is the desired system availability:

(a) Find an expression for the number of elements N that can be allowed to still meet the system availability.

(b) Determine the value for N, if $A = 0.99$ and $A_N = 0.9$.

2.10 In the testing of jitter effects on digital transmission, sinusoidal jitter is often specified as the type of jitter modulating the digital signal. Consider an expression for sinusoidal phase modulation, $\theta(t) = A \sin \omega t$, where A is the jitter amplitude and ω is the jitter frequency.

(a) Find an expression for the slope of $\theta(t)$ that will describe the rate of change of the phase.

(b) Find an expression for the maximum phase shift between pulses. Interpret this expression for a bit rate of 1.544 Mb/s with a time between pulses of 15 bits (that is, 15 bit times for which no pulses occur).

3

Analog-to-Digital Conversion Techniques

OBJECTIVES

- Explains the overall design, performance, and applications of analog-to-digital conversion techniques
- Discusses sampling, quantizing, and coding—the basis for pulse code modulation
- Covers linear and logarithmic PCM, with emphasis on international standards for voice transmission
- Describes techniques such as differential PCM and delta modulation that apply to highly correlated signals (speech or video)
- Compares the performance of pulse code modulation, delta modulation, and differential PCM—including the effects of transmission error rate and tandeming (multiple A/D coding)
- Describes the basic principles of vocoding techniques and hybrid coders
- Compares speech coding techniques with respect to transmission bit rate and quality

3.1 Introduction

Most communication systems are dominated by signals that originate in analog form, such as voice, music, or video. For such analog sources, the initial process in a digital transmission system is the conversion of the analog source to a digital signal. Numerous analog-to-digital (A/D) conversion techniques have been developed. Some have widespread use and are the subject of standards, such as pulse code modulation in voice networks, while others have very specialized and limited use, such as vocoders for secure voice. The type of A/D converter selected by the system designer thus depends on the application and the required level of performance.

Performance of A/D converters can be characterized by both objective and subjective means. The most significant parameter of objective assessment is the **signal-to-distortion (S/D) ratio.** Typical sources of distortion are quantization distortion and slope or amplitude overload distortion, both of which can be characterized mathematically in a straightforward manner. Subjective evaluation is more difficult to quantify because it involves human hearing or sight. In speech transmission, several tests have been devised for quantifying speech quality. They require listener juries who judge such factors as speech quality, word intelligibility, and speaker recognition.

Analog-to-digital coders can be divided into two classes: **waveform coders** and **source coders** [1]. Waveform coders are designed to reproduce the input signal waveform without regard to the statistics of the input. In speech coding, for example, pulse code modulation can provide adequate performance not only for speech but also for nonspeech signals such as signaling tones or voice-band data, which may appear on 3-kHz telephone circuits. Moreover, waveform coders provide uniform performance over a wide range of input signal level (dynamic range) and tolerate various sources of degradation (robustness). These advantages are offset somewhat by poor economies in transmission bit rates. The second class of A/D coders makes use of a priori knowledge about the source. If one can use certain physical constraints to eliminate source redundancy, then improved transmission efficiency can be realized. Human speech, for example, is known to have an information content of a few tens of hertz but requires a transmission bandwidth of a few thousand hertz. Source coders for speech, known as vocoders, are nearly able to match the transmission bandwidth to the information bandwidth. Vocoders tend to lack robustness, however, and produce unnatural sounding speech.

3.2   Pulse Code Modulation

Pulse code modulation (PCM) converts an analog signal to digital format by three separate processes: sampling, quantizing, and coding. The analog signal is first sampled to obtain an instantaneous value of signal amplitude at regularly spaced intervals; the sample frequency is determined by the Nyquist sampling theorem. Each sampled amplitude is then approximated by the nearest level from a discrete set of quantization levels. The coding process converts the selected quantization level into a binary code. If 256 ($= 2^8$) quantization levels are used, for example, then an 8-bit binary code is required to represent each amplitude sample.

The following step-by-step description of PCM is based on the illustration in Figure 3.1:

1. An analog input signal $s(t)$ is band-limited by a low-pass filter to F hertz.
2. The band-limited signal is sampled at a rate f_s that must equal or exceed the Nyquist frequency ($f_s \geq 2F$).
3. The sampled signal $s(iT)$ is held in a sample-and-hold circuit between two sampling instants (T seconds).
4. During this interval the sample is quantized into one of N levels. Quantization thus converts the sample into discrete amplitudes and in the process produces an error equal to the difference between input and quantized output. The greater the number of levels in the quantizer, the smaller the quantizing error introduced by the quantizer.
5. The coder maps the amplitude level selected by the quantizer into an n ($n = \log_2 N$) bit code designated $b_1, b_2, \ldots, b_n$, where $b_n = 1$ or 0. The format

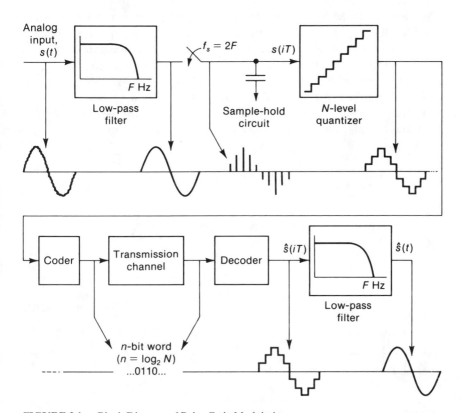

FIGURE 3.1   Block Diagram of Pulse Code Modulation

of the code is selected to facilitate transmission over a communication channel.

6. The decoder maps the PCM words back into amplitude levels, and the amplitude samples are low-pass filtered by a filter having a bandwidth of F hertz, which results in an estimate $\hat{s}(t)$ of the original sample.

The importance of PCM in digital transmission is based on its high level of performance for a variety of applications and its universal acceptance as a standard, especially for voice digitization. Pulse code modulation at 64 kb/s is an international standard for digital voice, based on a sampling rate of 8 kHz and an 8-bit code per sample. Two standards for 64-kb/s PCM voice have been developed, and these will be examined and compared. First, however, we will consider the processes of sampling, quantizing, and coding in more detail, with emphasis on voice applications.

3.2.1 Sampling

The basis for PCM begins with the sampling theorem, which states that a band-limited signal can be represented by samples taken at a rate f_s that is at least twice the highest frequency f_m in the message signal. An illustration of the sampling theorem is shown in Figure 3.2. A summarized proof of the sampling theorem is presented here, followed by a discussion of the application of sampling to a practical communication system.

Proof of the sampling theorem is contingent upon the assumptions that the input signal is band-limited to F hertz, that the samples are taken with

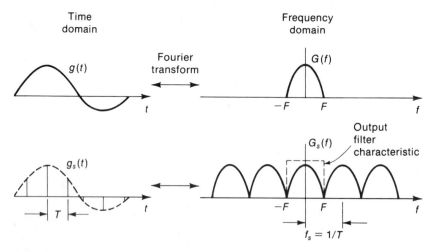

FIGURE 3.2   Illustration of Sampling Theorem

Proof of Sampling Theorem [2]

Consider a band-limited signal $g(t)$ that has no spectral components above F hertz. Suppose this signal is sampled at regular intervals T with an ideal impulse function $\delta_T(t)$ having infinitesimal width. The sampled signal $g_s(t)$ may be written as a sequence of impulses multiplied by the original signal $g(t)$:

$$g_s(t) = \sum_{k=-\infty}^{\infty} g(t)\,\delta_T(t - kT) \qquad (3.1)$$

Applying the Fourier transform to both sides of (3.1) we obtain

$$G_s(f) = G(f) * \delta_{f_s}(f) \qquad (3.2)$$

where * denotes the convolutional integral, $\delta_{f_s}(f)$ is a sequence of impulse functions separated by f_s hertz, and $f_s = 1/T$. Carrying out the convolution indicated in (3.2) we obtain

$$G_s(f) = \sum_{k=-\infty}^{\infty} G\left(f - \frac{k}{T}\right) \qquad (3.3)$$

The sampled spectrum is thus a series of spectra of the original signal separated by frequency intervals of $f_s = 1/T$. Note that $G(f)$ repeats periodically without overlap as long as

$$f_s \geq 2F$$

or equivalently

$$F \leq \frac{f_s}{2}$$

Therefore, as long as $g(t)$ is sampled at regular intervals $T \leq (2F)^{-1}$, the spectrum of $g_s(t)$ will be a periodic replica of $G(f)$ and will contain all the information of $g(t)$. We can recover the original signal $G(f)$ by means of a low-pass filter with a cutoff at F hertz. The ideal filter characteristic required to recover $g(t)$ from $g_s(t)$ is shown in Figure 3.2.

---

impulses of infinitesimal width, and that the low-pass filter used to recover the signal is ideal. In practice, these assumptions do not hold and therefore certain types of error are introduced. Because an ideal (rectangular) low-pass filter characteristic is not feasible, the spectrum of the "band-limited" analog signal contains frequency components beyond $f_s/2$ as shown in Figure 3.3a. The resulting adjacent spectra after sampling will overlap, causing an error in the reconstructed signal, as illustrated in Figure 3.3c. This error is called **aliasing** or **foldover distortion** (Figure 3.3d). The filtered spectrum is distorted due to the addition of tails from higher harmonics and the loss of tails for $|f| > f_s/2$. Design limitations on filter performance make it necessary to leave a guard band near the half sampling rate to minimize aliasing effects. For example, sampling a voice signal at 8 kHz provides a usable bandwidth of approximately 3.4 kHz. In this case the original signal can be recovered at the receiver by use of a low-pass filter with a gradual rolloff characteristic and a 3-dB cutoff around 3.4 kHz.

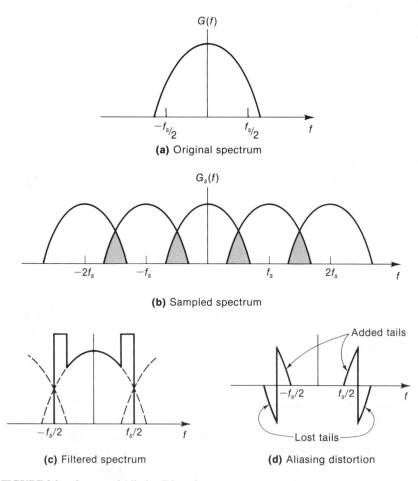

(a) Original spectrum

(b) Sampled spectrum

(c) Filtered spectrum

(d) Aliasing distortion

FIGURE 3.3   Spectra of Aliasing Distortion

3.2.2   Quantization

In contrast to the sampling process, which converts a continuous signal to one that is discrete in time, **quantizing** is the process of converting a continuous signal to one that is discrete in amplitude. As shown in Figure 3.4, the amplitude range of the analog input is converted into discrete steps. All samples of the input falling into a particular quantizing interval are replaced by a single value. An N-step quantizer may be defined by specifying a set of $N + 1$ decision thresholds $x_0, x_1, \ldots, x_N$ and a set of output values $y_1, y_2, \ldots, y_N$. When the value x of the input sample lies in the jth quantizing region, that is,

$$R_j = (x_{j-1} < x < x_j) \tag{3.4}$$

the quantizer produces the output y_j. The end thresholds x_0 and x_N are given values equal to the minimum and maximum values, respectively, expected of the input signal. A distinct n-bit binary word can then be associated with each output value if the set of output values contains $N = 2n$ members— hence the term n-bit quantizer.

The input/output characteristic $F(x)$ of a quantizer has a staircase form. For the simplest type of quantizer, the quantizing steps are all equal in size—hence the name *uniform* or *linear* quantization. If the allowed input range is bounded by $\pm V$, then the uniform step size q is

$$q = \frac{2V}{2^n} = \frac{2V}{N} \tag{3.5}$$

for an *n*-bit quantizer. The error introduced by the quantizing process is categorized into two types: **quantizing distortion** and **overloading** (or **clipping**), illustrated in Figure 3.4b. As long as the input signal lies within the quantizer-permitted range of $-V$ to $+V$, the only form of error introduced is quantizing distortion,* limited to a maximum or $\pm q/2$ for the linear quantizer. If the input signal exceeds this allowed range, the quantizer output will remain at the maximum allowed level ($\pm Nq/2$ for the linear quantizer), resulting in the input signal being clipped. Quantizing error is bounded in magnitude and tends to be sawtooth-shaped except at the maxima and at points of inflection, while overload noise is unbounded. A properly designed quantizer should match the particular input signal statistics (to

---

*The terms *noise* and *distortion* are used interchangeably here when describing quantization performance. It should be noted that quantizing and overload distortion are solely dependent on the particular quantization process and the statistics of the input signal.

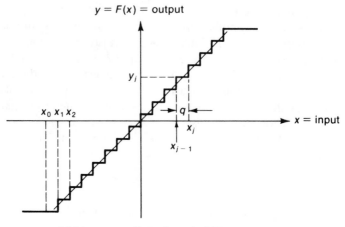

(a) Linear quantizer characteristic

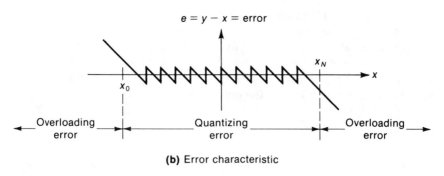

(b) Error characteristic

FIGURE 3.4   Linear Quantization

the extent such statistics are known). In particular, the choice of overload thresholds x_0 and x_N controls a tradeoff between the relative amounts of quantizing and overload distortion.

Quantizing Distortion
Performance of a quantizer can be described by a signal-to-distortion ratio that takes both quantizing and overload distortion into account. Distortion performance is stated in statistical terms, usually by mean square error, which will be used here. If we define the error due to the quantization process as

$$e = F(x) - x \qquad (3.6)$$

then the mean square error can be computed as

$$\overline{e^2} = \int_{-\infty}^{\infty} [F(x) - x]^2 p(x)\,dx \tag{3.7}$$

where $p(x)$ is the probability density function (pdf) of the input signal x. To compute $\overline{e^2}$ for a particular type of quantizer (linear or otherwise) we must divide the region of integration according to the decision thresholds $x_0, x_1, \ldots, x_N$ that define the quantizing regions:

$$\overline{e^2} = \sum_{j=1}^{N} \int_{x_{j-1}}^{x_j} (y_j - x)^2 p(x)\,dx$$

$$+ \int_{-\infty}^{x_o} (y_1 - x)^2 p(x)\,dx + \int_{x_N}^{\infty} (y_N - x)^2 p(x)\,dx \tag{3.8}$$

where $F(x) = y_j$ when x is in the region R_j. Note that the first term of (3.8) corresponds to quantizing distortion and the last two terms correspond to overload distortion. The regions R_j can be made arbitrarily small with large N, with the exception of the overload regions R_1 and R_N, which are unbounded. Then $p(x)$ can be assumed to be uniform over the inner regions and replaced by a constant $p(y_j)$. Further assuming that $p(x) \approx 0$ for x in the overload regions, Equation (3.8) simplifies to

$$\overline{e^2} = \sum_{j=1}^{N} \frac{(x_j - x_{j-1})^2}{12} (x_j - x_{j-1}) p(y_j)$$

$$= \frac{1}{12} \sum_{j} (\delta e_j)^2 \left[p(y_j) \delta e_j \right] \tag{3.9}$$

where $\delta e_j = x_j - x_{j-1}$.

For linear quantization, the decision thresholds are equally spaced so that the quantizing steps $(x_j - x_{j-1})$ are of constant length and equal to the step size q, as illustrated in Figure 3.4a. The mean square error then becomes

$$\overline{e^2} = \frac{q^2}{12} \sum_{j=1}^{N} p(y_j) q$$

$$= \frac{q^2}{12} \tag{3.10}$$

since $p(y_j)q$ describes the probability that the signal amplitude is in the region x_{j-1} to x_j and the sum of probabilities over index j equals 1. Thus the mean square distortion of a linear quantizer increases as the square of the step size. This result may also be obtained from the expression

$$\overline{e^2} = \int_{-\infty}^{\infty} e^2 p(e)de \tag{3.11}$$

by assuming the quantizing error pdf to be uniform over each quantization interval $(-q/2, +q/2)$ and neglecting overload distortion, so that

$$\overline{e^2} = \int_{-q/2}^{q/2} e^2 p(e)de = \int_{-q/2}^{q/2} \frac{e^2}{q} de = \frac{q^2}{12} \tag{3.12}$$

To compute the ratio of signal to quantizing distortion S/D_q, the input signal characteristics (such as power) must also be specified. Quite often performance for a quantizer is based on sinusoidal inputs, because S/D_q for speech and sinusoidal inputs compare favorably and use of sinusoidal inputs facilitates measurement and calculation of S/D_q [3]. For the case of a full range $(-V,V)$ sinusoidal input that has zero overload error, the average signal power is

$$S = \frac{V^2}{2} \tag{3.13}$$

From (3.5) we know that the peak-to-peak $(-V,V)$ range of the linear quantizer is $2^n q$. Therefore from (3.12)

$$\overline{e^2} = \frac{2}{3} \frac{V^2/2}{2^{2n}} \tag{3.14}$$

so that for the linear quantizer the signal to quantizing distortion ratio is

$$\frac{S}{D_q} = \frac{V^2/2}{(2/3)\left[(V^2/2)/2^{2n}\right]} = \left(\frac{3}{2}\right)2^{2n} \tag{3.15}$$

or, expressed in decibels,

$$\left(\frac{S}{D_q}\right)_{dB} = 6n + 1.8 \text{ dB} \tag{3.16}$$

This expression indicates that each additional quantization bit adds 6 dB to the S/D_q ratio.

For random input signals with root mean square (rms) amplitude σ, a commonly used rule for minimizing overload distortion is to select a suitable **loading factor** α, defined as $\alpha = V/\sigma$. A common choice is the so-called 4σ loading ($\alpha = 4$), in which case the total amplitude range of the quantizer is 8σ. The linear quantizing step then becomes

$$q = \frac{8\sigma}{2^n} \tag{3.17}$$

The S/D_q can then be expressed as

$$\frac{S}{D_q} = \frac{\sigma^2}{(2\alpha\sigma/2^n)^2/12} = \frac{\sigma^2}{(8\sigma/2^n)^2/12} = \frac{3}{16}(2^{2n}) \tag{3.18}$$

or, in decibels,

$$\left(\frac{S}{D_q}\right)_{dB} = 6n - 7.2 \text{ dB} \tag{3.19}$$

which is 9dB less than the expression given in (3.16) because of the difference in average signal power.

Example 3.1 _____

A sinusoid with maximum voltage $V/2$ is to be encoded with a PCM coder having a range of $\pm V$ volts. Derive an expression for the signal-to-quantization distortion and determine the number of quantization bits required to provide an S/D_q of 40 dB.

Solution   For the given sinusoid, we know that

$$\frac{S}{D_q} = \frac{V^2/8}{e^2}$$

From (3.14) we have

$$\frac{S}{D_q} = \frac{V^2/8}{(2/3)\left[(V^2/2)/2^{2n}\right]} = \frac{3}{8}(2^{2n})$$

or, in decibels,

$$\left(\frac{S}{D_q}\right)_{dB} = 6n - 4.3 \text{ dB}$$

The number of bits required to yield an S/D_q of 40 dB is

$$n = \frac{40 + 4.3}{6}$$

$$= 8 \text{ bits}$$

Overload Distortion
Overload distortion results when the input signal exceeds the outermost quantizer levels $(-V, V)$. The distortion power due to quantizer overload has been previously defined by the last two terms of (3.8). To compute the mean square error due to overload distortion (D_0), the input signal pdf must be specified. First, let us assume the pdf to be symmetric so that the last two terms of (3.8) can be written

$$D_0 = 2 \int_V^\infty (V - x)^2 \, p(x) dx \tag{3.20}$$

If the input signal has a noiselike characteristic, then a gaussian pdf can be assumed, described by

$$p(x) = \frac{1}{\sigma\sqrt{2\pi}} e^{-x^2/2\sigma^2} \tag{3.21}$$

An important example of signals that are closely described by a gaussian pdf, by virtue of the law of large numbers, is an FDM multichannel signal [4]. Speech statistics are often modeled after the laplacian pdf [5], given by

$$p(x) = \frac{1}{\sigma\sqrt{2}} e^{-\sqrt{2}|x|/\sigma} \tag{3.22}$$

where σ in (3.21) and (3.22) is the rms value of the input signal and σ^2 is the average signal power. Substituting the gaussian and laplacian pdf's into (3.20) we obtain, respectively,

$$D_0 = \sqrt{\frac{2}{\pi}} (V^2 + \sigma^2) \int_{V/\sigma}^\infty e^{-x^2/2} \, dx$$

$$- \sqrt{\frac{2}{\pi}} V\sigma e^{-V^2/2\sigma^2} \quad \text{(gaussian input)} \tag{3.23a}$$

and

$$D_0 = \sigma^2 e^{-\sqrt{2}V/\sigma} \qquad \text{(laplacian input)} \qquad (3.23b)$$

Figure 3.5 plots the S/D ratio for linear PCM versus the loading factor α for a gaussian input signal. For small α, performance is limited by an asymptotic bound due to amplitude overload given by (3.23a). Similarly for large α, performance is bounded by a quantization noise asymptote given by (3.18). The peak of each curve indicates an optimum choice of loading factor, which maximizes the S/D ratio by balancing amplitude overload and quantization noise. Code lengths of $n = 2, 4, 6, 8,$ and 10 are shown. These curves also illustrate that by increasing the code length by 1 bit, the S/D ratio is improved by 6 dB.

Effects of Transmission Errors
Transmission errors cause inversion of received bits resulting in another form of distortion in the decoded analog signal. The distortion power due to

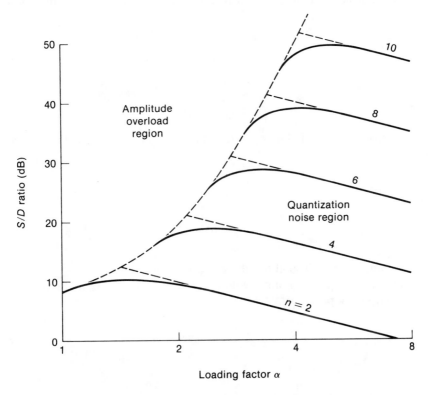

FIGURE 3.5   Linear PCM *S/D* Performance for Gaussian Signal Input

decoded bit errors can be computed on the basis of mean square value of the decoded errors. Assuming the use of linear quantization, each sample $s(iT)$ is transmitted by a weighted binary code that can be written

$$s(iT) = \sum_{j=1}^{n} qb_{ij}2^{j-1} \qquad (3.24)$$

and the reconstructed sample can be written

$$\hat{s}(iT) = \sum_{j=1}^{n} qc_{ij}2^{j-1} \qquad (3.25)$$

where n represents the number of bits used in the code and b_{ij}, c_{ij} are the transmitted and received bits, respectively, where the c_{ij} may have errors. An error in the jth bit will cause an error in the decoded sample value of $e_j = 2^{j-1}q$. If each error occurs with probability p_j, then the mean square value is

$$\overline{e^2} = \sum_{j=1}^{n} e_j^2 p_j$$
$$= \sum_{j=1}^{n} 4^{j-1}q^2 p_j \qquad (3.26)$$

If bit errors are assumed independent, then $p_j = p_e$ for all bits in the code so that

$$\overline{e^2} = q^2 p_e \sum_{j=1}^{n} 4^{j-1}$$
$$= q^2 p_e \frac{4^n - 1}{3} \qquad (3.27)$$

In computing the signal to bit error distortion ratio, S/D_e, two cases are calculated here: gaussian input with average signal power σ^2 and sinusoidal input with peak power V^2:

$$\frac{S}{D_e} \approx \frac{3\sigma^2}{4^n q^2 p_e} \qquad \text{(gaussian)} \qquad (3.28)$$

and

$$\frac{S}{D_e} \approx \frac{3}{4p_e} \qquad \text{(sinusoidal)} \qquad (3.29)$$

To compare the effects of transmission error versus quantization error, consider the ratio of signal to total distortion, $S/(D_q + D_e)$. For a sinusoidal input signal with peak power V^2, we have, using (3.15) and (3.29),

$$\frac{S}{D_q + D_e} = \frac{3}{4p_e + 2^{-2n}} \tag{3.30}$$

Rearranging terms, we may rewrite (3.30) as

$$\frac{S}{D_q + D_e} = \frac{(3)2^{2n}}{4p_e 2^{2n} + 1} \tag{3.31}$$

Two specific cases are of interest. First, for large p_e we have $4p_e 2^{2n} \gg 1$ and Equation (3.31) is dominated by transmission errors, in which case

$$\frac{S}{D_q + D_e} \approx \frac{3}{4p_e} \quad \text{(large } p_e) \tag{3.32}$$

Second, for small p_e we have $1 \gg 4p_e 2^{2n}$ and Equation (3.31) is dominated by quantization distortion, in which case

$$\frac{S}{D_q + D_e} \approx (3)2^{2n} \quad \text{(small } p_e) \tag{3.33}$$

3.2.3   Companding

For most types of signals such as speech and video, a linear quantizer is not the optimum choice in the sense of minimizing mean square error. Strictly speaking, the linear quantizer provides minimum distortion only for signals with a uniform probability density function and performs according to predictions only for stationary signals. Speech signals, however, exhibit nonuniform statistics with a wide dynamic range (up to 40 dB) and smaller amplitudes are more likely than larger amplitudes. Here a choice of linear quantization would result in a poorer S/D (up to 40 dB worse) for weak signals. An alternative approach is to divide the input amplitude range into nonuniform steps by increasing the number of quantization steps in the region around zero and correspondingly decreasing the number around the extremes of the input range. The result of this nonuniform code is an input/output characteristic that is a staircase with N steps of unequal width, as shown in Figure 3.6.

As indicated in Figure 3.7, nonuniform quantization can be achieved by first compressing the samples of the input signal and then linearly quantizing the compressed signals; at the receiver a linear decoder is followed by an

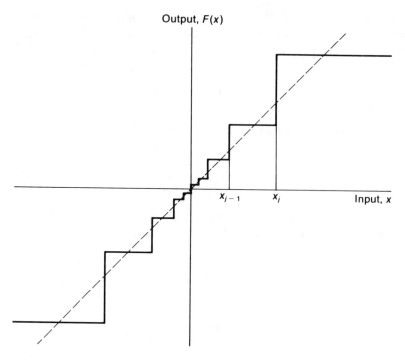

FIGURE 3.6   Nonuniform Quantizer Characteristic

expander that provides the inverse characteristic of the compressor. This technique is called **companding.**

The input signal is compressed according to the characteristic $F(x)$, illustrated in Figure 3.8 for positive values of x. The characteristic of the linear quantizer is once again a staircase but now preceded by the compressor, which provides N steps of unequal width as shown in Figure 3.6. The input

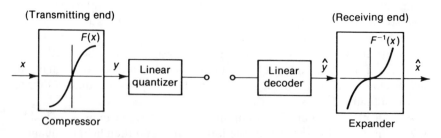

FIGURE 3.7   Companding Technique

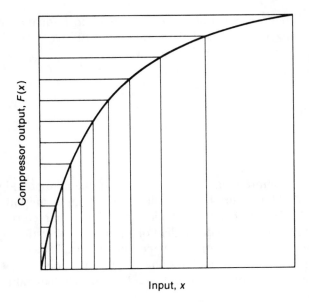

FIGURE 3.8   Compressor Characteristic (Positive Values Only)

signal x is recovered by applying the inverse characteristic $F^{-1}(x)$ to produce an estimate $\hat{x}$.

An approximate expression for the mean square error of a nonuniform quantizer can be derived by starting with (3.9). First assume a compression characteristic $y = F(x)$, with N nonuniform input intervals and N uniform output intervals each of step size q. Next define the compression slope to be $F'(x) = dF(x)/dx$. For large N, this curve of $F(x)$ in the jth quantizing interval can be approximated by

$$F'(x_j) \approx \frac{q}{x_j + x_{j-1}} \tag{3.34}$$

Substituting (3.34) into (3.9) and approximating the resulting sum by an integral (good for $N \gg 1$) yields

$$\overline{e^2} = \frac{1}{12} \sum_{j=1}^{N} \frac{q^2}{[F'(x_j)]^2} (x_j - x_{j-1}) p(y_j)$$

$$\approx \frac{q^2}{12} \int \frac{p(x)}{[F'(x)]^2} \, dx \tag{3.35}$$

The improvement in performance of the nonuniform quantizer over the uniform quantizer is expressed by the ratio of (3.10) to (3.35), which yields a factor C_l called the companding improvement:

$$C_l = \frac{1}{\displaystyle\int \frac{p(x)}{[F'(x)]^2}\,dx} \tag{3.36}$$

Logarithmic Companding
The choice of compression characteristic $F(x)$ is made to provide constant performance (S/D) over the dynamic range of input amplitudes for a given signal. For speech, a logarithmic compression characteristic is used, since the resulting S/D is independent of input signal statistics [6]. In practice, an approximate logarithmic shape must be used since a truly logarithmic representation would require an infinite code set. One such logarithmic compressor curve widely used for speech digitization is the μ-law curve given by

$$F(x) = \text{sgn}(x)V \frac{\ln\left(1 + \frac{\mu|x|}{V}\right)}{\ln(1 + \mu)} \qquad 0 \le |x| \le V \tag{3.37}$$

where sgn(x) is the polarity of x. As seen in Figure 3.9, $F(x)$ approaches a linear function for small μ and a logarithmic function for large μ. Specifically, by applying l'Hôpital's rule to (3.37), the case for $\mu = 0$ can be seen to correspond to no companding and therefore reduces to linear quantization. For $\mu \gg 1$ and $\mu x \gg V$, $F(x)$ approximates a true logarithmic form. The mean square error for the logarithmic compandor can be found by first differentiating (3.37) to obtain the compressor slope

$$\frac{dF}{dx} = \frac{\mu}{\ln(1 + \mu)}\left[1 + \frac{\mu|x|}{V}\right]^{-1} \tag{3.38}$$

Substituting (3.38) into the expression for mean square error (3.35) then yields

$$D_q = \frac{q^2[\ln(1 + \mu)]^2}{12\mu^2}\left(1 + \frac{2\mu E[|x|]}{V} + \frac{\mu^2 \sigma_x^2}{V^2}\right) \tag{3.39}$$

where the quantity $E[|x|]$ is the rms mean absolute value of the input signal and where σ_x is the rms value of the input signal. Now since $q = 2V/N$, Equation (3.39) can be simplified to

$$D_q = \frac{\ln^2(1+\mu)}{3N^2\mu^2}\left(V^2 + 2V\mu E[|x|] + \mu^2\sigma_x^2\right)$$ (3.40)

The signal to quantizing distortion ratio then becomes

$$\frac{S}{D_q} = \frac{3N^2}{\ln^2(1+\mu)}\frac{1}{1 + 2BC/\mu + C^2/\mu^2}$$ (3.41)

where we define

$$B = \frac{E[|x|]}{\sigma_x} = \frac{\text{mean absolute input}}{\text{rms input}}$$

$$C = \frac{V}{\sigma_x} = \frac{\text{compressor overload voltage}}{\text{rms input}}$$

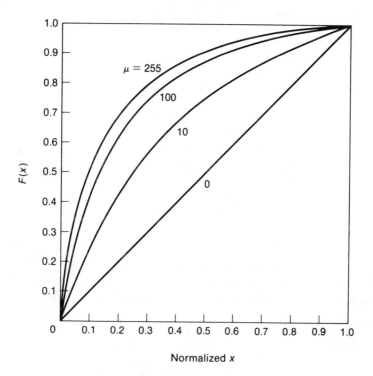

FIGURE 3.9   Normalized μ-Law Compressor Curve (Positive Values Only)

Example 3.2

Find an expression for S/D_q for the case of a speech input to a $\mu = 255$ companded PCM coder. Assume that speech may be modeled with laplacian statistics and that the rms amplitude of the input is equal to the compressor overload voltage. Repeat the calculation for $\mu = 100$.

Solution   To calculate S/D_q for μ-law companding, we will use (3.41). But first we note that from (3.22) for a laplacian pdf,

$$B = \frac{1}{\sqrt{2}}$$

and from the assumptions given

$$C = 1$$

Substituting these values into (3.41), we find

$$\left(\frac{S}{D_q}\right)_{dB} = 10 \log\left[\frac{3N^2}{\ln^2(1+\mu)} \frac{1}{1 + 1/\mu^2 + \sqrt{2}/\mu}\right] \tag{3.42}$$

For $\mu = 255$,

$$\left(\frac{S}{D_q}\right)_{dB} = 20 \log N - 10.1 \text{ dB} \tag{3.43}$$

$$= 6n - 10.1 \text{ dB}$$

where $N = 2^n$ so that n represents the number of bits required to code N levels. For $\mu = 100$ a similar calculation yields

$$\left(\frac{S}{D_q}\right)_{dB} = 6n - 8.5 \text{ dB} \tag{3.44}$$

Now let us consider the performance of u-law companding for the frequently used case of sinusoidal inputs. The parameter C, which indicates the range of the input, can be related to a full-load sinewave by noting that

$$10 \log\left[\frac{V^2/2}{\sigma_x^2}\right] \begin{array}{l} = 10 \log\left[\frac{C^2}{2}\right] \\[2ex] = 20 \log C - 3 \text{ dB} \end{array} \tag{3.45}$$

We can then plot the S/D_q ratio given by (3.41) versus input signal power referenced to a full-load sinusoid. Figure 3.10 is such a plot for $N = 256$ (8-bit quantizer) and for $\mu = 0,10,40,100,255$, and 400. The S/D_q ratio for $\mu = 0$ (linear quantization) is found by applying l'Hôpital's rule to (3.41) to obtain

$$\left(\frac{S}{D_q}\right)_{dB} = 10\log\left(\frac{3N^2}{C^2}\right) \tag{3.46}$$

The results of Figure 3.10 indicate that $\mu \geq 100$ is required in order to obtain a relatively flat S/D_q ratio over a 40-dB dynamic range. In practice the usual choices made are $\mu = 100$ for 7-bit PCM and $\mu = 255$ for 8-bit PCM. Another performance indicator is the companding improvement C_I, which for weakest signals (where the companding characteristic is nearly linear) is given by Equation (3.36):

$$C_I = [F'(0)]^2 = \left[\frac{\mu}{\ln(1 + \mu)}\right]^2 \tag{3.47}$$

Then for $\mu = 255$, we obtain $C_I = 33.25$ dB.

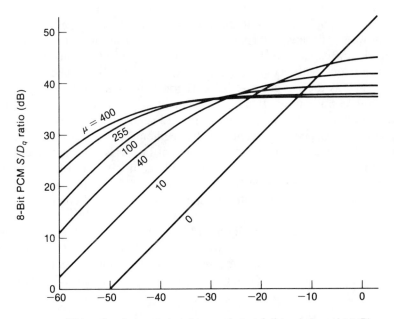

FIGURE 3.10   Quantizing Distortion Curves for μ-Law Compandor with Sinusoidal Input

Another widely used logarithmic characteristic for speech is the A-law curve where

$$F(x) = \begin{cases} \text{sgn}(x)\, \dfrac{A|x|}{1 + \ln A} & 0 \le |x| \le 1/A \qquad\qquad \text{(3.48a)} \\[4mm] \text{sgn}(x)\, \dfrac{1 + \ln\big(A|x|\big)}{1 + \ln A} & 1/A \le |x| \le 1 \qquad\qquad \text{(3.48b)} \end{cases}$$

The parameter A determines the dynamic range. A value of $A = 87.6$ is typical for 8-bit implementation. Over a 40-dB dynamic range, μ-law has a flatter S/D_q ratio than A-law, as shown in Figure 3.11.* This figure also reproduces the recommended CCITT specification [7]. Comparison of these curves indicates that 8-bit A-law and μ-law both meet the recommended limits. A similar comparison of 7-bit S/D_q ratios would show that neither A-law nor μ-law meets the recommended limits.

Piecewise Linear Segment Companding
In practice, logarithmic companding laws are approximated by nonlinear devices such as diodes or by piecewise linear segments. The D1 channel bank, for example, uses 7-bit, μ-law companding implemented with diodes [8]; however, because of characteristic diode variation and temperature dependency, experience has indicated difficulty in maintaining matching characteristics at the compressor and expander [9, 10]. In later PCM systems, such as the D2 channel bank, segment implementations have been more successfully used. Two of the segment law families have emerged as international (CCITT) standards—the 13-segment A-law and the 15-segment μ-law. The significance of the segment laws resides in the digital linearization feature, by which the segment intervals can be conveniently coded in digital form. Digitally linearizable compression also facilitates a variety of digital processing functions commonly found in digital switches such as call conferencing, filtering, gain adjustment, and companding law conversion [11].

As an example, consider Figure 3.12, which shows the piecewise linear approximation using 15-segment $\mu = 255$ PCM. The vertices of the connected segments lie on the logarithmic curve being approximated. For this μ-law curve, there are eight segments on each side of zero. The two seg-

---

*The term dBm is the absolute power level in decibels referred to 1 milliwatt; dBm0 is the absolute power level in dBm referred to a point of zero relative level.

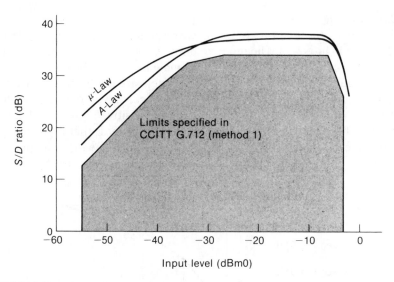

S/D ratio (dB)

Input level (dBm0)

FIGURE 3.11   Quantizing Distortion Curves with Noiselike (Gaussian) Input for 8-Bit PCM

ments about the origin are collinear and counted as a single segment, however, thus yielding a total of 15 segments. The encoding of the 8 bits—B1 (most significant bit) through B8 (least significant bit)—of each PCM word is accomplished as follows. The B1 position is the polarity bit encoded as a 0 if positive or 1 if negative. The encoding of bits B2, B3, and B4 is determined by the segment into which the sample value falls. Examination of the curve in Figure 3.12 shows that the linear segments have a binary progression (that is, segment 2 represents twice as much input range as segment 1, and segment 3 twice as much as segment 2, and so on) which provides the overall nonlinear characteristic. The remaining 4 bits are encoded using linear quantization with 16 equal-size intervals within each segment.

The corresponding curve for an 8-bit A-law characteristic with $A = 87.6$ uses 13 segments, where the center four segments around zero are made collinear and the other remaining segments are related by binary progression. The 15-segment $\mu = 255$ characteristic and 13-segment $A = 87.6$ characteristic have been adopted by the CCITT as two PCM standards; μ-law is used principally in North America and Japan and A-law primarily in Europe [7]. The S/D_q curve for the 15-segment and 13-segment laws are essentially as shown in Figure 3.11, although the curves would show a scalloping effect that is due to the use of segmented rather than continuous companding.

3.2.4 Coding

Table 3.1 lists the PCM code words corresponding to the magnitude bits (B2-B8) of Figure 3.12. The input amplitude range has been normalized to a maximum amplitude of 8159 so that all magnitudes may be represented by integer values. Note that the quantization step size doubles with each successive segment, a feature that generates the logarithmic compression characteristic. With 128 quantization intervals for positive inputs and likewise 128 levels for negative inputs, there are 256 code words that correspond to the relative input amplitude range of −8159 to +8159. The first and most significant bit indicates the sign and the remaining bits indicate the magnitude.

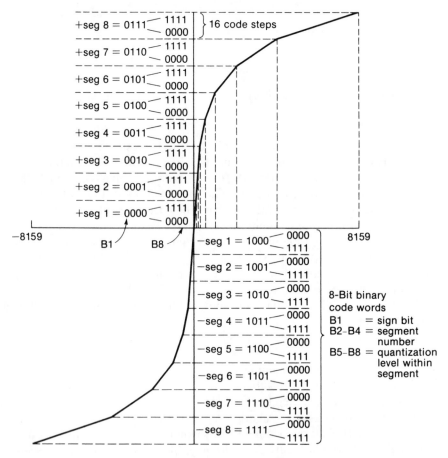

FIGURE 3.12   Piecewise Linear Segment Approximation to μ = 255 Logarithmic Compandor

TABLE 3.1   PCM Coding for μ = 255 Segment Companding

| | | | | | Coder Output | | |
Segment Number	Segment Endpoint	Step Size	Amplitude Input Range[a] (x_n to x_{n+1})	Code Value (n)	Segment Code (B2–B4)	Quantization Code (B5–B8)	Decoder[b] Output (y_n)
	0		0–1	0		0000	0
			1–3	1		0001	2
			3–5	2	000	0010	4
1		2	⋮	⋮		⋮	
	31		29–31	15		1111	30
	31		31–35	16		0000	33
2		4	⋮	⋮	001	⋮	
	95		91–95	31		1111	93
	95		95–103	32		0000	99
3		8	⋮	⋮	010	⋮	
	223		215–223	47		1111	219
	223		223–239	48		0000	231
4		16	⋮	⋮	011	⋮	
	479		463–479	63		1111	471
	479		479–511	64		0000	495
5		32	⋮	⋮	100	⋮	
	991		959–991	79		1111	975
	991		991–1055	80		0000	1023
6		64	⋮	⋮	101	⋮	
	2015		1951–2015	95		1111	1983
	2015		2015–2143	96		0000	2079
7		128	⋮	⋮	110	⋮	
	4063		3935–4063	111		1111	3999
	4063		4063–4319	112		0000	4191
8		256	⋮	⋮	111	⋮	
	8159		7903–8159	127		1111	8031

[a]Magnitude only: normalized to full-scale value of 8159. Sign bit B1 = 0 for positive and 1 for negative.
[b]Decoder output is $y_0 = x_0 = 0$ and $y_n = (x_n + x_{n+1})/2$ for $n = 1, 2, \ldots, 127$.

Each quantization interval is assigned a separate code word that is decoded as the midpoint of the interval.

Example 3.3

For μ = 255, 15-segment companded PCM, determine the code word that represents a 5-volt signal if the encoder is designed for a ±10-volt input range. What output voltage will be observed at the PCM decoder, and what is the resulting quantizing error?

Solution   Since a 5-volt signal is half the allowed maximum input, the corresponding PCM amplitude is represented by

$$\left(\frac{1}{2}\right)(8159) = 4080$$

From Table 3.1, the code word is found to be

$$01110000 \qquad \text{(decimal 112)}$$

The corresponding decoder output is

$$y_{112} = \frac{x_{112} + x_{113}}{2}$$

$$= \frac{4063 + 4319}{2} = 4191$$

The voltage associated with this output value is

$$\left(\frac{4191}{8159}\right)(10 \text{ volts}) = 5.14 \text{ volts}$$

Therefore the quantizing error is 5.14 - 5.0 = 0.14 volt.

The code set shown in Table 3.1, known as a *signed* or *folded binary* code, is superior in performance to ordinary binary codes in the presence of transmission errors. With a folded binary code, errors in the sign or most significant bit cause an output error equal to twice the signal magnitude. Since speech signals have a high probability of being equal to or near zero, however, the output error due to a transmission error also tends to be small. For ordinary binary codes, where negative signals are expressed as the complement of the corresponding positive signal, an error in the first or most significant bit always causes an output error of half the amplitude range.

Table 3.1 also reveals that choice of a natural folded binary code results in a high density of zeros due to the most probable speech input amplitudes. When using bipolar coding in repeatered line applications, this code set would result in poor synchronization performance, since good clock recovery from a bipolar signal is dependent on a high density of 1's. (See Chapter 5 for more details.) By complementing the folded binary code, a predominance of 1's is created, which provides good synchronization performance for bipolar coding. The transmitted PCM code words for the inverted folded binary code are shown in Table 3.2, which also shows suppression of the all-zero code. When a sample encodes at this value (−127), the all-zero code is replaced by 00000010, which corresponds to the value −125. This **zero code suppression** then guarantees that no more than 13 consecutive zeros can appear in two consecutive PCM code words, which further enhances clock recovery in repeatered line transmission. Since the maximum amplitude represented by the all-zero code occurs infrequently, the degradation introduced by zero code suppression is insignificant. The inverted folded binary code with zero code suppression shown in Table 3.2 is exactly the set of codes specified by CCITT for μ-law PCM in Rec. G.711.

TABLE 3.2   PCM Code Words as Prescribed by CCITT for μ-Law Companding

256 Transmission Code Words								
1	0	0	0	0	0	0	0	(127)
1	0	0	0	0	0	0	1	(126)
1	0	0	0	0	0	1	0	(125)
1	1	1	1	1	0	1	0	(5)
1	1	1	1	1	0	1	1	(4)
1	1	1	1	1	1	0	0	(3)
1	1	1	1	1	1	0	1	(2)
1	1	1	1	1	1	1	0	(1)
1	1	1	1	1	1	1	1	(+0)
0	1	1	1	1	1	1	1	(−0)
0	1	1	1	1	1	1	0	(−1)
0	1	1	1	1	1	0	1	(−2)
0	1	1	1	1	1	0	0	(−3)
0	1	1	1	1	0	1	1	(−4)
0	1	1	1	1	0	1	0	(−5)
0	0	0	0	0	0	1	0	(−125)
0	0	0	0	0	0	0	1	(−126)
0	0	0	0	0	0	0	0	(−127)*

*The all-zero code is not used and instead is replaced by the code word 00000010.

For the A-law coder prescribed by CCITT Rec. G.711, only the *even* bits (B2, B4, B6, and B8) are inverted from the natural folded binary sequence and zero code suppression is not used [7].

As a result of these differences between A-law and μ-law coders, conversion is required when interfacing these two coders as with intercountry gateways [12]. The CCITT has prescribed that any necessary conversion will be done by the countries using μ-law. A conversion algorithm prescribed by the CCITT (Rec. G.711, tables 3 and 4 [7]) provides direct digital conversion via a "look-up table" that is realizable in the form of read-only memory. The S/D performance of the converters is displayed in Figure 3.13 for an input with laplacian distribution to simulate speech. For purposes of comparison, the S/D performance for ideal A-law and μ-law (curves 1 and 2) plus the S/D performance for nonoptimum $\mu \rightarrow A$ and $A \rightarrow \mu$ conversion (curves 3 and 4) are also shown. Curves 3 and 4 show the effects of using a μ-law compressor with an A-law expander, and vice versa, assuming the use of a natural folded binary code. Because of the difference in assignment of signal level to PCM codes between A-law and μ-law, curves 3 and 4 indicate much worse degradation than curves 1 and 2. For high input signal level the S/D is only slightly degraded, due to the fact that high input levels are represented by nearly identical PCM words for μ-law and A-law. For lower input signal levels the S/D degrades rapidly, due to the fact that lower signal levels are represented by significantly different PCM words for μ-law versus A-law. Comparing curves 5 and 6 with curves 1 and 2, we see that the degradation attendant upon optimum conversion is only about 3 dB. Hence the optimized code conversions (curves 5 and 6) via the look-up table must be used to provide acceptable S/D performance.

3.3 Differential PCM and Delta Modulation

For highly correlated signals such as speech or video, the signal value changes slowly from one Nyquist sample to the next. This makes it possible to predict a sample value from preceding samples and to transmit the difference between the predicted and actual samples. Since the variation of this difference signal is less than that of the input signal, the amount of information to be transmitted is reduced. This technique is generally known as *differential PCM encoding*.

An implementation of differential pulse code modulation (DPCM) is shown in Figure 3.14. The input signal is first low-pass filtered to limit its bandwidth to one-half (or less) of the sampling rate f_s. The predicted input value is then subtracted from the analog input $s(t)$ and the difference is sampled and quantized. The predicted input signal is generated in a feedback

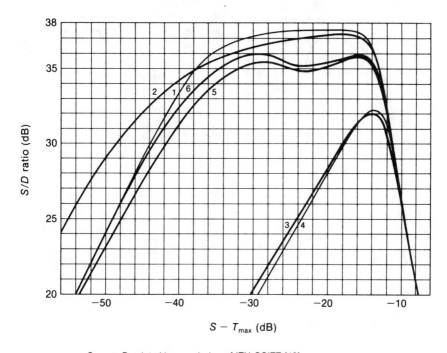

Source: Reprinted by permission of ITU-CCITT [12].
Curve 1: 13-segment A-law compandor
Curve 2: 15-segment μ-law compandor
Curve 3: 15-segment μ-law compressor with 13-segment A-law expander
Curve 4: 13-segment A-law compressor with 15-segment μ-law expander
Curve 5: same as curve 3 but with recoding
Curve 6: same as curve 4 but with recoding
$T_{max} = \begin{cases} 3.14 \text{ dBm0 for } A\text{-law} \\ 3.17 \text{ dBm0 for } \mu\text{-law} \end{cases}$

FIGURE 3.13   Signal-to-Distortion Ratio (S/D) as a Function of the Load $S - T_{max}$ for 8-Bit Coding with a Simulated (Laplacian) Speech Signal

loop that uses an integrator to sum past differences as the decoded sample estimate. The receiver contains a decoder identical to that used in the transmitter.

The simplest form of DPCM is the delta modulator, which provides 1-bit quantization of the difference signal. The output bits then represent only the polarity of the difference signal. If the difference signal is positive, a 1 is generated; if it is negative, a 0 is generated. The local decoder generates steps with amplitude $+\Delta$ or $-\Delta$ in accordance with the outputs 1 and 0. The decoded unfiltered signal is then a staircase waveform with uniform step size Δ, as illustrated in Figure 3.15. The integrator in the receiver is period-

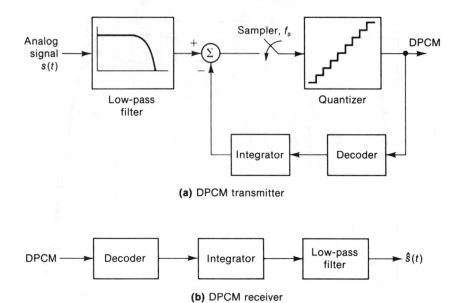

(a) DPCM transmitter

(b) DPCM receiver

FIGURE 3.14   DPCM Implementation

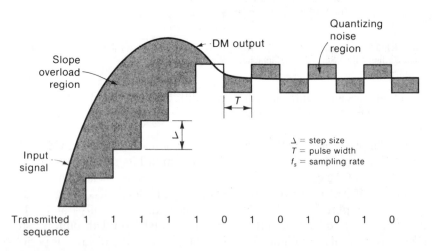

FIGURE 3.15   Delta Modulator Waveform

ically reset to limit the effect of transmission errors on the output, and the low-pass filter smoothes the staircase signal to recover the analog signal.

Figure 3.15 also illustrates the effects of slope overload distortion and quantizing distortion. If the slope of the input signal exceeds the slope of the staircase (Δ/T), the resulting error is known as **slope overload distortion.** For an input slope less than Δ/T, the errors are a form of quantizing distortion. For a fixed sampling rate, optimum S/D performance is obtained by selecting the step size to minimize the sum of slope overload and quantizing distortion. With a choice of small step size, slope overload distortion will dominate; with a choice of large step size, quantizing distortion will dominate.

3.3.1   Quantizing Distortion
Delta Modulation
Consider first the simplest measure of performance: quantizing distortion with no slope overload. Assume the input to be sinusoidal with the following characteristics:

$$\text{Input:} \qquad s(t) = A \sin 2\pi f_0 t$$

$$\text{Slope:} \qquad \frac{ds}{dt} = 2\pi f_0 A \cos 2\pi f_0 t$$

$$\begin{array}{l}\text{Maximum slope}\\ \text{of input:} \qquad \text{Slope max} = 2\pi f_0 A\end{array}$$

To prevent slope overload, the slope of the delta modulator must be greater than or equal to the maximum slope of the input, or

$$\Delta f_s \geq 2\pi f_0 A \tag{3.49}$$

The output signal power S can then be deduced from (3.49) as

$$S = \frac{A^2}{2} = \frac{1}{2}\left(\frac{\Delta f_s}{2\pi f_0}\right)^2 \tag{3.50}$$

The maximum quantizing error is $\pm\Delta$; assuming the quantizing error to be uniformly distributed (good approximation for small Δ), the mean square quantizing error is

$$\overline{e^2} = \int_{-\Delta}^{\Delta} e^2 p(e)de = \frac{1}{2\Delta}\int_{-\Delta}^{\Delta} e^2 de = \frac{\Delta^2}{3} \tag{3.51}$$

It is worth noting that for a PCM linear quantizer with $q = 2\Delta$, the same result as (3.51) is obtained for quantizing distortion. For a sinusoidal input the S/D_q ratio is given by

$$\frac{S}{D_q} = \frac{3A^2}{2\Delta^2} \tag{3.52}$$

The maximum S/D_q ratio for no slope overload is obtained from using the equality sign in (3.49):

$$\frac{S}{D_q} = \frac{3}{2} \frac{f_s^2}{(2\pi f_0)^2} \tag{3.53}$$

which indicates that doubling the sampling frequency (and hence doubling the bit rate) improves the S/D_q ratio by 6 dB.

The preceding calculation of the S/D_q ratio assumes no filtering of the output samples. Since the input bandwidth is less than the sampling frequency, we can enhance the output S/D_q ratio by proper filtering at the receiver. Assume the noise (error) power at the receiver input to be uniformly distributed over frequencies $(0, f_s)$. If an ideal low-pass or bandpass filter of bandwidth f_m is used at the receiver, the mean square output error becomes

$$\overline{e^2} = \frac{\Delta^2}{3}\left(\frac{f_m}{f_s}\right) \tag{3.54}$$

Assuming a sinusoidal input, the output S/D_q ratio is then

$$\frac{S}{D_q} = \frac{3A^2 f_s}{2\Delta^2 f_m} \tag{3.55}$$

The maximum S/D_q for no slope overload is obtained from (3.55) and (3.49):

$$\frac{S}{D_q} = \frac{3}{2} \frac{f_s^3}{f_m(2\pi f_0)^2} \tag{3.56}$$

or

$$\left(\frac{S}{D_q}\right)_{dB} = 10\log\left[\frac{f_s^3}{f_m f_0^2}\right] - 14\,dB \tag{3.57}$$

Notice that the low-pass filter yields an S/D_q ratio proportional to f_s^3 rather than to f_s^2 as found in (3.53), indicating a 9-dB improvement with doubling of the sample frequency.

Additional improvement in the S/D_q ratio is realized by increasing the order of integration in the delta modulator feedback loop. When double integration is incorporated, for example, the maximum S/D_q for a sinusoidal input has been derived [13]:

$$\left(\frac{S}{D_q}\right)_{dB} = 10 \log \left[\frac{f_s^5}{f_m^2 f_0^2}\right] - 32 \text{ dB} \tag{3.58}$$

Differential PCM (DPCM)
The maximum positive slope for l-bit DPCM is

$$\text{Slope max} = (2^l - 1) \, \Delta f_s \tag{3.59}$$

where the 2^l levels are equispaced and separated by 2Δ. The slope of the DPCM quantizer must be greater than or equal to the maximum slope of the input to prevent slope overload. For a sinusoidal input, this means that

$$(2^l - 1) \, \Delta f_s \geq 2\pi f_0 A \tag{3.60}$$

so that the output signal power S can be given as

$$S = \frac{A^2}{2} = \frac{1}{2} \left[\frac{(2^l - 1) \, \Delta f_s}{2\pi f_0}\right]^2 \tag{3.61}$$

The quantization distortion for filtered DPCM is identical to (3.54), so that the S/D_q ratio can be given as

$$\frac{S}{D_q} = \frac{2}{3} \frac{(2^l - 1)^2 f_s^3}{f_m (2\pi f_0)^2} \tag{3.62}$$

This yields an increase of $(2^l - 1)^2$ relative to the S/D_q ratio for two-level quantization (delta modulation). However, l bits per sample are transmitted with DPCM rather than 1 bit for delta modulation (DM). As shown in Table 3.3, a comparison of l-bit DPCM and DM, normalized to the same bit rate, indicates that DPCM provides significant improvement for $l \geq 3$.

TABLE 3.3   Comparison of Quantization Distortion for DPCM vs. DM (Normalized to the Same Bit Rate)

DM			DPCM		
Sample Frequency	D_q Improvement Factor		Bits/ Sample	D_q Improvement Factor	
(xf_s)	x^3	$10 \log x^3$	(l)	$(2^l - 1)^2$	$10 \log (2^l - 1)^2$
$1f_s$	1	0	1	1	0
$2f_s$	8	9	2	9	9.5
$3f_s$	27	14	3	49	17
$4f_s$	64	18	4	225	23

Expression (3.62) for peak S/D_q ratio of DPCM applies only to sinusoidal inputs. The main effect of a gaussian or speech input signal compared to a sinusoidal input is to reduce the peak S/D_q ratio. For example, Van de Weg [14] has derived an expression for S/D_q ratio with gaussian input statistics:

$$\frac{S}{D_q} \approx \left(\frac{3}{4\pi}\right)^2 \left(\frac{f_s}{f_m}\right)^3 \frac{(2^l - 1)^2}{l^3} \qquad l \geq 3 \tag{3.63}$$

Notice that the output S/D_q ratio is proportional to f_s^3 as obtained in (3.62). For DM ($l = 1$) systems the S/D_q ratio for gaussian input statistics is given by

$$\frac{S}{D_q} \approx 2\left(\frac{3}{4\pi}\right)^2 \left(\frac{f_s}{f_m}\right)^3 \tag{3.64}$$

We can compare the case for a gaussian input with sinusoidal inputs by re-arranging (3.56) to read

$$\frac{S}{D_q} = \frac{3}{2(2\pi)^2} \left(\frac{f_s}{f_m}\right)^3 \left(\frac{f_m}{f_0}\right)^2 \tag{3.65}$$

Obviously this result for sinusoidal inputs exceeds the result for a gaussian input for low-frequency input sinusoids with $f_0 \ll f_m$.

3.3.2   Slope Overload Distortion in Delta Modulation

Here we will outline a classic derivation of slope overload distortion, credited to S. O. Rice [15], and then interpret that result for a simple example.

Using an analysis similar to the study of fading statistics in radio transmission, Rice first calculated the average noise energy N_b for a single slope overload event and then calculated the expected number of slope overload events in 1 second, r_b. The product $N_b \cdot r_b$ is then an approximation of the average noise power due to slope overload, D_{so}, given as

$$D_{SO} = N_b \cdot r_b \approx \frac{1}{4\sqrt{2\pi}} \left(\frac{b_0^2}{b_2}\right) \left(\frac{3b_0^{1/2}}{\Delta f_s}\right)^5 \exp\left[\frac{-(\Delta f_s)^2}{2b_0}\right] \tag{3.66}$$

where
$$\Delta = \text{quantizer step size}$$
$$f_s = \text{sampling frequency}$$
$$b_0 = \text{variance of } f'(t)$$
$$b_2 = \text{variance of } f''(t)$$

$$b_n = \int_0^\infty (2\pi f)^{n+2} F(f)\, df, \quad n = 0, 2 \tag{3.67}$$

$$F(f) = \text{power spectrum of } f(t)$$
$$f(t) = \text{gaussian input signal}$$

Example 3.4

Let the gaussian input signal to a delta modulator be band-limited with spectrum

$$F(f) = \begin{cases} \dfrac{1}{f_m} & 0 < f < f_m \\ 0 & \text{otherwise} \end{cases}$$

For this case, find the expression for slope overload distortion.

Solution   From (3.67) the expressions for b_0 and b_2 are calculated as

$$b_0 = \frac{(2\pi f_m)^2}{3}$$
$$b_2 = \frac{(2\pi f_m)^4}{5}$$

These expressions allow calculation of the slope overload noise from (3.66), which becomes

$$D_{SO} = \frac{5}{4}\sqrt{\frac{3}{2\pi}\left(\frac{2\pi f_m}{\Delta f_s}\right)^5 \exp{-\left[\frac{3}{2}\left(\frac{\Delta f_s}{2\pi f_m}\right)^2\right]}} \qquad (3.68)$$

A plot of S/D ratio versus normalized step size shows a tendency to approach a slope overload asymptote for small step size and a quantization asymptote for large step size. To illustrate this effect, assume the total distortion power D to be approximated by the sum of D_{so} and D_q. This approximation is valid since quantizing noise does not occur due to a slope overload condition, and vice versa. Now if we normalize the rms value of the input to unity, the approximation for S/D is

$$\frac{S}{D} \approx \frac{1}{D_{SO} + D_q} \qquad (3.69)$$

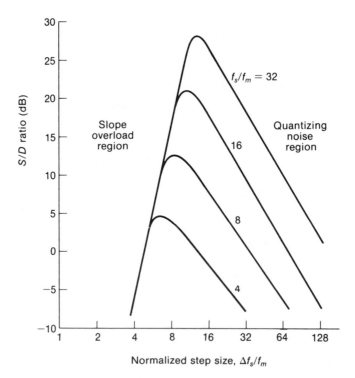

FIGURE 3.16   Delta Modulation Performance for a Flat Band-Limited Gaussian Input

Figure 3.16 plots the S/D ratio for a flat band-limited gaussian input signal. These theoretical curves were obtained by substituting expressions for slope overload noise (3.68) and quantization noise (3.64) into (3.69). These S/D ratios are plotted as a function of normalized step size, $\Delta f_s/f_m$. Sampling rates of 4, 8, 16, and 32 times the signal bandwidth are shown. For step sizes to the left of the peaks, slope overload noise dominates while to the right of the peaks quantization noise dominates. The peaks indicate an optimum value of normalized step size that maximizes the S/D ratio. These curves also demonstrate that by doubling the sampling frequency, the S/D ratio is improved by 9 dB in the quantization noise region.

3.3.3   Effects of Transmission Errors
on Delta Modulation
A delta modulator used with a digital transmission link suffers degradation from transmission errors. To characterize this measure of performance, consider Figure 3.17a, which shows an arbitrary two-level waveform produced at the transmitter. Due to channel disturbances, the recovered waveform is observed to contain errors. These errors are assumed to occur randomly with p being the probability of digit error. The resulting error

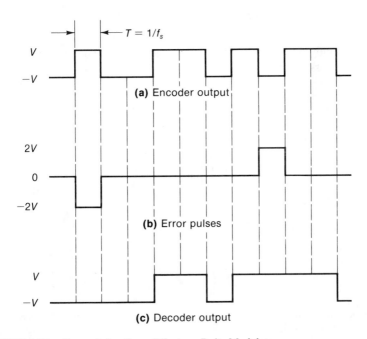

FIGURE 3.17   Transmission Error Effects on Delta Modulator

pulses have duration equal to one sampling period (T) and an amplitude which is double that of the transmitted signal, as shown in Figure 3.17b.

The receiver shown in Figure 3.18 is composed of a binary detector, an RC circuit that approximates an integrator,* and a bandpass filter. The effect of digit errors on the receiver output is determined by first computing the noise power due to the error pulses and then considering the effect of the integrator and bandpass filters. For sufficiently small p, the autocorrelation function $R_e(\tau)$ of the error pulses is maximum at $\tau = 0$ and effectively zero for $|\tau| > T$, as shown in Figure 3.19a. The power spectral density $S_e(f)$ is then of the form $(\sin x/x)^2$ with the first null at f_s, as shown in Figure 3.19b. If the maximum frequency f_m of the original analog signal is small compared to the sampling frequency f_s (that is, $f_m \ll f_s$), then the power density of the error noise spectrum is essentially constant over the range $(-f_m, f_m)$ and equal to

$$S_e(0) = \frac{4pV^2}{f_s} \tag{3.70}$$

If the RC integrator and bandpass filter have transfer function $H_1(f)$ and $H_2(f)$, respectively, then the spectral density at the receiver output is

$$S_0(f) = S_e(f)\left|H_1(f)\right|^2\left|H_2(f)\right|^2 \tag{3.71}$$

If $H_2(f)$ is an ideal filter with cutoff at f_1 and f_m, the mean square distortion due to transmission errors is

$$D_e = 2\int_{f_1}^{f_m}\left(\frac{4pV^2}{f_s}\right)\left(\frac{f_1^2}{f^2 + f_1^2}\right)df$$

$$= \frac{8pV^2f_1}{f_s}\left[\tan^{-1}\left(\frac{f_m}{f_1}\right) - \frac{\pi}{4}\right] \tag{3.72}$$

For the usual case of $f_m \gg f_1$,

$$D_e \simeq \frac{2\pi pV^2 f_1}{f_s} \tag{3.73}$$

---

*A perfect integrator cannot be realized in a delta modulator and is therefore replaced by an RC circuit. Since the "integration" of a constant input results in an output that follows an exponential curve, this configuration is called an exponential delta modulator [16].

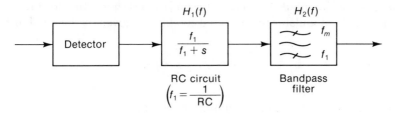

FIGURE 3.18   Receiver for Exponential Delta Modulator

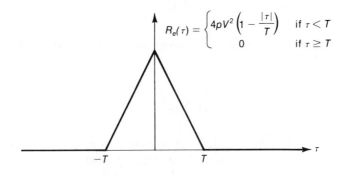

(a) Autocorrelation function

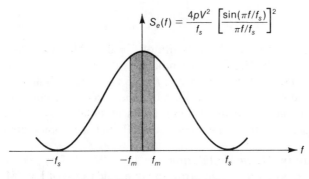

(b) Power spectral density

FIGURE 3.19   Error Waveform Characteristics for Exponential Delta Modulation

For a sinusoidal input $A \sin \omega_0 t$, Johnson [16] has shown that for the exponential delta modulator the overload condition is given by the relationship

$$A = \frac{V}{\sqrt{1 + \left(2\pi f_0 / 2\pi f_1\right)^2}} \tag{3.74}$$

The mean square value of the output signal is therefore

$$S = \frac{A^2}{2} = \frac{V^2}{2\left[1 + (2\pi f_0 / 2\pi f_1)^2\right]} \tag{3.75}$$

or for the usual case of $f_0 \gg f_1$,

$$S \approx \frac{V^2}{2} \frac{\left(2\pi f_1\right)^2}{\left(2\pi f_0\right)^2} \tag{3.76}$$

We can then write the signal to transmission error distortion ratio using (3.76) and (3.73):

$$\frac{S}{D_e} = \frac{f_s f_1}{4\pi f_0^2 p} \tag{3.77}$$

To consider the relative effects of transmission errors, compare (3.77) with the expression for the signal to quantizing error ratio (3.56). Solving for the value of p' for which $S/D_q = S/D_e$, we obtain

$$p' = \frac{2\pi f_1 f_m}{3 f_s^2} \tag{3.78}$$

For a typical voice channel bandwidth of $f_m = 3000$ Hz with low-end cutoff frequency of $f_1 = 300$ Hz, Table 3.4 indicates values of p' for various bit rates (f_s). These figures indicate that when $f_s = 16$ kb/s and with an error rate as high as 0.73 percent, the noise produced at the decoder output is of the same magnitude as the quantization noise. Delta modulation systems therefore offer good performance in the presence of a high transmission error rate.

3.3.4   Adaptive Delta Modulation

As in PCM, it is possible to improve the dynamic range of DPCM and DM coders by the use of companding. For DM coders, the most suitable form of companding consists of adapting the quantizer step size according to the

TABLE 3.4   Values of Transmission Error Rate p′ for $S/D_q = S/D_e$ in Delta Modulator Applied to Typical Voice Channel

Bit Rate (f_s)	p′
64 kb/s	4.6×10^{-4}
32 kb/s	1.8×10^{-3}
16 kb/s	7.3×10^{-3}
8 kb/s	2.9×10^{-2}

input signal. Unlike PCM where the companding characteristic is fixed, in companded DM systems the variable step sizes are derived dynamically; step sizes increase during steep segments of the input and decrease with slowly varying segments.

Linear (nonadaptive) DM and DPCM have the undesirable characteristic that there is only one input level that maximizes the signal-to-distortion ratio. Companding can greatly enhance the dynamic range of these coders, rendering them useful for widely varying signals such as speech. Whereas performance characterization for linear DM and DPCM could be done by straightforward derivation of quantization and slope overload distortion formulas, such derivations are not readily accomplished for adaptive techniques. Companding optimization has been done largely by experiment and computer simulation [17, 18] rather than by theory. Several versions of companded DM have been developed and given special names by their inventors. Here we will discuss a few representative examples and show applications.

In general, there are both *discrete* and *continuous* methods of adapting the DM system to changes in the slope of the input signal. In the discrete adaptive system, illustrated in Figure 3.20, the slope is controlled by a logic circuit that operates on the sequence of bits produced by the quantizer. When slope overload occurs, for example, the quantizer output tends toward all 1's or all 0's. In response to these bits of the same polarity, the logic selects a larger step size, continuing to do so until the largest discrete allowed value is reached. Conversely, the step size incrementally decreases when polarity reversals occur. The decoder senses the same bit sequences and controls slope values using logic identical to that in the coder, thus producing the same step sizes. Since step size changes are made at a rate equal to the sampling rate, this form of discrete adaptive delta modulation is known as **instantaneous companding.**

To illustrate one example, consider the set of weights {1,2,4,6,8,16} for the allowed step sizes and the logic given in Table 3.5. The step sizes to be used are determined by the last three transmitted bits. The direction (positive or negative) is controlled by the current bit. The logic rules have been

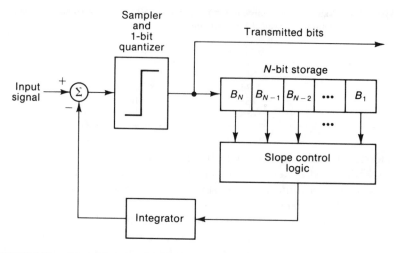

FIGURE 3.20   Block Diagram of Adaptive Delta Modulator

chosen to minimize slope overload by continuing to increase the step size for runs of three consecutive like bits. Figure 3.21 illustrates the adaptive DM output generated for a sample input using the logic of Table 3.5. The biggest design problems for discrete ADM are the choice of step sizes—both the number of allowed sizes and the gain per size—and the logic required to control changes in step sizes. Abate [17] has investigated this effect of the choice of step sizes on S/D performance of discrete ADM.

Winkler [19] proposed one of the earliest companded DM schemes in a system he termed **high-information delta modulation** (HIDM), so named because the system contained more information per pulse than linear DM.

TABLE 3.5   **Example of Slope Control Logic for Adaptive Delta Modulation**

Current Bit	Previous Bit	Pre-Previous Bit	Slope Magnitude Relative to Previous Slope Magnitude
0	0	0	Take next larger magnitude[a]
0	0	1	Slope magnitude unchanged
0	1	0	Take next smaller magnitude[b]
0	1	1	Take next smaller magnitude[b]
1	0	0	Take next smaller magnitude[b]
1	0	1	Take next smaller magnitude[b]
1	1	0	Slope magnitude unchanged
1	1	1	Take next larger magnitude[a]

[a]If slope magnitude is already at the maximum, this instruction is not obeyed.
[b]If slope magnitude is at the minimum, this instruction is not obeyed.

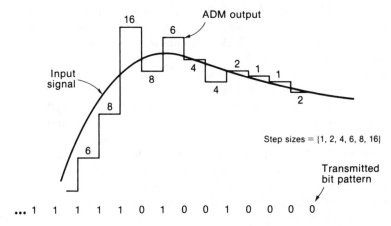

FIGURE 3.21   Sample Output of Adaptive Delta Modulator Using Logic of Table 3.5

His method consisted of doubling the step size whenever identical consecutive bits are produced at the coder output. Step sizes are divided in half after each polarity reversal. This scheme was applied to video signals and significantly improved the dynamic range over linear DM [20].

Bosworth and Candy [21] described a companded coder specifically designed for picture-phone transmission. It differs from HIDM in that its step response is more damped. The adaptation algorithm was designed to promptly increase the step size for transients (busy areas of picture and at edges) and to promptly decrease step size after a transient in order to minimize overload distortion and quantization noise, respectively. Subjective tests on picture-phone signals were used to pick the optimum weighting sequence (1,1,2,3,5), in which each weight is equal to the sum of the two previous weights. This sequence minimizes overshoot by converging to the smallest increment after a transient. Specifically, when three or more consecutive 1's or 0's are produced, the weighting logic increases the step size by 1,1,2,3,5, . . . ,5. Eventually a bit of opposite polarity is observed, and the step size returns to 1. The direction of any step is determined by the polarity of the current coder bit.

Jayant [22] has described an ADM design applicable to speech coding. Here the step size Δ_i at time i is related to the previous step size by

$$\Delta_i = \Delta_{i-1} \cdot P^{b_i b_{i-1}} \tag{3.79}$$

where b_i and b_{i-1} are the present and previous bits, respectively. Variation in step size is determined by the factor P, which for speech has an optimum value $1 < P < 2$ [22]. Note that a choice of $P = 1$ equates to linear delta modulation (LDM). For a choice of $P = 1.5$, a sampling rate of 60 kHz, and a speech signal band-limited to 3.3 kHz, computer simulation showed a 10-dB S/D improvement over LDM.

At low bit rates for digitized speech, instantaneously companded DM systems are seriously degraded by excessive quantization noise. One means of improving speech quality for low-rate coders is the use of **syllabic companding,** wherein the step sizes are adapted more smoothly in time, controlled by the syllabic rate of human speech. A syllabic filter has a typical bandwidth of 100 Hz, so that the step sizes are adapted with a time constant on the order of 10 ms. The slow adaptation of syllabic companding improves upon quantization noise at the expense of increased slope overload distortion. Even so, for speech bit rates below 25 kb/s, syllabic companding is preferable [23]. Moreover, syllabic companding provides resistance to bit errors by its slow-changing nature.

Continuously adaptive delta modulation, first described by Greefkes and DeJager [24], is another version of DM that can improve performance (that is, increase dynamic range) or, alternatively, reduce the bit rate required for digitized speech. This approach derives its name from the way the slope of the coder output varies in an almost continuous fashion, as opposed to discrete ADM where the step sizes are limited to a discrete set. Current implementations of this technique have been generally called **continuously variable slope delta modulation** (CVSD). A typical CVSD encoder/decoder is shown in Figure 3.22. Step size is controlled in a manner similar to discrete ADM, but with unlimited step size values and usually syllabic control to slow the rate of change of step size. For typical CVSD speech coders (32 kb/s) a dynamic range of 30–40 dB is possible, which represents approximately 10 dB improvement over discrete ADM coders.

To illustrate the operation of a CVSD coder, assume a 3-bit shift register in Figure 3.22 with the following logic applied:

- Let the smallest step size be Δ.
- For 3-bit combinations 000,111: Increase step size by 3Δ.
- For present bit reversals $0 \to 1$, $1 \to 0$: Decrease step size by 3Δ.
- For other 3-bit combinations: Leave slope unchanged.
- The direction of slope is determined by the present bit.

Figure 3.23 shows the resulting CVSD coder output for a sample analog input, assuming instantaneous companding (that is, no syllabic control).

3.4   Adaptive Differential Pulse Code Modulation

Similar to ADM, DPCM steps sizes can be adapted to provide increased dynamic range. Coders with these adaptive features have been referred to

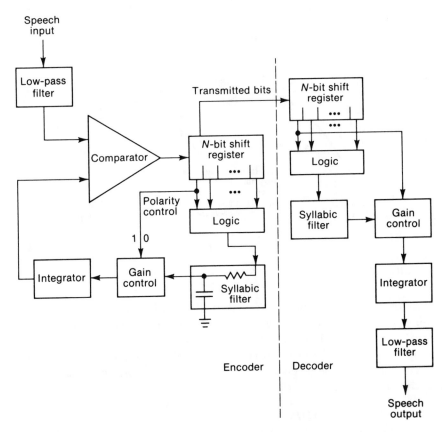

FIGURE 3.22   Block Diagram of Typical CVSD Modulator

as ADPCM coders. ADPCM at 32 kb/s has emerged as a digital transmission standard for telephone networks, rivaling 64-kb/s PCM in popularity. During the early 1980s, the CCITT developed, tested, and adopted a standard in Rec. G.721 for 32-kb/s ADPCM. Later, CCITT Rec. G.723 extended G.721 to rates of 24 and 40 kb/s. Even more recently, CCITT Rec. G.726 has consolidated and replaced G.721 and G.723 to form a single standard for 24-, 32-, and 40-kb/s ADPCM. The algorithms of G.726 require that the two endpoints and any intermediate points of each telephone channel operate at the same fixed coding rate. Another form of coding, called **embedded ADPCM,** uses a variable coding rate that can be reduced at any point in the telephone network without the need for coordination between the transmitter and receiver. This embedded ADPCM algorithm is standardized in CCITT Rec. G.727, which is an extension of the algorithm described in G.726 to variable rates of 16, 24, 32, and 40 kb/s. The motiva-

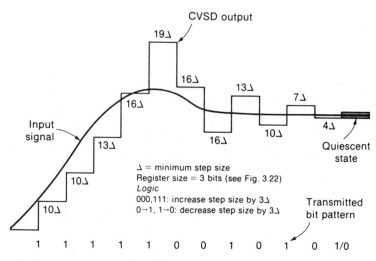

FIGURE 3.23   Sample Output of CVSD Coder

tion for use of lower rate ADPCM is of course improved transmission efficiency, but higher rate coding is required to obtain speech fidelity equivalent to 64-kb/s PCM or to carry high-speed voice-band data. An additional motivation for embedded, variable-rate ADPCM is the need to relieve occasional telephone network congestion by temporarily reducing the ADPCM coding rate.

3.4.1   Fixed-Rate ADPCM

The following description of fixed-rate 24-, 32-, and 40-kb/s ADPCM is based on two identical standards, CCITT Rec. G.726 and ANSI T1.303-1989. For further details, the reader is referred to these standards [25,26]. Both standards detail the characteristics necessary to convert 64-kb/s PCM to and from 24-, 32-, or 40-kb/s ADPCM. The encoding algorithm is based on a combination of a dynamic locking quantizer and an adaptive predictor applied to the difference signal [27]. The difference signal is first scaled and then quantized by a fixed quantization scale. The scaling factor is adapted to accommodate both fast-changing signals, like speech, and slow-changing signals, like voice-band data and tones. Quantization is performed by a 3-, 4-, or 5-bit nonuniform adaptive quantizer with 7-, 15-, or 31 levels (like μ-law PCM, the all-zero code is not allowed). At both the encoder and the decoder, the difference signal is recreated from these three, four, or five bits using the inverse of the quantizer. The reconstructed signal is obtained at both the encoder and decoder by adding the difference signal to the current estimate of the input signal, which is produced by identical adaptive predictors at both ends. Each adaptive predictor has two sections, one with two taps and the other with six taps, both having adaptive tap coefficients. The decoder includes the functions of the encoder plus a feature that prevents distortion accumulation when multiple PCM/ADPCM/PCM conversions (called **synchronous** tandem codings) are performed. The encoder and decoder are shown in Figures 3.24 and 3.25, respectively, and a more detailed description of their functions follows.

As shown in Figure 3.24, the first operation is conversion of a standard μ-law or A-law PCM signal, $s(k)$, to a linear PCM signal, $s_l(k)$. Next, the difference signal $d(k)$ is calculated by subtracting the signal estimate $s_e(k)$ from $s_l(k)$,

$$d(k) = s_l(k) - s_e(k) \qquad (3.80)$$

Quantization of the difference signal is performed in three steps. First, the linear difference signal $d(k)$ is converted to a base 2 logarithmic form. A scale factor $y(k)$ is then subtracted to yield

$$d_{ln}(k) = log_2 |d(k)| - y(k) \qquad (3.81)$$

This normalized signal $d_{ln}(k)$ is then quantized by table look-up to yield $I(k)$. Each sample of the output $I(k)$ contains three, four, or five bits, one bit for the sign and the remainder for the magnitude.

The quantization scale factor $y(k)$ is determined from past values of $I(k)$ and a speed control factor $a_l(k)$:

$$y(k) = a_l(k)y_u(k-1) + \left(1 - a_l(k)\right)y_l(k-1) \qquad (3.82)$$

where $y_u(k)$ is the fast (unlocked) scale factor for speech and $y_l(k)$ is the slow (locked) scale factor for voice-band data and tones. The basic idea here is to adapt the quantizer to the fluctuations in the difference signal $d(k)$. The speed control factor $a_l(k)$ varies between 0 for voice-band data and 1 for speech and therefore determines how the fast and slow scale factors are combined. The fast and slow scale factors are computed as

$$y_u(k) = (1 - 2^{-5})y(k) + 2^{-5}W[I(k)] \qquad (3.83)$$

$$y_l(k) = (1 - 2^{-6})y_l(k-1) + 2^{-6}y_u(k) \qquad (3.84)$$

where $y_u(k)$ is constrained to the limits $1.06 \leq y_u(k) \leq 10.00$, and $W[I(k)]$ is determined from a look-up table. The factor $(1 - 2^{-5})$ in (3.83) introduces

128

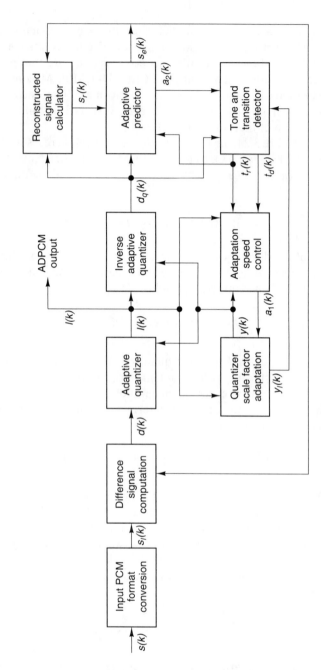

FIGURE 3.24 Fixed-Rate ADPCM Encoder Specified by CCITT Rec. G.726 (Courtesy CCITT [25])

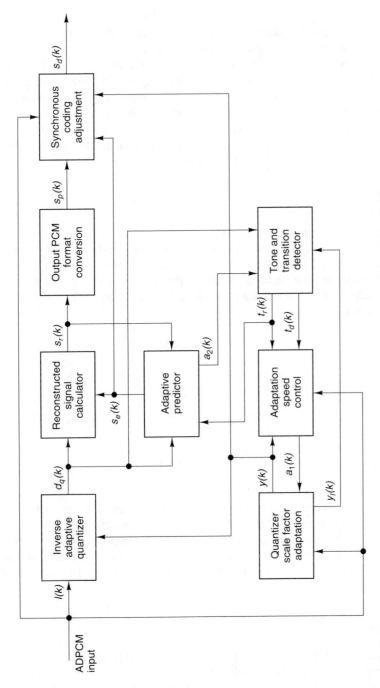

FIGURE 3.25 Fixed-Rate ADPCM Decoder Specified by CCITT Rec. G.726 (Courtesy CCITT [25])

129

finite memory into the adaptive process so that $y(k)$ will converge to its true value following the occurrence of transmission errors.

The inverse adaptive quantizer converts the signal I(k) into the signal $d_q(k)$, the quantized version of the difference signal. This operation is performed as the inverse of the three steps of the quantizer: (1) the magnitude of $I(k)$ is mapped into one of the four, eight, or sixteen mid-range output values $d_m(k)$ via table look-up; (2) the scale factor $y(k)$ is added to the output value $d_m(k)$; and (3) the antilog is taken to produce the magnitude of the quantized difference signal $d_q(k)$. In equation form,

$$|d_q(k)| = 2^{[d_m(k) + y(k)]}$$

(3.85)

Finally, $d_q(k)$ is assigned the same sign as $I(k)$.

The function of the adaptive predictor is to compute the signal estimate, $s_e(k)$, using the past history of the quantized difference signal $d_q(k)$. The predictor consists of a sixth-order section that models zeroes and a second-order section that models poles in the input signal, described by the following equations:

$$s_e(k) = s_{e2}(k) + s_{e6}(k)$$

(3.86)

where

$$s_{e2}(k) = \sum_{i=1}^{2} a_i(k-1)s_r(k-i)$$

(3.87)

$$s_{e6}(k) = \sum_{i=1}^{6} b_i(k-1)d_q(k-i)$$

(3.88)

in which $a_i(k)$ are the predictor coefficients for the second-order predictor and $b_i(k)$ are the predictor coefficients for the sixth-order predictor. This choice of a pole-zero predictor provides decoder stability in the presence of transmission errors. The predictor coefficients $a_i(k)$ and $b_i(k)$ are updated for each sample using a gradient search algorithm. The reconstructed signal $s_r(k)$ is calculated by adding the quantized difference signal $d_q(k)$ to the signal estimate $s_e(k)$,

$$s_r(k-i) = s_e(k-i) + d_q(k-i)$$

(3.89)

This reconstructed signal is also the output of the ADPCM decoder.

The final block in the encoder detects tones and transitions between tones. When a tone is detected, a tone detection signal $t_d(k)$ causes the quantizer to be driven into the fast mode of adaptation. The signal $t_d(k)$ also sets the tone transition detect signal $t_r(k)$ when a transition between tones occurs. This latter signal causes the fast adaptation mode to take effect immediately.

As can be seen by a comparison of Figures 3.24 and 3.25, the ADPCM encoder and decoder are nearly identical in terms of the blocks. Both perform the same algorithms for inverse quantization, linear prediction, tone detection, and adaptation. The reconstructed signal $s_r(k)$ is converted to standard μ-law or A-law PCM by table look-up to produce the PCM signal $s_d(k)$. This signal is passed through a synchronous coding adjustment, which ensures that samples remain unchanged with successive encodings, so that quantization distortion is limited to that for a single encoding.

3.4.2   Embedded ADPCM

ADPCM as prescribed by CCITT Rec. G.726 requires that both transmitter and receiver be configured beforehand to use the same bit rate. **Embedded ADPCM,** as standardized in CCITT Rec. G.727 and ANSI T1.310-1990 [28,29], allows bit rate reduction anywhere in the network without the need to make any changes to the transmitter or the receiver. A variable-rate, embedded algorithm is used in which the quantization levels in lower rate quantizers are a subset of the quantizer used at the highest rate. In contrast, the quantization levels of fixed-rate ADPCM algorithms are not subsets of one another, meaning that both ends of the telephone channel must operate at the same bit rate and must coordinate any change in bit rate. The embedded ADPCM algorithm thus allows the coder and decoder to operate at different rates. The decoder, however, must be informed of the number of bits to be used for each incoming sample. Code words produced by this algorithm consist of core bits and enhancement bits. Core bits cannot be discarded but enhancement bits can be to relieve network congestion. CCITT Rec G.727 and ANSI T1.310-1990 allow any combination of coding rates (core and enhancement bits) of 40, 32, 24, and 16 kb/s and core rates of 32, 24, and 16 kb/s, but of course the coding rate must be equal to or greater than the core rate. Embedded ADPCM is often described by (x,y) pairs where x refers to the core plus enhancement bits per sample and y refers to the core bits. For example, (4,2) represents 32-kb/s embedded ADPCM with two core bits, so that the minimum allowed bit rate is 16 kb/s.

Figure 3.26 is a simplified block diagram of the embedded ADPCM encoder and decoder. Since the principle of operation is an extension to that already described for fixed-rate ADPCM, here we shall focus on the features that make this algorithm variable in bit rate. The adaptive quantizer uses 4, 8, 16, or 32 levels to convert the difference signal d(k) into a code word of 2, 3, 4, or 5 bits. (Note that these quantizers produce the all-zero

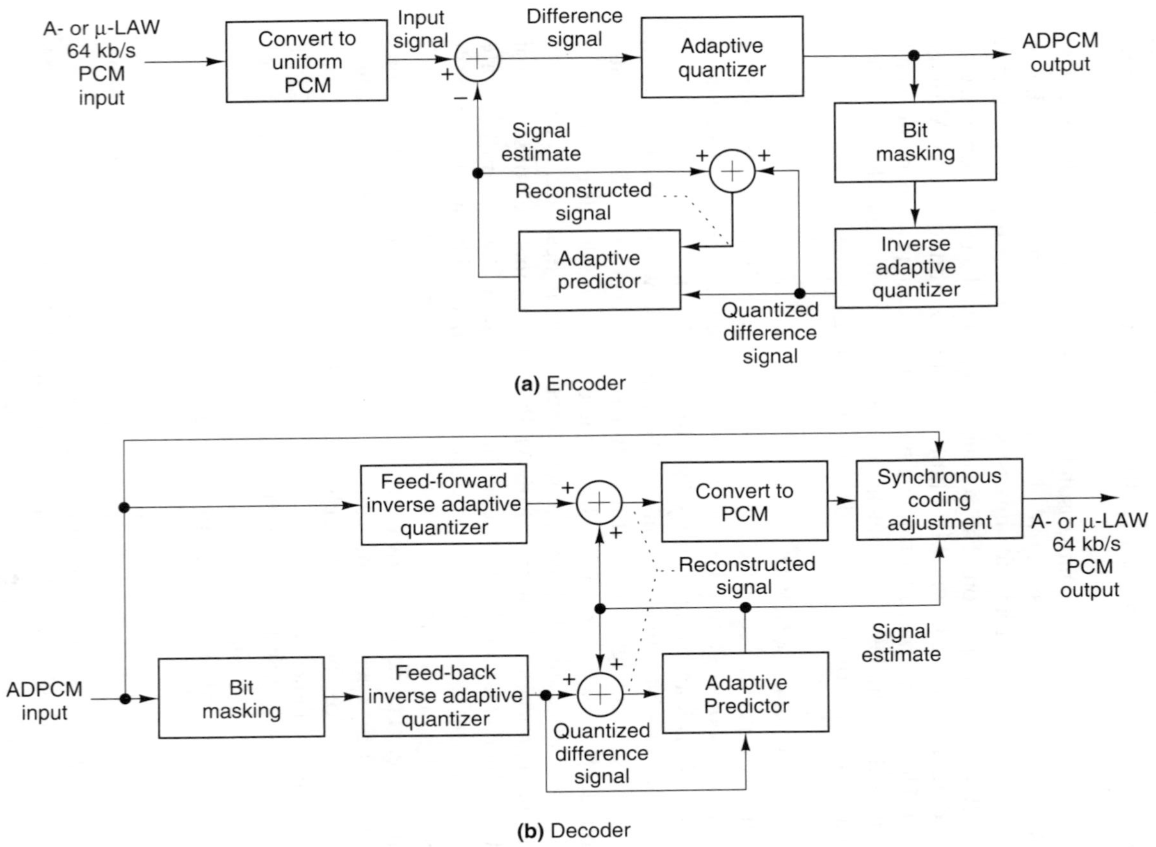

(a) Encoder

(b) Decoder

FIGURE 3.26 Embedded ADPCM Specified by CCITT Rec. G.727 (Courtesy CCITT [28])

132

code word that may violate one's density requirements in some networks.) These code words, containing both core and enhancement bits, are transmitted by the coder intact, but the feedback path uses only the core bits. In the feedback loop, the bit-masking block drops the enhancement bits so that the inverse adaptive quantizer and the adaptive predictor operate on only the core bits. At the decoder the feedforward path contains the core bits and possibly the enhanced bits, while the feedback path contains only the core bits. The penalty paid if enhancement bits are lost is an increase in quantization noise due to the use of larger quantization step sizes.

The embedded ADPCM algorithm specified by CCITT Rec. G.727 is suitable for use in packetized speech systems such as that prescribed by CCITT Rec. G.764. The multiplexing schemes that use embedded ADPCM in conjunction with packetized speech are further described in Chapter 4.

3.5   Subband Coding

The waveform coders described to this point have all applied the coding algorithm to a time-domain representation of the signal. Another class of coders, of which **subband coding** is an example, operate on the frequency-domain signal. The principle behind frequency-domain coders is that frequency bands may be assigned a variable number of coding bits, depending on the energy content of each band. The input signal is divided into a number of frequency bands by a bank of parallel filters. In theory, the subbands may be equal width or variable width, and the number of bits allocated to each subband may be fixed or adaptive [30]. In practice, most implementations use equal-width subbands with adaptive bit allocation, a version we shall now describe in more detail.

A block diagram of subband coding is shown in Figure 3.27. Each band is allocated a number of bits for quantization based on the energy content in that band. Bands with more energy are allocated more bits to accommodate the greater dynamic range in that band. This allocation process is done dynamically to ensure that the changing characteristics of the signal are followed. If the number of subbands is N, then each bandpass filter limits its output to approximately W/N of the full-signal bandwidth W. Each filter output is subsampled at a rate that is reduced by the factor N, which results in a per-band sampling rate of $2(W/N)$. These filtered signals are then fed to a bank of coders that convert each subband to a quantized and then digitized version. Various coding techniques have been used, including nonlinear PCM, DPCM, and ADPCM. However, the use of the quantizer for any of these coders usually follows the same basic strategy, namely, the number of quantization levels varies according to the number of bits assigned to that band. Furthermore, the step sizes of each quantizer are adjusted to match

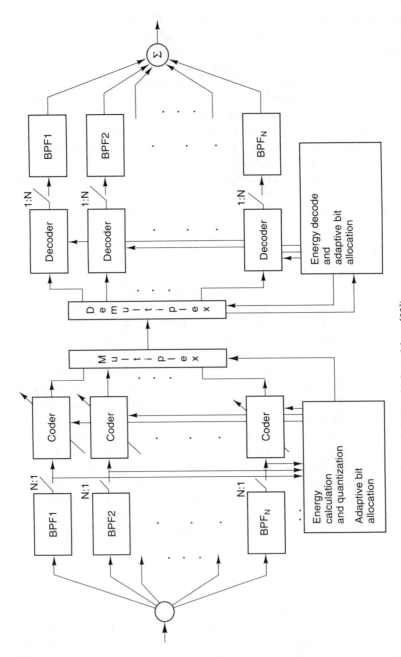

FIGURE 3.27 Block Diagram of Subband Coding (Adapted from [32])

134

the range of sample values for each corresponding subband. The transmitted data contain two types of information, codes describing the amount of energy in each band and codes describing the sampled and quantized values in each band. The receiver first decodes the energy values, which are then used to determine the bit allocation per band. The data corresponding to the sample codes can then be demultiplexed and sent to the appropriate band decoder. Each band decoder output is restored to its original level by using the corresponding band energy value. The band samples are returned to the original sampling rate by interpolation in which $(N - 1)$ zero values are inserted before the sample is sent to its corresponding bandpass filter. The zero values cause the band samples to be replicated across the full spectrum of the original signal, but the corresponding bandpass filter picks off energy only at the correct frequency. The outputs of the N filters are finally summed to obtain the reconstructed signal at the original sampling rate.

Subband coders of the type just described have been popular as speech coders for a variety of bit rates and applications. In digital mobile radio systems, subband coding has been used at coding rates of 12 to 15 kb/s [31,32]. Because of the fading nature of these radio channels, error protection is typically included, which adds another 1 to 3 kb/s of overhead. Error correction is usually reserved for just the so-called side information, which is the code describing the energy content in each band. (Unless the side information is received correctly, the remainder of the encoded frame cannot be decoded.) Testing has shown that 16-kb/s subband coding is roughly equivalent to 16-kb/s ADPCM in speech quality for a single encoding but is actually superior for tandem encodings [33].

CCITT Rec. G.722 prescribes subband coding for use in the 64-kb/s coding of 7-kHz audio [34]. The primary applications for this coder are in teleconferencing and loudspeaker telephones, with secondary applications to music. The coding algorithm uses two equal-width subbands, with the lower band coded with 6 bits per sample and the higher band with 2 bits per sample, resulting in code rates of 48 and 16 kb/s for the two bands. For the lower band, embedded ADPCM is used, which allows simultaneous data transmission at 8 or 16 kb/s by dropping one or two least significant bits out of the 6 bits per sample. This capability for coding rate reduction leads to three transmission modes: Mode 1 at 64 kb/s for speech only, Mode 2 at 56 kb/s for speech and 8 kb/s for data, and Mode 3 at 48 kb/s for speech and 16 kb/s for data. Embedded ADPCM coding, as described earlier in Section 3.4.2, allows the transmitter and receiver to operate at different coding rates without interruption. The least significant bit(s) can be dropped after 64-kb/s encoding to permit multiplexing of the auxiliary data channels without any change needed in the encoder. For proper operation of the decoder, however, the mode of operation must be transmitted to select the appropriate decoder version. Operation of the lower-band encoder is based on a 60-level adaptive quantizer that produces 6-bit ADPCM; in the feedback loop, bit masking of the two least significant bits produces a 4-bit signal, which is applied to a 15-level inverse adaptive quantizer. The lower-band decoder operates in one of three modes, using a 60-level, 30-level, or 15-level inverse adaptive quantizer corresponding to Modes 1, 2, and 3, respectively. The higher-band encoder uses a 4-level adaptive quantizer for both feedforward and feedback loops to produce two code bits and a corresponding 16-kb/s coding rate; the decoder is identical to the feedback portion of the encoder. Performance comparisons have shown that the CCITT G.722 algorithm operating at 64 kb/s has a S/D_q gain of 13 dB over the 64-kb/s PCM coder of G.712, of which 6 dB are due to the expanded bandwidth and 7 dB to more precise encoding. Even at 48 kb/s, the G.722 algorithm had a 6-dB improvement over G.712 coding [35].

Subband coding has also been applied to video signals typically using 4 or 16 subbands. The lower bands have higher semantics and therefore are given greater bit allocation and may be provided error protection as well. In addition to bit allocation among the subbands, bits can be allocated within a single subband to remove redundancy in the spatial domain. Those areas that have large changes are assigned more bits than quiet areas of the subband. The form of coding used with the subbands has varied depending on the application but has included DPCM [36] and transform coding [37]. Subband coding has been proposed as the compression technique for all-digital HDTV coding at rates from 45 Mb/s [38] to 135 Mb/s [37], and for video conferencing at rates of $384 \times m$ kb/s ($m = 1$ to 6) [39,40].

3.6   Transform Coding

Like subband coding, **transform coding** operates on the frequency-domain representation of the signal. For correlated signals such as speech and video, the redundancy in quiet areas of the signal may be efficiently removed by first transforming the signal and then processing it in its transform domain. Several discrete transforms have been proposed, including Fourier, Karhunen-Loeve, Hadamard, Cosine, Sine, and others. The signal to be encoded is first converted into blocks of N samples, which are then transformed into N transform coefficients. The coefficients are then quantized, coded, and transmitted independently and separately for each block. Quantization is usually made adaptive by using variable-bit allocation from block to block to account for varying degrees of activity in the signal. Applications of transform coding have included one-dimensional (1-D) transforms to speech coding, 2-D transforms to image processing, and 3-D transforms to television transmission.

Although adaptive transform coding has been used with speech coding [30], its most popular application has been to video systems. Discrete cosine transform (DCT) coding is commonly used for video applications because of its superior performance and implementation simplicity. Here the image is divided into nonoverlapping blocks of $N \times N$ picture elements (pixels), and a discrete cosine transform is performed on each block. The resulting N^2 coefficients then represent the frequency contents of the given block. Quantization bits are distributed according to the degree of activity in the block, with more bits assigned to busy areas than to quiet areas. Transform coefficients are assigned quantization bits according to their variance, so that coefficients corresponding to active picture areas will be transmitted with greater quantization while remaining coefficients may receive as few as zero bits. The overall measure of coding efficiency is the number of bits per pixel. Further compression is possible in moving pictures by extending transform coding to sequential frames, a technique known as **interframe coding,** which takes advantage of the temporal correlation between successive frames. Here the data are processed in $N \times N \times N$ blocks (N frames with N^2 pixels each) in which a three-dimensional transform is used to reduce both spatial and temporal redundancy.

To illustrate and further describe transform coding, consider Figure 3.28, which shows a typical 2-D cosine transform encoder and decoder applied to image processing. The digitized image represented by $M \times M$ pixels S_{ij} $(i,j = 1 \ldots M)$ is first partitioned into a number of blocks of size $N \times N$. Two-dimensional transform is performed on each block, producing transform coefficients $S_{u,v}$ $(u,v = 1 \ldots N)$. The activity level of the image in its transform domain next determines the bit allocations within each block for quantization. The energy in the (m,n)th block can be defined

$$E_{m,n} = \sum_{u=0}^{N-1} \sum_{v=0}^{N-1} \left[S_{m,n}(u,v) \right]^2 - \left[S_{m,n}(0,0) \right]^2 \tag{3.90}$$

with $m,n = 1,2, \ldots, M/N$. Each dc sample in the transform domain, $S_{m,n}(0,0)$, is subtracted since it determines only the brightness level. The sum of the ac energy of the transform samples from each block is calculated and used as a measure of activity for that block. The transformed blocks may then be classified according to the level of activity. The variances for each activity class are computed, from which an $N \times N$ bit allocation matrix can be determined for each class. The number of bits, $n(u,v)$, available for quantization of each element $S(u,v)$ is subject to a constraint on the total number of bits available for each $N \times N$ block. The value of $n(u,v)$ is proportional to the variance $\sigma^2(u,v)$ of the corresponding transform coefficient. Iterative calcu-

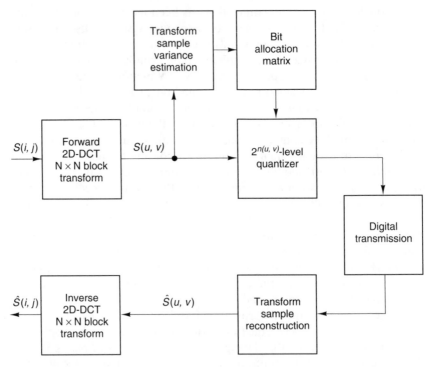

FIGURE 3.28   Block Diagram of Two-Dimensional Transform Encoder/Decoder (Adapted from [41])

lations can be used to find the individual $n(u,v)$'s and satisfy the constraint on total number of bits. Matrix elements for which $n(u,v) = 0$ are simply discarded, which results in significant signal compression. Once the bit-allocation matrix is set, quantization for each matrix element (that is, transform coefficient) is done with a quantizer having $2^{n(u,v)}$ levels. At the receiver, the quantized coefficients are inverse quantized and then inverse transformed block by block to reconstruct the image. When used with a 2-D DCT, typical block sizes are 16×16 extracted from a 256×256 digitized image, with an average rate of 1 bit/pixel [41,42].

Specific applications of transform coding DCT to video have included teleconferencing, color TV, and HDTV. CCITT Rec. H.261 specifies use of 2-D DCT for video conferencing at rates of $384 \times n$ kb/s ($n = 1$ to 6) [40]. Here blocks of 8 pixels by 8 lines are transformed, resulting coefficients are quantized, and variable-length coding is used. For color images, transform coding has been used to code the individual components (for instance, red, green, blue) of the signal or to code a composite signal that contains both luminance and chrominance. Transform coding of the three color components requires three separate, parallel transforms, which may be 2-D for color images or 3-D for color TV. Although HDTV standards are not yet mature, transform coding has been proposed in conjunction with subband coding [39,43,44], or with intraframe DPCM [45] or interframe (temporal) DPCM [46].

3.7 Comparison of Waveform Coders

Because DM, DPCM, and PCM have emerged as leading standards for digital transmission of analog information, a comparison of their performance is given here to assist the system engineer in the choice of applicable techniques. Since PCM has been shown to provide adequate performance for a variety of analog sources, the kernel of this comparison lies in the potential for reduced bit rate capabilities via some alternative form of coding. In light of this objective, issues of subjective quality, coder performance (quantizing and overload distortion), transmission performance (effects of error rate and multiple A/D-D/A conversion and ability to handle voice-band and nonspeech signals), and implementation complexity are examined using the example of speech coding.

3.7.1 Subjective Quality
Performance of A/D coders has so far been described in terms of various objective parameters, which can be precisely analyzed or measured. The quality of a coder, however, is ultimately determined by the human viewer or listener. The introduction of human perception into coder evaluation has led to the development of many subjective scoring schemes, generally based on the use of a jury that views or listens to standardized video or audio samples. The most widely used subjective scoring method for digital speech is the Mean Opinion Score (MOS), which has a five-level rating scale (5 = excellent, 4 = good, 3 = fair, 2 = poor, and 1 = unsatisfactory). On the MOS scale, a score of 4.0 is considered to be network quality, 3.5 to be a tolerable level of degradation for communications quality, and 3.0 to be a synthetic quality characterized by high intelligibility but inadequate level of naturalness or speaker recognition. For low-speed digital speech coders, other scoring systems have also been used, primarily the Diagnostic Rhyme Test (DRT) and the Diagnostic Acceptability Measure (DAM), which tend to measure intelligibility rather than quality.

The plot in Figure 3.29 compares the MOS for 64-kb/s PCM with the MOS values for ADPCM at several bit rates. The PCM coder tested was that specified in CCITT Rec. G.711, while the ADPCM coders were those specified by Rec. G.726 at fixed rates of 24, 32, and 40 kb/s and by Rec. G.727 at embedded rates of 16, 24, 32, and 40 kb/s with two core bits. These

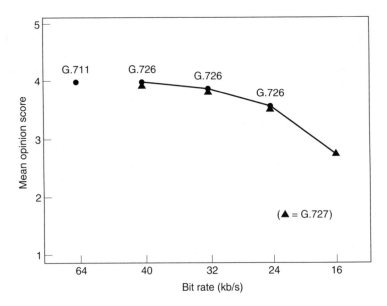

FIGURE 3.29   Mean Opinion Scores for PCM and ADPCM (Adapted from [47])

results indicate that embedded ADPCM provides essentially the same performance as fixed-rate ADPCM, and that ADPCM meets communications quality (MOS = 3.5 or better) for all rates except 16 kb/s.

3.7.2   Quantizing Distortion

Figure 3.30 shows the output S/D_q ratio for logarithmic (μ-law) PCM, linear PCM, and delta modulation as a function of input power level. A sinusoidal input of frequency $f_0 = 800$ Hz is assumed. The results for linear and logarithmic PCM were obtained from (3.41) by setting $\mu = 0$ and $\mu = 255$, respectively; $B = 1$; $N = 256$ for a transmission bit rate of 64 kb/s; and by varying the parameter C according to (3.45). The corresponding results for delta modulation were obtained from (3.55) and (3.56) by letting $f_s = 64$ kb/s and $f_m = 3400$ Hz. For all three curves, the input is normalized to a full-load sinusoid that exactly matches the quantizer range so that no overload distortion is introduced. Note that only μ-law PCM meets CCITT Rec. G.712.

Figure 3.31 shows the variation in S/D_q versus bit rate for μ-law PCM and delta modulation. Again a sinusoidal input is assumed. The curve for μ-law PCM was obtained from (3.41), where $\mu = 255$, $B = C = 1$, and N is varied according to the selected bit rate. The maximum S/D_q curves for delta modulation were obtained from (3.56) with $f_m = 3400$ Hz; $f_0 = 800$, 1600, and 2600 Hz, respectively, for curves a, b, and c; and f_s varied according to the bit rate.

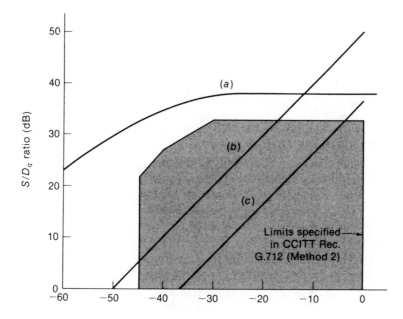

FIGURE 3.30 S/D_q Versus Input Signal Power: (a) μ-Law PCM with μ = 255 and n = 8 Bits; (b) Linear 8-Bit PCM; (c) Linear DM with f_s = 64 kb/s, f_m = 3400 Hz, f = 800 Hz
A comparison of the three curves in Figure 3.30 indicates the limited dynamic range of the delta modulator as compared to linear or logarithmic PCM. A similar comparison for gaussian or speech input signals would yield curves having the same general shapes as those in Figure 3.30 [48,23].

Curves *a, b,* and *c* illustrate the characteristic of delta modulation that the peak S/D_q decreases by 6 dB for every octave increase in f_0 and increases by 9 dB for every doubling of the sampling rate (bit rate) f_s. Curve *d* illustrates the PCM characteristic that the S/D_q improves by 6 dB for each additional bit of coding. A comparison of PCM versus DM performance curves shown in Figure 3.31 indicates that at high bit rates PCM outperforms DM while the reverse holds true for low bit rates.

3.7.3   Overload Distortion

Delta modulation overload noise results when the input signal *slope* exceeds the maximum slope capability of the DM quantizer; PCM overload noise results when the input signal *amplitude* exceeds the maximum levels of the PCM quantizer. Analytic expressions for both DM overload noise (3.68) and PCM overload noise (3.23a) have been derived for gaussian input signals. These two expressions can be conveniently compared as a

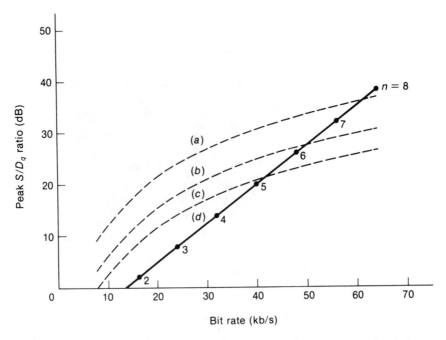

FIGURE 3.31   Peak S/D_q Versus Bit Rate for Sinusoidal Inputs: (a), (b), (c) Are DM with $f_0 = 800$, 1600, and 2600 Hz Respectively; (d) is μ-Law PCM with $\mu = 255$

function of a bandwidth expansion factor B, defined as the ratio of the bandwidth of the transmission channel to that of the signal. For DM systems, B is simply one-half the ratio of sampling rate to signal bandwidth, or $f_s/2f_m$. For PCM systems, B is identical to the number of coding bits. A comparison can now be made by plotting peak performance (that is, maximum S/D ratio obtained from Figures 3.5 and 3.16) versus the bandwidth expansion factor. It is clear from Figure 3.32 that for gaussian signals, linear PCM provides superior performance to that of linear DM. A comparison of the results of Figure 3.31 with Figure 3.32 indicates the significance of slope overload in DM systems. With only quantization noise taken into account, Figure 3.31 indicates that DM has superior performance over μ-law PCM for low rate coders. With *both* slope overload and quantization noise taken into account, however, Figure 3.32 indicates that linear PCM provides superior performance at all bit rates.

A similar comparison involving adaptive delta modulation would show that the maximum S/D (where D is the sum of quantizing and slope overload distortion) remains approximately the same as that of linear DM, thus allowing us to extend the results of Figures 3.31 and 3.32 to ADM [17]. The

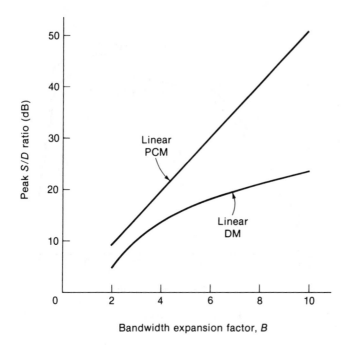

FIGURE 3.32   Peak S/D Versus Bandwidth Expansion Factor for Gaussian Inputs

companding improvement inherent in ADPCM and ADM does increase the dynamic range, however, so that the results of Figure 3.30 are not applicable to ADPCM or ADM. Measured results of log PCM and ADPCM S/D performance are shown in Figure 3.33 to illustrate the relative dynamic ranges achievable with these two companding techniques.

3.7.4   Transmission Errors

Delta modulation and DPCM are affected differently by bit errors than is PCM. The feedback action in DPCM leads to a propagation of errors in the decoder output. The cumulative effect of this error propagation becomes a significant problem for error bursts, producing a "streaking" effect in video transmission. For speech transmission, however, the natural gaps and pauses in speech effectively mitigate this effect. Subjective evaluations show that DPCM is more tolerant of bit errors than PCM for voice transmission systems. A bit error in PCM transmission can cause an error spike that has magnitude on the order of the peak-to-peak value of the input signal. Corresponding error spikes in a DPCM transmission systems are of smaller magnitude, since DPCM operates on the difference signal rather

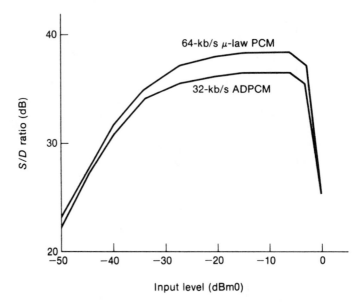

than absolute magnitude of the input signal. Therefore, even though errors do not propagate in a PCM system, the greater magnitude of error makes PCM more vulnerable to bit errors.

Speech intelligibility scores indicate that CVSD suffers negligible loss at an error rate of 10^{-3} as compared to error-free transmission [50]. Similarly, at error rates in the vicinity of 10^{-3}, delta modulation suffers only a loss that is comparable to quantization noise, as shown in Table 3.4. In Figure 3.34, the average MOS is plotted as a function of transmission line BER for 64-kb/s PCM and 32-kb/s ADPCM coders. The results shown in Figure 3.34 are from tests done in several countries in the native language, with the individual MOSs averaged to obtain world averages [51]. By comparison, PCM suffers significant degradation at an error rate of 10^{-3}, while ADPCM maintains communications quality at the same error rate. In a separate test, 32-kb/s ADPCM degraded to an MOS of 1.3 in the presence of a 10^{-2} BER [52].

3.7.5   Effect of Multiple A/D and D/A Conversions
An analog signal passed through multiple analog-to-digital and digital-to-analog conversions is degraded because of the accumulation of quantization noise and slope overload distortion. Jayant and Shipley [53] have investigated this effect for DM and PCM and have expressed results in terms of *S/D* degradation after *M* successive conversions. For a 60-kb/s

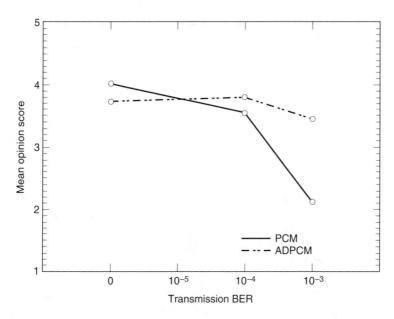

FIGURE 3.34   Average Mean Opinion Score versus BER for PCM and ADPCM
(Adapted from [51])

delta modulation system a computer simulation using real speech was used
to obtain the following empirical relationships [22,53]:

$$\text{LDM:} \quad \left(\frac{S}{D}\right)_M = 30 - 10\log(3.15 + 0.35M^2) \tag{3.91}$$

$$\text{ADM:} \quad \left(\frac{S}{D}\right)_M = 40 - 10\log(0.9 + 3.5M + 1.4M^2) \tag{3.92}$$

Additionally, by assuming that successive PCM conversions have the effect
of adding equal amounts of quantization noise, the following recursive formula results for S/D in a PCM system

$$\text{PCM:} \quad \left(\frac{S}{D}\right)_M = \left(\frac{S}{D}\right)_1 - 10\log(M) \tag{3.93}$$

Inspection of (3.93) and (3.92) indicates that S/D degradation in a PCM and
ADM system decreases monotonically with M, while (3.91) indicates that
LDM exhibits a constant rate of degradation, as illustrated in Table 3.6. As
a consequence, PCM and ADM systems provide superior performance over
LDM for networks employing multiple conversions.

TABLE 3.6   Degradation in S/D Due to Multiple (M) Analog-To-Digital and Digital-to-Analog Conversions

	Values of $\|(S/D)_M - (S/D)_{M+1}\|$ (dB)				
M	1	2	4	6	8
PCM	3.0	1.8	1.0	0.7	0.5
ADM	3.7	2.4	1.6	1.1	0.9
LDM	1.1	1.3	1.4	1.1	0.9

Source: After Jayant and Shipley [53].

The CCITT has used the quantizing distortion produced by a single 8-bit PCM conversion (A-law or μ-law) as a basic reference, assigning it a value of 1 qd (quantizing distortion) unit. Other A/D converters are assigned values referenced to 8-bit PCM, as shown in Table 3.7. For example, 7-bit PCM is known to produce 6 dB more quantizing distortion and is therefore assigned 4 qd units. Since quantizing distortion is uncorrelated from one converter to the next, individual qd units can be added algebraically to yield the overall distortion for an end-to-end connection. CCITT Rec. G.113 provisionally assigns a limit of 14 qd units for an international telephone connection [54]. This value corresponds to the quantizing distortion produced by 14 consecutive PCM conversions and to a S/D degradation of 11.5 dB relative to a single PCM conversion.

The effect of multiple conversions is dependent on the coding technique and bit rate. For example, Figure 3.35 shows MOS values for PCM at 64 kb/s and embedded ADPCM at 32, 24, and 16 kb/s as a function of the number of conversions. In general, 64-kb/s PCM maintains acceptable quality with up to 14 conversions, while other coders at ≤ 32 kb/s, including ADM, CVSD, and ADPCM, can tolerate at most three or four conversions. This limitation on *asynchronous tandeming,* which reproduces an analog representation with

TABLE 3.7   Quantizing Distortion for Various A/D Conversions [49, 54]

A/D Conversion	Quantizing Distortion (qd units)	S/D Degradation in dB Relative to 8-bit PCM (10 log qd)
8-bit PCM	1	0
7-bit PCM	4	6
A/μ law or μ/A law conversion	1	0
32-kb/s ADPCM	3.5	5.4
International telephone connection (CCITT Rec. G113)	14	11.5

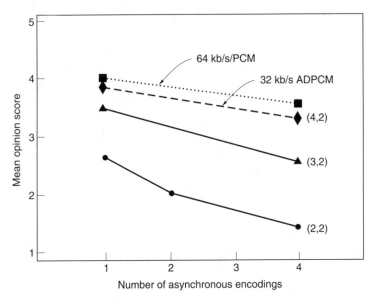

FIGURE 3.35   Mean Opinion Scores for PCM and Embedded ADPCM versus Number of Asynchronous Encodings
Source: Reprinted from [47] by permission of IEEE

each conversion, can be avoided by use of *synchronous tandeming,* which provides a direct digital-to-digital mapping between dissimilar coders.

3.7.6   Transmission of Voice-Band Nonspeech Signals

When DM and PCM coders are designed for transmission of speech-generated analog signals, there may also exist requirements to pass quasi-analog signals (voice-band data modems) or other nonspeech signals (signaling information). Logarithmic PCM used in telephony provides the wide dynamic range required for speech but leads to nonoptimum performance for voice-band data. The quantization noise of logarithmic PCM is largest at the peak of the data signal envelope, the point at which data detection is done. Hence the average quantization noise cannot be used as a parameter to predict voice-band data performance accurately. Similarly, the adaptation of step size used in ADM enhances the dynamic range for speech but is nonoptimum for modem signals. The average power level of modem signals remains fairly constant so that companding is not required if the proper (fixed) step size is selected. For DM systems to work properly with data signals, the adaptive feature must be disabled.

Performance of modem signals transmitted via various voice coders has been reported in analytic studies and simulation [55–57], but empirical

TABLE 3.8   Modem BER Performance over Multilink ADPCM, PCM, and DM Coders [56,60]

Coder Type and Channel Bit Rate	Modem Bit Rates											
	1200 b/s			2400 b/s			4800 b/s			9600 b/s		
	1 Link	2 Links	4 Links	1 Link	2 Links	4 Links	1 Link	2 Links	4 Links	1 Link	2 Links	4 Links
DM												
32 kb/s	O	O	X	X	X	X	X	X	X	X	X	X
48 kb/s	O	O	O	O	X	X	X	X	X	X	X	X
64 kb/s	O	O	O	O	O	O	X	O	X	X	X	X
PCM												
64 kb/s		O		O	O	O	O	O	X	X	X	X
ADPCM												
32 kb/s	Y	Y	Y	Y	Y	Y	Y	Y	Z	Z	Z	Z

Note: O = error free; X = error-free operation not possible; Y = BER $\leq 10^{-5}$; Z = BER $> 10^{-5}$.

results are not as widely published. Notably, May and colleagues have experimentally studied performance of five different modems transmitted through tandem connections of both DM and PCM [56]. Significant results are summarized in Table 3.8. PCM and DM are seen to provide comparable modem BER performance for a 64-kb/s channel rate. With errors inserted into the transmission line, however, the BER performance of a modem was superior for DM channels vis-à-vis PCM channels; both were operated at a channel rate of 64 kb/s. Results indicated that the modem BER falls off exponentially as the DM transmission BER decreases, whereas it decreases linearly as the PCM transmission BER decreases. Table 3.8 indicates that 32-kb/s DM will support modem data rates up to 2400 b/s, a conclusion reached by similar experiments [50, 58, 59] and analyses [55]. Moreover, analysis [55] and experiment [50] have shown that 16-kb/s DM will not support modem data rates of 1200 b/s and above.

ADPCM at 32 kb/s has also been extensively tested for performance of voice-band data modems [61]. No significant differences in modem performance were observed between one and several *synchronous* codings of ADPCM. For multiple *asynchronous* codings, PCM outperformed ADPCM for all modem rates tested. For a single coding, however, performance of standard 2400- and 4800-b/s modems were approximately the same for PCM and ADPCM. For modem rates above 7200 bp/s there were large differences in performance observed, with PCM providing acceptable performance but ADPCM generally not. Testing of 9600-b/s modems through 32-kb/s ADPCM indicates that V.32 modems provide marginally better performance than V.29 modems, but neither modem can meet a 1×10^{-5} error objective even for a single coding [60].

For switched telephone networks, in-band signaling information may be transmitted through the voice coder. Testing of single-frequency (SF) and dual-tone multifrequency (DTMF) signaling transmitted over CVSD, PCM, and ADPCM channels indicated the following [49, 50, 60, 61]:

1. For SF signaling at 2.6 kHz, 64-kb/s CVSD provided acceptable performance; 16-kb/s and 32-kb/s CVSD did not.
2. For DTMF signaling at frequencies from 697 to 1477 Hz, 32-kb/s CVSD provided acceptable performance but 16 kb/s CVSD did not.
3. Log PCM at 64 kb/s and ADPCM at 32 kb/s provided adequate performance for transmission of SF or DTMF signaling.

The increasing use of telephone channels for transmission of data and facsimile has led to the development of voice-band signal classification techniques [62] that can distinguish voice from various other signals that may be present in a telephone network. Such classifiers can then assign the appropriate coding type and bit rate to accommodate the voice or voice-band data signal. After separation of voice from nonvoice signals, the non-voice signals are further classified by identifying the baud rate and modulation type. Voice-band data signals are separated into several cate-gories, say, low, medium, and high speed, and then given a coding rate that will support that modem signal with an acceptable error rate. If the modem signal is a standard one, it is even possible to identify and demodulate the signal, transport the data signal at its original bit rate through the trans-mission system, and reconstruct the voice-band data signal at the terminat-ing location. This approach has been used with Group Three facsimile, which has a bit rate of 9.6 kb/s, thus saving bandwidth and improving per-formance [63].

3.7.7   Implementation Complexity

Coders
Coder implementation for DM and DPCM is potentially much simpler than for PCM, although a PCM coder may be shared among a number of sub-scribers while a DM coder cannot. With the advent of large-scale integra-tion techniques, however, coder costs have become less important. Current-generation channel banks in fact utilize one coder per subscriber to improve overall reliability and eliminate crosstalk caused by switching the coder from one input to the next.

Filters
Filtering in a PCM coder and decoder requires sharp cutoff to $f_s/2$ Hz in accordance with the Nyquist sampling frequency. This requirement for sharp cutoff filtering is in contrast to a DPCM or DM coder and decoder where the sampling frequency is typically much larger than the message bandwidth so that a gradual rolloff can be used for the filter characteristic. As a consequence filter design for DPCM and DM is simpler than for PCM.

3.8   Voice Coders (Vocoders)

Vocoders are based on the speech model shown in Figure 3.36, in which the source of sound is separated from the vocal tract filter. Speech sound is assumed to be either voiced ("buzz"), corresponding to the periodic flow of air generated by the vocal cords, or unvoiced ("hiss"), corre-sponding to a turbulent flow of air past a constriction in the vocal tract. Unvoiced sound is an acoustic noise source that can be represented by a random noise generator; voiced sound is an oscillation of the vocal cords that can be represented by a periodic pulse generator. Sound, voiced or unvoiced, passes through the vocal tract, which includes the throat,

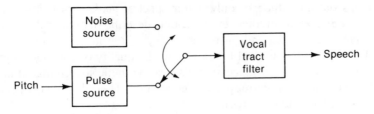

FIGURE 3.36   Model of Human Speech

tongue, mouth, and nasal cavity. The moving parts of the vocal tract, called the articulators, modulate the sound, thereby creating speech signals. During voiced sounds the articulators assume different positions causing resonances in the vocal tract that impose spectral peaks called *formants* (Figure 3.37). In Figure 3.37 the speech signal is assumed to be voiced, resulting in a spectrum with equally spaced harmonics at the fundamental pitch of the speaker. The spectral envelope is determined by the vocal tract shape, which is determined by the spoken sound. Should the sound be unvoiced, the periodic structure would disappear leaving only the spectral envelope.

All vocoders work on the same basic principles. Speech is analyzed to determine whether the excitation function (sound) is voiced or unvoiced; if it is voiced, the vocoder then estimates the fundamental pitch. Simultaneously, the vocal tract filter transfer function (spectral envelope) is also estimated. Since speech is a nonstationary process, this voiced/unvoiced decision, pitch extraction, and spectral estimate must be updated with each successive speech segment. These principles have been implemented in vocoders in a variety of forms: In the **channel vocoder,** values of the short-time spectrum are evaluated by specific frequencies; the **formant vocoder**

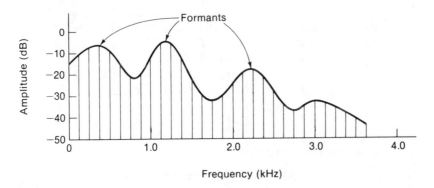

FIGURE 3.37   Spectral Content of Typical Speech

evaluates the amplitude of only major spectral peaks (formants); in the **LPC vocoder,** linear predictive coefficients describe the spectral envelope. All such vocoders depend on the accuracy of the model in Figure 3.36, which places a fundamental limitation on the quality of vocoders. Typical low-bit-rate vocoders produce an unnatural synthetic speech that although intelligible tends to obscure speaker recognition. The value of vocoders lies in the inherent bandwidth reduction.

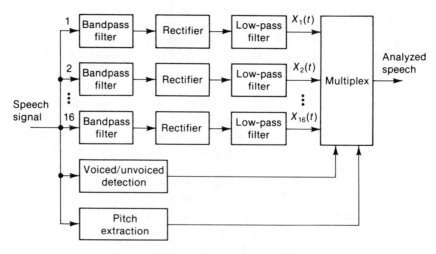

(a) Channel vocoder analyzer

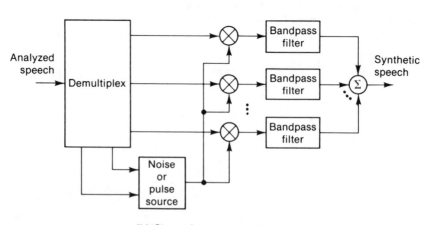

(b) Channel vocoder synthesizer

FIGURE 3.38   Block Diagram of Channel Vocoder

3.8.1   Channel Vocoders

The strategy of the channel vocoder is to represent the spectral envelope by samples taken from contiguous frequency bands within the speech signal. In parallel, the voiced/unvoiced decision and pitch extraction are performed; these operations determine the fine-grain structure of the speech spectrum. A block diagram of a channel vocoder analyzer and synthesizer is shown in Figure 3.38. The outputs of a bank of bandpass filters are connected to a rectifier and low-pass filter. The signal $X_i(t)$ represents the smoothed amplitude of the input speech for the ith frequency band. The vocal tract filter can be reasonably well described by 16 channel values taken every 20 ms. Low-pass filtering limits the bandwidth of each spectral channel to about 25 to 30 Hz. The excitation signals require another 50 Hz of bandwidth, so that vocoded speech may be transmitted in a bandwidth of about 500 Hz for a bandwidth savings of 7:1.

At the synthesizer, the speech signal is regenerated by modulating the excitation function (noise or pulse source) with the samples of the vocal tract filter. After bandpass filtering and summation, the result is synthetic speech that reproduces the short-term spectrum of the original speaker.

Despite some 40 years of research in channel vocoders [64], no optimum design exists today and channel vocoders still produce the unnatural machinelike speech that was characteristic of early vocoders [65]. Their chief application was in early attempts at providing digital voice to facilitate encryption in military applications. A digital channel vocoder operating at 2400 b/s could be readily encrypted and transmitted by existing 3-kHz telephone channels. In a 2400-b/s vocoder there are nominally 3 bits per spectral channel, or 48 bits for a 16-channel vocoder, plus 6 bits for pitch information resulting in a total of 54 bits per speech sample. A frame synchronization bit and a voiced/unvoiced decision bit usually occupy a least-weight bit position of one of the spectral channels. A transmission rate of 2400 b/s results if the speech signal is sampled every 22.5 ms. Further bit rate (or bandwidth) reduction was realized with the formant vocoder, which specified only the frequencies of the spectral peaks (formants) and their amplitudes. However, the added complexity and even poorer voice quality compared to the channel vocoder have limited the application of formant vocoders. Voice-excited vocoders (VEV) [66] provide improved voice quality at the expense of higher data rates (typically 9.6 kb/s). The VEV performs PCM on the frequency band below 900 Hz, which replaces the voiced/unvoiced decision and pitch extraction of the channel vocoder. The spectral information is handled in a similar way to the channel vocoder, except that the synthesizer modulators are excited by the PCM information below 900 Hz.

3.8.2 Linear Predictive Coding (LPC) Vocoders

The linear predictive coding (LPC) vocoder is based on the concept that a reasonable prediction of a speech sample can be obtained as a linear weighted sum of previously measured samples. In the block diagram of Figure 3.39, the LPC analyzer and synthesizer both use a predictor filter of order P that provides an estimate $\hat{S}_n$ expressed as

$$\hat{S}_n = \sum_{m=1}^{P} a_m S_{n-m} \qquad (3.94)$$

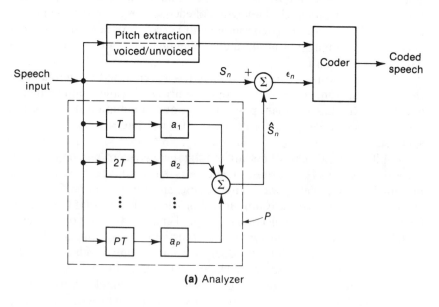

(a) Analyzer

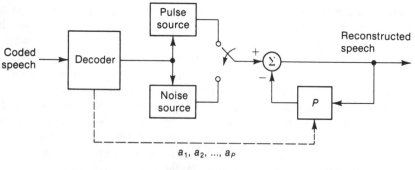

(b) Synthesizer

FIGURE 3.39   Linear Predictive Coding (LPC) Vocoder

where the a_m are known as predictor coefficients. The order of the predictor filter is typically 8 to 12, where, for example, the spectral resolution provided by $P = 8$ matches that of the 16-channel vocoder. The predicted value $\hat{S}_n$ of the speech signal is next subtracted from the signal value S_n to form the difference ε_n, which is coded and transmitted to the receiver.

Voicing and pitch extraction are estimated as they are in a channel vocoder. At the synthesizer the inverse filter to the analyzer is excited by the pulse (voiced) or noise (unvoiced) source. Predictor coefficients are generally updated every 10 to 25 ms (approximately the syllabic rate). The difference between methods of implementing LPC vocoders lies within the method of determining the predictor coefficients [67–68]. The implementation of LPC has proved simpler than channel vocoders, one of the significant factors in the growing popularity of LPC. A United States standard used for secure voice applications of LPC is based on a 2400-b/s data rate, 10 predictor coefficients, and a frame length of 54 bits. Each frame has duration 22.5 msec and contains 7 bits for pitch, 5 bits for gain, 41 bits for the 10 coefficients, and one framing bit [69].

3.8.3 Adaptive Predictive Coding (APC) Vocoders

Because of the nonstationary nature of speech signals, a fixed predictor cannot efficiently predict the signal values at all times. Ideally the predictor coefficients must vary with changes in the spectral envelope and changes in the periodicity of voiced speech. Adaptive prediction of speech is a deconvolving process that is done conveniently in two separate stages as shown in Figure 3.40a. One prediction (P_1) is with respect to the vocal tract filter and the other (P_2) is with respect to pulse excitation [70]. The short-time spectral envelope predictor can be characterized in Z-transform notation as

$$P_1(Z) = \sum_{K=1}^{P} a_K Z^{-K} \tag{3.95}$$

where Z^{-1} represents a delay of one sample interval and $a_1, a_2, \ldots a_p$ are the P predictor coefficients. The signal ε_1 that remains after the first prediction exhibits the periodicity associated with the excitation signal when the speech is voiced. The second predictor is simply a delay of M speech samples and a weighting factor β, so that

$$P_2(Z) = \beta Z^{-M} \tag{3.96}$$

In most cases, the delay M is chosen to be equivalent to the pitch period or an integral number of pitch periods [71, 72]. The residual signal E_2 that remains after the second predictor is quantized prior to transmission. The receiver structure for the APC vocoder is depicted in Figure 3.40b and is the inverse of the transmitter. This receiver filter is determined by the values of β, M and the a_i's, each of which has been quantized, coded, transmitted, and decoded at the receiver. Speech is reconstructed at the receiver by feeding the quantized residual through the inverse predictor filters.

There are two limiting factors in APC vocoder performance: One is the existence and significance in listener perception of small variations in the periodicity of voiced speech; the other is degradation due to incorrect decoding of the vocal tract filter parameters. Although the first limitation is nontrivial, the second problem is readily solved via error correction coding. As to implementations of APC, Atal and Schroeder [70] claim that 10-kb/s APC compares favorably with logarithmic ($\mu = 100$) PCM at 40 kb/s. Detailed testing of 8-kb/s APC indicated performance equivalent in voice quality to 16-kb/s CVSD.

3.9 Hybrid Coders

At bit rates below 8 kb/s, the number of bits available for encoding the residual signal in APC is less than 1 bit/sample, assuming that the sampling frequency is 8 kHz. Moreover, for low bit rate APC, coarse (for instance, two-level) quantization is usually the major source of audible distortion in the reconstructed speech signal. Improved quantization of the residual sig-

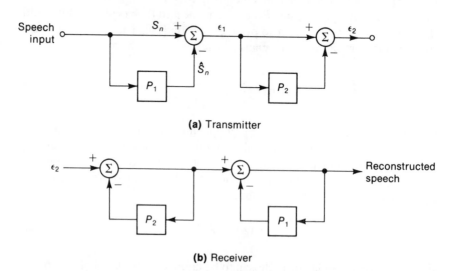

(a) Transmitter

(b) Receiver

FIGURE 3.40   Adaptive Predictive Coding (APC)

nal for low bit rates has been achieved by actually increasing the number of quantization levels but using variable-length codes so that the net output rate of the quantizer is less than 1 bit/sample. In this technique, a number of samples are grouped together in a block, and the resulting block is used as an input to the coder rather than individual samples. Unlike APC, which treats samples independently, this scheme assumes that a block of the residual signals can be matched with one out of a set of code words or by a small set of parameters. In addition to minimizing the error between the residual signal and its quantized version, these low bit rate coders also seek to minimize the error between the synthesized signal and the original signal. This additional feature requires a minimization procedure (for instance, minimum mean-squared error) that operates on a weighted version of the error signal and feeds back changes to the coder parameters. This form of coding has been called **analysis-by-synthesis adaptive predictive coding** [73, 74], but is better known as **hybrid coding** [75]. Hybrid coders in general combine the high-quality advantage of waveform coders with the bandwidth efficiency of vocoders.

The basic structure of a hybrid coder is shown in Figure 3.41. The two filters labeled Pitch and Vocal Tract synthesis are essentially those used in APC. Updating of the filter coefficients can be done either from the speech signals or from the error minimization procedure. Once the filter coefficients have been determined, the excitation function is found in a blockwise fashion. For every block of samples (typically, 5–30 ms), the excitation function is selected such that the weighted error between the reconstructed and original speech signal is minimized. The error-weighting filter incorporates a model of how human perception treats the differences between the original and reconstructed speech. Many variations of this basic scheme have been proposed. Schemes in which the coefficients for both predictors are transmitted are called **forward adaptive prediction,** while those schemes that do not transmit the predictor coefficients are called **backward adaptive**

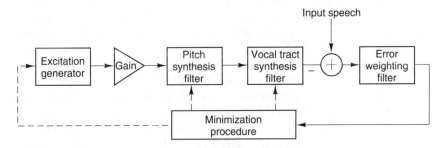

FIGURE 3.41   Block Diagram of Hybrid Coder

prediction. Other variations include elimination of the pitch synthesis filter, and an interchange of the location of the pitch and vocal tract filters. The most fundamental difference among these variations, however, lies in the choice of the excitation function. We shall now describe two of the most popular choices in hybrid coders, whose names designate the choice of excitation function.

3.9.1   Code-Excited Linear Prediction

The excitation generator of Figure 3.41 may be a collection of possible sequences or code words. Each incoming block of residual signals is matched to the closest fitting code word from a code book. By storing the collection of code words at both the transmitter and receiver, only the index of the code word that produced the minimum error needs to be sent. Because a large number of code words is required to adequately represent the residual sequence, a sequential search approach is impractical. To render this technique practical, the code book is structured as a tree or trellis. Each sequence traces a different path through the tree or trellis. Code words with similar elements are grouped together to speed up the search process. This form of coding is known as **code-excited linear prediction (CELP).** Because a group of residual samples can be thought of as a vector, this coding technique is also referred to as **vector quantization.**

CELP has been used at rates from 4.8 to 16 kb/s, with the lower rate versions applied to cellular radio [76], mobile satellite [77], and secure voice [78], and the 16-kb/s version proposed as a telephone network standard [52]. The 16-kb/s version has received considerable attention by the CCITT in proposed Rec. G.7XY and therefore merits describing in more detail. A primary objective in the development of this 16-kb/s coder was low delay combined with high quality. The delay objective forced the use of backward adaptive prediction to avoid the buffering of tens of milliseconds otherwise required for forward adaption. The only source of coding delay remaining is that required for selection of the excitation signal. The excitation vector has a block size of five samples, which is equal to a 0.625-ms delay at an 8-kHz sampling rate. This coder has been named the Low-Delay CELP, or LD CELP, and has less than 2 ms of total delay. In the LD CELP, only the excitation sequence is transmitted, which means that at a rate of 16 kb/s, there are 10 bits available to encode each vector of five samples. The pitch predictor found in conventional CELP is eliminated, and a fiftieth-order LPC predictor is used. Except for the excitation generator, all other parameters are backward adapted, with the excitation gain updated every vector, and the LPC predictor coefficients and weighting filter coefficients updated every four vectors. Performance tests indicate that 16-kb/s LD CELP meets the CCITT requirement for 16-kb/s coders of less than 4 qd units for a single encoding and 14 qd units for three tandem encoding, passes 2400-b/s modems error free, and provides satisfactory performance for DTMF signals [52].

3.9.2   Multipulse and Regular-Pulse Excitation

Rather than select the excitation sequence from a set of codewords, it is possible to use a sequence of pulses that are either uniformly or nonuniformly spaced. The coder selects the appropriate pulse sequence and quantizes that sequence for transmission at low bit rates. **Multipulse excitation** represents the residual signals by a small number of pulses with nonuniform spacing. The position and amplitudes of these pulses are selected to minimize a weighted-error criterion. **Regular-pulse excitation** represents the residual signal by a pulse train, selected with regular spacing but with more pulses per frame than the multipulse scheme. The position of the first pulse in the frame and the amplitudes of the pulses are determined by the encoder. For either type of pulse excitation, quantization of the pulses is accomplished by an adaptive quantizer whose levels are adjusted to the maximum absolute value within a frame. This maximum value and each of the pulses within the frame are quantized before transmission. For regular-pulse excitation the position of the first pulse must also be encoded, while for multipulse excitation all pulse positions must be encoded.

Compared to CELP, these pulse-excitation coders have the advantage of providing adequate performance without the need for a pitch predictor. The pulse-excitation approach is most efficient at rates around 10 kb/s, while code book excitation offers better performance at lower rates. Concerning applications, a CEPT working group has selected a regular-pulse excitation coder combined with long-term LPC for the European digital mobile radio system. A multipulse excitation coder operating at 9.6 kb/s is used in Skyphone, an airline telephone service [75].

3.10   Summary

The summary for this chapter is presented in the form of a comparison of digital coders for speech and video signals. Comparative evaluations of speech and video processing algorithms must be tentative because of ongoing developments. Nevertheless, some observations can be made based on past experience, current technology, and future trends. The choice of a speech or video coder for a given application involves a tradeoff of transmission bit rate and quality. Each of these elements is addressed in the following paragraphs.

3.10.1   Transmission Bit Rate

Figure 3.42 compares transmission bit rates for speech coders currently of interest. The lowest practical rates are 1.2 to 2.4 kb/s, used with formant and channel vocoders. Bit rates of 2.4 and 4.8 kb/s are most typical for LPC, while APC rates range from 8 to 16 kb/s. Common bit rates for CVSD and ADPCM are 16 and 32 kb/s. Linear DM or DPCM typically requires 32 to perhaps 64 kb/s. PCM is at the high end of the spectrum, ranging from 48-kb/s (6-bit) logarithmic to 96-kb/s (12-bit) linear PCM. Of course, 64-kb/s logarithmic PCM is the internationally recognized standard for digital speech. To segregate transmission bit rate requirements for digital speech, two general categories are commonly used. **Narrowband coders** utilize rates up to and including 9.6 kb/s that can be handled by 3-kHz telephone channels using modems. **Wideband coders** exceed the transmission rate (generally 9.6 kb/s) that can be handled by a nominal telephone channel.

Figure 3.43 indicates transmission rates for various applications of video coding. Lower coding rates shown are applicable to video phone and video conferencing at rates of $m \times 64$ kb/s and $n \times 384$ kb/s, with a upper bound of about 2.048 Mb/s [40]. The combination of DCT and DPCM coding has been popular for such video services. The trend in coding of conventional television such as NTSC or PAL is toward the use of hybrid techniques such as intraframe and interframe coding combined with DPCM and variable-length coding [79]. HDTV transmission rates are determined by the intended application. HDTV subscriber quality is achievable at transmission rates of 15 to 30 Mb/s using various hybrid coding techniques. Distri-

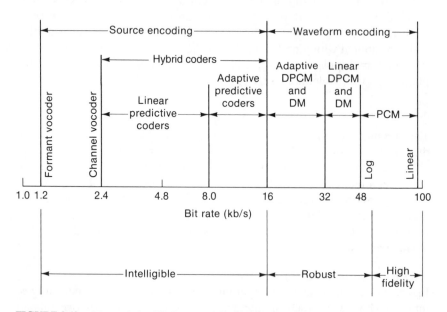

FIGURE 3.42   Transmission Bit Rates and Quality for Speech Coders

bution of HDTV signals requires rates of 30 to 600 Mb/s, depending on the exact application. Subband coding, usually in combination with DCT or DPCM, is a standard choice for HDTV distribution. For highest quality, such as that used in television studio production, 8- to 10-bit PCM is a standard choice because of its quality and simplicity, with resulting rates of 600 Mb/s to 2.4 Gb/s [80].

3.10.2   Speech Quality

Figure 3.42 also indicates the quality of speech reproduction that can presently be attained by a prescribed bit rate. The quality indication is roughly divided into three categories: intelligible, robust, and high fidelity. Vocoders, having the lowest bit rates, fall into the intelligible range; they are characterized as highly intelligible but suffering from some loss of naturalness and reduced speaker recognition. At rates above 16 kb/s, robust speech is obtainable by using techniques such as CVSD, ADPCM, or logarithmic PCM. These schemes depend less on individual speaker characteristics and maintain adequate performance with transmission impairments (tandeming or transmission errors). Hence most coders operating in this range provide adequate telephone toll quality. High-fidelity speech can be provided at rates above 64 kb/s, where input signal bandwidth may exceed the nominal telephone channel. This grade of quality is appropriate for radio broadcast material, including music.

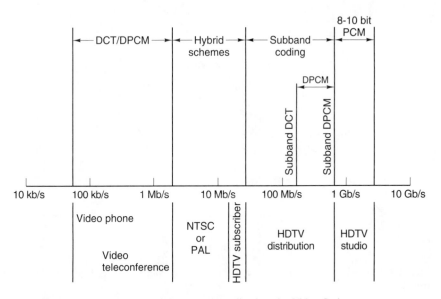

FIGURE 3.43   Transmission Bit Rates and Applications for Video Coders

Numerous comparisons of speech coders have been made, both objective and subjective, some of which have been referenced here. Figure 3.44 summarizes speech quality using MOS for the general classes of speech coders and their respective range of bit rates considered in this chapter. The solid curves represent actual performance for generic coders, and the broken curve represents a research objective realizable within practical bounds. The curve for PCM represents the standard form of nonadaptive, logarithmic pulse code modulation popular in telephone networks worldwide. Reduced bit rates without loss of performance are observed with waveform coders, which remove redundancy inherent in speech signals but may be limited in their ability to accommodate nonspeech signals. Even greater reduction in bit rates is possible by use of vocoders, which model speech as a excitation function modulated by the vocal tract filter. However, vocoders produce artificial speech that lacks naturalness and speaker recognition. The hybrid curve represents a combination of vocoder and waveform principles, with a resulting improvement in quality over vocoders and

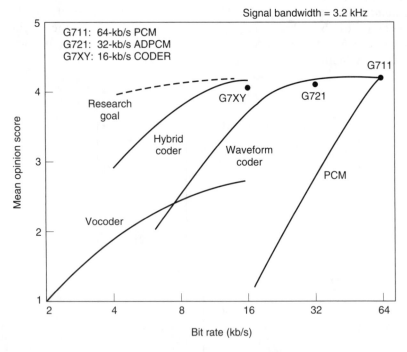

FIGURE 3.44   Mean Opinion Scores versus Bit Rate for Various Speech Coders (Adapted from [75])

no increase in bit rates, making 16-kb/s coding with network quality possible. The dots in Figure 3.44 refer to specific coders, namely, G.711 64-kb/s PCM, G.721 32-kb/s ADPCM, and G.7XY 16-kb/s hybrid coding, all of which provide high-quality performance. For further subjective evaluation, the reader is referred to speech recordings included on plastic records furnished with papers on speech coding [1, 23].

References

1. J. L. Flanagan and others, "Speech Coding," *IEEE Trans. on Comm.*, vol. COM-27, no. 4, April 1979, pp. 710–736.
2. B. P. Lathi, *An Introduction to Random Signals and Communication Theory* (Scranton: International Textbook Company, 1968).
3. Bell Telephone Laboratories, *Transmission Systems for Communication*, 4th ed. rev. (Winston-Salem: Western Electric Company, 1971).
4. B. D. Holbrook and J. T. Dixon, "Load Rating Theory for Multi-Channel Amplifiers," *Bell System Technical Journal* 43(1939):624–644.
5. M. D. Paez and T. H. Glisson, "Minimum Mean-Squared Error Quantization in Speech PCM and DPCM Systems," *IEEE Trans. on Comm.*, vol. COM-20, no. 2, April 1972, pp. 225–230.
6. B. Smith, "Instantaneous Companding of Quantized Signals," *Bell System Technical Journal* 36(1957):653–709.
7. CCITT Yellow Book, vol. III.3, *Digital Networks–Transmission Systems and Multiplexing Equipment* (Geneva: ITU, 1981).
8. H. Mann, H. M. Stranbe, and C. P. Villars, "A Companded Coder for an Experimental PCM Terminal," *Bell System Technical Journal* 41(1962):1173–1226.
9. S. M. Schreiner and A. R. Vallarino, "48-Channel PCM System," 1957 *IRE Nat. Conv. Rec.*, pt. 8, pp. 141–149.
10. A. Chatelon, "Application of Pulse Code Modulation to an Integrated Telephone Network, pt. 2—Transmission and Encoding," *ITT Elec. Comm.* 38(1963):32–43.
11. H. Kaneko, "A Unified Formulation of Segment Companding Laws and Synthesis of Codecs and Digital Compandors," *Bell System Technical Journal* 49(1970):1555–1588.
12. CCITT Green Book, vol. III. 3, *Line Transmission* (Geneva: ITU, 1973).
13. F. DeJager, "Delta Modulation: A Method of PCM Transmission Using a 1-Unit Code," *Philips Research Report,* December 1952, pp. 442–466.
14. H. Van de Weg, "Quantization Noise of a Single Integration Delta Modulation System with an N-digit Code," *Philips Research Report,* October 1953, pp. 367–385.
15. J. B. O'Neal, Jr., "Delta Modulation Quantizing Noise Analytical and Computer Simulation Results for Gaussian and Television Input Signals," *Bell System Technical Journal* 45(1966):app. A, pp. 136–139.
16. F. B. Johnson, "Calculating Delta Modulation Performance," *IEEE Trans. on Audio and Electroacoustics,* vol. AU-16, March 1968, pp. 121–129.

17. J. E. Abate, "Linear and Adaptive Delta Modulation," *Proc. IEEE* 55(3)(1967):298–308.

18. H. R. Schindler, "Delta Modulation," *IEEE Spectrum,* October 1970, pp. 69–78.

19. M. R. Winkler, "High Information Delta Modulation," *IEEE Inter. Conv. Record,* pt. 8, 1963, pp. 260–265.

20. M. R. Winkler, "Pictorial Transmission with HIDM," *IEEE Inter. Conv. Record,* pt. 1, 1965, pp. 285–291.

21. R. H. Bosworth and J. C. Candy, "A Companded One-Bit Coder for Television Transmission," *Bell System Technical Journal* 48(1)(1969):1459–1479.

22. N. S. Jayant, "Adaptive Delta Modulation with a One-Bit Memory," *Bell System Technical Journal* 49(1970):321–342.

23. N. S. Jayant, "Digital Coding of Speech Waveforms: PCM, DPCM, and DM Quantizers," *Proc. IEEE* 62(May 1974):611–632.

24. J. A. Greefkes and F. DeJager, "Continuous Delta Modulation," *Philips Research Report* 23(2)(1968):233–246.

25. CCITT Recommendation G.726, "40, 32, 24, and 16 kbit/s Adaptive Differential Pulse-Code Modulation (ADPCM)," CCITT Study Group XV, 1990.

26. ANSI T1.303–1989, "Digital Processing of Voice-Band Signals—Algorithms for 24-, 32-, and 40-kbit/s Adaptive Differential Pulse-Code Modulation (ADPCM)."

27. N. Benvenuto and others, "The 32-kb/s ADPCM Coding Standard," *AT&T Technical Journal,* 65 (September/October 1986):12–22.

28. CCITT Recommendation G.727, "5-, 4-, 3-, and 2-bits/sample Embedded Adaptive Differential Pulse Code Modulation," CCITT Study Group XV, 1990.

29. ANSI T1.310-1990, "Digital Processing of Voice-Band Signals—Algorithms for 5-, 4-, 3-, and 2-bits/sample Embedded Adaptive Differential Pulse-Code Modulation (ADPCM)."

30. N. S. Jayant and P. Noll, *Digital Coding of Waveforms,* (Englewood Cliffs, NJ: Prentice Hall, 1984).

31. J. E. Natvig, "Evaluation of Six Medium Bit-Rate Coders for the Pan-European Digital Mobile Radio System," *IEEE Journal on Sel. Areas in Comm.* 6 (February 1988):324–331.

32. M. J. McLaughlin and P. D. Rasky, "Speech and Channel Coding for Digital Land-Mobile Radio," *IEEE Journal on Sel. Areas in Comm.* 6 (February 1988):332–345.

33. R. V. Cox and others, "New Directions on Subband Coding," *IEEE Journal on Sel. Areas in Comm.* 6 (February 1988):391–409.

34. CCITT Recommendation G.722, "7 kHz Audio-Coding within 64 kbit/s," CCITT Blue Book, vol. III.4 (Geneva: ITU, 1989).

35. P. Mermelstein, "G.722, A New CCITT Coding Standard for Digital Transmission of Wideband Audio Signals," *IEEE Comm. Mag,* January 1988, pp. 8–15.

36. J. W. Woods and S. D. O'Neil, "Subband Coding of Images," *IEEE Trans. on Acoustics, Speech, and Signal Proc.,* vol. ASSP-34, No. 5, October 1986, pp. 1278–1288.

37. A. Fernandez and others, "HDTV Subband/DCT Coding: Analysis of System Complexity," *1990 International Conference on Communications*, pp. 343.1.1–343.1.5.
38. M. Vetterli and D. Anastassiou, "A Multiresolution Approach for All-Digital HDTV," *1990 International Conference on Communications*, pp. 320.2.1–320.2.5.
39. H. Gharavi, "Subband Based CCITT Compatible Coding for HDTV Conferencing," *1990 GLOBECOM*, pp. 508.4.1–508.4.4.
40. CCITT Recommendation H.261, "Codec for Audiovisual Services at n × 384 kbit/s," CCITT Blue Book, vol. III.6 (Geneva: ITU, 1989).
41. J. W. Modestino, D. G. Daut, and A. L. Vickers, "Combined Source-Channel Coding of Images Using the Block Cosine Transform," *IEEE Trans. on Comm.*, vol. COM-29, No. 9, September 1981, pp. 1261–1274.
42. W. H. Chen and C. H. Smith, "Adaptive Coding of Monochome and Color Images," *IEEE Trans. on Comm.*, vol. COM-25, No. 11, November 1977, pp. 1285–1292.
43. Y. Okumura, K. Irie, and R. Kishimoto, "High Quality Transmission System Design for HDTV Signals," *1990 International Conference on Communications*, pp. 325.B.2.1–325.B.2.5.
44. K. Sawada, "Coding and Transmission of High-Definition Television—Technology Trends in Japan," *1990 International Conference on Communications*, pp. 320.3.1–320.3.7.
45. I. Tamitani, H. Harasaki, and T. Nishitani, "A Real-Time HDTV Signal Processor: HD-VSP System and Applications," *1990 International Conference on Communications*, pp. 325.B.3.1–325.B.3.5.
46. M. Cominetti and F. Molo, "A Codec for HCTV Signal Transmission Through Terrestrial and Satellite Digital Links," *1990 GLOBECOM*, pp. 508.6.1–508.6.6.
47. M. H. Sherif and others, "Overview of CCITT Embedded ADPCM Algorithms," *1990 International Conference on Communications*, pp. 324.5.1–324.5.5.
48. R. Steele, *Delta Modulation Systems* (New York: Wiley, 1975).
49. AT & T Contribution to CCITT Question 7/Study Group XVIII, "32 kb/s ADPCM-DLQ Coding," April 1982.
50. E. D. Harras and J. W. Presusse, *Communications Performance of CVSD at 16/ 32 Kilobits/Second*, U.S. Army Electronics Command, Ft. Monmouth, NJ, January 1974.
51. G. Williams and H. Suyderhoud, "Subjective Performance Evaluation of the 32-kbit/s ADPCM Algorithm," *1984 GLOBECOM*, pp. 23.2.1–23.2.8.
52. J. H. Chen and R. V. Cox, "LD-CELP: A High Quality 16 Kb/s Speech Coder with Low Delay," *1990 GLOBECOM*, pp. 405.4.1–405.4.5.
53. N. S. Jayant and K. Shipley, "Multiple Delta Modulation of a Speech Signal," *Proc. IEEE* 59(September 1971):1382.
54. CCITT Yellow Book, vol. III.1, *General Characteristics of International Telephone Connections and Circuits* (Geneva: ITU, 1981).

55. J. B. O'Neal, "Delta Modulation of Data Signals," *IEEE Trans. on Comm.*, vol. COM-22, no. 3, March 1974, pp. 334–339.

56. P. J. May, C. J. Zarcone, and K. Ozone, "Voice Band Data Modem Performance over Companded Delta Modulation Channels," *Conf. Rec. 1975 International Conference on Communications*, June 1975, pp. 40.16–40.21.

57. J. B. O'Neal, "Waveform Encoding of Voiceband Data Signals," *Proc. IEEE* 68(February 1980):232–247.

58. E. V. Stansfield, "Limitations on the Use of Delta Modulation Links for Data Transmission," SHAPE Technical Centre Report STC CR-NICS-39, January 1979.

59. J. Evanowsky, "Test and Evaluation of Reduced Rate Multiplexers," Rome Air Development Center, Rome, NY, April 1981.

60. ANSI T1.501-1988, "Network Performance—Tandem Encoding Limits for 32 kbit/s Adaptive Differential Pulse Code Modulation (ADPCM)," 1988.

61. J. M. Raulin, W. R. Belfield, and T. Nishitani, "Objective Test Results for the 32 kb/s ADPCM Coder," *1984 GLOBECOM*, pp. 23.3.1–23.3.5.

62. N. Benvenuto and W. R. Daumer, "Classification of Voiceband Data Signals," *1990 International Conference on Communications*, pp. 324.4.1–324.4.4.

63. W. J. Giguere, "New Applications of Wideband Technology," *1990 International Conference on Communications*, pp. 324.1.1–324.1.3.

64. B. Gold, "Digital Speech Networks," *Proc. IEEE* 65(December 1977):1636–1658.

65. H. Dudley, "The Vocoder," *Bell Labs Record* 17(1939):122–126.

66. B. Gold and J. Tierney, "Digitized Voice-Excited Vocoder for Telephone Quality Inputs Using Bandpass Sampling of the Baseband Signal," *J. Acoust. Soc. Amer.* 37(April 1965):753–754.

67. B. Atal and S. L. Hanauer, "Speech Analysis and Synthesis by Linear Prediction of the Speech Wave," *J. Acoust. Soc. Amer.* 50(1971):637–655.

68. J. Marhoul, "Linear Prediction: A Tutorial Review," *Proc. IEEE* 63(1975):561–580.

69. U.S. Federal Standard 1015, *Analog to Digital Conversion of Voice by 2400 Bit/Second Linear Predictive Coding*, National Communications System, Washington, D.C., July 1983.

70. B. S. Atal and M. R. Schroeder, "Adaptive Predictive Coding of Speech Signals," *Bell System Technical Journal* 49(1970):1973–1986.

71. E. E. David and H. S. McDonald, "Note on Pitch Synchronous Processing of Speech," *J. Acoust. Soc. Amer.*, November 1956, pp. 1261–1266.

72. A. H. Frei, H. R. Schindler, P. Vettiger, and E. Von Felten, "Adaptive Predictive Speech Coding Based on Pitch-Controlled Interruption/Reiteration Techniques," *1973 International Conference on Communications*, pp. 46.12–46.16.

73. B. S. Atal, "Predictive Coding of Speech at Low Bit Rates," *IEEE Trans. on Comm.*, vol. COM-30, no. 4, April 1982, pp. 600–614.

74. P. Kroon and E. F. Deprettere, "A Class of Analysis-by-Synthesis Predictive Coders for High Quality Speech Coding at Rates Between 4.8 and 16 kbit/s," *IEEE Journal on Sel. Areas in Comm.* 6 (February 1988):353–363.

75. N. S. Jayant, "High-Quality Coding of Telephone Speech and Wideband Audio," *IEEE Comm. Mag.*, January 1990, pp. 10–20.

76. E. S. K. Chien, D. J. Goodman, and J. E. Russell, "Cellular Access Digital Network (CADN): Wireless Access to Networks of the Future," *IEEE Comm. Mag.*, June 1987, pp. 22–27.
77. N. S. Jayant, V. B. Lawrence, and D. P. Prezas, "Coding of Speech and Wideband Audio," *AT&T Technical Journal* 69 (September/October 1990):25–41.
78. FED-STD 1016, "Telecommunications: Analog to Digital Conversion of Radio Voice by 4800 bit/sec Code Excited Linear Prediction (CELP)," National Communications System, Washington, D.C., November 1989.
79. CCIR Report 646-4, "Digital or Mixed Analogue-and-Digital Transmission of Television Signals," Annex to Vol. XII, *Television and Sound Transmission (CMTT)*, (Geneva: ITU, 1990).
80. R. Kishimoto and I. Yamashita, "HDTV Communication Systems in Broadband Communication Networks," *IEEE Comm. Mag.*, August 1991, pp. 28–35.

Problems

3.1 A signal $s(t) = s_1(t) + s_2(t)$ is to be transmitted as a digital signal, where $s_1(t)$ is band-limited to B_1 Hz and $s_2(t)$ is band-limited to $4B_1$.

   (a) What is the maximum allowable time between samples to ensure that the signal will be faithfully reconstructed by the receiver?

   (b) How does the time between samples for $s(t)$ compare to those respective times for $s_1(t)$ and $s_2(t)$?

3.2 Determine the minimum (Nyquist) sampling rate for the signal $s(t) = 2 \cos (50t + \pi) + 5 \cos (100t + \pi)$.

3.3 A signal with duration 15 μ's is to be sampled at the Nyquist rate. What is maximum frequency allowed in this signal?

3.4 Using sinusiods as your example, show graphically that

   (a) A signal band-limited to 2 kHz is uniquely represented by samples taken every 250 μs.

   (b) If a signal band-limited to 2 kHz is sampled every 750 μs, signals other than the original can be represented by the samples.

3.5 A signal given by the sum of two sinusoids, one at 4 kHz and the other at 8 kHz, is to be filtered and sampled. Draw the waveform

   (a) at the input to the filter,

   (b) at the output of the filter, assuming the filter has a cutoff frequency of 6 kHz,

   (c) and at the output of the sample and hold, assuming a sampling rate of 24 kHz.

3.6 Using a frequency domain representation, show the effect of sampling a signal at $f_s = (4/3) f_{max}$ where f_{max} is the maximum frequency in the signal.

3.7 Consider a signal consisting of three tones: one at 5 MHz, one at 12 MHz, and one at 34 MHz. What frequencies will be present at the output of a PCM decoder if the sampling rate is 20 MHz and the input is unfiltered? (Assume the output filter cutoff frequency is 10 MHz.)

3.8 A signal to be quantized has a range normalized to ±1 and a probability density function $p(x) = 1 - |x|$ with $-1 \leq x \leq 1$.
  (a) Find the quantizer step size and levels for a uniform quantizer with eight levels.
  (b) Find the eight levels for a nonuniform quantizer necessary to make the quantizer levels equiprobable.
  (c) Plot the compressor characteristic for part (b).

3.9 A signal with maximum frequency 12 MHz is to be transmitted with a linear PCM system. The quantization distortion is to not exceed ±0.5 percent of the peak-to-peak signal.
  (a) How many bits per sample are required?
  (b) What is the resulting bit rate if the Nyquist sampling rate is used?

3.10 A linear dequantizer has its output samples offset from the center of a quantization interval by a distance equal to 10 percent of the interval. Find the resulting degradation (in dB) relative to the optimum dequantizer.

3.11 A linear quantizer is to be designed to operate with a gaussian signal.
  (a) Derive an expression for the S/D_q in dB as a function of the number of coding bits n and the loading factor α.
  (b) Plot S/D_q versus loading factor for $n = 6, 7, 8$, and 9, and for S/D_q values from 0 to 50 dB.

3.12 In a compact-disc (CD) digital audio system, 16-bit linear PCM is used with a sampling frequency of 44.1 KHz for each of two stereo channels. (a) What is the resulting data rate? (b) What is the maximum frequency allowed on the input signal? (c) What is the maximum S/D_q ratio in dB? (d) If music has a loading factor $\alpha = 20$, find the average S/D_q in dB. (e) If the total playing time of the CD is 70 minutes, find the total number of bits stored on the disc. Assume that error correction coding, synchronization, and other overhead bits make up one-half of the total capacity of the disc with the remaining one-half dedicated to PCM bits.

3.13 The bandwidth of a TV video plus audio signal is 4.5 MHz. This signal is to be converted into linear PCM with 1024 quantizing levels. The sampling rate is to be 20 percent above the Nyquist rate. (a) Determine the resulting bit rate. (b) Determine the S/D_q if the quantizer loading factor is $\alpha = 6$.

3.14 Improved S/D_q in linear PCM systems is sometimes achieved by oversampling the signal and using a low-pass filter at the receiver to reduce the noise bandwidth. If B_m and B_s are the message and sampled bandwidths, respectively,

   (a) Derive an expression for the distortion power D_q as a function of the maximum input amplitude A, number of coding bits n, B_m and B_s.

   (b) For a sinewave of bandwidth 20 kHz and amplitude A, a sampling frequency of 125 kHz, and 12-bit PCM, find the S/D_q.

3.15 A 1-kHz signal is to be transmitted by linear PCM. The maximum tolerable error in sample amplitudes is 0.5 percent of the peak signal amplitude. The signal must be sampled 20 percent above the Nyquist rate. Determine the minimum possible data rate that must be used.

3.16 A 12-bit linear PCM coder is to be used with an analog signal having range ±10 volts. (a) Find the size of the quantizing step. (b) Find the mean square error of the quantizing distortion. (c) Find the S/D_q for a full-scale sinusoidal input. (d) Find the S/D_q for a ±1 volt sinusoidal input.

3.17 A compressed HDTV signal has a bandwidth of about 6 MHz. What bit rate is required if this signal is to be digitized with uniform PCM at signal-to-quantizing distortion ratio of 35 dB? Assume a sampling rate that is 10 percent greater than the Nyquist rate.

3.18 For a uniform PCM coder with sinewave input signal:
   (a) Derive an expression for the S/D_q where the range of the coder is $\pm A_{max}$ and the range of the sinewave is $\pm A$.
   (b) How much dynamic range is provided with 12 bits per sample and a minimum S/D_q of 33 dB?

3.19 North American PCM channel banks use "zero code suppression" wherein the next to least significant bit of every all-zero code (00000000) is overwritten with a 1. (This 1's insertion is done to maintain good timing content in the PCM code.) Assuming that all codes are equally likely, determine the relative increase in overall quantization noise produced by this process under the following conditions:
   (a) The decoder treats the 1's insertion bit as a voice bit and decodes the PCM sample as eight bits.
   (b) The decoder ignores the 1's insertion bits and generates an output sample corresponding to the middle of the quantization interval defined by the six most significant bits.

3.20 A sinusoidal signal given by $s(t) = 5 \cos 20\pi t$ is to be quantized into 32 uniformly spaced levels over the input voltage range ±5 volts.
   (a) Assuming sampling at the Nyquist rate, what is the resulting bit rate?
   (b) Show a table of code words corresponding to the first eight samples of s(t). Assume the use of a signed binary code.

3.21 For a signal with the same range and probability density function as given in Problem 3.8, find the average number of one's transmitted per code word when using 4-bit linear PCM and (a) a natural binary code; (b) a signed binary code; and (c) an inverted signed binary code.

3.22 A signal with amplitude range of ±100 volts is to be linearly quantized with a step size of 0.5 volts.
   (a) How many bits are required?
   (b) What is the peak signal to quantizing distortion ratio, assuming a uniformly distributed signal?

3.23 Consider a linear 8-level PCM coder designed to operate with a range of ±10 volts. A periodic sawtooth waveform with peak amplitude of ±12 volts is to be sampled and quantized by this coder. Assuming that this signal is sampled eight times in a period, sketch: (a) the coder input and output waveforms; and (b) the resulting quantization and overload distortion.

3.24 A linear PCM system is to be used on signals having the same range as that shown in Figure 3.12. Assume a quantization step size of 2 and use of signed binary coding. Determine the signal-to-quantization noise ratio of each of the following signal values: 423.4, 6908.7, −233.5, and −14.9.

3.25 For a gaussian input to a PCM coder, with a loading factor $\alpha = 4$: (a) Determine the probability that overload will occur; (b) Find the resulting overload distortion.

3.26 Repeat Problem 3.25 but for a laplacian input.

3.27 For an 8-bit linear PCM coder, determine the dynamic range if a minimum S/D_q of 30 dB is required.

3.28 Plot the A-law compressor characteristic of Equation (3.48) versus normalized input $0 \leq x \leq 1$ for $A = 10, 100, 1000$. Compare your curves with Figure 3.9 for μ-law.

3.29 For an A-law compander:
   (a) Derive an expression for the quantizing distortion (D_q) as a function of step size q, A, and average signal power σ^2.

(b) Find an expression for S/D_q as a function of the number of coding bits n and loading factor α. Assume $A = 87.6$ and the compressor overload voltage V is normalized to 1.

3.30 An 8-bit μ-law coder with $\mu = 100$ is designed to operate over a full-scale range of ± 100 volts. Determine the S/D_q: (a) for a full-scale sinusoid; (b) for a 2-volt sinusoid.

3.31 An A-law coder with $A = 10$ is designed to operate over a voltage range of ± 5 volts.

(a) Sketch the complete compressor.

(b) Sketch the complete expander.

(c) On the drawing in part (a), superimpose a 16-level nonlinear characteristic that corresponds to the A-law characteristic.

3.32 Using the μ 255 encoding table given in Table 3.1 for an 8-bit PCM coder with a sampling rate of 8 kHz,

(a) Determine the sequence of code words representing a 2-kHz sinewave with one-fourth maximum power. Assume the phase of the first sample is 45°.

(b) Determine the signal to quantizing distortion ratio (S/D_q) for part (a).

(c) Construct an encoding table, similar to Table 3.1, for an 8-bit linear quantizer with an 8-kHz sampling rate that covers the same range as the μ 255 quantizer. Using that table, determine the sequence of codewords representing the 2-kHz sinewave with one-fourth maximum power.

(d) Determine the S/D_q for part (c).

3.33 Consider a 6-bit logarithmic PCM compressor implemented with piecewise linear segments, with 3 bits used for segment identification and 3 bits for linear quantization within the segment. Let σ (rms value of input) be equal to 1 volt and assume the compressor is exactly matched to $\pm 7 \sigma$.

(a) Draw to scale the resulting compressor characteristic that approximates a logarithmic shape, with successive segments related by powers of 2. Label axes.

(b) Generate a table to show the assignment of signal levels to 6-bit PCM words, using a natural binary code.

(c) For the following input signal, which has a filtered bandwidth of 4 kHz, write the 6-bit PCM words, using the table of part (b), corresponding to periodic samples taken at the minimum required sampling rate.

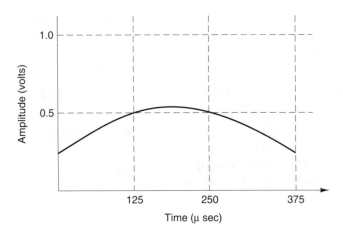

FIGURE p3.33

3.34 The conversion of a linear PCM code to a μ-law PCM code and vice versa can be accomplished by first biasing the linear code. This bias amounts to adding the value 33 to the magnitude of all linear samples. The position of the first one in all linear biased codes indicates the value of the segment (bits $B2$ through $B4$) in the μ-law code. The value of the segment is equal to seven minus the number of leading zeros before the first one. The values of bits $B5$ through $B8$ are equal to the four bits immediately following the leading one. All trailing bits, signified by an X, are ignored when generating the logarithmic code. The general form of linear-biased codes and their corresponding logarithmic codes can be represented in the form of a table.

(a) The linear biased code 0 0 0 0 0 0 0 1 B5 B6 B7 B8 X corresponds to the logarithmic code 0 0 0 B5 B6 B7 B8. Complete the other seven entries of a table to describe the encoding algorithm.

(b) Using the table in part (a), convert the unbiased linear code 0001110001111 (magnitude only) to its equivalent logarithmic code.

(c) The logarithmic code 0 0 0 B5 B6 B7 B8 corresponds to the linear biased code 0 0 0 0 0 0 0 1 B5 B6 B7 B8 1. Complete the other seven entries of a table to describe the decoding algorithm.

(d) Using the table in part (c), convert the logarithmic code 0111001 (magnitude only) to its equivalent unbiased linear code.

3.35 A-law PCM with 8 bits per sample and $A = 87.6$ has a coder input range of ±4096 and step sizes of [2, 2, 4, 8, 16, 32, 64, 128] corresponding to the 8 segments.

(a) Construct a table such as Table 3.1 to show the coding of input signals into A-law PCM and the decoding of A-law PCM into output signals.

(b) If the coder is matched to an input having a maximum peak value of 10 volts, determine the code word generated by a sample value of 1 volt.

3.36 For the A-law PCM coder, show that if the first segment has a ratio of $F(x)/x = 16$, then $A = 87.6$. (Hint: Begin by differentiating Equation (3.48a) to obtain the gradient of the slope within the first segment.)

3.37 Equation (3.56) specifies the S/D_q of delta modulation for a sinewave with a fixed amplitude. The following relationship expresses the S/D_q also as a function of a dynamic range factor R

$$S/D_q = \frac{3\,(f_s)^3}{2\,Rf_m\,(2\pi f_0)^2}$$

where

$$R = (A_{max}/A_{min})^2$$

(a) Plot S/D_q in dB versus R in dB for $f_s = 32$ kHz, $f_m = 3.4$ kHz, and $f_o = 800$ Hz.

(b) Using an 800-Hz sinewave as an input signal, determine the step size and sampling rate of a delta modulator to provide voice encoding with an S/D_q of 398 (26 dB) and a dynamic range of 1000 (30) dB. Assume an output filter cutoff at 3.4 kHz.

3.38 In a delta modulation system, a voice signal is to be sampled at 64,000 samples per second. The maximum signal amplitude is $A_{max} = 1$.

(a) Determine the value of the step size to avoid slope overload.

(b) Determine the quantization distortion D_q if the voice signal bandwidth is 3.5 kHz.

(c) Assuming that the voice signal is a sinusoid, determine the average signal power and S/D_q.

3.39 A delta modulation system operates with a sampling frequency of $f_s = 1/T$ and a fixed step size q. If the input to the system is $s(t) = k^2 t$, (a) Determine the value of k above which a slope overload will occur. (b) Determine the value of k which minimizes the mean-square quantization noise.

3.40 Consider the following binary sequence, which represents the input to a digital to analog converter:

101011111000000101010101

————————> time

For each of the following three decoders, reconstruct the analog waveform represented by this binary sequence.

(a) Log PCM. Assume 8 bits/sample, μ-law companding ($\mu = 255$), with a piecewise linear approximation to $\mu = 255$ using 15 linear segments. Also assume the above binary sequence to exactly represent three 8-bit PCM words. By use of a table, show your assignment of signal levels to PCM words, and then use that table to reconstruct the three analog samples.

(b) Variable Slope Delta. Assume the allowed slope values in the encoder to be 1, 1, 2, 3, 5 (Fibonacci sequence), and assume the encoder examines three bits at a time (current, previous, and pre-previous) to determine the slope magnitude. Using the slope control logic shown in Table 3.5, reconstruct the analog waveform.

(c) Continuous Variable Slope Delta (CVSD). Assume that the allowed change in slope is 3Δ per sample, that the encoder examines three bits at a time to determine slope magnitude, and that Δ is the minimum step size. Using the slope control logic shown in Figure 3.23, reconstruct the analog waveform.

3.41 Assume the input x(t) to a delta modulator is gaussian, white noise, bandlimited from 0 to f_m Hz and σ^2 is the mean square value of $x(t)$. Let the sampling frequency be f_s and the delta mod step size be s. The delta modulator is to be designed to accept a maximum input signal slope of

$$\left(\frac{dx}{dt}\right)_{\text{Max}} = 4\sigma_d$$

where

$$\sigma_d = \frac{d\sigma}{dt}$$

(a) Determine the sf_s necessary to prevent slope overload. (Express answer in terms of f_m and σ.) Hint: The power spectral density $S_{\dot{x}}(f)$

of $\dot{x} = \dfrac{dx}{dt}$ is found by

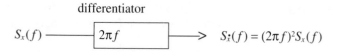

differentiator

$$S_x(f) \longrightarrow \boxed{2\pi f} \longrightarrow S_{\dot{x}}(f) = (2\pi f)^2 S_x(f)$$

(b) Assuming that the quantization error is uniformly distributed over $(0, f_s)$, and assuming that the receiver is bandlimited over $(0, f_m)$, calculate the quantization noise for the delta modulator.

(c) Using results of parts (a) and (b), calculate the signal to quantizing noise ratio (expressed in terms of f_s and f_m only).

3.42 Consider the exponential delta modulator shown in Figure 3.18.

(a) Plot the waveform at the output of the detector, for an input 111100110000.

(b) Consider a linear PCM system with coder range $\pm A$ and average input signal power $A^2/2$. Find the value of P_e, the transmission error rate, for which the quantization distortion (D_q) is equal to the distortion caused by bit errors (D_e). Express your answer in terms of n, the number of bits per PCM sample. Assuming an 8-kHz sampling rate, indicate by table the values of P_e for $D_q = D_e$ corresponding to bit rates of 64, 32, 16, and 8 kb/s. Comment on the relative performance of D_q versus D_e for PCM versus the exponential delta modulator.

3.43 A linear delta modulator is designed to operate with a sampling rate of 32 kHz and a step size Δ of 0.1 volts.

(a) Determine the maximum amplitude of a 2-kHz sinusoid for which the delta modulator is not in a slope overload situation.

(b) Determine the prefiltered S/D_q for the signal described in part (a).

(c) Now determine the S/D_q for the signal described in part (a) if the output signal is filtered with an ideal low-pass filter of bandwidth 4 kHz.

3.44 The use of a double integrator in delta modulation is known to improve the maximum S/D_q over the use of a single integrator. Determine this improvement in dB for sampling rates of 8, 16, 32, and 64 kHz, assuming input of a 2-kHz sinusoidal signal, no slope overload, and a receiver filter bandwidth of 4 kHz.

3.45 Consider delta modulation (DM), differential PCM (DPCM), and linear PCM (LPCM) coders that are all perfectly matched to a sinusoidal input signal.

(a) Find an expression for the ratio $(S/D_q)_{\text{LPCM}}/(S/D_q)_{\text{DM}}$.

(b) Find an expression for the ratio $(S/D_q)_{\text{LPCM}}/(S/D_q)_{\text{DPCM}}$.

3.46 Consider embedded ADPCM as defined by CCITT Rec. G.727.

(a) List all allowed combinations of core (x) and enhancement (y) bits by giving them as (x,y) pairs.

(b) Given the following table of quantizer input/output values for 40-kb/s embedded ADPCM, create the corresponding tables for 32-, 24-, and 16-kb/s embedded ADPCM:

Normalized Quantizer Input Range $d_{ln}(k)$	Quantizer Output $I(k)$
$(-\infty, -1.05)$	0
$(-1.05, -0.05)$	1
$(-0.05, 0.54)$	2
$(0.54, 0.96)$	3
$(0.96, 1.30)$	4
$(1.30, 1.58)$	5
$(1.58, 1.82)$	6
$(1.82, 2.04)$	7
$(2.04, 2.23)$	8
$(2.23, 2.42)$	9
$(2.42, 2.60)$	10
$(2.60, 2.78)$	11
$(2.78, 2.97)$	12
$(2.97, 3.16)$	13
$(3.16, 3.43)$	14
$(3.43, \infty)$	15

3.47 This problem considers the number of quantization levels used in fixed-rate versus embedded ADPCM for like bit rates.

(a) Explain why fixed-rate ADPCM (CCITT Rec. G.726) uses a quantizer with a number of levels given by 2^N-1. What effect does this choice have on the transmission line?

(b) Explain why embedded ADPCM (CCITT Rec. G.727) uses a quantizer with a number of levels given by 2^N. What effect does this choice have on the transmission line?

3.48 Consider a speech subband coder that has the following five subbands:

Subband	Frequency Range (Hz)
1	0–500
2	500–1000
3	1000–2000
4	2000–3000
5	3000–4000

Determine the coder bit rates if the bit allocations for subbands 1 to 5 are: (a) 4,4,2,2,0 bits/sample; (b) 5,5,4,3,0 bits/sample; and (c) 5,5,4,4,3 bits/sample.

4

Time-Division Multiplexing

OBJECTIVES

- Explains the basis of multichannel digital communications systems: time-division multiplexing (TDM)
- Describes the techniques and performance of TDM frame synchronization
- Discusses the multiplexing of asynchronous signals by means of pulse stuffing and transitional coding
- Summarizes the CCITT recommendations that apply to the three different digital multiplex hierarchies
- Describes the multiplex equipment specified by these digital hierarchies
- Considers statistical multiplexing for voice and data transmission
- Summarizes the essential TDM design and performance considerations

4.1 Introduction

The basis of multichannel digital communications systems is **time-division multiplexing (TDM)** in which a number of channels are interleaved in time into a single digital signal. Each channel input is periodically sampled and assigned a certain time slot within the digital signal output. An application of TDM to voice channels is illustrated in Figure 4.1. At the transmit end (multiplex), four voice channels are sequentially sampled by a switch resulting in a train of amplitude samples. The coder then sequentially converts each sample into a binary code using an A/D technique as described in Chapter 3. The coder output is a string of binary digits representing channel 1, channel 2, and so on. These voice bits are combined with framing bits for multiplexer/demultiplexer synchronization and signaling bits for telephone

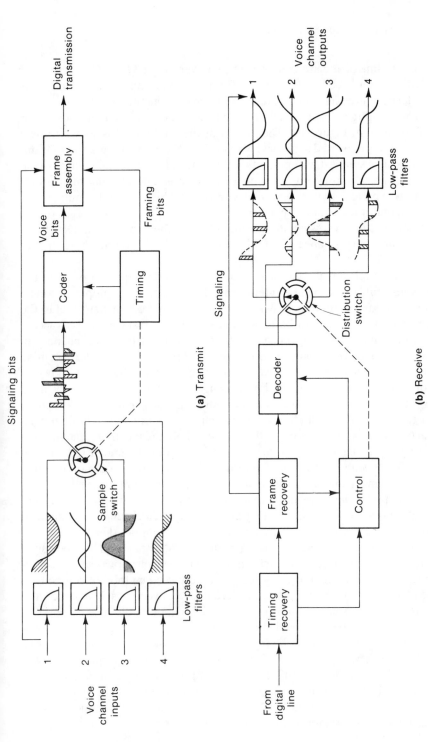

FIGURE 4.1 Block Diagram of Time-Division Multiplexing

179

network control and are then transmitted over a digital line. The receiver or demultiplexer recovers timing and framing synchronization, which allows the decoder to assign each sample to the proper channel output.

The framing operation of a time-division multiplexer/demultiplexer pair is illustrated in Figure 4.2. The output of the multiplexer is shown to be a string of bits allocated successively to each of K channel inputs and to framing bits. The smallest group of bits containing at least one sample from each channel plus framing bits is known as a **frame**. The framing bits form a repetitive pattern that is combined with channel bits within the multiplexer. The demultiplexer must then successfully recognize the contents of the framing pattern to distribute incoming bits to the proper channels. Techniques and performance of TDM frame synchronization are described in Section 4.2.

If a multiplexer assigns each channel a time slot equal to one bit, the arrangement is known as **bit interleaving.** In this case, the multiplexer can be thought of as a commutator switching sequentially from one channel to the next without the need for buffer storage. A second arrangement is to accept a group of bits, making up a word or character, from each channel in sequence. This commonly used scheme is applicable where the incoming channels are character- or word-oriented; hence this technique is known as **word interleaving** or **character interleaving.** This scheme introduces the need for buffering to accumulate groups of bits from each channel while waiting for transfer to the multiplexer. The sampling sequence shown in Figure 4.2 can easily be extended to cover those cases where the incoming channels are not all at the same rate but a fixed relationship exists among the channel rates. The frame length is then determined by the lowest common multiple of the incoming channel rates, as shown in the example of Figure 4.3.

Thus far we have assumed that incoming channels are continuous and **synchronous**—that is, with timing provided by a clock that is common to both the channels and the multiplexer. Channel inputs that are not synchronous with multiplexer timing require additional signal processing. When a digital source is **plesiochronous** (nearly synchronous) with the multiplexer, for example, buffering is a commonly used method of compensating for the small frequency differences between each source and multiplexer. This use of buffering is described in more detail in Chapter 10 where we consider independent clock operation for network synchronization. For digital signals with larger frequency offsets, buffering may prove to be unacceptable because of the large buffer lengths required or, conversely, because of high buffer slip (reset) rate. For **asynchronous** signals, whose transitions do not necessarily occur at multiples of a unit interval, another multiplex interface is more appropriate, most commonly pulse stuffing or less commonly transitional coding.

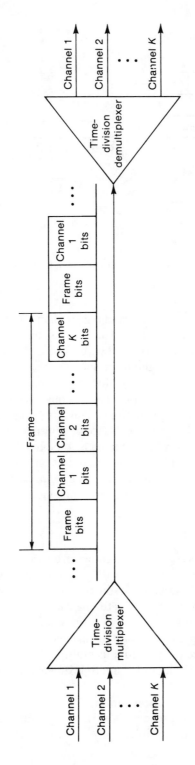

FIGURE 4.2 TDM Frame Structure

181

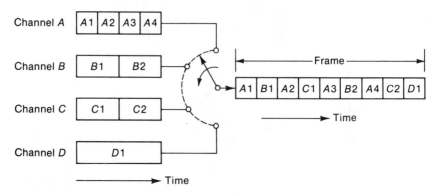

FIGURE 4.3   Multiplexing of Channels Having Different Bit Rates (Adapted from Bylanski and Ingram [1])

The use of TDM to build bigger and bigger channel capacities has led to different digital hierarchies in North America, Japan, and Europe. The different multiplex equipment associated with these digital hierarchies is now standardized, however. Throughout this chapter we will consider examples based on these standards, and Section 4.4 summarizes the applicable CCITT recommendations. In Section 4.5, we consider the case where channel inputs are inactive part of the time. This case suggests a multiplexing scheme where a channel is given access to the multiplexer only during periods of activity, thus allowing more channels to be handled than by conventional TDM techniques. Multiplex designs of this type have been developed for voice transmission, which is called **speech interpolation,** and for data transmission, which is called **statistical multiplexing.**

4.2   Synchronous Multiplexing

A time-division multiplexer may be categorized as providing either synchronous or asynchronous interface with each information channel. A synchronous interface implies that each source that interfaces with the TDM is timed by the same clock that provides timing within the TDM. For this case, there is no ambiguity between the arrival time of each channel bit and the time to multiplex each channel bit. In this section we will describe the design and performance of synchronous multiplexers, and in the following section we will cover techniques for multiplexing asynchronous channels.

4.2.1   Types of Frame Synchronization

The frame synchronization scheme illustrated in Figure 4.4a is characteristic of most time-division multiplexers. A specific code generated by the

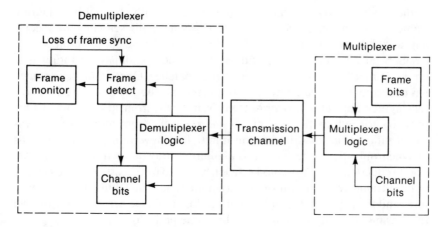

(a) Forward-acting frame synchronization

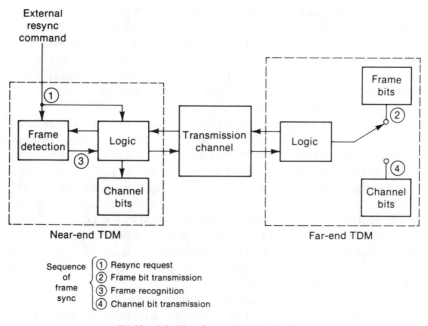

(b) Handshaking frame synchronization

FIGURE 4.4   Types of Frame Synchronization

frame bits is interleaved and transmitted along with channel bits. During frame synchronization the demultiplexer sequentially searches the incoming serial bit stream for a match with the known frame code. Frame search circuitry makes an accept/reject decision on each candidate position based on the number of correct versus incorrect matches. Once frame acceptance has occurred, the selected frame position is continuously monitored for correct agreement. This *forward-acting* frame synchronization scheme thus provides for automatic recognition and recovery from loss of frame synchronization, since the frame monitor initiates a new frame search upon loss of synchronization. There are three inherent drawbacks to this technique: slight reduction in transmission efficiency due to overhead, finite possibility of false frame synchronization due to random data matching the frame pattern, and finite possibility of inadvertent loss of frame synchronization due to transmission errors that corrupt the frame pattern.

A second type of frame synchronization involves a sequence of actions known as *handshaking* (Figure 4.4b). To initiate frame synchronization, a request for synchronization or resynchronization is transmitted by the near-end TDM to the far-end TDM. A preamble of 100 percent framing bits is then transmitted back to the near-end TDM until a frame recognition signal is activated. The transmission of frame bits is then terminated and transmission of 100 percent channel bits is commenced. As long as frame synchronization is maintained, no further action is required. However, loss of synchronization at other levels in a transmission system will also disrupt frame synchronization. Since the demultiplexer has no frame to monitor continuously, this loss in frame synchronization would go unrecognized unless signaled via external means. The advantage of this frame strategy is that once frame synchronization is established, no additional overhead bits are required to maintain synchronization. Moreover, frame acquisition is accomplished more quickly than forward-acting schemes since there exists no ambiguity between frame bits and channel bits.* Once frame is acquired, no accidental loss of synchronization can be caused by the TDM itself, since no frame bits exist to monitor for loss of synchronization. If loss of synchronization occurs due to any other cause, however, this type of TDM must rely on external equipment to recognize the loss of synchronization and command a resynchronization.

An application of handshaking (also called *cooperative*) frame synchronization is a synchronous TDM in which standard low-speed rates (say 1.2

---

*Note that for satellite applications of the handshaking synchronization scheme, propagation delay becomes the dominant contributor to resynchronization times due to the two round trips required through the transmission channel.

kb/s) are combined into a standard high-speed rate (say 4×1.2 kb/s $\rightarrow$ 4.8 kb/s). By not using continuous overhead, the combined rate can be constrained to standard transmission rates (say 4.8-kb/s modem operation over a telephone circuit). Such applications of synchronous multiplexing are an exception, however; the predominant application of TDM warrants the use of forward-acting frame synchronization. The following sections that deal with frame synchronization assume the use of forward-acting schemes, but the techniques of analysis can be applied to the simpler handshaking synchronization scheme as well.

4.2.2 Frame Structures

Frame synchronization bits may be inserted one at a time (**distributed**) or several at a time (**burst** or **bunched**) to form an N-bit frame pattern. Figure 4.5 represents a distributed frame structure where frame bits are inserted one at a time at the beginning or end of the frame. Each frame contains M bits for each of K channels plus one frame bit for a total of $F = KM + 1$ bits. To transmit the complete N-bit frame pattern, a total of NF transmission bits are required, which forms a **multiframe** (sometimes called superframe or major frame). The advantage of distributing frame bits is immunity from the error bursts that might occur, for example, in a radio channel. An error burst of length F bits can affect at most one synchronization bit rather than several or all of the frame bits. The disadvantage of the distributed frame structure is the additional time required to acquire frame synchronization as compared to a bunched frame structure. An example of a multiplexer that employs the distributed frame structure is the 24-channel PCM multiplexer specified by CCITT Rec. G.733 [2].

Figure 4.6 illustrates the bunched frame structure, with the N-bit frame code appearing as contiguous bits at the beginning or end of the multiframe. Note that the total multiframe length is NF bits, so that the ratio of frame bits to total bits per multiframe is identical to that of the distributed frame. The advantage of the bunched frame is that less time is required to acquire frame synchronization when compared to the time required with the distributed frame. One disadvantage lies in the fact that error bursts may effect most or all of a particular N-bit pattern. Another disadvantage is that during the insertion of the frame pattern in N consecutive bit positions, the TDM must store incoming bits from each channel, which results in a design penalty for the extra buffering required in the multiplexer and a performance penalty for residual jitter from demultiplexer smoothing of the reconstructed data. The 2.048-Mb/s PCM multiplexer and the 8.448-Mb/s second-level multiplexer recommended by CCITT Recs. G.732 and G.742, respectively, are examples of multiplexers using bunched frame structure [2].

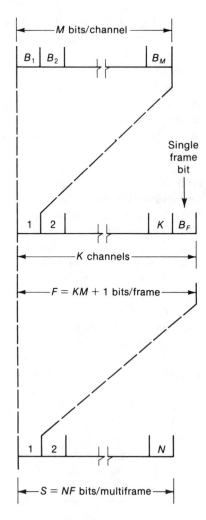

FIGURE 4.5   Distributed Frame Structure

These two definitions of frame structure represent the two extremes of frame organization. In practice, a TDM may use a combination of the two types in organizing frame structure—as, for example the 6.312-Mb/s second-level multiplexer recommended by CCITT Rec. G.743 [2].

4.2.3   Frame Synchronization Modes
TDM frame synchronization circuits commonly employ two modes of operation. The initial mode, called the **frame search** or **frame acquisition mode,**

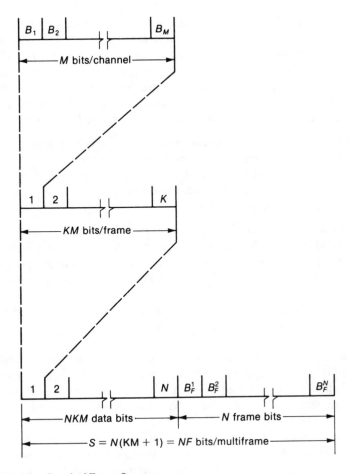

FIGURE 4.6   Bunched Frame Structure

searches through all candidate framing bits until detection occurs. Frame acquisition is declared when a candidate frame alignment position meets the specified acceptance criterion, which is based on correlation with the known frame pattern. With acquisition, the frame synchronization circuit switches to a second mode of operation, called the **frame maintenance mode,** which continuously monitors for correlation between the received framing pattern and the expected framing pattern. If the maintenance mode detects a loss of synchronization as indicated by loss of required correlation, a loss of frame synchronization is declared and the search mode is reentered. A third and sometimes fourth mode of frame synchronization are occasionally used as intermediate modes between search and maintenance.

In general, these additional modes only marginally enhance frame synchronization performance [3, 4]. We limit our treatment here to the more typical two-mode operation, which is depicted in a state transition diagram (Figure 4.7). The strategies employed with the search and maintenance modes, described in the following paragraphs, are considered typical of current TDM design. Variations on these basic approaches will also be discussed.
    Frame synchronization performance is characterized by specifying:

- The time to acquire initial synchronization, or **frame acquisition time**
- The time that frame synchronization is maintained as a function of BER, or **time to loss of frame alignment**
- The time to reacquire frame synchronization after a loss of synchronization, or **frame reacquisition time**

    To be complete, a specification must state the transmission bit error rate and the required probabilities of acquisition, maintenance, or reacquisition within the specified times. Frame synchronization times are usually specified by the statistical mean and sometimes also the variance [4], which will be used here.
    Before proceeding with a description and analysis of frame synchronization modes, it is necessary to state the assumptions:

1. Sequential search is assumed in which only a single frame alignment candidate is under consideration at any given time. This is the most commonly used method of frame search. However, the time to acquire frame synchronization can be reduced with parallel search of all candidate positions in a frame.
2. In the demultiplexer, bit timing is locked to timing of the transmit TDM, so that the time to acquire or maintain frame synchronization is independent of other synchronization requirements (such as bit synchronization).

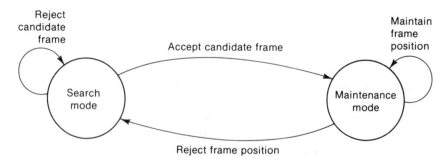

FIGURE 4.7   State Transition Diagram of Two-Mode Frame Synchronization

3. Channel bits are assumed equally likely and statistically independent—that is, for binary transmission, $P(0) = P(1) = \frac{1}{2}$. Therefore the probability that a channel bit will match a frame synchronization bit is equal to $\frac{1}{2}$. Bit errors are also assumed statistically independent, thus allowing use of Bernoulli trials formulation where the bit error probability p_e is constant.

Frame Search Mode
The search mode provides a sequential search comparing each candidate frame alignment position with the known frame pattern until a proper match is found. For the distributed frame structure each candidate position is scanned one bit at a time at F-bit intervals for up to N times, where F is the frame length and N is the number of frames in the pattern. For the bunched frame structure each candidate is scanned by examining the last N bits. The criterion for acceptance of a candidate position is that the N-bit comparison must yield ε or fewer errors, while rejection occurs for more than ε errors. When acceptance occurs, the frame maintenance mode is activated; for rejection the search mode shifts one bit position and tests the next candidate position using the same accept/reject criterion. This operation of the search mode can be conveniently described in terms of a set of states (candidate positions) and probabilities of transitions between these states (accept/reject probabilities), as shown in Figure 4.8. Thus there are four transition probabilities [5]:

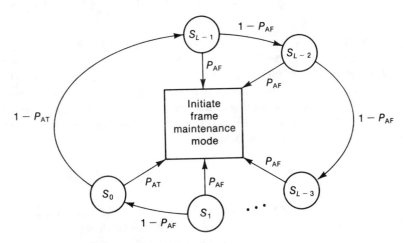

P_{AF} = Prob [accept false candidate]
P_{AT} = Prob [accept true candidate]
S_i = search mode state, i bit positions away from true frame alignment
($i = 0 \rightarrow$ true frame alignment; $i \neq 0 \rightarrow$ false frame alignment)
L = number of candidate alignment positions

FIGURE 4.8   State Transition Diagram of Frame Search Mode

1. Probability of accepting a false position (P_{AF}):

$$P_{AF} = \sum_{i=0}^{\varepsilon} \binom{N}{i}(0.5)^N \qquad (4.1)$$

2. Probability of rejecting a false position (P_{RF}):

$$P_{RF} = 1 - P_{AF}$$

$$= \sum_{i=\varepsilon+1}^{N} \binom{N}{i}(0.5)^N \qquad (4.2)$$

3. Probability of accepting the true position (P_{AT}):

$$P_{AT} = \sum_{i=0}^{\varepsilon} \binom{N}{i}(1 - p_e)^{N-i}p_e^i \qquad (4.3)$$

4. Probability of rejecting the true position (P_{RT}):

$$P_{RT} = 1 - P_{AT}$$

$$= \sum_{i=\varepsilon+1}^{N} \binom{N}{i}(1 - p_e)^{N-i}p_e^i \qquad (4.4)$$

The probability of acquiring true alignment during the first search of all possible candidates is the probability of rejecting all random bit positions and accepting the true frame position. Should no match be found during the first search, even at the true frame position, additional searches are made until a match is found. Assuming that search begins with the first (channel) bit of the frame, the probability that the synchronization position selected by the search mode is the true position is given by

$$P_T = P_{RF}^{L-1}P_{AT} + (P_{RF}^{L-1})^2 P_{AT}(1 - P_{AT}) + (P_{RF}^{L-1})^3 P_{AT}(1 - P_{AT})^2 + \cdots \qquad (4.5)$$

where L is the number of candidate positions. For the distributed frame structure the number of candidates is $L = F$; for the bunched frame structure the number of candidates is $L = NF$. The geometric series in (4.5) reduces to

$$P_T = P_{RF}^{L-1}P_{AT}(1 + c + c^2 + \cdots)$$

$$= \frac{P_{RF}^{L-1}P_{AT}}{1 - c} \qquad (4.6)$$

where

$$c = P_{\text{RF}}^{L-1}(1 - P_{\text{AT}}) \tag{4.7}$$

Similarly, the probability of the search mode inadvertently acquiring a false frame position is found to be

$$P_{\text{F}} = \frac{1 - P_{\text{RF}}^{L-1}}{1 - c} \tag{4.8}$$

From (4.6) and (4.8), we find that the total probability of acquisition in the search mode is, as expected,

$$P_{\text{F}} + P_{\text{T}} = 1 \tag{4.9}$$

Frame Maintenance Mode
Two commonly used schemes for the frame maintenance mode are examined here. In both schemes, each frame pattern is first tested against a pass/ fail criterion, similar or identical to the accept/reject criterion used in the search mode. Either scheme must be able to recognize actual loss of frame alignment and initiate frame reacquisition. If a maintenance mode scheme were employed that rejected the frame position based on a single failed test, however, the time to hold true frame synchronization would probably be unacceptably small in the presence of transmission bit errors. For this reason, the maintenance mode reject criterion is based on more than one frame pattern comparison, which results in lower probability of inadvertent loss of frame synchronization. This protection against false loss of synchronization is known as the *hysteresis* or *flywheel* provided by the maintenance mode. The first scheme to be considered requires that r successive tests fail before the search mode is initiated. The second scheme is based on an up-down counter, which is initially set at its highest count, $M + 1$. For each test of the received frame pattern, agreement results in an increment of 1 and disagreement results in a decrement of D. A loss of frame is declared and the search mode initiated if the counter reaches its minimum (zero) state.

For the first scheme, we have to calculate the number of tests required before r successive misalignments are identified. This calculation can be done using results from Feller, who derived expressions for the mean μ_n and variance σ_n^2 of the number of tests n required to observe runs of length r [6]:

$$\mu_n = \frac{1 - p^r}{qp^r} \tag{4.10}$$

$$\sigma_n^2 = \frac{1}{(qp^r)^2} - \frac{2r + 1}{qp^r} - \frac{p}{q^2} \tag{4.11}$$

Here p is the probability of misalignment detected in any single test and $q = 1 - p$.

For the second scheme employing an up-down counter, the calculation of number of tests to detect misalignment is facilitated with the state diagram of Figure 4.9. This figure shows that if the counter value is within D counts of zero, then the counter moves to state zero when frame disagreement occurs. Similarly, if the counter is already at its maximum value, succeeding frame agreements leave the counter at its maximum value. Once the counter arrives at its zero state, the search mode is activated. In general, the calculation of the number of tests n required before the zero state is reached is difficult and requires numerical techniques. For $D = 1$, however, the problem reduces to the classic simple random walk, which has been analyzed by several authors [6, 7, 8]. These results may be used to find an analytic expression for the expected value of n:

$$\mu_n = \frac{M}{p - q} + \frac{q^{M+1}}{p^M(q - p)^2}\left[1 - (p/q)^M\right] \quad (p \neq q) \tag{4.12}$$

$$= M(M + 1) \quad (p = q = 0.5) \tag{4.13}$$

where q is the probability of an up count (frame agreement) and $p = 1 - q$ is the probability of a down count (frame disagreement). For a general D, recursive methods exist that are suitable for numerical solution by computer. One such approach is to note that the up-down counter as shown in Figure 4.9 is a finite Markov chain and take advantage of the well-known properties of finite Markov chains to obtain the mean, variance, and even probability density function of n for general choice of p, M, and D [9].

We are also interested in specifying the probability of detection. Since the number of tests is a sum of independent random variables, the central limit theorem may be invoked, which asserts that the number of tests is gaussian-distributed with mean μ_n and variance σ_n^2. Hence the number of tests to detect frame misalignment with 90 percent probability, for example, is given by

$$n_{0.9} = \mu_n + 1.28\sigma_n \tag{4.14}$$

and for 99 percent probability

$$n_{0.99} = \mu_n + 2.33\sigma_n \tag{4.15}$$

To specify the number of bit times to detection of misalignment, we note that each maintenance mode test of a frame pattern occurs at regularly

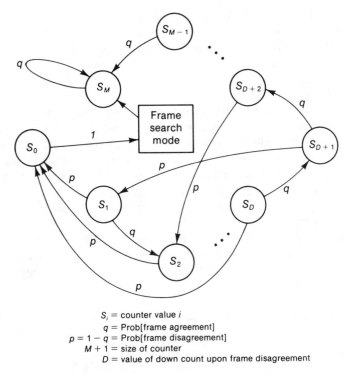

S_i = counter value i
q = Prob[frame agreement]
$p = 1 - q$ = Prob[frame disagreement]
$M + 1$ = size of counter
D = value of down count upon frame disagreement

FIGURE 4.9   State Diagram of Maintenance Mode Using Up-Down Counter

spaced intervals. For the distributed and bunched frame structures of Figures 4.5 and 4.6, respectively, each test occurs every NF bits, since there are exactly NF bits between successive frame words for either the distributed or bunched frame structure. Thus Expressions (4.10) through (4.15) given here for number of tests to misalignment detection may be converted to bit times by using the multiplying factor of NF bit times per test.

Detection of Misalignment
Suppose a true loss of frame synchronization has occurred. The maintenance mode will then be examining random data rather than the true frame pattern. The probability of misalignment detection in a particular frame pattern test depends on the length of the frame pattern and the accept/reject criterion and is specified in Equation (4.2). The number of such tests required before the maintenance mode declares loss of frame is then dependent on the scheme employed. The necessary calculations are made by using appropriate expressions selected from (4.10) through (4.15), where p is equivalent to P_{RF} and n is the number of tests to detect misalignment.

Table 4.1 presents selected examples of performance for the two maintenance mode schemes considered here. To convert these results from number of tests to number of bit times, recall that there are NF bit times per test in both the bunched and distributed frame structure.

Frame Alignment Time

From Figure 4.7 we see that the time to true frame synchronization involves time spent in both the search and maintenance modes. Each time the search mode accepts a candidate position, the maintenance mode is entered. If false alignment is acquired by the search mode, the maintenance mode will eventually reject that false alignment and return to the search mode. The search mode then accepts another candidate position and reenters the maintenance mode. This cycle continues until true frame alignment is found in the search mode.

As indicated earlier, the time to true alignment is usually specified by its mean value, which we will derive here for both the bunched and distributed frame structures. First suppose that true frame alignment has been lost and

TABLE 4.1   Performance of Frame Misalignment Detection

(a) Scheme 1: r successive frame word errors result in reframe.

Detection Probability in Each Test (P_{RF})	Successive Tests Resulting in Frame Rejection (r)	Number of Tests to Detect Misalignment			
		Mean (μ_n)	Variance (σ_n^2)	90% Probability	99% Probability
0.5	2	6	22	12.0	16.9
0.5	4	30	734	64.7	93.1
0.5	6	126	14,718	281.3	408.7
0.5	8	510	257,790	1160	1693

(b) Scheme 2: Count up 1 for correct frame word, down D for incorrect frame word. Counter size M = 15. Reframe occurs at state zero.

Detection Probability in Each Test (P_{RF})	Value of Down Count upon Frame Disagreement (D)	Number of Tests to Detect Misalignment			
		Mean (μ_n)	Variance (σ_n^2)	90% Probability	99% Probability
0.5	2	27.5	191	45.2	59.7
0.5	4	10.1	23.8	16.4	21.5
0.5	6	6.4	10.2	10.5	13.8
0.5	8	4.5	7.4	8.0	10.8

that detection of loss of frame alignment has taken place in the maintenance mode. We are now interested in the time for the combined action of the search and maintenance modes to *acquire* true frame alignment. The time to *reacquire* frame alignment can then be obtained by summing the time to detect misalignment with the time to acquire true alignment.

Bunched Frame Structure.   Assuming a random starting position, the search mode examines an average of $(NF - 1)/2$ false positions, each with a probability P_{AF} of being accepted, before reaching the true alignment position. For those positions that are rejected, the search mode shifts one bit position and examines the next candidate position, so that eventually all $(NF - 1)/2$ positions are examined. On average, however, there will be $(P_{AF})(NF - 1)/2$ false candidates accepted. For each of these there will be a mean number of tests to reject false synchronization, μ_n, in the maintenance mode, with NF bits per maintenance mode test for the bunched frame structure. The mean number of bits to reach the true frame alignment position, μ_T, is then the sum of the time to shift through $(NF - 1)/2$ positions and the time for the maintenance mode to reject all false frame alignment positions:

$$\mu_T = \left(\frac{NF - 1}{2}\right)\left[1 + (P_{AF})(\mu_n)(NF)\right] \text{ bit times} \tag{4.16}$$

The *maximum average frame acquisition time* is also sometimes used as a specification, in which case the search mode is assumed to start at maximum distance from the true frame position. In this case

$$(\mu_T)_{\max} = (NF - 1)\left[1 + (P_{AF})(\mu_n)(NF)\right] \text{ bit times} \tag{4.17}$$

Distributed Frame Structure.   Assuming a random starting position, the search mode examines an average of $(F - 1)/2$ false positions before reaching the true alignment position. The time to test each candidate is NF bit times, so that $NF(F - 1)/2$ bit times are required to test $(F - 1)/2$ candidates. For the distributed frame structure, we assume that any cyclic version of the N-bit frame pattern will be accepted by the search mode. Therefore each false alignment position has a probability NP_{AF} of being accepted. Then, on the average, there will be $(NP_{AF})(F - 1)/2$ false candidates accepted. For each of these there will be a mean number of tests in the maintenance mode to reject false positions, μ_n, with NF bits per maintenance mode test. Thus the mean number of bits to reach true frame alignment is

$$\mu_T = NF\left(\frac{F - 1}{2}\right)(NP_{AF}\mu_n + 1) \tag{4.18}$$

The maximum average frame acquisition time is then

$$(\mu_T)_{max} = NF(F-1)(NP_{AF}\mu_n + 1) \tag{4.19}$$

Example 4.1 _____

The 2.048-Mb/s PCM multiplex equipment specified by CCITT Rec. G.732 contains a 7-bit bunched frame alignment word that is repeated every 512 bits (see Figure 4.31). Frame alignment requires the presence of the correct frame alignment word without error. Frame alignment is assumed lost when three consecutive frame words are received in error. Assuming a random starting position, find the mean time for the demultiplexer to acquire frame alignment. Assume the transmission error rate to be zero.

Solution   For each test of a false frame alignment word, the probability of acceptance is given by (4.1):

$$P_{AF} = \left(\frac{1}{2}\right)^7 = 7.8125 \times 10^{-3}$$

The number of tests for rejection of false alignment by the maintenance mode is found from (4.10) with $p = P_{RF} = 1 - P_{AF}$ and $q = P_{AF}$:

$$\mu_{RF} = \frac{1 - (0.9921875)^3}{(7.8125 \times 10^{-3})(0.9921875)^3} = 3.05 \text{ tests}$$

Then from (4.16),

$$\mu_T = \frac{511}{2}\left[1 + (7.8125 \times 10^{-3})(3.05)(512)\right]$$

$$= 3373 \text{ bit times}$$

or

$$\mu_T = \frac{3373 \text{ bit times}}{2.048 \times 10^6 \text{ b/s}} = 1.65 \text{ ms}$$

Note that from (4.17) the maximum average frame acquisition time would then be 3.3 ms.

_____

False Rejection of Frame Alignment
In the presence of transmission errors, frame bits may be corrupted, causing the maintenance mode to declare loss of frame and to initiate realignment in the search mode. Calculation of the time between such false rejections begins with the probability of false rejection in a particular test, given by

(4.4). For the maintenance mode scheme that requires r consecutive test rejections, the mean and variance of the number of tests to false rejection are given by (4.10) and (4.11), with $p = P_{RT}$. For the maintenance mode scheme employing an up-down counter, the statistics of the number of tests to false rejection can be obtained by recursive techniques using the Markov chain model as previously referenced. Once again invoking the central limit theorem, we note that the number of tests is gaussian-distributed so that (4.14) and (4.15) apply for the 90 percent and 99 percent probabilities. Because false resynchronization occurs infrequently, however, the mean is commonly the only statistic used as a specification. Using the statistical mean, false resynchronization performance is illustrated in Figures 4.10 and 4.11 for the two maintenance mode schemes.

Example 4.2 _____

For the 2.048-Mb/s multiplexer described in Example 4.1, find the mean time between loss of frame alignment due to a transmission error rate of 10^{-3}.

Solution   During operation of the maintenance mode, the probability of rejecting the true alignment word is found from (4.4):

$$P_{RT} = 1 - P_{AT}$$
$$= 1 - (1 - p_e)^7 = 7 \times 10^{-3}$$

The mean number of tests to reject the true alignment pattern is found from (4.10) with $p = P_{RT}$:

$$\mu_{RT} = \frac{1 - (P_{RT})^3}{(1 - P_{RT})(P_{RT})^3}$$

$$\approx \frac{1}{(7 \times 10^{-3})^3} = 2.9 \times 10^6 \text{ tests}$$

Since there are 512 bits between frame words,

$$\mu_{RT} \text{(bits)} = (2.9 \times 10^6 \text{ tests})(512 \text{ bits/test})$$
$$= 1.5 \times 10^9 \text{ bits}$$

The mean time in seconds to loss of frame alignment is then

$$\mu_{RT} \text{(seconds)} = \frac{1.5 \times 10^9 \text{ bits}}{2.048 \times 10^6 \text{ b/s}} = 732 \text{ s}$$

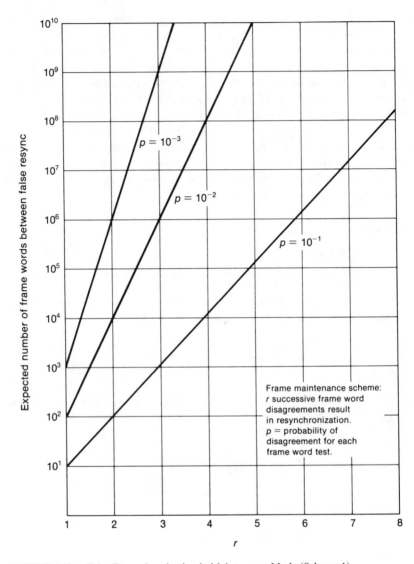

FIGURE 4.10   False Resynchronization in Maintenance Mode (Scheme 1)

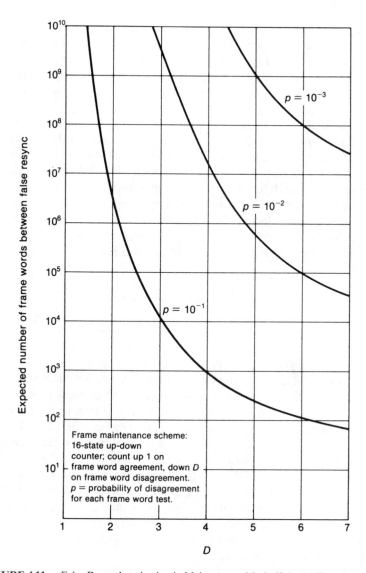

FIGURE 4.11   False Resynchronization in Maintenance Mode (Scheme 2)

Variations of Search and Maintenance Modes
Variations of the basic search or maintenance mode strategy described above have been utilized to improve performance—that is, to minimize frame synchronization times or occurrences of spurious rejections. The choice of frame synchronization strategy is determined by a tradeoff between complexity and performance.

One technique that reduces frame acquisition time is a parallel search of all candidate frame alignment positions. In parallel, each of the candidates is continuously checked against the accept/reject criterion; failed candidates are eliminated until only one candidate remains. Parallel search is clearly faster than serial search but requires additional storage and comparator circuitry. The increased speed in frame acquisition depends on how quickly all false candidates are rejected. The probability of any one false candidate being rejected in n or less comparisons is $1 - P_{AF}{}^n$. For an F-bit frame, the probability of rejecting all $(F - 1)$ false candidates in n or less comparisons is

$$\text{Prob[rejection time} \leq n] = (1 - P_{AF}^n)^{F-1} \qquad (4.20)$$

Of course, after all false candidates have been eliminated, the remaining position is accepted as the true frame alignment, so that (4.20) is also the probability of true frame acquisition in n or less comparisons.

To minimize false (spurious) losses of synchronization, a maintenance mode can lock to the old frame position while the search mode looks for the actual frame position. Upon declaration of loss of frame, frame is held at the old frame position while in parallel a search mode is initiated. The old frame position is held until the search mode locates the correct frame position. By this process, if a spurious loss of frame occurs the search mode will return to the same position that was held when loss of frame was declared. No loss of synchronization occurs for this process, since the true frame position is held during the search mode. Moreover the time to reacquire true frame alignment is reduced if the search mode starts with examination of bit positions immediately adjacent to the old frame position, since these are the most likely true frame positions.

Finally, the choice of type of search and maintenance mode and parametric values for each mode involves a tradeoff between quick synchronization times and immunity to transmission errors and false synchronization patterns. In the search mode, performance is determined by the frame structure (distributed versus burst), by the frame pattern length (and code itself as shown in Section 4.2.4.), and by the number of errors allowed in each comparison. Expressions (4.1) to (4.9) can be used to evaluate these tradeoffs for the search mode. Maintenance mode performance is usually determined by the same criteria as the search mode plus the number of frame miscompares required before loss of frame is declared. Table 4.1 and Figures 4.10 and 4.11 can be used to trade off misalignment detection with false resynchronization in the maintenance mode.

4.2.4   Choice of Frame Pattern

For the distributed frame structure of Figure 4.5, each test of a candidate frame position is independent of previous tests. Further, each candidate position contains either all channel bits or all frame bits. In this case, the probability of random bits exactly matching an N-bit framing pattern is 2^{-N}. This probability can be made arbitrarily small simply by lengthening the pattern length, but at the penalty of reduced transmission efficiency and increased synchronization times. If the channel bits are assumed to be random with equally likely levels—that is, $P(0) = P(1) = \frac{1}{2}$—then any choice of frame pattern is equally good for a selected length N.

In contrast, for the bunched frame structure of Figure 4.6, successive tests for the frame pattern are not independent and certain tests will contain a mix of channel bits and framing bits. If the overlapping positions are in agreement, the probability of accepting a false position is increased, thus degrading overall performance. Consider Figure 4.12, for example, which shows a TDM that is out of frame alignment by exactly one bit time. For the candidate 7-bit pattern being examined, six of the positions overlap with the true alignment pattern while one does not. Since the code selected for this example consists of all 1's, the six overlapping positions all match with the true frame pattern, and the seventh (channel bit) position will also match if

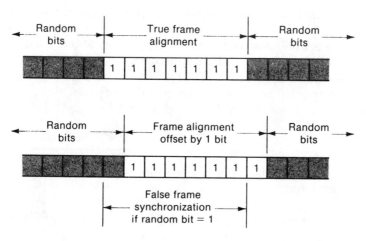

FIGURE 4.12   False Synchronization for Bunched Frame

it is a 1. Thus the probability of frame pattern imitation has been reduced to ½ rather than $(½)^N$ as would be the case for purely random bits. This argument can easily be extended to other degrees of overlap and for other poor choices of frame alignment pattern.

Certain framing patterns have been discovered that minimize the degree of agreement for shifted versions of the frame pattern. The autocorrelation function $R_x(k)$ measures this property of a frame pattern. If the pattern is represented by $x_1, x_2, \ldots, x_m$, then it is desired that the autocorrelation

$$R_x(k) = \sum_{i=1}^{m-k} x_i x_{i+k} \qquad \text{for } x_i = \pm 1 \qquad (4.21)$$

be small for $k \neq 0$. R. H. Barker investigated codes that have the property

$$|R_x(k)| \leq 1 \qquad k \neq 0 \qquad (4.22)$$

and found three such codes of length 3, 7, and 11. When represented with 1's and 0's, these are [5]:

110
1110010
11100010010

For each Barker code, the complement, reflected (mirror image), and reflected complement versions also obey Expression (4.22) and thus are also good synchronization codes. M. W. Williard has also derived a set of codes that exhibit similar good synchronization properties, although they do not meet the Barker property of (4.22); however, Williard has shown that the probability of false synchronization in any overlap condition is less than the probability of false synchronization due to random bits [10].

Example 4.3 _____

Show that 7-bit code 1110010 meets the autocorrelation property of (4.22) necessary for Barker codes.

Solution   To determine its autocorrelation, this 7-bit binary code must first be rewritten as the sequence $x_1, x_2, \ldots, x_7$:

$$1110010 \rightarrow + + + - - + -$$

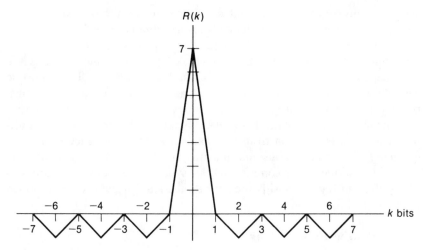

FIGURE 4.13   Autocorrelation of 7-bit Barker Code

Now applying (4.21) we find the autocorrelation as a function of k:

$k = 0$ $R(0) = 7$
$k = 1$ $R(1) = 0$
$k = 2$ $R(2) = -1$
$k = 3$ $R(3) = 0$
$k = 4$ $R(4) = -1$
$k = 5$ $R(5) = 0$
$k = 6$ $R(6) = -1$

which is sketched in Figure 4.13. Since the autocorrelation $R(k)$ is an even function, this code is in fact a Barker code.

4.3   Asynchronous Multiplexing

If each digital source is timed by its own internal clock, the interface to a TDM is said to be asynchronous. For this case, slight differences between the source clock and TDM clock arise due to clock inaccuracies and instabilities; the two clocks then wander apart and eventually cause a bit to be added or deleted in the TDM, resulting in loss of synchronization. To compensate for this lack of synchronization between each source and the TDM, the technique known as **pulse stuffing*** is commonly used to convert each

---

*Other terms also used are **justification** (by the CCITT) and **bit stuffing.**

asynchronous source to a rate that is synchronous with the TDM clock frequency. Each digital source is clocked at some nominal rate f_i that ranges to $f_i \pm \Delta f_i$, where the tolerance $\pm \Delta f_i$ is dependent on clock performance parameters but is always much smaller than f_i. A typical tolerance value is ± 50 parts per million (ppm) for rates of 1.544 or 2.048 Mb/s. The TDM must be capable of accepting each nominal input rate over its full range and must provide a conversion of each channel input to a rate that is synchronous with the internal clock of the TDM. As illustrated in Figure 4.14, pulse stuffing provides a conversion from the nominal source frequency f_i to a frequency f_i' for synchronous multiplexing in the TDM. Two commonly applied techniques for accomplishing this synchronous conversion are **positive pulse stuffing** and **positive-negative pulse stuffing,** which are described in the following sections.

4.3.1   Positive Pulse Stuffing
With positive pulse stuffing each channel input rate f_i is synchronized to a TDM channel rate f_i' that is higher then the maximum possible frequency of the channel input, $f_i + \Delta f_i$. Figures 4.15 and 4.16 show typical block diagrams of the transmit and receive sections, respectively, of a positive pulse stuffing TDM. Channel input bits are written into a digital buffer at a rate of f_i and read out of the buffer at a rate f_i'. This buffer is often referred to as an **elas-**

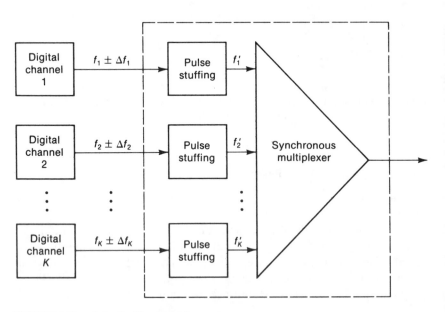

FIGURE 4.14   Pulse Stuffing Multiplexer

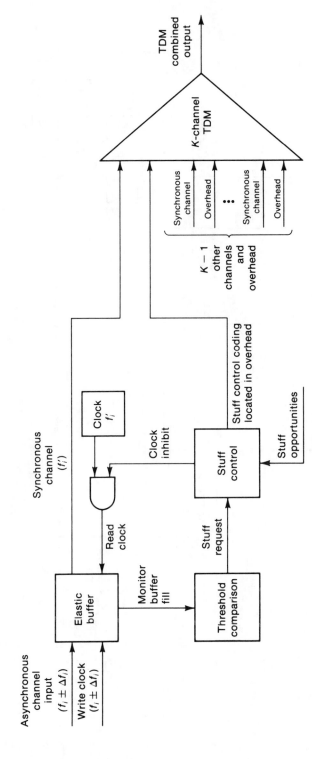

FIGURE 4.15 Block Diagram of Positive Pulse Stuffing Multiplexer

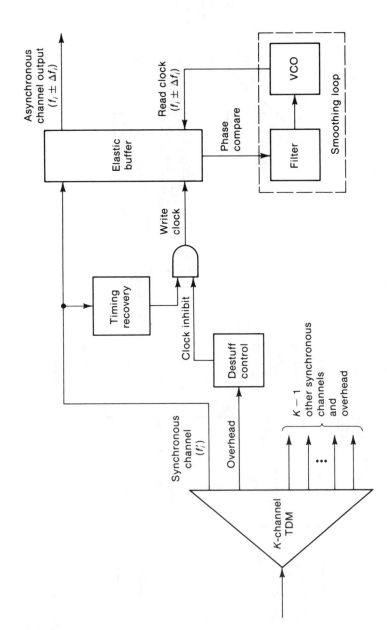

FIGURE 4.16   Block Diagram of Positive Pulse Stuffing Demultiplexer

tic buffer, since the length of its contents varies with differences between the write and read clocks. Since f_i' is at a rate slightly higher than f_i, there is a tendency to deplete the buffer contents. To avoid emptying the buffer, the buffer fill is monitored and compared to a preset threshold. When this threshold is reached, a request is made to stuff a time slot with a stuff (dummy) bit. At the next available stuff opportunity, the read clock is inhibited for a single clock pulse, allowing a stuff bit (or bits) to be inserted into a designated time slot within the synchronous channel while the asynchronous channel input continues to fill the buffer. Prior to the actual stuffing operation, the precise location of the stuffed time slot is coded into the overhead channel and transmitted to the receive end (demultiplexer) of the TDM. Here the received channel bits are written into an elastic buffer, but stuff bits are prevented from entering the buffer by the overhead channel, which drives a clock inhibit circuit. Bits are read out of the buffer by a smoothed clock. This smoothed clock is derived by using the buffer fill, which is a measure of the difference between phases of the write and read clocks and thus can be used as a phase detector in a clock recovery circuit. The smoothing loop is achieved by using the buffer fill to drive a phase-locked loop, which consists of a filter plus voltage-controlled oscillator.

The timing relationship between the asynchronous channel input and synchronous TDM channel is illustrated in Figure 4.17. In this figure, which

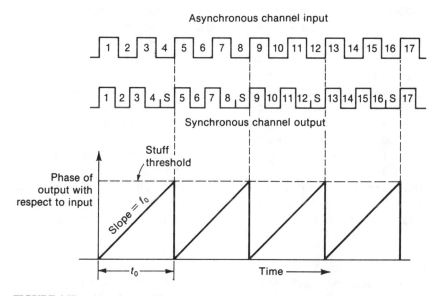

FIGURE 4.17   Asynchronous/Synchronous Timing Relationships for Positive Pulse Stuffing

exaggerates the typical frequency difference for the sake of illustration, each synchronous output bit has a period that is four-fifths the period of the asynchronous input bit so that the instantaneous output rate is 1.25 times the input rate. Every fifth bit of the synchronous output is converted to a stuff (dummy) bit. Because of the rate difference, the phase of the output signal increases linearly with respect to the input signal. When this phase difference accumulates to 1 bit, the phase is reset by inserting a dummy bit within the synchronous channel output.

4.3.2 Positive-Negative Pulse Stuffing

In a positive-negative pulse stuffing technique, each channel input rate f_i is synchronized to a TDM channel rate that is identical to the nominal channel input rate. Buffering of the input is provided just as shown in Figure 4.15. Now the buffer contents may deplete or spill, however, since both the channel input rate and TDM channel rate may have either negative or positive offset and drift about the nominal frequency f_i. If the channel input rate is lower than the multiplexer channel rate, the buffer tends to deplete and for this case stuff bits are added (positive stuff); if the input rate is higher, the buffer tends to spill and here information bits are subtracted (negative stuff) from the channel input and transmitted as part of the overhead channel. For positive-negative pulse stuffing, two actions may occur with each stuff opportunity: positive stuff or negative stuff (or spill). In a similar scheme, called **positive-zero-negative pulse stuffing,** three actions may occur with each stuff opportunity: no stuff, positive stuff, or negative stuff. Stuff/spill locations are coded and spilled information bits are added to an overhead channel to allow the receiving end to properly remove positive stuff bits and add negative stuff information bits.

Because the multiplexer channel rate is equal to the nominal asynchronous channel rate, synchronous operation is easily facilitated by providing common clock to the TDM and digital source and ignoring the stuff/spill code since it will be inactive. Moreover, since stuffing circuitry is independent for each channel, synchronous channels can be mixed with asynchronous channels. These advantages are not found with positive pulse stuffing because of the frequency offset employed with the multiplexer channel rate. Although positive pulse stuffing is prevalent among higher-order multiplexers, there are positive-zero-negative stuffing schemes standardized by the CCITT [2].

4.3.3 Stuff Coding

Pulse stuffing multiplexers provide each channel with one time slot for stuffing in each frame or multiframe. This time slot contains either a dummy bit when stuffing has occurred or an information bit when no stuffing is required. Each time slot that is available for stuffing has an associated stuff control bit or code that allows the demultiplexer to interpret each stuffing time slot properly. Incorrect decoding of a stuffing time slot due to transmission errors causes loss of bit count integrity in the associated channel since the demultiplexer will have incorrectly added or deleted a bit in the derived asynchronous channel. To protect against potential loss of synchronization, each stuffing time slot is redundantly signaled with a code, where two actions (stuff/no stuff) are signaled for a positive pulse stuffing TDM and three actions (no stuff/positive stuff/ negative stuff) are signaled for a positive-zero-negative pulse stuffing TDM. At the receiver, majority logic applied to each code word determines the proper action. As an example, a 3-bit code is commonly used for positive pulse stuffing in which the code 111 signals stuffing and 000 signals no stuffing. In this case, if two or more 1's are received for a particular code, the associated stuffing time slot is detected as a stuffing bit whereas two or more 0's received results in detection of an information bit in the stuffing time slot.

Performance is specified by the probability of decoding error per stuff code, and from this probability the time to loss of synchronization due to incorrect decoding of a stuff code may be obtained. If we assume independent probability of error from code bit to code bit, then the probability of incorrect decoding for a particular stuff code is given by the binominal probability distribution:

$$\text{Prob[stuff code error]} = P(\text{SCE}) = \sum_{i=(N+1)/2}^{N} \binom{N}{i} p_e^i (1 - p_e)^{N-i} \qquad (4.23)$$

where

N = code length
P_e = average probability of bit error over transmission channel

For small probability of bit error, this expression can be approximated by

$$P(\text{SCE}) \approx \binom{N}{x} p_e^x \qquad (4.24)$$

where $x = (N + 1)/2$. For the case of a 3-bit code,

$$P(\text{SCE}) \approx 3p_e^2 \qquad (4.25)$$

Knowing the rate of stuffing opportunities, f_s, given in stuffs per unit time, the mean time to loss of BCI due to stuff code error can be calculated as

$$\text{Mean time to loss of BCI} = \frac{1}{f_s P(\text{SCE})} \qquad (4.26)$$

Stuff code bits are also distributed within the frame to protect against error bursts and to minimize the number of overhead bits inserted in the information channel at any one time. By distributing the code bits, the probability of an error burst affecting all bits of a particular code is reduced, which therefore increases the effectiveness of the redundancy coding. Distributed code bits also result in smaller buffer sizes required to store information bits during multiplexing and demultiplexing and in improved jitter performance. In the demultiplexer, the clock recovery circuit for each channel must smooth the gaps created by deletion of overhead bits. The jitter resulting from this smoothing process is minimized when overhead bits are distributed rather than bunched.

4.3.4   Pulse Stuffing Jitter
The removal of the stuffed bits at the demultiplexer causes timing jitter at the individual channel output, which has been called **pulse stuffing jitter.** As shown in Figure 4.17, pulse stuffing jitter is a sawtooth waveform with a peak-to-peak value of 1 bit interval and a period t_0 seconds. The *actual stuffing rate* occurs at average frequency $f_0 (= 1/t_0)$ stuffs per second and is equal to the difference between the multiplexer synchronous rate and the channel input rate. The slope of the sawtooth in Figure 4.17 thus equals the actual stuffing rate. The *maximum stuffing rate* f_s is a design parameter whose value is selected to minimize pulse stuffing jitter. Moreover, the maximum stuffing rate sets a limit on allowable frequency difference between the channel input and multiplexer clocks. This relationship is described by the stuffing ratio ρ, defined as the ratio of the actual stuffing rate to the maximum stuffing rate:

$$\rho = \frac{f_0}{f_s} \leq 1 \qquad (4.27)$$

For pulse stuffing jitter to be the periodic sawtooth waveform shown in Figure 4.18a, stuff bits need to be inserted at the specific instant that the phase error crosses threshold. Only discrete time slots are available for stuff bits at the maximum stuffing rate, however, and phase threshold crossings do not always occur at these times. The frequency offset f_0 tends to change slowly in time due to small differences between the multiplexer and channel

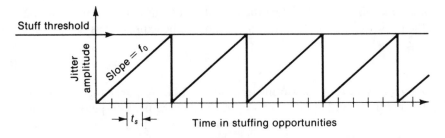

(a) Pulse stuffing jitter without waiting time jitter ($\rho = \frac{1}{5}$)

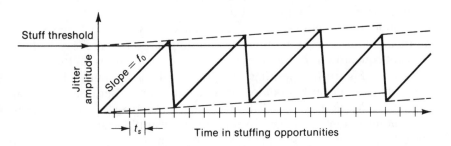

(b) Pulse stuffing jitter with superimposed waiting time jitter as indicated by dashed envelope ($\rho = \frac{1}{5} +$)

FIGURE 4.18   Waiting Time Jitter

input clocks. With these changes in f_0, the interval between threshold crossings also varies. In the case of Figure 4.18b, stuffing usually occurs once in every five opportunities, but occasionally at every four opportunities. The corresponding stuffing ratio ρ is slightly greater than 1:5 rather than exactly 1:5 as in Figure 4.18a. This effect is called **waiting time jitter** because of the delay or waiting time between the initiation of a stuff request and the time at which the stuff is actually accomplished. The result of waiting time jitter is a low-frequency sawtooth-shaped jitter superimposed on high-frequency pulse stuffing jitter.

The peak waiting time jitter occurs when the phase error crosses the threshold immediately after an available stuff opportunity; then the error accumulates until the next possible stuff opportunity. The peak-to-peak phase error that can accumulate during this interval equals the frequency offset times the interval between stuff opportunities:

$$\phi_\varepsilon = f_0 t_s \text{ bits} \tag{4.28}$$

where $t_s = 1/f_s$ seconds. From the definition of stuffing ratio given in (4.27), the peak-to-peak waiting time jitter can also be expressed as

$$\phi_\varepsilon = \rho \text{ bits} \tag{4.29}$$

At $\rho = 1$, peak-to-peak waiting time jitter is 1 bit. At $\rho = \frac{1}{2}$, the peak-to-peak jitter equals $\frac{1}{2}$ bit; likewise, at $\rho = \frac{1}{3}$, $\frac{1}{4}$, and $\frac{1}{5}$, the peak-to-peak jitter equals $\frac{1}{3}$, $\frac{1}{4}$, and $\frac{1}{5}$ bits. In general, for $\rho = 1/n$ the peak-to-peak jitter equals $1/n$. Figure 4.19 illustrates this relationship in a plot of jitter amplitude versus stuffing ratio. This plot shows that waiting time jitter peaks when ρ is a simple fraction $1/n$. Hence waiting time jitter can be minimized by selecting a nonrational number n for the stuffing ratio.

In practice the pulse stuffing recommended by the CCITT for second, third, and fourth-order multiplex has stuffing ratios that are less than 0.5 and not equal to simple fractions in order to avoid significant jitter peaks. As seen in Figure 4.19, waiting time jitter would be minimized by selecting a value of ρ near but not exactly equal to $\frac{1}{3}$ or $\frac{2}{5}$. Because the actual stuffing ratio varies due to drift in system clocks, the system designer is also interested in the range of values that the stuffing ratio may experience. By judiciously selecting the maximum stuffing rate based on expected clock frequency variations and known jitter characteristics (Figure 4.19), the stuffing ratio can be contained within a range that avoids large peaks of waiting time jitter. However, D. L. Duttweiler has

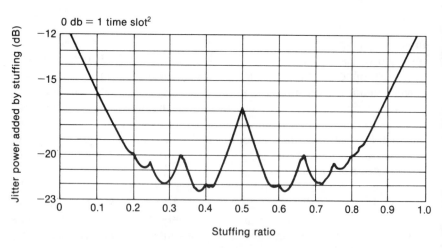

FIGURE 4.19   Waiting Time Jitter due to Pulse Stuffing (Courtesy CCITT [12]) (Smoothing Loop Bandwidth = $0.15f_s$)

shown that much of this advantage is lost if jitter is already present on a multiplexer input. As the amount of input jitter is increased, the peaks and valleys in Figure 4.19 become less pronounced and the choice of stuffing ratio becomes less critical [11].

Example 4.4 _____

The 6.312-Mb/s second-order multiplexer described in CCITT Rec. G.743 combines four 1.544-Mb/s channels using positive pulse stuffing with the following characteristics (also see Table 4.5):

Frame length: 294 bits
Information bits per frame: 288 bits
Maximum stuffing rate per channel: $f_s = 5367$ stuffs per second
Stuff code length: 3 bits
Tolerance on 1.544-Mb/s channels: $\Delta f = 50$ ppm

(a) Determine the nominal, maximum, and minimum stuffing ratios.
(b) Find the mean time to loss of BCI per channel due to stuff decoding errors, with an average bit error rate of 10^{-4}.

Solution
(a) The multiplexer synchronous rate per channel is given by

$$f_i' = \left(\frac{6.312 \times 10^6}{4}\right)\left(\frac{288}{294}\right) = 1,545,796 \text{ b/s}$$

The nominal, maximum, and minimum channel input rates are

Nominal: $f_i = 1,544,000 \text{ b/s}$

Maximum: $f_i + \Delta f = 1,544,000 + \left(\frac{50}{10^6} \times 1.544 \times 10^6\right)$

$= 1,544,077 \text{ b/s}$

Minimum: $f_i - \Delta f = 1,544,000 - \left(\frac{50}{10^6} \times 1.544 \times 10^6\right)$

$= 1,543,923 \text{ b/s}$

The corresponding stuffing ratios, defined by (4.27), are then

$$\text{Nominal:} \quad \rho = \frac{f_i'-f_i}{f_s} = \frac{1796}{5367} = 0.3346$$

$$\text{Maximum:} \quad \rho = \frac{f_i'-(f_i-\Delta f)}{f_s} = \frac{1873}{5367} = 0.3490$$

$$\text{Minimum:} \quad \rho = \frac{f_i'-(f_i+\Delta f)}{f_s} = \frac{1719}{5367} = 0.3203$$

(b) From (4.25), the probability of decoding error per stuffing time slot is

$$3(10^{-4})^2 = 3 \times 10^{-8}$$

From (4.26), the mean time to loss of BCI per 1.544-Mb/s channel is

$$\frac{1}{(5367)(3 \times 10^{-8})} = 6.2 \times 10^3 \text{ s} = 1.7 \text{ hr}$$

4.3.5 Smoothing Loop

The purpose of the phase-locked loop (PLL) shown in the demultiplexer of Figure 4.16 is to derive the original channel input clock. The timing signal at the input to the PLL will have an instantaneous frequency of f_i', the multiplexer channel rate, but with occasional timing cycles deleted so that the long-term average frequency is f_i. This gapped version of the original channel timing signal is then smoothed by the PLL to provide an acceptable recovered timing signal.

The effect of a smoothing loop on stuffing jitter is illustrated in Figure 4.20. First-order phase-locked loops have been found adequate to attenuate stuffing jitter properly, as will be verified here. This loop has a cutoff fre-

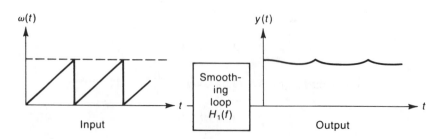

FIGURE 4.20   Model of Smoothing Loop Effect on Jitter

quency that is low compared to the stuffing rate. In this way, jitter on the output signal due to removal of stuffing bits can be controlled to acceptable levels. The closed-loop transfer function of a first-order loop is given by

$$H_1(f) = \frac{f_c}{f_c + jf}$$ (4.30)

The corresponding impulse response is then

$$h_1(t) = 2\pi f_c e^{-2\pi f_c t}$$ (4.31)

where f_c is the 3-dB cutoff frequency of the smoothing filter normalized to the actual stuffing rate f_0. The periodic sawtooth waveform representing unsmoothed pulse stuffing jitter can be expressed as

$$\omega(t) = t - n \qquad n - 0.5 < t < n + 0.5; n = 0, \pm 1, \pm 2, \cdots$$ (4.32)

where the period has been normalized by the reciprocal of the stuffing rate, f_0, and the sawtooth peak-to-peak amplitude has been normalized by 1 bit. Note that the waveform given in (4.32) neglects the presence of waiting time jitter. This turns out to be a reasonable assumption, since waiting time jitter is a low-frequency jitter and is not significantly attenuated by the smoothing loop, whereas pulse stuffing jitter is at a higher frequency and can be significantly attenuated by the loop. The output of the first-order smoothing loop with the periodic sawtooth input is obtained by applying the convolution theorem, which results in

$$y(t) = \int_{-\infty}^{t} \omega(\alpha) 2\pi f_c e^{-2\pi f_c (t - \alpha)} \, d\alpha$$ (4.33)

Upon carrying out the integration, the output jitter waveform over the period $-0.5 < t < 0.5$ is

$$y(t) = t - \frac{1}{2\pi f_c} + e^{-2\pi f_c t} \left(\frac{e^{-\pi f_c}}{1 - e^{-2\pi f_c}} \right)$$ (4.34)

The phase jitter $y(t)$ given by (4.34) is plotted in Figure 4.21. Note that the smoothing loop with narrower bandwidth provides greater jitter attenuation. In Figure 4.21, as before, f_c is the 3-dB cutoff frequency of the smoothing loop normalized to the stuffing frequency.

To determine the peak-to-peak output jitter from Figure 4.21 note that the waveform of (4.34) has a maximum value at the two extremes ($t = \pm 0.5$)

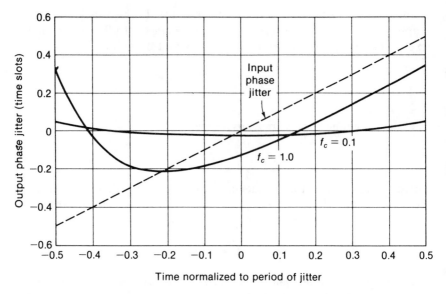

FIGURE 4.21   Stuffing Jitter Waveform at Output of First-Order Smoothing Loop

and minimum value in between. The minimum point is found by differentiating $y(t)$, setting this equal to zero, and solving for t. From the minimum and maximum values, the peak-to-peak output jitter can be expressed as

$$\Delta y_{p-p} = \frac{1}{1 - e^{-2\pi f_c}} - \frac{1}{2\pi f_c}\left[1 + \ln\left(\frac{2\pi f_c}{1 - e^{-2\pi f_c}}\right)\right] \qquad (4.35)$$

A plot of this peak-to-peak stuffing jitter is shown in Figure 4.22 as a function of f_c, the smoothing loop bandwidth normalized to stuffing rate. As expected, stuffing jitter is significantly attenuated with narrow loop bandwidths.

Example 4.5 _____

For the 6.312-Mb/s multiplexer described by CCITT Rec. G.743, the two specifications pertaining to pulse stuffing jitter are:

1. With no jitter at the input to the multiplexer and demultiplexer, the jitter at the demultiplexer output should not exceed one-third of a time slot peak-to-peak.
2. The gain of the jitter transfer function should not exceed the limits given in Figure 4.23.

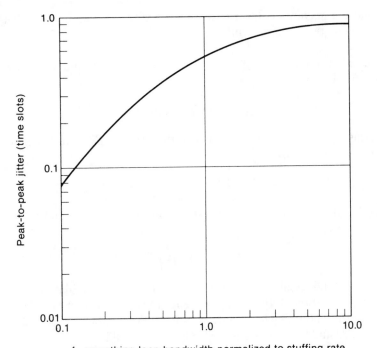

FIGURE 4.22   Peak-to-Peak Jitter at Output of First-Order Smoothing Loop

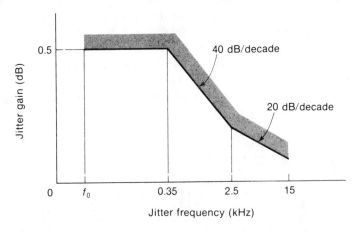

Note: The frequency f_0 should be as low as possible
taking into account the limitations of measuring equipment.

FIGURE 4.23   Required Jitter Attenuation for 6.312-Mb/s Pulse Stuffing Multiplexer
(Courtesy CCITT [2])

Find the maximum cutoff frequency f_c for the smoothing loop that will satisfy these two requirements.

Solution   From Figure 4.22 we see that the first requirement is met with a cutoff frequency

$$f_c = 0.45 \text{ Hz/stuffing rate}$$

In terms of hertz and for a nominal stuffing rate of 1796 stuffs per second (from Example 4.4),

$$f_c \text{ (Hz)} = (0.45)(1796) = 808 \text{ Hz}$$

while for the minimum stuffing rate of 1719 stuffs per second (also from Example 4.4),

$$f_c \text{ (Hz)} = (0.45)(1719) = 773 \text{ Hz}$$

For the second requirement on pulse stuffing jitter, examination of Figure 4.23 shows that the 3-dB cutoff frequency must be at about 625 Hz or less. Further, the smoothing loop must exhibit a rolloff characteristic as shown in Figure 4.23 and limit jitter gain to 0.5 dB below 350 Hz.

A potential problem in introducing a pulse stuffing multiplexer into a digital transmission system is the effect that pulse stuffing jitter has on bit synchronizers in the system. Excessive jitter can cause a phase-locked loop used for bit synchronization to lose lock or slip bits. Since timing derived by the PLL is used to make bit decisions, it is not enough that the loop merely maintain lock. It must also track instantaneous bit timing, or bit decisions will be made with a timing error. The ability of the bit

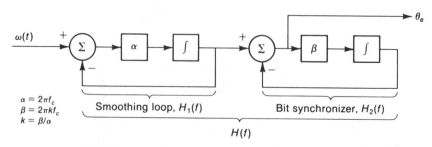

FIGURE 4.24   Block Diagram of First-Order Smoothing Loop Cascaded with First-Order Bit Synchronizer

synchronizer to track input jitter is improved by increasing the loop bandwidth but at the expense of added noise bandwidth, so that bit synchronizer bandwidths are generally no greater than necessary for proper timing recovery.

To analyze the effect of pulse stuffing jitter on bit synchronizers, a model of a first-order smoothing loop cascaded with a first-order bit synchronizer is needed, as shown in Figure 4.24. This model represents a first-order loop as a phase detector, a multiplier of constant gain (α or β), and an integrator. The transfer function of this cascade from input signal to bit synchronizer phase detector output is

$$H(f) = H_1(f) \cdot H_2(f) = \frac{f_c(jf)}{(f_c + jf)(kf_c + jf)} \tag{4.36}$$

By partial fractions this expression can be expanded to

$$H(f) = \frac{1}{(1-k)} \left(\frac{f_c}{f_c + jf} - \frac{kf_c}{kf_c + jf} \right) \tag{4.37}$$

We can convolve the sawtooth waveform representing the jitter input as given in (4.32) with $H(f)$, similar to that done earlier in (4.33), to arrive at

$$\theta_e(t) = \frac{1}{2\pi k f_c} + \frac{1}{1-k} \left(e^{-2\pi f_c t} \cdot \frac{e^{-\pi f_c}}{1 - e^{-2\pi f_c}} - e^{-2\pi k f_c t} \cdot \frac{e^{-\pi k f_c}}{1 - e^{-2\pi k f_c}} \right) \tag{4.38}$$

This is an expression for phase error over one stuffing jitter period, $-0.5 < t < 0.5$. This waveform has a maximum value at the extremes ($t = \pm 0.5$) and a minimum point inbetween. The minimum point is found by differentiating $\theta_e(t)$, setting this equal to zero, and solving for t.

From the maximum and minimum values of (4.38), the peak-to-peak phase error is found to be

$$\Delta\theta_{e_{p-p}} = \frac{1}{1-k} \cdot \frac{1}{1 - e^{-2\pi f_c}} \left[1 - \frac{1 - e^{-2\pi f_c}}{1 - e^{-2\pi k f_c}} + \left(\frac{1-k}{k} \right) \right.$$
$$\left. \cdot \left(k \cdot \frac{1 - e^{-2\pi f_c}}{1 - e^{-2\pi k f_c}} \right)^{1/(1-k)} \right] \tag{4.39}$$

Using (4.39), curves are shown in Figure 4.25 for peak timing error when a first-order smoothing loop is interfaced with a first-order bit synchronizer. This figure presents the results for a constant ratio of bit synchronizer loop bandwidth to smoothing loop filter bandwidth. The curves only consider

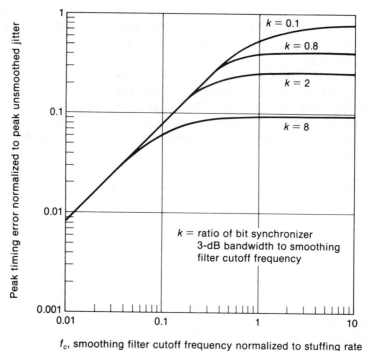

f_c, smoothing filter cutoff frequency normalized to stuffing rate

FIGURE 4.25   Effect of Pulse Stuffing Jitter on Timing Recovery for a First-Order Bit Synchronizer

the effects of stuffing jitter since Expression (4.32) does not include waiting time jitter, which is so low in frequency that its effects are negligible for bit synchronizers. To illustrate the significance of these curves, first note that when the smoothing loop filter has a bandwidth equal to the stuffing rate ($f_c = 1$), the timing error is significantly decreased only when the bit synchronizer has a bandwidth approximately 10 times the stuffing rate. When the bit synchronizer bandwidth and the smoothing filter bandwidth are both equal to the stuffing rate ($k = 1, f_c = 1$), the timing error is approximately 30 percent of a bit, which corresponds to an 8.0-dB loss in signal-to-noise ratio. Figure 4.25 points out the importance of minimizing pulse stuffing jitter by proper smoothing loop design to avoid excessive degradation in downstream bit synchronizers.

Another concern in the application of pulse stuffing multiplexers is the degree to which jitter accumulates along a digital path. Waiting time jitter has been shown to accumulate according to $\sqrt{N}$ where N is the number of multiplex/demultiplex pairs connected in tandem [11]. Proper choice of the stuffing ratio in each multiplexer/demultiplexer pair limits the accumulation of waiting time jitter. As observed earlier, waiting time jitter has little effect on bit synchronizers, although it does produce distortion in PCM systems due to the irregular spacing of reconstructed analog samples. A CCITT study of jitter accumulation along a digital path indicates that PCM multiplexers and pulse stuffing multiplexers tolerate high input jitter and cause low output jitter [13]. This jitter-reducing effect of digital multiplexers means that limitations in jitter amplitude such as those shown in Figure 4.23 will not be exceeded in a path of tandemed multiplexers. The same study also pointed out, however, that digital repeatered lines tolerate only low input jitter and generate relatively high output jitter. This question of jitter and its accumulation in repeatered lines is considered in Chapter 7.

4.3.6 Transitional Coding

Many low-speed terminals transmit characters asynchronously, where characters have a uniform length but do not necessarily occur at a uniform rate. Teletypewriters and other keyboard terminals produce alphanumeric symbols, one at a time, in character format. Each character is made up of several information bits plus 2 or 3 bits for character synchronization; for example, the CCITT International Alphabet No. 2 uses five information bits. In the United States, most devices use the 7-bit ASCII code (American Standard Code for Information Interchange), which is compatible with CCITT Alphabet No. 5. As shown in the example of Figure 4.26, character synchronization for the ASCII code is provided by a start bit of value 0 (space) and two stop bits of value 1 (mark). The ASCII code also provides a parity bit for error detection, thus resulting in an 11-bit character.

Several techniques are used in the digital multiplexing of asynchronous, character-oriented data. A form of synchronous multiplexing is possible by using buffers and some means of character detection to assemble characters in blocks of data. Each block of data is delimited or framed by a synchronization code. Moreover, each character's start and stop pulses are dropped

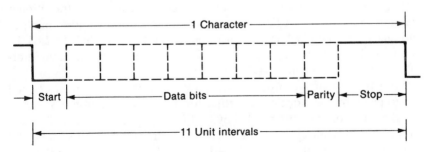

FIGURE 4.26   Format of Character from CCITT Alphabet No. 5 or ASCII Code

at the multiplexer and reinserted by the demultiplexer before interface with the receiving data terminal. This technique results in high transmission efficiency, since only the character's data bits are transmitted, and avoids the problem of handling stop pulses that are not an integer number of bit intervals. A form of pulse stuffing can also be used to compensate for variation in arrival time of characters and to absorb any fraction of a bit used as part of the stop pulse. Pulse stuffing is also a highly efficient transmission technique, but the system designer must be aware of the effects of pulse stuffing jitter and stuff code errors.

A low-speed asynchronous data signal can also be converted to a synchronous signal by simply sampling the data at a very high rate, a technique that is inexpensive to implement. **Transitional coding** is a similar technique that uses extra bits to code the data transition times. Decoding in the receiver provides reconstruction of the transition time and binary value for each data bit. The disadvantage of both techniques is the high transmission bit rate required by the sampling or encoding process and the jitter resulting from reconstruction of the data transitions. Of the two schemes, transitional coding is the more commonly used, since it results in more efficient transmission.

The simplest form of transitional coding is the 2-bit version illustrated in Figure 4.27a. Shown are the data signal and corresponding 2-bit code. When a data transition occurs, the two coded bits indicate the transition time and data bit value. The first bit following the transition (indicated as X) is a timing bit whose value indicates whether the data transition occurred in the first or second half of the sampling clock period. After this first half/second half bit has been transmitted, the coded bits are set equal to the data bit until another transition occurs. Since 2 bits of code are required for each data transition, the code rate is twice that of the data—one of two disadvantages of transitional coding. The second disadvantage is the jitter introduced when reconstructing the data transitions at the receiver, as illustrated in Figure 4.27b. The clock is shown partitioned into the first and second half of a cycle. In the receiver, the transition is reconstructed in the center of the first or second half-clock cycle. Peak jitter occurs when the data transition is precisely at the boundary between the two half-clock cycles and amounts to one-quarter of the clock cycle. Since the data rate is one-half that of the sampling clock, the peak jitter is therefore one-eighth of the data bit interval, which is tolerable in most data terminals.

Jitter performance can be improved by increasing the number of code bits to partition the clock cycle further for data reconstruction. For example, the 3-bit coder recommended by CCITT Rec. R.111 [14] uses 2 bits to identify the quarter cycle in which each data transition occurs. Table 4.2 shows the assignment of the 3 bits—B_1, B_2, and B_3—according to the data

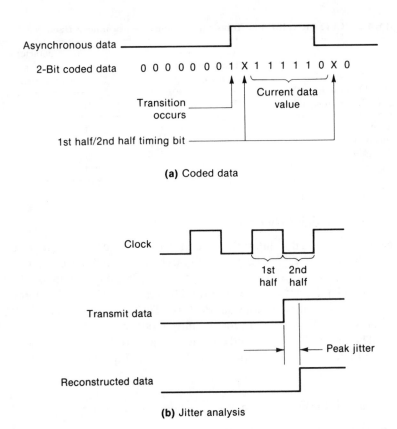

(a) Coded data

(b) Jitter analysis

FIGURE 4.27   Two-Bit Transitional Coding

bit value and position of the transition. This scheme clearly reduces the jitter compared to 2-bit coding but increases the transmission rate by 50 percent since 3-bit coding requires a code rate three times that of the data rate. The peak jitter J_p due to transitional coding or simple sampling can be expressed in percentage by

$$J_p = \frac{1}{2^N} = \left(\frac{f_D}{f_C}\right) \times 100 \tag{4.40}$$

where N is the number of code bits ($N = 1$ for simple sampling), f_D is the data rate, and f_C is the clock rate. A typical limit imposed on transitional coders is a maximum of 10 percent peak jitter, which is acceptable to most asynchronous data terminals.

TABLE 4.2   CCITT Rec. R.111 for Transitional Coding of Asynchronous Data

Position of Transition in a Group of Four Sampling Pulses	Code Character for Transition from 1 to 0 in Data Signal			Code Character for Transition from 0 to 1 in Data Signal		
	B_1	B_2	B_3	B_1	B_2	B_3
First quarter	0	0	0	1	1	1
Second quarter	0	0	1	1	1	0
Third quarter	0	1	0	1	0	1
Fourth quarter	0	1	1	1	0	0

4.4   Digital Multiplex Hierarchies

Three different digital multiplex hierarchies have emerged and been adopted by the CCITT. The three hierarchies reflect choices made by AT&T and followed by the United States and Canada, by Nippon Telegraph and Telephone (NTT) and followed by Japan, and by the Conférence Européene des Administrations des Postes et Télécommunications (CEPT) and followed by Europe and most countries outside North America and Japan. The bit rate structure and channel capacity of these hierarchies are shown in Figure 4.28. This section describes the multiplex equipment spec-

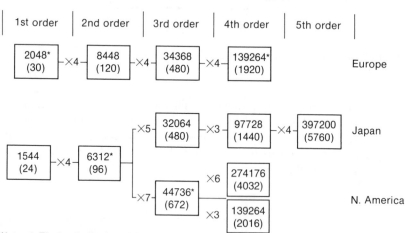

Notes: 1. The top figures in each box indicate bit rates in kb/s.
2. The figures in parentheses indicate capacity in number of 64 kb/s time slots.
3. The figures ×*N* indicate the number of multiplexed digital channels.
4. The * indicates a hybrid hierarchy used for interworking of dissimilar digital hierarchies.

FIGURE 4.28   Multiplex Hierarchies Recommended by CCITT

ified by these digital hierarchies. As will be seen, some common features exist among the various levels in the digital hierarchies, such as the multiplexing method and pulse stuffing technique. However, the divergence of essential features such as channel and aggregate bit rates has generally precluded interface between the various hierarchies except at the lowest level (zero order: voice channel).

4.4.1   North American Digital Hierarchy

As illustrated in Figure 4.29, the North American digital hierarchy is based on four levels, in which the digital signals (DS) at rates of 1.544, 6.312, 44.736, and 274.176 Mb/s (or alternatively 139.264 Mb/s) are denoted DS-1, DS-2, DS-3, and DS-4 (or alternatively DS-4NA). Transmission (T) lines corresponding to these rates are denoted T1, T2, T3, and T4, respectively. Multiplex (M) equipment is denoted by the channel and aggregate level—for example, an M13 combines 28 DS-1 signals into a DS-3 signal.

North American First-Level Multiplex

The first-level multiplexer of each CCITT digital hierarchy is a PCM multiplexer operating at a bit rate of 1.544 Mb/s in the North American and Japanese hierarchies and 2.048 Mb/s in the CEPT hierarchy. Table 4.3 lists characteristics of these two PCM standards as prescribed by the CCITT.

TABLE 4.3   CCITT Characteristics of PCM Multiplex Equipment [2]

Characteristic	1.544 Mb/s PCM Standard (CCITT Rec. G.733)	2.048 Mb/s PCM Standard (CCITT Rec. G.732)
Encoder		
Compander law	μ-law	*A*-law
No. of segments	15	13
Zero code suppression	Yes	No
Bit Rate	1.544 Mb/s ± 50 ppm	2.048 Mb/s ± 50 ppm
Timing signal	Internal source or incoming digital signal or external source	Internal source or incoming digital signal or external source
Frame structure	See Figure 4.30	See Figure 4.31
No. of bits per channel	8	8
No. of channels per frame	24	32
No. of bits per frame	193	256
Frame repetition rate	8000	8000
Signaling	Least significant bit of each channel every sixth frame	Time slot 16 used; rate ≤ 64 kb/s
*Coding**	Bipolar or B8ZS	HDB3

*See Chapter 5 for a description of baseband transmission codes.

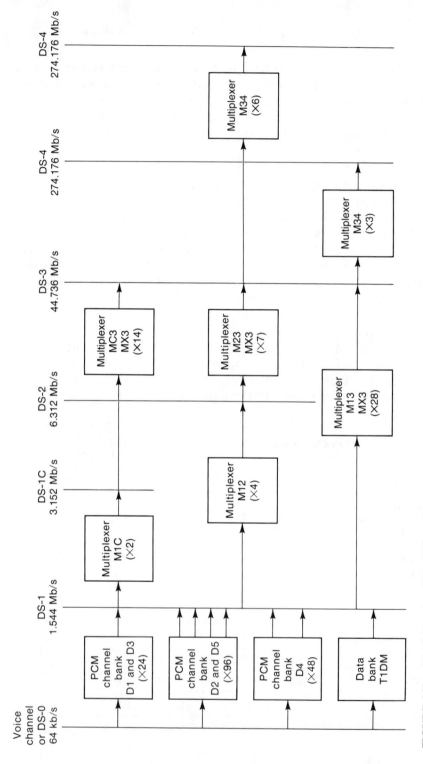

FIGURE 4.29 North American Digital Hierarchy

Figure 4.30 illustrates the 1.544-Mb/s frame structure, which provides 24 channels of 64-kb/s PCM speech signals. This grouping of 24 voice channels with a transmission rate of 1.544 Mb/s is known as a **digroup.** Each frame consists of 24 eight-bit words and one synchronization bit for a total of 193 bits. Each channel is sampled 8000 times per second; thus there are 8000 frames per second and a resulting aggregate rate of (8000 frames) × (193 bits/frame) = 1,544,000 b/s. Each superframe consists of 12 frames. The 12 synchronizing bits per superframe are used for frame and superframe synchronization. This 12-bit sequence is subdivided into two sequences. The frame alignment pattern is 101010 and is located in odd-numbered frames; the superframe alignment pattern is 001110 and is located in even-numbered frames. The superframe structure also contains signaling bits that are used to carry supervisory information (on-hook, off-hook) and dial pulses for each voice channel. This signaling function is accomplished by time-sharing the least significant bit (B8) between speech and signaling. The B8 bits carry speech sample information for five frames, followed by signaling information in every sixth frame. Two separate signaling channels

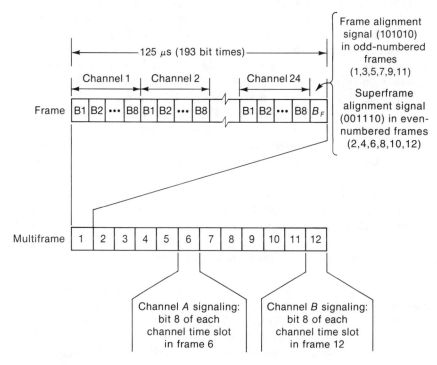

FIGURE 4.30   Frame and Superframe Format for 24-Channel PCM Multiplex

are provided, which alternate between the sixth and twelfth frame of each superframe. Signaling bits carried by frame 6 are called *A* and those carried by frame 12 are called *B* signaling bits. This form of signaling is known as *robbed bit* signaling.

As an alternative to the superframe (SF) format shown in Figure 4.30, an **extended superframe (ESF)** format, which improves performance and facilitates maintenance, has been developed in the United States and adopted in CCITT Rec. G.704. (Chapter 11 describes the advantages of ESF in monitoring and testing DS-1 signals.) The ESF format extends the superframe structure from 12 to 24 frames and redefines the 8-kb/s overhead in the 1.544-Mb/s DS-1 signal, as shown in Table 4.4. Whereas the superframe 8-kb/s overhead shown in Figure 4.30 is used exclusively for frame synchronization, the 8-kb/s overhead channel in ESF provides three separate functions:

- *2-kb/s Framing:* Beginning with frame 4, the framing bit of every fourth frame forms the 2-kb/s pattern 001011 . . . 001011, which is used to obtain frame and superframe synchronization. Frame synchronization is used to identify each 64-kb/s PCM channel of each frame. Superframe synchronization is used to locate each frame within the superframe in order to extract the CRC and data link information and identify the frames that contain signaling information. Signaling bits carried by frames 6, 12, 18, and 24 are called, respectively *A, B, C,* and *D* signaling bits.
- *2-kb/s Cyclic Redundancy Check (CRC):* This is contained within the framing bit position of frames 2, 6, 10, 14, 18, and 22 of every superframe. The check bits (CB) that make up the CRC are used to detect errors within the extended superframe. The CRC can be monitored at any DS-1 access point and can be used for end-to-end performance monitoring, false framing protection, protection switching, automatic restoral after alarms, line verification in maintenance actions, and the measurement of error-free seconds.
- *4-kb/s Data Link:* Beginning with frame 1 of every superframe, every other 193rd bit is part of the 4-kb/s data link. Using X.25 level 2 protocol, this data link has several possible applications, including transmission of alarms, status signals, supervisory signaling, network configuration, performance indicators, and maintenance information.

The extended superframe format also offers four signaling options, as shown in Table 4.4. Three of the options use robbed bit signaling, with the capability to signal two states using the *A* bits, four states using the *A* and *B* bits, and 16 states using the *A, B, C,* and *D* bits. The fourth option offers transparent (T) signaling, which uses a common channel rather than robbed bits. With the T option, the twenty-fourth DS-0 channel is used for common channel signaling as specified, for example, in CCITT Rec. I.431.

TABLE 4.4   Extended Superframe Format

Frame Number	Overhead			Signaling Channel			
	Framing	Data	Error Detection	16-State	4-State	2-State	Transparent
1	—	D	—				
2	—	—	CB_1				
3	—	D	—				
4	0	—	—				
5	—	D	—				
6	—	—	CB_2	A	A	A	—
7	—	D	—				
8	0	—	—				
9	—	D	—				
10	—	—	CB_3				
11	—	D	—				
12	1	—	—	B	B	A	—
13	—	D	—				
14	—	—	CB_4				
15	—	D	—				
16	0	—	—				
17	—	D	—				
18	—	—	CB_5	C	A	A	—
19	—	D	—				
20	1	—	—				
21	—	D	—				
22	—	—	CB_6				
23	—	D	—				
24	1	—	—	D	B	A	—

AT&T uses the term **digital channel bank** to denote a PCM multiplexer. The D1 channel bank (D for digital, 1 for first generation) was placed in service in 1962. The D1 used 7-bit logarithmic (μ-law) PCM coding and combined 24 channels into the DS-1 rate [15]. In 1969, the D2 channel bank was introduced, which provided the 8-bit coding scheme used in all later channel banks [16]. The D2 combines up to 96 channels into four separate DS-1 signals. The third-generation channel bank, the D3, was introduced in 1972 for 24-channel (1.544 Mb/s) application and to replace the D1 [17]. Because of differences in signaling format and companding characteristic, the D1 is not end-to-end compatible with the later-generation channel banks. However, the D1D, a modified D1 that is compatible with a D3, was also introduced to upgrade existing D1 channel banks [18]. In 1976, the versatile D4 channel bank was introduced; it provides up to 48-channel operation [19]. The D4 has the following options for transmission line interface: two independent T1 lines, a single T1C line (3.152 Mb/s),* and two D4's ganged and interfaced with a T2 line. It should also be noted that AT&T has developed a D4E that is compatible with the 2.048-Mb/s multiplexer of CCITT Rec. G.732 and operates over a T1E (2.048 Mb/s) transmission line. Finally, the D5 channel bank was announced in 1982 and provides up to 96-channel operation. A comparison of these PCM channel banks is presented in Table 4.5

Many other versions of DS-1 multiplexers exist, some *channelized* into DS-0 channels and others *nonchannelized* without DS-0 boundaries. The designation *M*24 specifies a DS-1 multiplexer that combines 24 DS-0 channels into a DS-1 using either SF or ESF overhead. CCITT Rec. G.734 specifies a DS-1 synchronous multiplexer capable of 23 DS-0 digital data signals [2]. Other channelized DS-1 multiplexers that are standardized are described in the following subsections. Nonchannelized DS-1 multiplexers commonly used in network applications are described in Chapter 13.

Subrate Data Multiplexing.   Synchronous digital multiplexers operating at the DS-1 level are used to provide data channels at 64 kb/s (designated the DS-0 level) and at various subrates such as 2.4, 4.8, 9.6, 19.2, and 56 kb/s (designated subrate data channels). The generic term for such multiplexers is *subrate data multiplexers* (SRDM). Subrate data multiplexing is accomplished within a channel bank using special dataport cards or within *T1 data multiplexers* (T1DM). Within AT&T's Digital Data System (DDS) [20], these subrate channels are built up to DS-0 signals, which have interface characteristics identical to a DS-0 but have varying formats depending on the exact subrate and use. A 56-kb/s channel is converted to a DS-0 signal by adding a control or *C* bit to seven data bits to create an 8-bit byte; the control bit is set to 1 when the byte contains data and is set to 0 when the byte contains network control codes. Individual subrate channels at 2.4, 4.8, and 9.6 kb/s are converted to DS-0 by adding a framing bit before a group of six data bits and a control bit after the six data bits to form an 8-bit byte; using a technique known as *byte stuffing,* this byte is repeated 20, 10, or 5 times, respectively, to create a DS-0 signal. When only a single subrate signal is carried within the DS-0, the format is referred to as DS-0A. A DS-0B format, conversely, has the same byte structure as DS-0A but can accommodate multiple subrate channels, up to twenty 2.4 kb/s, ten 4.8 kb/s, five 9.6 kb/s, or a mix of these DS-0A formatted signals. (Subrates within DS-0A and DS-0B signals are distinguishable by the choice of framing bits.) The SRDM removes the stuffed bytes from incoming DS-0A signals and multiplexes the resulting data channels from each DS-0A into a DS-0B channel.

---

*T1C and M1C are designations for the transmission line and multiplexer for DS-1C (3.152 Mb/s) operation.

TABLE 4.5 Comparison of PCM Channel Banks

Characteristic	D1	D2	D1D	D3	D4	D4E	D5
No. of voice channels	24	24, 48, 72, or 96	24	24	24 or 48	30	24, 48, 72, or 96
Sampling sequence	1, 13, 2, 14, 3, 15, 4, 16, 5, 17, 6, 18, 7, 19, 8, 20, 9, 21, 10, 22, 11, 23, 12, 24	12, 13, 1, 17, 5, 21, 9, 15, 3, 19, 7, 23, 11, 14, 2, 18, 6, 22, 10, 16, 4, 20, 8, 24	Same as D1	1, 2, 3, 4, 5, 6, 7, 8, 9, 10, 11, 12, 13, 14, 15, 16, 17, 18, 19, 20, 21, 22, 23, 24	Options for three sampling sequences corresponding to D1, D2, D3	1, 2, 3, 4, . . . , 30	Options for D1, D2, D3, D4
No. of coding bits per channel	7	8	8	8	8	8	8
Signaling bit location	Eighth bit of every channel	Eighth bit of every channel of every sixth frame	Same as D2	Same as D2	Same as D2	Time slot 16	Same as D2
Companding characteristic	$\mu = 100$ nonlinear diode characteristic	$\mu = 255$ 15-segment piecewise linear approximation	Same as D2	Same as D2	Same as D2	$A = 87.6$ 13-segment piecewise linear approximation	Same as D2
Transmission line	T1	Up to four independent T1's	T1	T1	T1, T1C, or T2	T1E (2.048 Mb/s)	T1, T1C, or T2

Once formatted, these DS-0B signals can be multiplexed into a DS-1 signal using a T1DM, ESF, or SF format.

Fractional T1. Transmission of data signals at speeds greater than subrates can be accomplished via fractional T1 (FT1) rates, which are generally available in increments of 64 kb/s. For example, AT&T's fractional T1 service, called the Accunet Spectrum of Digital Services [21], provides circuits at 56, 64, 128, 256, 512, and 768 kb/s. Fractional T1 circuits are usually embedded in a T1 multiplexer at the customer premise. For a T1 multiplexer to carry fractional T1 circuits, data must be formatted as 24 DS-0s with standard superframe or extended superframe overhead. The customer's data are placed into the frame in alternating DS-0 slots, and unused slots are filled with a pattern that ensures proper 1's density is maintained. At the carrier's facility, the fractional T1 signal is extracted and multiplexed with other fractional T1 circuits. The fractional T1 circuit may be carried through the carrier's network in a variety of ways, but at the far end the fractional T1 signal is restored to its original form.

Transcoders. The popularity of 32-kb/s ADPCM for telephone network applications has led to the development and standardization of **transcoding,** a technique that allows direct digital conversion of PCM to ADPCM and vice versa. Standard transcoders used in the North American digital hierarchy interface with two PCM channel banks, convert 64-kb/s PCM to 32-kb/s ADPCM, and thus compress up to 48 voice channels into a single 1.544-Mb/s signal [22]. The compressed 1.544-Mb/s signal is organized into 193-bit frames, with each frame comprising 48 4-bit time slots plus a framing bit. Transcoders also provide the ability to pass through 64-kb/s channels, for example, when selected channels are to retain PCM coding or are to be used for digital data. Two basic transcoder versions are available, one that provides up to 48 voice channels plus the option of four signaling channels and one that provides 48 voice channels with robbed bit signaling. In the first version, called the *bundle format,* the frame is further organized into 4 independent 384-kb/s bundles. Each bundle thus consists of 12 32-kb/s time slots capable of carrying 12 voice channels. As an option, channel-associated signaling may be conveyed via the twelfth time slot of each bundle, which reduces the voice channel capacity to 11 for each bundle. These 32-kb/s signaling channels are referred to as *delta channels* and, if used, are located at time slots 12, 24, 32, and 48 of each frame. (This type of transcoder is sometimes referred to as a M44.) Use of the delta channel is selected on a per-bundle basis, independent of other bundles. The delta channel contains 16-state *a-b-c-d* signaling, alarm information, signaling alignment, and error detection bits.

The second transcoder version uses robbed bit signaling, similar to PCM channel banks, in which the least significant bit (that is, the fourth bit) is used for signaling every sixth frame. In frames containing robbed bit signaling, 7-level quantization is used in lieu of 15-level quantization.

ADPCM multiplexing may also be done directly in a first-level multiplex, which combines 44 or 48 voice channels and has the same options as the transcoder. The ADPCM transcoder and multiplexer are standardized in CCITT Recs. G.762 and G.724, respectively [2].

North American Higher Level Multiplex
Higher level multiplexers as standardized in North America are shown in Figure 4.29 and Table 4.6. Original versions of these multiplexers were all based on the use of cyclic bit interleaving and positive pulse stuffing. The M1C and M12 combine two and four DS-1 signals into a DS-1C and DS-2 signal, respectively. The complete frame and multiframe structure of the M12 is shown in Table 4.7. Both the M1C and M12 multiplexers were developed principally for metallic cable applications, at T1C and T2 rates, respectively. These lower level multiplex have been superseded by higher level multiplex as transmission rates have increased with the availability of fiber optics and digital radio. The DS-3 level can be attained through several types of multiplexers, the M23, M13, MC3, or MX3. The M13 combines 28 DS-1's into a DS-3, the MC3 combines 14 DS-1Cs into a DS-3, and the M23 combines 7 DS-1s into a DS-3. The MX-3 is capable of multiplexing these same combinations, or any mix of DS-1s, DS-1Cs, or DS-2s up to an aggregate of DS-3. The fourth level is attained by the M34 multiplexer. The original fourth level of 274.176 Mb/s in the North American hierarchy has

TABLE 4.6   Characteristics of North American Digital Multiplex

	Second-Level	Third-Level		Fourth-Level
Multiplex Characteristic	CCITT Rec. G.743	CCITT Rec. G.752	DS-4	CCITT Rec. G.755
Aggregate bit rate (Mb/s)	6.312	44.736	274.176	139.264
Tolerance on aggregate rate (ppm)	±30	±20	±10	±15
Aggregate digital interface*	B6ZS	B3ZS		CMI
Channel bit rate (Mb/s)	1.544	6.312	44.736	44.736
No. of channels	4	7	6	3
Channel digital interface*	Bipolar/B8ZS	B6ZS	B3ZS	B3ZS
Frame structure	Table 4.7	Table 4.8		CCITT Rec. G.755
Pulse stuffing technique	Positive pulse stuffing with 3-bit stuff code and majority vote decoding			
Multiplexing method	Cyclic bit interleaving in channel numbering order			
Timing signal	Derived from external as well as internal source			

*See Chapter 5 for a description of baseband transmission codes.

TABLE 4.7 Frame Structure of 6.312-Mb/s Second-Order Multiplex

Frame Structure

M_0	I_{1-48}	C_{11}	I_{1-48}	F_0	I_{1-48}	C_{12}	I_{1-48}	C_{13}	I_{1-48}	F_1	S_1	I_{2-48}			
M_1	I_{1-48}	C_{21}	I_{1-48}	F_0	I_{1-48}	C_{22}	I_{1-48}	C_{23}	I_{1-48}	F_1	I_1	S_2	I_{3-48}		
M_1	I_{1-48}	C_{31}	I_{1-48}	F_0	I_{1-48}	C_{32}	I_{1-48}	C_{33}	I_{1-48}	F_1	I_1	I_2	S_3	I_{4-48}	
X	I_{1-48}	C_{41}	I_{1-48}	F_0	I_{1-48}	C_{42}	I_{1-48}	C_{43}	I_{1-48}	F_1	I_1	I_2	I_3	S_4	I_{5-48}

Frame length: 294 bits
Multiframe length: 1176 bits
Bits per channel per multiframe: 288 bits
Maximum stuffing rate per channel: 5367 b/s
Nominal stuffing ratio: 0.334

F_i = frame synchronization bit with logic value i
M_i = multiframe synchronization bit with logic value i
I_i = information bits from channels
C_{ij} = jth stuff code bit of ith channel
S_i = bit from ith channel available for stuffing
X = alarm service bit

been rarely used, has not been adopted by the CCITT, and has been superseded by the CEPT fourth level, which is called DS-4NA in North America. Any one of these four levels in the North American hierarchy may also be created by a single source, such as commercial color television at the DS-3 rate. These nonmultiplexed signals are commonly referred to as *nonsubrated* signals and contain the same overhead as multiplexed signals.

M13 Signal Format [23].   The multiplexing of DS-1 signals into a DS-3 is done in two steps in a M13 multiplexer. In the first step, 4 DS-1 signals are multiplexed using pulse stuffing to obtain the DS-2 rate within the multiplexer. At this point, the DS-2 is identical to that signal produced by an M12. In the second step, 7 DS-2 signals are multiplexed using dedicated pulse stuffing synchronization to generate the DS-3 signal. The resulting DS-3 is identical to the signal produced by an M23. Both stages of multiplexing interleave bits according to the input channel numbering order. These two stages of multiplexing thus ensure compatibility of an M13 with both the M12 and M23. Table 4.8 provides a view of the M13 subframe (85 bits), frame (680 bits), and multiframe or *M-frame* (4760 bits) structure. Each group of 84 information bits consists of the 7 DS-2 signals interleaved one bit at a time, done 12 times. The frame (F) and multiframe (M) bits are used for synchronization, the signaling (X) bits for network control, and the parity (P) bits for parity check on the 4704 information bits of each multiframe. The stuff control (C) bits in the multiframe provide 3-bit words, one for each of the 7 input DS-2 signals, to signal the presence or absence of a stuff bit (S) at the stuffing opportunity. Since the final stage of multiplexing from DS-2 to DS-3 is done within the multiplexer, all 7 of the DS-2 signals are timed off a common clock in the multiplexer and are therefore already synchronous. Hence the C bits used for pulse stuffing, necessary only to preserve backward compatability with older multiplexers, could be used for other purposes.

C-bit Parity Format [24].   The C-bit parity format was developed to add an in-band data link and to enhance monitoring and control of the DS-3 signal. Multiplexing of DS-1 signals to a DS-3 signal is accomplished in two stages, similar to the M13, but here the first stage of multiplexing produces a pseudo DS-2 rate. In the second stage, 7 pseudo DS-2 signals are multiplexed to form a DS-3 signal. Both stages use bit interleaving for multiplexing of the input channels, but while the first stage uses conventional pulse stuffing, the second stage inserts stuff bits into every stuffing opportunity, which forces the DS-2 rate to a pseudo rate of 6.306 Mb/s. These "stuff" bits are reserved for network use. The 21 C bits normally used for stuff indications are thus unnecessary and are used for other purposes, which means that the C-bit format is incompatible with the M13 format. Table 4.9 shows the C-bit frame and multiframe, which have the same basic structure as the M13, including identical use of the frame (F), multiframe (M), and parity (P) bits. The C bits, however, are used for different purposes; for example, the C

TABLE 4.8   M13 M-Frame Structure for DS-3 Multiplexer

X	I_{1-84}	F_1	I_{1-84}	C_{11}	I_{1-84}	F_0	I_{1-84}	C_{12}	I_{1-84}	F_0	I_{1-84}	C_{13}	I_{1-84}	F_1		S_1	I_{2-84}
X	I_{1-84}	F_1	I_{1-84}	C_{21}	I_{1-84}	F_0	I_{1-84}	C_{22}	I_{1-84}	F_0	I_{1-84}	C_{23}	I_{1-84}	F_1	I_1	S_2	I_{3-84}
P	I_{1-84}	F_1	I_{1-84}	C_{31}	I_{1-84}	F_0	I_{1-84}	C_{32}	I_{1-84}	F_0	I_{1-84}	C_{33}	I_{1-84}	F_1	I_{1-2}	S_3	I_{4-84}
P	I_{1-84}	F_1	I_{1-84}	C_{41}	I_{1-84}	F_0	I_{1-84}	C_{42}	I_{1-84}	F_0	I_{1-84}	C_{43}	I_{1-84}	F_1	I_{1-3}	S_4	I_{5-84}
M_0	I_{1-84}	F_1	I_{1-84}	C_{51}	I_{1-84}	F_0	I_{1-84}	C_{52}	I_{1-84}	F_0	I_{1-84}	C_{53}	I_{1-84}	F_1	I_{1-4}	S_5	I_{6-84}
M_1	I_{1-84}	F_1	I_{1-84}	C_{61}	I_{1-84}	F_0	I_{1-84}	C_{62}	I_{1-84}	F_0	I_{1-84}	C_{63}	I_{1-84}	F_1	I_{1-5}	S_6	I_{7-84}
M_0	I_{1-84}	F_1	I_{1-84}	C_{71}	I_{1-84}	F_0	I_{1-84}	C_{72}	I_{1-84}	F_0	I_{1-84}	C_{73}	I_{1-84}	F_1	I_{1-6}	S_7	I_{8-84}

X = signaling bit for network control
I_i = information bit from channels
F_i = frame synchronization bit with logic value i
C_{ij} = jth stuff code bit of ith channel
S_i = bit from ith channel available for dedicated stuffing
P = parity bit
M_i = multiframe synchronization bit with logic value i

TABLE 4.9   C-Bit Parity M-Frame Structure for DS-3 Multiplexer

X	I_{1-84}	F_1	AIC	I_{1-84}	F_0	I_{1-84}	NA	I_{1-84}	F_0	I_{1-84}	FEA	I_{1-84}	F_1	S_1	I_{2-84}	
X	I_{1-84}	F_1	DL	I_{1-84}	F_0	I_{1-84}	DL	I_{1-84}	F_0	I_{1-84}	DL	I_{1-84}	F_1	I_1	S_2	I_{3-84}
P	I_{1-84}	F_1	CP	I_{1-84}	F_0	I_{1-84}	CP	I_{1-84}	F_0	I_{1-84}	CP	I_{1-84}	F_1	I_{1-2}	S_3	I_{4-84}
P	I_{1-84}	F_1	FEBE	I_{1-84}	F_0	I_{1-84}	FEBE	I_{1-84}	F_0	I_{1-84}	FEBE	I_{1-84}	F_1	I_{1-3}	S_4	I_{5-84}
M_0	I_{1-84}	F_1	DL	I_{1-84}	F_0	I_{1-84}	DL	I_{1-84}	F_0	I_{1-84}	DL	I_{1-84}	F_1	I_{1-4}	S_5	I_{6-84}
M_1	I_{1-84}	F_1	DL	I_{1-84}	F_0	I_{1-84}	DL	I_{1-84}	F_0	I_{1-84}	DL	I_{1-84}	F_1	I_{1-5}	S_6	I_{7-84}
M_0	I_{1-84}	F_1	DL	I_{1-84}	F_0	I_{1-84}	DL	I_{1-84}	F_0	I_{1-84}	DL	I_{1-84}	F_1	I_{1-6}	S_7	I_{8-84}

X = signaling bit for network control
I_i = information bit from channels
F_i = frame synchronization bit with logic value i
S_i = bit from ith channel, which is always stuffed
P = parity bit
M_i = multiframe synchronization bit with logic value i
AIC = application identification channel
NA = reserved for network application
FEA = far-end alarm
DL = data link
CP = C-bit parity
$FEBE$ = far-end block error

bits found in the third frame are used to carry path parity (CP) bits. At the transmitter, the CP bits are set equal to the P bits. The 3 CP bits carry parity information and are not altered anywhere along the path from DS-3 source to DS-3 sink. At any intermediate point in the path, or at the far end, wherever the DS-3 signal is terminated, the receiver can determine if a parity error has occurred up to that point. Conversely, the P bits (for both the M13 and C-bit parity formats) are corrected at each intermediate point in the path so that there is no end-to-end monitoring, just section-to-section monitoring. The occurrence of a C-bit parity error on a block of data (4704 bits) received at the near end is transmitted back to the far end by use of the far-end block error ($FEBE$) bits located in the fourth frame. Monitoring of both the $FEBE$ and CP bits allows determination of performance for both directions of transmission, from either end of the path. The X bits are also redefined, to transmit an indication of "degraded seconds" from the far end to the near-end terminal. Degraded seconds are defined as one-second intervals in which an incoming out-of-frame (OOF) or alarm-indication-signal (AIS) is detected at the far-end terminal.

SYNTRAN Format [25].   SYNTRAN (Synchronous Transmission) redefines the DS-3 to provide a synchronous hierarchy. This reformatting of the DS-3 eliminates two stages of pulse stuffing to allow synchronous digital cross-connect, switching, add, and drop of DS-1 and DS-0 signals without the need for demultiplexing and remultiplexing. Although SYNTRAN is compatible with DS-3 transmission facilities, it is incompatible with both M13 and C-bit parity multiplexers. The basic frame structure of the asynchronous DS-3 format is retained in SYNTRAN in order to use the parity bits for monitoring and control. However, the 4704 information bits per multiframe are reorganized into 588 8-bit bytes, and the 21 C bits are reassigned to other functions such as restoration, provisioning, maintenance, and testing. Furthermore, the 4760-bit multiframe of the standard DS-3 becomes part of a 699-multiframe structure called the "synchronous superframe." The length of this synchronous superframe is selected to coincide with an integral number of 125-μsec DS-1 frames. Each of the 699 multiframes is identified, verified, and numbered by use of the C bits, which also perform the function of superframe synchronization. The SYNTRAN format accommodates three classes of signals: (1) byte synchronous, which permits direct access to both DS-1 and DS-0 signals; (2) bit synchronous, which permits direct access to DS-1 signals; and (3) asynchronous DS-1, as done in the M13 and C-bit parity formats. These signal classes may be transported singly or mixed on the same DS-3. The chief advantage to SYNTRAN stems from its ability to add and drop DS-0 or DS-1 signals in an *Add-Drop Multiplexer (ADM)*. One example is the ADM 3/X, which interfaces two DS-3 signals and can add or drop any number of DS-1s, a function that otherwise would require back-to-back M13 multiplexers [26]. Although the SYNTRAN format has been standardized [25], it has not gained popularity because of its basic incompatibility with existing asynchronous digital multiplexer, cross-connect, and switching equipment.

4.4.2 CEPT Digital Hierarchy

The CEPT digital hierarchy employs four levels of multiplex as shown in Figure 4.28 and described in CCITT Rec. G.702. These levels are sometimes referred to as European- or E- levels; for example, E-1 refers to the 2.048-Mb/s rate. All four levels are standardized by the CCITT. The 2.048-Mb/s frame format and multiplex characteristics are given by G.704 and G.732, respectively. Standards for the upper three levels exist for both positive and positive-zero-negative pulse stuffing. For the second level, these CCITT standards are G.742 and G.745, respectively; for the third level, G.751 and G.753, respectively; and for the fourth level, G.751 and G.754, respectively. The following descriptions are based on these CCITT standards [2].

CEPT First-Level Multiplex

The CEPT standard for PCM multiplexing is illustrated in Figure 4.31. Each frame consists of 30 8-bit PCM words (time slots 1–15 and 17–31), an eight-bit frame synchronization word (time slot 0), and an eight-bit signaling word (time slot 16)—a total of 32 8-bit words or 256 bits. Since there are 8000 frames transmitted per second, the resulting aggregate rate is 2,048,000 b/s. The multiframe structure shown in Figure 4.31b comprises 16 frames, with signaling information inserted in time slot 16 of each frame. The first 8-bit signaling time slot is actually used for multiframe alignment. The next 15 time slots for signaling provide 4 signaling bits per channel for all 30 speech channels.

A 60-channel transcoder can be used to convert two 30-channel, 2.048-Mb/s PCM signals to one 60-channel, 2.048-Mb/s ADPCM signal. The frame structure is identical to that shown in Figure 4.31. Time slots 1 to 15 and 17 to 31 may contain 64-kb/s PCM or data signals as in Figure 4.31 or two 4-bit samples corresponding to two 32-kb/s ADPCM signals. Time slots 0 and 16 are used in the same manner as Figure 4.31, except that signaling in time slot 16 is provided for up to 60 channels. The 60-channel transcoder is standardized in CCITT Rec. G.761 [2].

CEPT Higher Level Multiplex

All CEPT higher level multiplex are based on a cyclic bit interleaving multiplexing scheme in which the four tributaries are sampled in order of their port number. However, multiframe size and organization differ among the

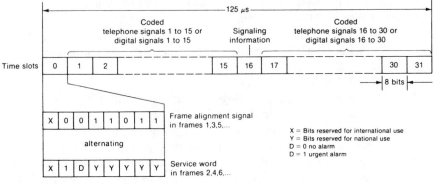

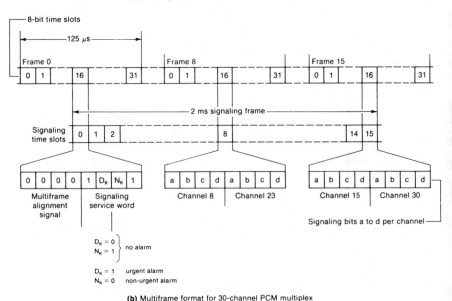

FIGURE 4.31   Frame Structure for 30-Channel PCM (Reprinted by Permission of Siemens AG)

various levels and between the positive and positive-zero-negative stuffing multiplexers for each level. In the case of the second-level multiplex, which operates at an aggregate rate of 8.448 Mb/s, Rec. G.742 specifies an 848-bit multiframe, while Rec. G.745 specifies a 1056-bit multiframe. In a similar manner, for the third-level multiplex, which has an aggregate rate of 34.368 Mb/s, Rec. G.751 prescribes a 1536-bit multiframe and Rec. G.753 prescribes a 2148-bit multiframe. Finally, for the fourth-level multiplex, with a

139.264 Mb/s aggregate rate, Rec. G.751 prescribes a 2928-bit multiframe and Rec. G.754 prescribes a 2176-bit multiframe. Rec. G.751 also specifies two methods to achieve the 139.264-Mb/s level, by multiplexing 4 level-three signals at 34.368 Mb/s or by directly multiplexing 16 level-two signals at 8.448 Mb/s. Table 4.10 provides a more detailed picture of the frame structure for the 8.448-Mb/s second-level multiplex based on CCITT Rec. G.742. Table 4.11 summarizes general characteristics of all CEPT higher-level digital multiplexers.

4.4.3   Japanese Digital Hierarchy

Figure 4.28 shows the basic Japanese digital hierarchy, which has five levels, the first two of which have the same data rate as the North American digital hierarchy. The first-level multiplex combines 24 voice channels using 64-kb/s μ-law PCM. The Japanese second-level multiplexer is based on the synchronous frame structure specified by CCITT Rec. G.704, which is incompatible with the North American M12, although both have the same aggregate rate.

Whereas the M12 interleaves information, control, and synchronization bits in a 1176-bit multiframe, the Japanese second-level multiplex has a 789-bit multiframe which is organized into 96 octets of information, 16 bits of signaling, and 5 bits of overhead, as shown in Table 4.12. The information octets correspond to samples from each of the 96 DS-0 channels, which constitute a DS-2 signal. According to CCITT Rec. G.704, the 16 ST bits reserved for signaling are allocated to 16 different DS-0 channels in each of 6 frames, preceded by a 16-bit multiframe synchronization code and followed by 16 bits for alarm indication. As specified in Rec. G.704, the 5 F bits are allocated to frame synchronization, a data link, a remote-end alarm, a CRC-5 code, and spare bits, as shown in Table 4.12(b).

The third and fourth levels of multiplex are as prescribed in CCITT Rec. G.752. Table 4.13 summarizes characteristics of the higher-level multiplex in the Japanese digital hierarchy.

4.4.4   Hybrid Digital Hierarchy

To facilitate interworking of dissimilar digital hierarchies, a hybrid hierarchy has been developed and standardized in CCITT Rec. G.802 [27]. As shown in Figure 4.28, the hybrid hierarchy comprises the rates of 2.048, 6.312, 44.736, and 139.264 Mb/s. Three stages of multiplexers are required to develop these rates. The 2.048-6.312–Mb/s multiplexer combines three E-1 signals into a DS-2 signal, as specified in CCITT Rec. G.747. The 6.312-44.736–Mb/s multiplexer is identical to the M23 as specified in Rec. G.752. And the 44.736-139.264 Mb/s multiplexer combines three DS-3 signals into a European fourth-level signal, as specified in CCITT Rec. G.755.

TABLE 4.10 Frame Structure of 8.448-Mb/s Second-Order Multiplex

Frame Structure

F_1	F_1	F_1	F_1	F_0	F_1	F_0	F_0	F_0	F_0	X	Y	I_{13}	$\cdots$	I_{212}
C_{11}	C_{21}	C_{31}	C_{41}	I_5			·							I_{212}
C_{12}	C_{22}	C_{32}	C_{42}	I_5		·								I_{212}
C_{13}	C_{23}	C_{33}	C_{43}	S_1	S_2	S_3	S_4	I_9			·			I_{212}

Frame length: 212 bits
Multiframe length: 848 bits
Bits per channel in a multiframe: 206 bits
Maximum stuffing rate per channel: 9962 b/s
Nominal stuffing ratio: 0.424

F_i = frame synchronization bit with logic value i
I_i = information bit from channels
X = alarm indication to remote multiplexer
Y = reserved for national use
C_{ij} = jth stuff code bit of ith channel
S_i = bit from ith channel available for stuffing

TABLE 4.11   Characteristics of CEPT Digital Multiplex

Multiplex Characteristic	Second-Level CCITT Rec.		Third-Level CCITT Rec.		Fourth-Level CCITT Rec.	
	G.742	G.745	G.751	G.753	G.755	G.754
Aggregate bit rate (Mb/s)	8.448	8.448	34.368	34.368	139.264	139.264
Tolerance on aggregate rate (ppm)	±30	±30	±20	±20	±15	±15
Aggregate digital interface*	HDB3	HDB3	HDB3	HDB3	CMI	CMI
Channel bit rate (Mb/s)	2.048	2.048	8.448	8.448	34.368	34.368
No. of channels	4	4	4	4	4	4
Channel digital interface*	HDB3	HDB3	HDB3	HDB3	HDB3	HDB3
Frame structure	Table 4.10	CCITT G.745	CCITT G.751	CCITT G.753	CCITT G.751	CCITT G.754
Pulse stuffing technique	+	+/0/–	+	+/0/–	+	+/0/–
	Stuff code is 3-bit for + stuffing and 6-bit for +/0/– stuffing, with majority vote decoding used for all versions					
Multiplexing method	Cyclic bit interleaving in channel numbering order					
Timing signal	Derived from external as well as internal source					

*See Chapter 5 for a description of baseband transmission codes.

Other provisions of this CCITT standard for interworking deal with the first-level multiplex and speech encoding laws. International digital links between countries that have adopted different coding laws (A-law or μ-law) are specified to use A-law coding. Any conversion necessary between two countries is the responsibility of the country using μ-law coding, using the A-law/μ-law conversion tables given in Rec. G.711. Newly introduced first-level multiplex is required to provide 64-kb/s *clear channel capability* (also called *bit sequence independence*), which means that any binary sequence is allowed without the need to change any bit value. Existing 1.544-Mb/s multiplex is required to support only 56-kb/s bit sequence independent channels, in order to meet existing requirements for 1s density of at least one in every 8 bits.

4.4.5 SONET Digital Hierarchy

The Synchronous Optical Network (SONET) digital hierarchy has been developed to provide common standards and technology for applications of high-speed digital transmission. Initially, SONET standards were developed in North America under the auspices of Bellcore [28] and the ANSI

TABLE 4.12   Frame Structure of Japanese Second-Level Multiplex

(a) 789-bit Frame Format

I_{1-8}	I_{9-16}	I_{17-24}	- - - - - -	$I_{753-760}$	$I_{761-768}$	ST_1	ST_2	- - -	ST_{16}	F_1	- - -	F_5

I = information bits
ST = signaling bits
F = frame bits

(b) Allocation of F Bits

		Bit Number			
Frame number	785	786	787	788	789
1	1	1	0	0	M
2	1	0	1	0	0
3	X	X	X	A	M
4	E_1	E_2	E_3	E_4	E_5

M = data link bit
X = spare bits
A = remote-end alarm bit
E_i = ith bit of CRC-5

(c) Allocation of ST Bits

Frame number	Use of ST bits
1	Multiframe synchronization
2	Signaling for channels 1–16
3	Signaling for channels 17–32
4	Signaling for channels 33–48
5	Signaling for channels 49–64
6	Signaling for channels 65–80
7	Signaling for channels 81–96
8	Alarm indication signal

T1X1 Committee [29], but the CCITT has followed suit with its own international standard under the name Synchronous Digital Hierarchy (SDH) [2]. The basic building block of SONET and SDH is a 51.840-Mb/s signal, which is termed the Level 1 Synchronous Transport Signal (STS-1). A family of higher-speed rates, formats, and electrical interfaces is defined at a rate of $N \times 51.840$ Mb/s, or STS-N, where N is an integer ranging up to 255. The basic STS-1 is divided into a portion for the payload and a portion for overhead. The payload is designed to carry DS-3 signals or a number of subrates, including DS-1, DS-1C, E-1, and DS-2. To accommodate rates below DS-3, a virtual tributary (VT) is used to transport and switch pay-

TABLE 4.13   Characteristics of Japanese Digital Multiplex

Multiplex Characteristic	Second-Level CCITT Rec. G.704	Third-Level CCITT Rec. G.752	Fourth-Level CCITT Rec. G.752
Aggregate bit rate (Mb/s)	6.312	32.064	97.728
Tolerance on aggregate rate (ppm)	±30	±10	±10
Aggregate digital interface*	B6ZS	Scrambled AMI	Scrambled AMI
Channel bit rate (Mb/s)	0.064	6.312	32.064
No. of channels	96	5	4
Channel digital interface*	Bipolar	B6ZS	Scrambled AMI
Frame structure	Table 4.12	CCITT Rec. G.752	CCITT Rec. G.752
Pulse stuffing technique	Not used	Positive pulse stuffing with 3-bit stuff code and majority vote decoding	
Multiplexing method	Byte interleaving	Cyclic bit interleaving in channel numbering order	
Timing signal	Derived from external as well as internal source		

*See Chapter 5 for a description of baseband transmission codes.

loads in a uniform manner. Rates above DS-3 may also be mapped into payloads, for example, the 139.264-Mb/s CEPT level-four signal is mapped into a STS-3 signal. Also, CCITT Recs. G.781, G.782, and G.783 have been developed for multiplexing a number of asynchronous digital hierarchy signals to an STS-M rate [30].

The optical counterpart of the STS-N is the Optical Carrier–Level N (OC-N), which is generated by a direct optical conversion of the electrical STS-N signal. The OC-N rates are identical to the STS-N rates, which are the result of synchronous multiplexing of the basic 51.840-Mb/s signals. Standard rates are identified for $N = 1, 3, 9, 12, 18, 24, 36,$ and 48. Further discussion of SONET formats and applications is given in Chapter 8 on fiber optic transmission.

4.5   Statistical Multiplexing and Speech Interpolation

With conventional time-division multiplexing, each channel is assigned a fixed time slot. Fixed frames are transmitted continuously between the multiplexer and demultiplexer. This means that each of the time slots belonging to an inactive channel must be filled with an "idle" character in order to maintain proper framing. Since human speech and most data terminals are not continuously active, the average activity tends to be low compared to the available bandwidth. **Statistical multiplexing** removes the idle time and fills it with active samples from another channel. Thus time slots are no longer dedicated to a particular channel but rather can be shared among all active sources. Relative to conventional TDM, the advantage of statistical multiplexing can be thought of as an increase in the number of sources that can be handled, or as a decrease in the transmission bit rate or bandwidth required to multiplex a certain number of sources. The degree of concentration, sometimes referred to as the **gain,** depends on the signal activity and the number of signals being multiplexed. This section will describe statistical multiplexing of data and voice signals, using both circuit and packet multiplexing techniques.

4.5.1   Statistical Multiplexing of Data

Figure 4.32 illustrates the improved transmission efficiency that can be provided with statistical time-division multiplexing (STDM) by dynamically allocating transmission bandwidth to whichever terminals are active at any one time. Statistical multiplexing results in transmission efficiency improvements by a factor of 2 to 1 or more, as in the example of Figure 4.32.

As data arrive at the statistical multiplexer, they are placed in a buffer that has been dynamically assigned for each active channel. Frames of data are then assembled as the data are systematically removed from these buffers. However, buffer overflow can occur during prolonged peak traffic activity or excessive retransmission of data unless the terminal equipment is signaled to cease transmission of data. Transmission flow control on incoming channels can be invoked by using either in-band or out-of-band traffic control signals. In-band signals are generally nonprinting control characters, which are transmitted to the terminal by the multiplex in-band as if they were data characters. Out-of-band traffic control is achieved by placing control signals on the separate control lines available in data terminals.

Should the terminal equipment continue sending data during overload, buffer overflow will occur. The buffer recovers from overflow by dumping data of one or more terminals, transmitting a "data lost" message to the affected terminal, and possibly disconnecting the terminal. When total buffer utilization drops below the preset overload condition, flow control will be released by either in-band or out-of-band signals to allow transmission to be resumed.

Most statistical multiplexers also utilize the CCITT Rec. X.25 level 2 link protocol procedure. This procedure calls for transmission of information in variable-length frames. The fields within the frame are:

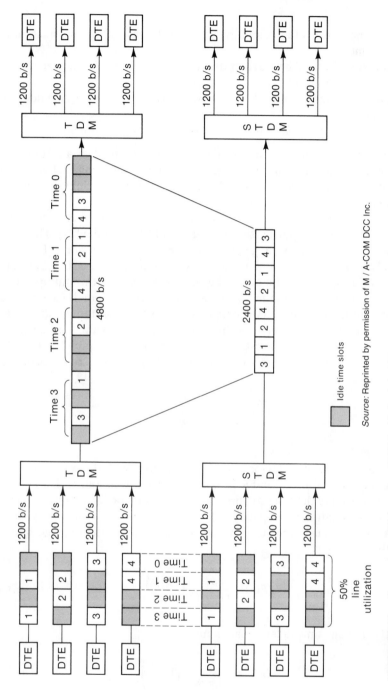

FIGURE 4.32 Comparison of Conventional TDM with Statistical TDM

Source: Reprinted by permission of M / A-COM DCC Inc.

247

- *Flag* synchronizes the demultiplexer with the rest of the information in the frame.
- *Address* indicates the source or destination of a particular message.
- *Control* identifies the type of message.
- *Data* is a variable-length field into which data from input channels are packed.
- *Cyclic redundancy check* (CRC) performs error detection.

The error-checking function of the link protocol involves a retransmission of the frame in error. This is made possible by retaining each frame in buffer storage until an acknowledgment is received from the far end of the link.

The performance of statistical multiplexers may be evaluated using queueing models, in particular, the so-called M/D/1 model [31]. The multiplexer is modeled as a single-server queue with random (Poisson) arrivals of the user data and a constant service time (that is, fixed bit rate and continuous transmission at the aggregate side of the multiplexer). Queueing occurs if the number of active input data channels exceeds the capacity of the multiplexer, forcing some users to be buffered and to wait for service. The delay incurred by a particular user is the waiting time, which depends on the statistics of the arriving traffic, plus the time to serve the user. The average arrival rate (in b/s) for the sum of all incoming channels is

$$\lambda = \sum_{i=1}^{N} \alpha_i R_i \qquad (4.41)$$

where α_i is the activity factor of each data source, with $0 < \alpha < 1$, N is the number of input sources, and R_i is the data rate for each source. The service time (s) is simply the time required to transmit one bit, given by the reciprocal of the multiplexer capacity (C). For this queueing model of a STDM, the average queue occupancy, which is equivalent to the average buffer length required, is

$$E(n) = \frac{\rho}{(1-\rho)} \left(1 - \frac{\rho}{2}\right) \qquad (4.42)$$

and the average time delay for an input is

$$E(T) = \frac{s}{(1-\rho)} \left(1 - \frac{\rho}{2}\right) \qquad (4.43)$$

where ρ is the utilization, or fraction of the multiplexer capacity being used, defined as

$$\rho = \lambda \, s \qquad (4.44)$$

E(n) given by (4.42) can be interpreted as the average number of users (bits, packets, characters, frames, and so on) appearing in the queue. Thus (4.42) provides the average buffer length required in a STDM. See Problem 4.29 for further interpretation of (4.42) and (4.43). It should be noted here that these equations describe only average values. Greater buffer sizes are required and longer time delays will occur with variance in the queue length. Variance is proportional to the utilization so that a higher level of utilization will require longer buffer lengths to avoid buffer overflow.

4.5.2   Statistical Multiplexing of Voice

A form of statistical multiplexing can also be applied to voice channels in order to realize concentrations of 2:1 or more. The technique is known as **time assignment speech interpolation (TASI)** for analog transmission and **digital speech interpolation (DSI)** for digital transmission. The principle involved is the filling of the silent gaps and pauses occurring in one voice channel with speech from other channels. The receiver reorders the bursts of speech into the required channel time slots. The technology consists of digital speech processing (such as PCM), buffer memory for speech storage, and microprocessors for switching and diagnostics. Speech interpolation devices have been employed by AT&T [32], COMSAT [33], and European telephone companies [34]. Common configurations include 48 voice channels concentrated onto 24 transmission channels (abbreviated 48:24) for North American systems and 60:30 for European systems.

The operation of speech interpolation devices is similar to that of statistical multiplexers for data. As shown in Figure 4.33, incoming speech is recognized by speech activity detectors and simultaneously is stored in a buffer dynamically assigned to that voice channel. Data modem signals are also recognized and provided a dedicated connection, which then reduces the number of channels available for effective speech interpolation. Stored speech bursts are transmitted as soon as transmission bandwidth becomes available. The buffer memory allocated to a channel is of variable length to allow storage of speech until transmission is possible. After transmission of a speech burst the buffer is returned to a pool for future use. Periodically a control signal is transmitted that informs the receiver where to place each incoming speech burst. With sufficiently high

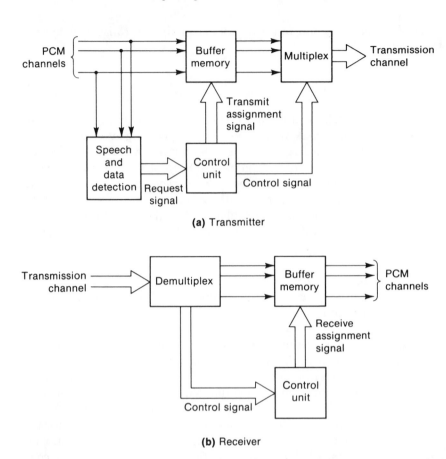

(a) Transmitter

(b) Receiver

FIGURE 4.33   Block Diagram of a Digital Speech Interpolation Device

speech activity on incoming channels or a large number of data signals, buffer overflow can result in loss of speech segments. Overload conditions can be controlled by blocking any unseized channels from being seized, by dropping segments on a controlled basis [35], or by reducing the code rate per voice channel for DSI systems [36]. The loss of speech segments, due to overload or delay in detecting a speech burst, is known as **clipping** or **freezeout** and is the limiting factor in the performance of speech interpolation systems.

DSI Performance
The key performance parameters of a DSI system are the probability of clipping and the probability of blocking. In order to satisfy the limits on

clipping and blocking, the DSI concentration or gain must not exceed certain levels. DSI gain is defined as

$$g = \frac{n-k}{c-k} \tag{4.45}$$

where n is the number of subscriber channels, c is the number of transmission channels, and k is the number of data channels that require full-period, unrestricted operation. The choice of a gain factor is also a function of speech statistics. The speech activity factor (α) for typical two-way conversations rarely exceeds 40 percent, measured as the ratio of talk bursts to the sum of talk bursts plus silent periods for a given speaker. As the number of subscribers increases, the number of simultaneous speakers approaches a limit set by the reciprocal of the average speech activity factor. Thus, for α = 0.4 the DSI gain will approach 2.5 when n becomes large. As the number of subscribers is reduced, the DSI gain that can achieved for given limits on clipping and blocking will become smaller.

The clipping probability or freezeout fraction can be described by the binomial distribution

$$B(c,n,p) = \sum_{x=c}^{n} \frac{n!}{x!(x-n)!}\, p^x (1-p)^{n-x} \tag{4.46}$$

which is the probability that the number of simultaneous speakers on n subscriber channels with activity α will equal or exceed the number of transmission channels c. Here p is equal to the speech activity factor α. In actual DSI systems, some clipping is allowed because it has negligible effect on the listener. An exponential distribution for p can be used that accounts for allowed clipping,

$$p(T) = \alpha e^{-(T/T_s)} \tag{4.47}$$

where T is the allowed clipping duration and T_s is the average speech burst duration. For given values of T, T_s, and $B(c,n,p)$, (4.46) can be used to plot c versus n for various values of α. Figure 4.34 is such a plot for one industry standard: the probability of a clip of duration ≥ 50 ms should be less than 0.02 with T_s assumed to be 1.2 s [37,38]. The gain is determined by noting the number of transmission channels (c) required for a given number of subscriber channels (n).

Calculations of the probability of blocking due to buffer overflow in a DSI system can be made using the Erlang delay formula [39]. In a DSI system, the maximum number of users served is the sum of the number of

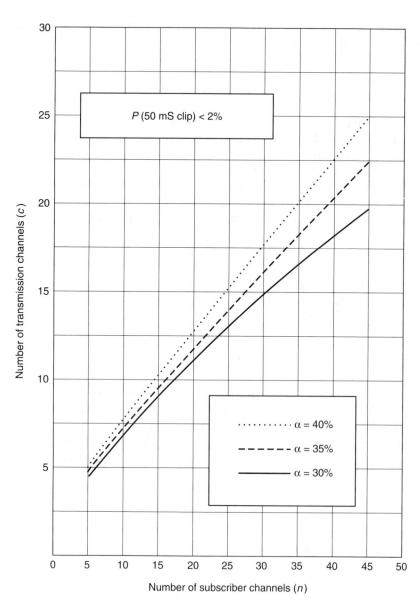

FIGURE 4.34   DSI Gain (n/c) Allowed by Clipping Criterion

available transmission channels (c) plus the number of buffer waiting positions (b). Defined in terms of total buffer length T_b and average speech burst T_s, the number of buffer waiting positions is

$$b = \frac{T_b}{T_s} \tag{4.48}$$

The probability of the buffer being in an overflow condition is equal to the probability that $c + b$ speech bursts await service simultaneously, which is given by

$$P_{c+b} = \frac{a^{c+b}}{c! \; c^b} P_0 \tag{4.49}$$

where

$$P_0 = \left[\sum_{k=0}^{c-1} \frac{a^k}{k!} + \frac{a^c}{c!} \sum_{i=0}^{b} \left(\frac{a}{c} \right)^i \right]^{-1} \tag{4.50}$$

$$a = (T_s + T_d) \frac{n\alpha}{T_s} \tag{4.51}$$

and T_d is the average delay time for a speech burst. Using (4.49), (4.50), and (4.51), a calculation of c versus n can be done as a function of buffer overflow probability. As an example, Figure 4.35 plots c versus n for three different probabilities of buffer overflow and typical values of the design parameters ($\alpha = 0.4$, $T_s = 1.2$ s, $T_d = 40$ ms, $T_b = 8.2$ s). Equations (4.46) and (4.49) can be used to determine the total effect of clipping plus blocking objectives on the DSI gain factor. In some cases, one objective will be more stringent than the other, allowing the designer to ignore the less stringent criterion. For example, if an α of 0.4 and an overflow probability of 0.01 are assumed, a comparison of Figures 4.34 and 4.35 reveals that the clipping objective results in a smaller allowed gain, meaning that only the clipping criterion needs to be used in system design [40].

DSI Standards

DSI is often combined with low rate encoding of voice to provide increased concentration of subscriber channels to transmission channels. One version that has been standardized, in CCITT Rec. G.763, is the **Digital Circuit Multiplication Equipment (DCME),** which combines 32 kb/s ADPCM with DSI [2]. DCME is specified to accommodate speech, voice-band data, and unrestricted 64-kb/s signals. The DCME transcoding algorithm converts 64-kb/s PCM into 24-kb/s ADPCM during overload conditions, 32-kb/s ADPCM during normal conditions, and 40-kb/s ADPCM for voice-band

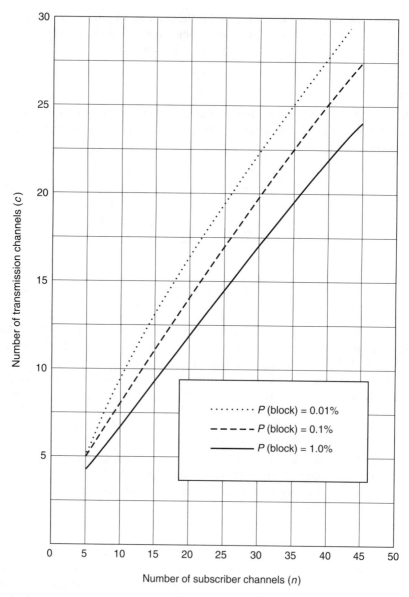

FIGURE 4.35   DSI Gain (n/c) Allowed by Blocking Criterion

data. By varying the coding rate, increased quantization noise is traded against clipping of speech. To maintain acceptable speech performance, an average of 3.7 bits per sample is maintained and the probability of clipping exceeding 50 ms is kept below 2 percent. Compression of 5:1 is typical for DCME, for example, 120 PCM channels in a single 2.048-Mb/s channel.

DCME is available in both 1.544-Mb/s and 2.048-Mb/s versions. Data rates on the subscriber and transmission sides are normally the same. In the case of 1.544-Mb/s systems, the subscriber side is specified to meet the characteristics shown in Table 4.3, while the transmission side uses a 193-bit frame consisting of one framing bit and 192 bits of DCME traffic and control signals. Likewise, in the case of 2.048-Mb/s systems, the subscriber side is specified to meet the characteristics of Table 4.3, and the transmission side uses time slot 0 as in Figure 4.31 and the remaining 31 time slots to carry traffic and control signals.

Operational modes of DCME include point-to-point, multibundle, and multidestination. Point-to-point applications have a single destination and all channels are grouped without separation. Applications of the multibundle, also called multiclique, mode divide the pool of transmission channels into bundles, each associated with a different destination. Subscriber channels destined for a certain route are bundled together and kept separate from other bundles, which limits the use of digital speech interpolation to just those channels in a given bundle. Operation in the multidestination mode permits any subscriber channel to be assigned to any transmission channel over any of a number of possible routes. The receiving unit uses the preassignment channel to select those subscriber channels destined for that location. This mode allows greater DSI gain than the multibundle mode.

4.5.3 Packet Multiplexing

The popularity of packetized data communications has led to the application of the same principles to voice communications. Such applications have usually included digital speech interpolation and embedded ADPCM as means to compress the bit rate and adapt to network congestion. The choice of packetization protocol must overcome the problems that characterize a packet network, such as variable delay in packet arrival times, lost packets, and network congestion. To minimize the effects of delay and lost packets, voice packet lengths of 50 ms or less are used. A time stamp can be assigned to each packet so that the receiver may place incoming packets in their correct order. Packets are transmitted only during talk bursts, and at the receiver, silent periods are reconstructed and filled with a background noise. To distinguish lost packets from silent periods, each packet is flagged to indicate whether it is part of a burst or the last packet in a burst. A maximum allowable delay is specified, and late packets are dropped at the terminating end. Under overload conditions, analysis and tests have shown that bit dropping offers a more graceful degradation to voice than dropping of packets. Furthermore, dropped packets in circuits carrying voice-band data may cause loss of synchronization in the receiving modem [41]. The use of embedded ADPCM allows bit dropping at any point in the network, but the packet protocol must provide a means to signal any intermediate nodes and the terminating node of any bit dropping that has occurred up to that point in the network.

Standards of Packetized Voice Protocol have been published by ANSI in T1.312-1990 [42] and the CCITT in G.764 [2] and will form the basis of our discussion here. Before packetization, voice signals are sampled at 8000 Hz and coded by one of several allowed techniques, including PCM, fixed-rate ADPCM, and variable-rate (embedded) ADPCM. Coded speech is transformed into packets that are 16-ms long and contain 128 voice samples from a single voice channel. The sample bits are arranged so that the least significant bits are all located in the same block, followed by the next least significant bits in the next block, and so on. For sake of illustration, we shall assume that embedded ADPCM with a 32-kb/s coding rate is to be used for all samples, so that there are four blocks corresponding to the four code bits. Figure 4.36 shows the resulting organization of the packet, which consists of a 10-byte header and four blocks of speech samples. All the least significant bits from the 128 samples are located in block 1, the next most significant bits are found in block 2, and the most significant bits are in blocks 3 and 4. The packet header contains information necessary for the terminating end to reconstruct the speech signals. This header information indicates the packet destination, the coding type (here assumed to be embedded ADPCM), the time stamp for proper ordering, and the number of droppable blocks.

The block-dropping algorithm used in Packetized Voice Protocol is based on the embedded ADPCM use of core and enhanced bits, of which the core bits must be retained but the remaining bits can be discarded with network congestion. To describe this algorithm, consider the classic queueing problem in which a packet is buffered while waiting to be transmitted. Now let L denote the buffer fill in packets and assume two thresholds of buffer fill, Q_1 and Q_2, are to be used to determine when to drop blocks. If L is smaller than Q_1, no block dropping is needed. When L exceeds Q_1 but is smaller than Q_2, block 1 containing the least significant bits is dropped. When L exceeds Q_2, blocks 1 and 2 are both dropped, thus reducing the information in the packet to the two most significant bits. The header indicates the initial and current number of droppable blocks in the packet. This header information allows packet delineation and block dropping at intermediate nodes, and provides for decoding and speech reconstruction at the terminating end. Since the

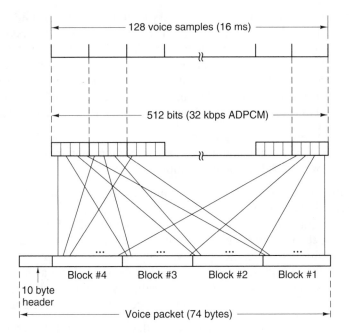

128 voice samples (16 ms)

512 bits (32 kbps ADPCM)

Block #4   Block #3   Block #2   Block #1

10 byte
header

Voice packet (74 bytes)

FIGURE 4.36   Structure of Voice Packet with Embedded ADPCM [43] (© 1989 IEEE)

packet interval (16 ms) is fixed, differences in coding type or block dropping, either at the originating node or intermediate nodes, lead to variable-length packets rather than fixed-length packets [43].

4.6   Summary

Here we will summarize essential TDM design and performance considerations. These criteria can be used by the system designer in specifying and selecting TDM equipment.

Organization of a TDM frame depends on the number of channels to be multiplexed, the channel bit rates, the relationship between channel input timing and multiplexer timing, and the number and location of overhead bits. These overhead bits are required for frame alignment and possibly other functions. Their location within the frame can be bunched to facilitate frame alignment in the demultiplexer or distributed to protect against bursts of transmission errors. A fixed timing relationship between each channel and the multiplexer permits synchronous multiplexing, the simplest form of TDM. For asynchronous channels, more complex TDM schemes such as pulse stuffing or transitional encoding are required to synchronize each channel to multiplexer timing. Asynchronous forms of multiplexing require additional overhead bits that must be transmitted within the frame to allow reconstruction of the original asynchronous channels by the demultiplexer. The choice of frame alignment pattern, both its length and content, and the frame synchronization strategy, which typically includes a search mode, maintenance mode, and some accept/reject criteria for each mode, are selected in accord with the following performance characteristics:

- Time to acquire frame alignment
- Time to detect misalignment and reacquire frame alignment
- Time between false resynchronizations

Calculation of these synchronization times is possible by use of the equations given in Section 4.2.

Pulse stuffing is the most efficient and commonly used synchronization scheme for the multiplexing of asynchronous channels. Two basic forms of pulse stuffing are available: positive only or positive-negative. Positive stuffing is less complex, but positive-negative stuffing facilitates the multiplexing of synchronous channels along with asynchronous channels. Four design parameters are of primary importance to the system designer:

- *Coding of stuff decisions:* Errors in the decoding of stuff decisions result in loss of synchronization in the affected channel. Therefore simple redundancy coding of stuff decisions is usually employed to protect each channel against transmission bit errors. The choice of code length determines a key performance characteristic—the time to loss of synchronization due to stuff decoding error, as given by Equation (4.26).
- *Maximum stuffing rate:* This parameter determines the upper limit on the allowed frequency offset between the timing signal used with the channel input and the timing signal internal to the multiplexer. Setting this parameter arbitrarily high is undesirable, however, since an increase in the maximum stuffing rate raises the percentage of overhead bits, lowers the time between stuff decoding errors, and may increase waiting time jitter.
- *Stuffing ratio:* This parameter is given by the ratio of actual stuffing rate—which is determined by the difference in frequency between the asynchronous channel input and the synchronous multiplexer—to the maximum stuffing rate, which is a design parameter. By careful choice of the stuffing ratio, waiting time jitter can be controlled to within specified levels.
- *Smoothing loop:* In the demultiplexer, the timing associated with each asynchronous channel is recovered by using a clock smoothing loop. Design of the loop involves a choice of loop type (a first-order phase-locked loop is typical) and loop bandwidth. This loop also must attenuate

the jitter caused by the insertion and deletion of stuff bits. Residual jitter in the recovered clock must be within the range of input jitter tolerated by the interfacing equipment.

Transitional encoding is commonly used with asynchronous data channels because of its simplicity and transparency to data format. The transmission overhead is high, however, typically equal to or greater than the data rate itself. The system designer must also consider the input jitter requirements for the data terminal equipment, since transitional encoding introduces jitter in the process of reconstructing data transitions at the receiver. The magnitude of this jitter can be reduced by increasing the number of coding bits, as indicated in Equation (4.40), but at the expense of increased overhead.

Statistical multiplexing and speech interpolation are two multiplexing techniques that apply to data and voice channels, respectively, using the same principle of filling inactive time of one channel with active time from another channel. Several parameters are of interest to the system designer:

- The *gain* or concentration ratio—that is, the ratio of the number of channel inputs to the number of channel time slots available for transmission
- The allowed blocking of incoming channels or clipping of existing channels due to an overload of channel activity
- The buffer capacity required and user time delay allowed
- The allowed reduction in voice coding rate due to an overload
- The allowed number of channel inputs that require a dedicated, full-period channel for transmission

References

1. P. Bylanski and D. G. W. Ingram, *Digital Transmission Systems* (London: Peter Peregrinus Ltd., 1980).
2. CCITT Blue Book, vol. III 4, *General Aspects of Digital Transmission Systems; Terminal Equipments* (Geneva: ITU, 1989).
3. V. L. Taylor, "Optimum PCM Synchronizing," *Proceedings of the National Telemetering Conference,* 1965, pp. 46–49.
4. M. W. Williard, "Mean Time to Acquire PCM Synchronization," *Proceedings of the National Symposium on Space Electronics and Telemetry,* Miami Beach, October 1972.
5. R. H. Barker, "Group Synchronizing of Binary Digital Systems." In *Communications Theory* (New York: Academy Press, 1953).
6. W. Feller, *An Introduction to Probability Theory and Its Applications* (New York: Wiley, 1957).
7. D. R. Cox and H. D. Miller, *The Theory of Stochastic Processes* (New York: Wiley, 1965).

8. B. Weesakul, "The Random Walk Between a Reflecting and Absorbing Barrier," *Ann. Math. Statistics* 32(1971):765–769.

9. J. G. Kemeny and J. L. Snell, *Finite Markov Chains* (New York: Van Nostrand, 1960).

10. M. W. Williard, "Optimum Code Patterns for PCM Synchronization," *Proceedings of the National Telemetering Conference*, Washington D.C., May 1962.

11. D. L. Duttweiler, "Waiting Time Jitter," *Bell System Technical Journal* 51(January 1972):165–207.

12. CCITT Green Book, vol. III.3, *Line Transmission* (Geneva: ITU, 1973).

13. Federal Republic of Germany, *Jitter Accumulation on Digital Paths and Jitter Performance of the Components of Digital Paths*, vol. COM XVIII, no. 1-E (Geneva: CCITT, 1981), pp. 77–83.

14. CCITT Orange Book, vol. VII, *Telegraph Technique* (Geneva: ITU, 1977).

15. C. G. Davis, "An Experimental Pulse Code Modulation System for Short-Haul Trunks," *Bell System Technical Journal* 41(January 1962):1–24.

16. H. H. Henning and J. W. Pan, "The D2 Channel Bank-System Aspects," *Bell System Technical Journal* 51(October 1972):1641–1657.

17. J. B. Evans and W. B. Gaunt, "The D3 PCM Channel Bank," *1973 International Conference on Communications,* June 1973, pp. 70-1–70-5.

18. H. A. Mildonian, Jr., and D. A. Spires, "The DID PCM Channel Bank," *1974 International Conference on Communications,* June 1974, pp. 7E-1–7E-4.

19. C. R. Crue and others, "D4 Digital Channel Bank Family: The Channel Bank," *Bell System Technical Journal* 61(November 1982):2611–2664.

20. P. Benowitz and others, "Digital Data System: Digital Multiplexers," *Bell System Technical Journal* 54(May/June 1975):893–918.

21. AT&T Technical Reference, TR 62421, *Accunet Spectrum of Digital Services,* December 1989.

22. ANSI T1.302-1989, "Digital Processing of Voice-Band Signals—Line Format for 32-kbit/s Adaptive Differential Pulse-Code Modulation (ADPCM)," December 1987.

23. ANSI T1.107-1988, "Digital Hierarchy—Formats Specifications," April 1989.

24. AT&T Technical Reference, PUB 54014, *Accunet T45 Service Description and Interface Specifications,* June 1987.

25. ANSI T1.103-1987, "Digital Hierarchy—Synchronous DS3 Format Specifications," August 1987.

26. Bellcore Technical Reference, TR-TSY-000010, *Synchronous DS3 Add-Drop Multiplex (ADM 3/X) Requirements and Objectives,* February 1988.

27. CCITT Blue Book, vol. III.5, *Digital Networks, Digital Sections and Digital Line Systems* (Geneva: ITU, 1989).

28. Bellcore Technical Reference, TR-TSY-000253, *Synchronous Optical Network (SONET) Transport Systems: Common Generic Criteria,* September 1989.

29. ANSI T1.105-1988, "Digital Hierarchy Optical Interface Rates and Formats Specifications," 1988.

30. R. Balcer and others, "An Overview of Emerging CCITT Recommendations for the Synchronous Digital Hierarchy: Multiplexers, Line Systems, Management, and Network Aspects," *IEEE Comm. Mag.,* (August 1990): 21–25.

31. M. Schwartz, *Telecommunication Networks: Protocols, Modeling and Analysis* (Reading, MA: Addison-Wesley, 1987).

32. R. L. Easton and others, "TASI-E Communications System," *IEEE Trans. on Comm.*, vol. COM-30, no. 4, April 1982, pp. 803–807.

33. J. H. Rieser, H. G. Snyderhoud, and Y. Yatsueuka, "Design Considerations for Digital Speech Interpolation," *1981 International Conference on Communications*, June 1981, pp. 49.4.1–49.4.7.

34. D. Lombard and H. L. Marchese, "CELTIC Field Trial Results," *IEEE Trans. on Comm.*, vol. COM-30, no. 4, April 1982, pp. 808–814.

35. P. A. Vachon and P. G. Ruether, "Evolution of COM 2, a TASI Based Concentrator," *1981 International Conference on Communications*, June 1981, pp. 49.6.1–49.6.5.

36. INTELSAT Earth Station Standard, IESS-501, "Digital Circuit Multiplication Equipment Specification: 32 kbit/s ADPCM with DSI," March 1989.

37. S. J. Campenella, "Digital Speech Interpolation," *COMSAT Technical Review* (Spring 1976): pp. 127–158.

38. J. M. Elder and J. F. O'Neil, "A Speech Interpolation System for Private Networks," *1978 National Telecommunications Conference*, pp. 14.6.1–14.6.5.

39. R. Cooper, *Introduction to Queueing Theory* (New York: Macmillan, 1972).

40. D. R. Smith and R. A. Orr, "Application of Reduced Rate Coding and Speech Interpolation to the DCS," *1988 MILCOM*, pp. 3.3.1–3.3.6.

41. M. H. Sherif, R. J. Clark, and G. P. Forcina, "CCITT/ANSI Voice Packetization Protocol," *International Journal of Satellite Communication* (1990): 429–436.

42. ANSI T1.312-1990, "Voice Packetization—Packetized Voice Protocol," 1990.

43. K. Sriram and D. M. Lucantoni, "Traffic Smoothing Effects of Bit Dropping in a Packet Voice Multiplexer," *IEEE Trans. on Comm.*, vol. 37, no. 7, July 1989, pp. 703–712.

44. F. G. Stremler, *Introduction to Communication Systems* (Reading, MA: Addison-Wesley, 1982).

45. CCITT Green Book, vol. VI-4, supplement No. 2/XI, *TASI Characteristics Affecting Signalling* (Geneva: ITU, 1973).

Problems

4.1 In a certain telemetry system, there are four analog signals, $m_1(t)$, $m_2(t)$, $m_3(t)$, and $m_4(t)$, with bandwidths of 1200, 700, 500, and 200 Hz, respectively. A TDM is to be designed in which each signal is sampled at exactly the Nyquist rate, with each sample to be represented with 8 bits.

(a) Show a proposed TDM frame which includes a single framing bit.

(b) Determine the number of bits per frame, the frame rate, the frame duration, and the TDM aggregate bit rate.

4.2 Repeat Problem 4.1 with four analog signals to be multiplexed, $m_1(t)$, $m_2(t)$, $m_3(t)$, and $m_4(t)$, with bandwidths of 3.6 kHz, 1.4 kHz, 1.4 kHz, and 1.4 kHz, respectively.

4.3 Consider a Time-Division Multiple Access (TDMA) system (used with INTELSAT V satellites) that has a total transmission rate of 120.832 Mb/s and a frame period of 2 ms.

(a) How many frames are there per second and how many bits are found in a frame?

(b) If each earth station is transmitting at 2.048 Mb/s, how many bits are contained in the frame from each station and what is the duration of the data burst from each station?

4.4 Consider the D2 channel bank that combines 24 voice channels using 64-kb/s μ255 PCM. With an additional 8 kb/s of overhead for frame synchronization, the aggregate rate is 1.544 Mb/s. The framing bits are distributed, one per 193-bit frame, and form a 12-bit framing pattern. The frame acceptance criterion requires a perfect match with the 12-bit pattern. Loss of frame synchronization is declared if one or more errors occur in four consecutive frame patterns.

(a) Assuming an error-free channel and a worst-case starting point, determine the probability that true frame synchronization is found during initial frame synchronization.

(b) Assuming an error-free channel, equally likely 1's and 0's in the data channels, and a worst-case starting point, determine the maximum average frame alignment time.

(c) For a transmission error rate of $p = 10^{-2}$, determine the mean time (in seconds) to false rejection of true frame synchronization in the maintenance mode.

4.5 Consider the same D2 channel bank described in Problem 4.4, but where the transmission error rate is zero and the data to be multiplexed has 70 percent 1's and 30 percent 0's.

(a) In the search mode, what is (1) the probability of accepting false synchronization, and (2) the probability of accepting true synchronization?

(b) What is the average frame synchronization time (in seconds)?

4.6 A DS-1 multiplexer using the superframe format is designed to examine 15 candidate frame positions in parallel during the frame search mode. What is the resulting average frame alignment time (from a random starting point)?

4.7 Consider a TDM system operating at 10 Mbps, which is specified to have a maximum average frame alignment time of 10 ms.

(a) Determine the maximum possible frame length with a distributed frame structure and a frame pattern of length $N = 10$ bits. (Assume that 1s and 0s in data channels are equally likely.)

(b) Calculate the mean time to false rejection of frame synchronization with a bit error rate of 10-3. Assume use of a 16-state up-down counter in the maintenance mode with no errors allowed in the 10-bit frame pattern for frame acceptance.

4.8 The CEPT third-level multiplex with an aggregate rate of 34.368 Mb/s uses a 10-bit frame alignment code bunched at the beginning of the 1536-bit frame. Frame alignment is assumed with the presence of three consecutive frame alignment codes received without error. Loss of frame alignment is assumed when four consecutive frame alignment signals are received in error.

(a) What is the probability of random data exactly matching the 10-bit frame alignment pattern? What is the probability of not recognizing this pattern with a bit error rate of 1×10^{-2}?

(b) What is the maximum average frame alignment time, assuming an error-free transmission channel, worst-case starting position, and equally likely 1's and 0's in the data channels?

(c) What is average time to loss of frame alignment with a bit error rate of 1×10^{-3}?

4.9 Consider a frame maintenance mode that declares loss of sync when it observes frame misalignment in two out of three frames. If p is the probability of misalignment in any one test, what is the average number of frames before loss of sync is declared? Compare this answer with a scheme that requires two successive misalignments before declaring loss of sync.

4.10 Consider a parallel frame search algorithm for the D1 frame format applied to all 193 bits in the frame. If candidate frame positions are examined one bit at a time, what is the probability that all false candidates have been rejected in twelve or less comparisons? Assume that the error threshold ε is 0 and that the transmission error rate is 0.

4.11 The frame word length for a 1.544-Mb/s multiplexer is to be designed to minimize the average reframe time. The number of errors allowed in the frame word is to be set at zero in the search mode and three in the maintenance mode. The maintenance mode strategy is to be based on a loss of sync declared with two successive misaligned frame words.

(a) For a fixed frame length of 193 bits, plot the maximum average frame acquisition time (in ms) versus the frame word length (N) for N up to 32. Create two plots, corresponding to a probability of transmission error $p_e = 0$ and 10^{-2}. Based on these plots, select an optimum value of N for $p_e = 0$ and 10^{-2}.

(b) Now plot the maximum average frame acquisition time (in ms) versus frame word length for N up to 32 for frame lengths (F) of 50, 100, 200, 400, and 800, assuming $p_e = 0$. Based on these plots, pick an optimum frame word length for $F = 193$ bits and $F = 772$ bits.

4.12 By plotting the autocorrelation, show that the following codes are Barker sequences:

(a) $+ -$

(b) $+ + - +$

(c) $+ + + - +$

(d) $+ + + + + - - + + - + - +$

4.13 List all 7-bit codes that meet the Barker code characteristic, and show a calculation of the autocorrelation function for each code.

4.14 Determine the rate at which DS-2 signals in a M23 multiplex will lose framing because stuff codes in the M23 signal are incorrectly interpreted, if the transmission channel bit error rate is 10-3. Hint: the M23 structure is shown in Table 4.8 if one assumes that the stuff bits (S) are actually used.

4.15 Consider the 8.448-Mb/s second-level multiplex described in Table 4.10.

(a) For a tolerance of 50 ppm on each 2.048-Mb/s channel, find the nominal, maximum, and minimum stuffing ratios.

(b) For a transmission error rate of 10^{-3}, determine the mean time (in seconds) between losses of pulse stuffing synchronization.

4.16 Determine the minimum ($\rho = 1$) and maximum ($\rho = 0$) input channel rates accommodated by the CEPT second-level multiplex (8.448-Mb/s aggregate rate), where ρ is the stuffing ratio.

4.17 A 1.544 Mb/s PCM signal is to be input to a second-level multiplex using positive pulse stuffing. The clock rate of the second-level multiplexer per signal input is 1.545 Mb/s.

(a) Determine the required stuffing opportunity rate if the peak to peak waiting time jitter is to be less than ½ bit.

(b) For the stuffing opportunity rate found in part (a), assume that a 3-bit code is used to denote stuff or no stuff for each opportunity and that majority vote is applied at the demultiplexer to decode each stuff opportunity. For a transmission bit error rate of 10^{-3}, find the mean time between losses of pulse stuffing synchronization. What would the mean time be for the same bit error rate if only one bit were used to denote stuffing?

4.18 Consider an asynchronous multiplexer that combines 143 DS-1 signals and operates at an output rate of 224 Mb/s. The frame contains
145 bits—143 data bits plus two overhead bits, called the F bit and S
bit, as shown below. The F bits alternate between 0 and 1, and occupy
bit position 145 in the frame. The S bit occupies bit position 73 in the
frame and has a format as shown below. The synchronization bits M_i
($i = 1$ to 16) form the 16-bit pattern 0110101001011010. Each 1.544-
Mb/s channel is synchronized using pulse stuffing with a 3-bit stuff
control code: C_{i1}, C_{i2}, C_{i3}, for the ith channel, where 000 = no stuff,
111 = stuff [44].

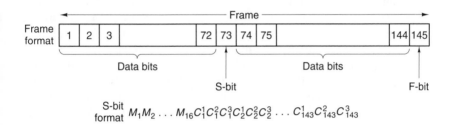

S-bit
format $M_1 M_2 \ldots M_{16} C_1^1 C_1^2 C_1^3 C_2^1 C_2^2 C_2^3 \ldots C_{143}^1 C_{143}^2 C_{143}^3$

(a) What is the number of frames in a multiframe? (Here a multiframe
is defined as the number of frames required to send a stuff control
code once for each of the input channels.)
(b) Multiframe synchronization is established when a perfect match (no
errors) is found with the 16-bit M_i pattern. What is the probability
that random data will match this pattern? What is the probability of
rejecting this 16-bit pattern with an error rate of 10^{-2}.
(c) Calculate the maximum stuffing rate.
(d) Calculate the stuffing ratio for a nominal input rate of 1544000.

4.19 This problem concerns the M12 multiplex that uses positive pulse
stuffing, has an aggregate bit rate of 6.312 Mb/s, and a framing format
shown in Table 4.7.
(a) What is the duration (in seconds) of a frame? A superframe?
(b) What is the probability of false synchronization in the search mode,
assuming random data from the tributaries, if a perfect match is
required with the 8 F bits of the frame pattern? What is the probability of rejecting true frame synchronization in the search mode for
a bit error rate of 10^{-3}?
(c) What is the mean time in seconds to loss of frame synchronization in
the maintenance mode if rejection occurs when five consecutive
framing patterns (consider only F bits) fail to provide a perfect

match with the expected frame pattern? Assume a probability of bit error of 10^{-3}.

4.20 An asynchronous TDM is to be designed with four channels multiplexed using bit interleaving with positive pulse stuffing, to operate over a random error channel transmission channel with average $p_e = 10^{-4}$. Each input channel is at a rate of 1 Mb/s ± 100 b/sec. The multiplexer overhead shall provide the necessary framing bits, stuffing bits, and pulse stuffing control words, as explained below, and meet a 98 percent efficiency requirement. Frame bits are to be inserted in a distributed fashion, one every frame, using a 10-bit pattern. The frame length is to be 100 bits. The search mode of the frame synchronization algorithm is to reject any candidate if one or more errors occur. The maintenance mode will accept zero or one errors in the frame pattern, but declares loss of sync if any one pattern fails the comparison.

(a) The stuff opportunity should be selected such that the stuffing ratio ρ never exceeds 0.25 for any input frequency within the allowed offset of ± 100 b/s. Each stuffing opportunity is to be coded with a redundant M-bit code, with majority vote used at the demultiplexer for decoding. For your choice of stuffing opportunity rate, show what length (M) code is required to meet the requirement that the mean time to stuff decode error due to transmission errors be no less than twenty-four hours for each channel.

(b) Plot pulse stuffing and waiting time jitter, prior to any smoothing in the demultiplexer, by showing phase error versus time. Estimate the peak jitter for both.

(c) Show the construction of the multiplexer overhead, including frame and stuff code bits, which meets the 98 percent efficiency requirement.

(d) Calculate the mean time to acquire frame synchronization and the mean time to spurious rejection of the frame pattern due to transmission errors.

4.21 In the CCITT recommendations for positive-zero-negative pulse stuffing multiplexers, positive stuffing is indicated by the code 111 transmitted in each of two consecutive frames, negative stuffing is indicated by 000 transmitted in each of two consecutive frames, and no stuffing is indicated with the 111 code in one frame followed by 000 in the next frame. Each 3-bit code is decoded with majority vote. Each frame contains one opportunity per input channel for positive stuffing and one opportunity for negative stuffing. For both the 34.368-Mb/s third-level multiplexer and the 139.264-Mb/s fourth-level multiplexer:

(a) Find the frame duration in μsec.

(b) Find the maximum stuffing rate per channel.

(c) Calculate the mean time to stuff decoding error for a transmission channel with BER = 10^{-4}.

4.22 For the M12 multiplexer, create a table that shows mean time to stuff decoding error for bit error rates of 10^{-2}, 10^{-3}, 10^{-4}, and 10^{-5}. Now assume that the existing 3-bit code is to be made more robust by using a longer-length code, without changing the maximum stuffing rate per channel. Create additional tables for 5-, 7-, and 9-bit codes to show mean time to decoding error for the same bit error rates.

4.23 A 3-bit transitional encoder has its 3 bits defined as follows:

> Bit 1 = 0 for a transition
> 1 for no transition
>
> Bit 2 = 0 if transition occurs in first half
> 1 if transition occurs in second half
>
> Bit 3 = 0 if data bit equals 0
> 1 if data bit equals 1

(a) For the data signal 110001111100, show the corresponding transitional encoder waveform. Assume that the data rate is equal to exactly one-third of the transitional encoding rate.
(b) What is the peak jitter that can occur, assuming that the data rate is equal to exactly one-third of the transitional encoding rate?
(c) What are the advantages and disadvantages of this 3-bit transitional encoder compared to that of CCITT Rec. R.111?

4.24 CCITT Rec. G.706 [2] specifies that for DS-1 multiplex the maximum average reframe time should not exceed 50 ms for the 12-frame superframe structure and 15 ms for the 24-frame extended superframe structure.

(a) Propose a strategy to meet the CCITT recommendation for the DS-1 superframe structure. Substantiate your choice with appropriate calculations. Assume that frame and multiframe alignment are done simultaneously.
(b) Propose a strategy to meet the CCITT recommendation for the DS-1 extended superframe structure. Substantiate your choice with appropriate calculations.
(c) One strategy for declaring loss of frame in the ESF is based on any two of four consecutive frame bits received in error. Calculate the mean time to loss of frame synchronization for a bit error rate of 1×10^{-3}.

4.25 For the M23 multiplexer, calculate the maximum stuffing rate per channel and the nominal stuffing ratio. Hint: the M23 structure is shown in Table 4.8 if one assumes that the stuff bits (S) are actually used.

4.26 Consider the C-bit parity DS-3 format shown in Table 4.9.
    (a) Show that with the 100 percent stuffing rate, the frequency of the pseudo DS-2 rate is equal to 6.306 Mb/s.
    (b) For the pseudo DS-2 rate, find the stuffing ratio of the M12 stage, assuming a nominal 1.544000-Mb/s DS-1 signal input.

4.27 Consider the SYNTRAN DS-3 format. How many DS-1 frames are contained in the "synchronous superframe"? What time interval is occupied by the synchronous superframe?

4.28 A fifth-level multiplex used in submarine fiber optic systems combines three 139.264-Mb/s channels into a 442.059-Mb/s signal. The frame structure has a length of 2100 bits with 662 bits per channel per frame. Each channel has one stuffing opportunity per frame.
    (a) Find the maximum stuffing rate per channel and the nominal stuffing ratio.
    (b) Each stuffing opportunity is indicated with a 5-bit code, which is decoded by the demultiplexer with majority vote. Find the mean time between stuff decode errors for a transmission error rate of 1×10^{-4}.
    (c) The frame alignment signal is 12 bits long and is repeated every frame. Loss of frame alignment is assumed to have taken place when four consecutive frame alignment signals are incorrectly received (one or more bits in error per signal). For a transmission error rate of 1×10^{-4}, find the average time between losses of frame alignment.

4.29 A statistical TDM with fixed-length frames (100 bits) is to be designed by trading off link utilization against buffer capacity and time delay. (a) Plot the buffer size (in frames) versus utilization using Equation (4.42). (b) Plot the time delay (in ms) versus utilization using Equation(4.43). (c) Using results of parts (a) and (b), suggest a maximum utilization.

4.30 The tandeming of speech interpolation systems is known to experience accumulation of clipping and blocking, which places a limit on the number of allowed tandems. From data published by the CCITT [45], it is known that two TASI systems in tandem increase the clipping probability by a factor of four over a single system. For two TASI systems in tandem:
    (a) Define a suitable clipping probability for each individual system so that the end-to-end objective specified by the CCITT is met.

(b) Plot the number of transmission channels c versus the number of subscriber channels n for $\alpha = 0.4, 0.35$, and 0.3, using the clipping criterion found in part (a). Compare your results with Figure 4.34.

4.31 The packetized voice protocol described in Section 4.5.3 uses a block-dropping algorithm in which blocks are dropped according to the congestion state. Assuming the algorithm to be defined by the buffer fill (L) and two thresholds of buffer fill (Q_1 and Q_2),

(a) Find the size (in bits) of the transmitted voice packet for $L \leq Q_1$, $Q_1 < L \leq Q_2$, and $L > Q_2$.

(b) Find the time (in ms) to transmit the voice packet for $L \leq Q_1$, $Q_1 < L \leq Q_2$, and $L > Q_2$.

5

Baseband Transmission

OBJECTIVES

- Describes techniques for transmitting signals without frequency translation
- Discusses the commonly used codes for binary transmission, including presentation of waveforms, properties, and block diagrams for the coder and decoder
- Compares bandwidth and spectral shape for commonly used binary codes
- Compares error performance for binary codes in the presence of noise and other sources of degradation
- Describes block codes, which introduce redundancy to gain certain advantages over binary transmission
- Explains how pulse shaping can be used to control the form of distortion called intersymbol interference
- Describes how multilevel transmission increases the data rate packing or spectral efficiency over binary transmission
- Discusses several classes of partial response codes that permit performance superior to that for binary and multilevel transmission
- Explains how the eye pattern shows the effects of channel perturbations and thus indicates overall system "health"
- Discusses the techniques of automatic adaptive equalization of signals distorted by a transmission channel
- Describes data scrambling techniques that minimize interference and improve performance in timing and equalization
- Describes spread spectrum techniques, both direct sequence and frequency hopping, and shows applications to anti-jamming and code division multiple access

5.1 Introduction

The last two chapters have described techniques for converting analog sources into a digital signal and the operations involved in combining these signals into a composite signal for transmission. We now turn our attention to transmission channels, starting in this chapter with baseband signals—signals used for direct transmission without further frequency translation. Transmission channels in general impose certain constraints on the format of digital signals. Baseband transmission often requires a reformatting of digital signals at the transmitter, which is removed at the receiver to recover the original signal. This signal formatting may be accomplished by shaping or coding, which maintains the baseband characteristic of the digital signal. The techniques described here are applicable to cable systems, both metallic and fiber optics, since most cable systems use baseband transmission. Later in our treatment of digital modulation techniques, we will see that many of the techniques applicable to baseband transmission are also used in radio frequency (RF) transmission.

There are a number of desirable attributes that can be achieved for baseband transmission through the use of shaping or coding. These attributes are:

- *Adequate timing information:* Certain baseband coding techniques increase the data transition density, which enhances the performance of timing recovery circuits—that is, bit and symbol synchronization.
- *Error detection/correction:* Many of the codes considered here have an inherent error detection capability due to constraints in allowed transitions among the signal levels. These constraints in level transitions can be monitored to provide a means of performance monitoring, although error correction is not possible through this property of baseband codes.
- *Reduced bandwidth:* The bandwidth of digital signals may be reduced by use of certain filtering and multilevel transmission schemes. The penalty associated with such techniques is a decrease in signal-to-noise ratio or increase in the amount of intersymbol interference.
- *Spectral shaping:* The shape of the data spectrum can be altered by scrambling or filtering schemes. These schemes may be selected to match the signal to the characteristics of the transmission channel or to control interference between different channels.
- *Bit sequence independence:* The encoder must be able to encode any source bit sequence, that is, other attributes of the code must be realized independently of the source statistics.

In describing the characteristics of various transmission codes, first a number of definitions are needed. Most of the transmission codes considered here do not use all possible combinations of allowed signal levels, and the resulting redundancy can be used to achieve certain properties such as spectrum control or error detection. **Redundancy** is the difference or ratio between the unused codes and total possible codes. Codes that use all possible combinations of the code are said to be **saturated.** The requirement to provide adequate timing information depends on the transition density, defined as the average number of transitions per bit. To control the frequency content near dc, the transmitted code should be **balanced,** that is, contain an equal number of 1's (marks) and 0's (spaces). **Disparity** is the difference between the number of transmitted marks and spaces. In a **zero-disparity code,** the transmitted code is constrained such that each block of symbols must have an equal number of 1's and 0's. Timing and spectral control attributes also place limits on the lengths of strings of like symbols. A bound on the length of such strings is conveniently specified by the **running digital sum (RDS)** given by

$$
\text{RDS}\,(k) = \sum_{i=-\infty}^{k} a_i \tag{5.1}
$$

where a_i are the weights assigned to the transmitted symbols. The **digital sum variation (DSV)** is given by the difference between the largest and smallest RDS. An unbalanced code means that the DSV is infinite, which leads to a power spectrum that does not roll off to zero near dc. Conversely, a zero-disparity code maintains an RDS = 0 at the end of each block, which eliminates a dc component in the coded signal. Elimination of the dc component becomes necessary for ac or transformer coupling of the signal to the medium. DC balance can be achieved by using an appropriate transmission code or by balancing each transmitted symbol. Any drift in the transmitted signal from the center baseline level, due to an uncontrolled RDS or the effects of ac coupling, will create a dc component, an effect known as **baseline wander.** This effect degrades performance at the receiver since a shift in the baseline will reduce the margin against noise and increase the BER.

5.2 Types of Binary Coding

Binary codes simply condition binary signals for transmission. This signal conditioning provides a square wave characteristic suitable for direct transmission over cable. Here we discuss commonly used codes for binary transmission, including presentation of waveforms, properties, and block diagrams of the coder and decoder. Later we will compare bandwidth and spectral shape (Section 5.3) and error performance (Section 5.4) for these same binary codes.

5.2.1   Nonreturn-to-Zero (NRZ)

With **nonreturn-to-zero (NRZ),** the signal level is held constant at one of two voltages for the duration of the bit interval T. If the two allowed voltages are 0 and V, the NRZ waveform is said to be **unipolar,** because it has only one polarity. This signal has a nonzero dc component at one-half the positive voltage, assuming equally likely 1's and 0's. A **polar NRZ** signal uses two polarities, $\pm V$, and thus provides a zero dc component.

Various versions of NRZ are illustrated in Figure 5.1. With NRZ(L), the voltage level of the signal indicates the value of the bit. The assignment of bit values 0 and 1 to voltage levels can be arbitrary for NRZ(L), but the usual convention is to assign 1 to the higher voltage level and 0 to the lower voltage level. NRZ(L) coding is the most common mode of NRZ transmission, due to the simplicity of the transmitter and receiver circuitry. The coder/decoder consists of a simple line driver and receiver, whose characteristics have been standardized by various level 1 (physical) interface standards, as discussed in Chapter 2.

In the NRZ(M) format, a level change is used to indicate a mark (that is, a 1) and no level change for a space (that is, a 0); NRZ(S) is similar except that the level change is used to indicate a space or zero. Both of these formats are examples of the general class NRZ(I), also called **conditioned**

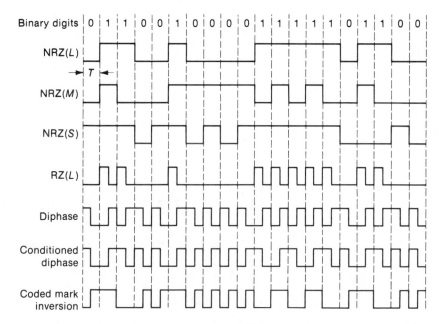

FIGURE 5.1   Binary Coding Waveforms

NRZ, in which level inversion is used to indicate one kind of binary digit. The logic required to generate and receive NRZ(*I*) is shown in the coder and decoder block diagrams of Figure 5.2. The chief advantage of NRZ(*I*) over NRZ(*L*) is its immunity to polarity reversals, since the data are coded by the presence or absence of a transition rather than the presence or absence of a pulse, as indicated in the waveform of Figure 5.2c.

As indicated in Figure 5.1, the class of NRZ signals contains no transitions for strings of 1's with NRZ(*S*), strings of 0's with NRZ(*M*), and strings of 1's or 0's with NRZ(*L*). Since lack of data transitions would result in poor clock recovery performance, either the binary signal must be precoded to eliminate such strings of 1's and 0's or a separate timing line must be transmitted with the NRZ signal. This characteristic has limited the

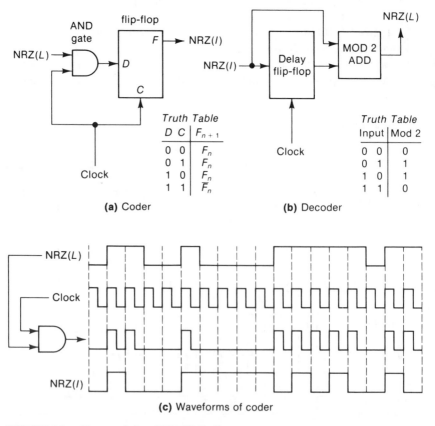

FIGURE 5.2   Characteristics of NRZ(*I*) Coding

application of NRZ coding to short-haul transmission and intrastation connections.

5.2.2   Return-to-Zero (RZ)

With **return-to-zero (RZ),** the signal level representing bit value 1 lasts for the first half of the bit interval, after which the signal returns to the reference level (0) for the remaining half of the bit interval. A 0 is indicated by no change, with the signal remaining at the reference level. Its chief advantage lies in the increased transitions vis-à-vis NRZ and the resulting improvement in timing (clock) recovery. The RZ waveform for an arbitrary string of 1's and 0's is shown in Figure 5.1 for comparison with other coding formats. Note that a string of 0's results in no signal transitions, a potential problem for timing recovery circuits unless these signals are eliminated by precoding.

The RZ coder, waveforms, and decoder are shown in Figure 5.3. The RZ code is generated by ANDing NRZ(L) with a clock operating at the system bit rate. Decoding is accomplished by delaying the RZ code ½ bit and applying the delayed version with the original version to an EXCLUSIVE OR (also known as MOD 2 ADD).

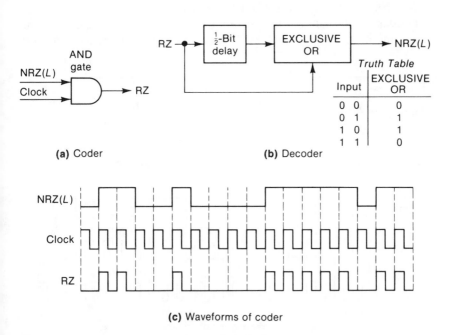

(a) Coder   **(b)** Decoder

(c) Waveforms of coder

FIGURE 5.3   Characteristics of Return-to-Zero Coding

5.2.3 Diphase
Diphase (also called biphase, split-phase, and Manchester) is a method of two-level coding where

$$f_1(t) = \begin{cases} V & 0 \leq t \leq T/2 \\ -V & T/2 < t \leq T \end{cases}$$

$$f_2(t) = -f_1(t)$$

(5.2)

This code can be generated from NRZ(L) by EXCLUSIVE OR or MOD 2 ADD logic, as shown in Figure 5.4, if we assume 1's are transmitted as $+V$ and 0's as $-V$. From the diphase waveforms shown in Figure 5.4, it is readily apparent that the transition density is increased over NRZ(L), thus providing improved timing recovery at the receiver—a significant advantage of diphase. Data recovery is accomplished by the same logic employed by the coder.

Diphase applied to an NRZ(I) signal generates a code known as **conditioned diphase.** This code has the properties of both NRZ(I) and diphase signals—namely, immunity to polarity reversals and increased transition density, as shown in Figure 5.1.

5.2.4 Bipolar or Alternate Mark Inversion
In **bipolar** or alternate mark inversion (AMI), binary data are coded with three amplitude levels, 0 and $\pm V$. Binary 0's are always coded as level 0;

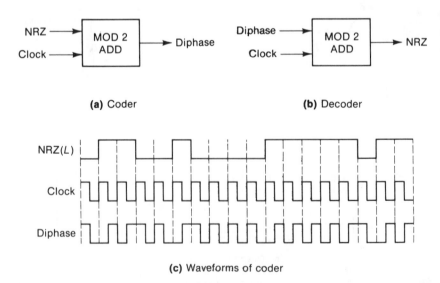

(a) Coder

(b) Decoder

(c) Waveforms of coder

FIGURE 5.4   Characteristics of Diphase Coding

binary 1's are coded as $+V$ or $-V$, where the polarity alternates with every occurrence of a 1. Bipolar coding results in a zero dc component, a desirable condition for baseband transmission. As shown in Figure 5.5, bipolar representations may be NRZ (100 percent duty cycle) or RZ (50 percent duty cycle). Figure 5.6 indicates a coder/decoder block diagram and waveforms for bipolar signals. The bipolar signal is generated from NRZ by use of a 1-bit counter that controls the AND gates to enforce the alternate polarity rule. Recovery of NRZ(L) from bipolar is accomplished by simple full-wave rectification [1].

The numerous advantages of bipolar transmission have made it a popular choice, for example, by AT&T for T1 carrier systems that use 50 percent duty cycle bipolar. Since a data transition is guaranteed with each binary 1, the clock recovery performance of bipolar is improved over that of NRZ. An error detection capability results from the property of alternate mark inversion. Consecutive positive amplitudes without an intervening negative amplitude (and vice versa) are a **bipolar violation** and indicate that a trans-

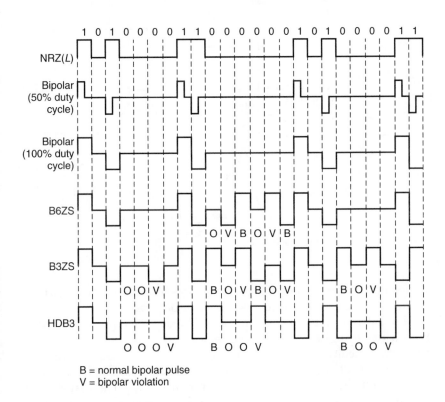

B = normal bipolar pulse
V = bipolar violation

FIGURE 5.5   Bipolar Coding Waveforms

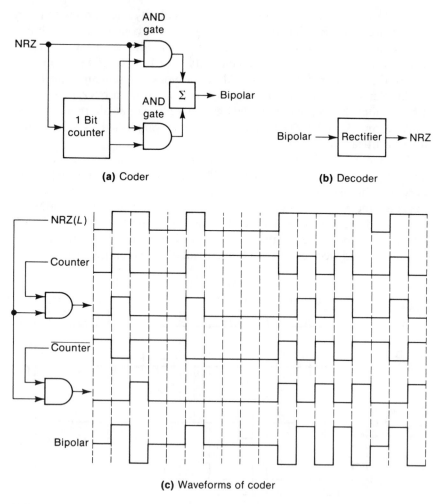

(a) Coder

(b) Decoder

(c) Waveforms of coder

FIGURE 5.6   Characteristics of Bipolar Coding

mission error has occurred. This property allows on-line performance monitoring at a repeater or receiver without disturbing the data.

Although clock recovery performance of bipolar is improved over that of
NRZ, a long string of 0's produces no transitions in the bipolar signal, which
can cause difficulty in clock recovery. For T1 carrier repeaters or receivers,
the maximum allowed string of consecutive 0's is 14. The common practice
in PCM multiplex is to design the encoder in such a way that it satisfies this
code restriction. In data multiplex equipment, however, a restriction of the

binary data is not practical. The answer to this problem is to replace the string of 0's with a special sequence that contains intentional bipolar violations, which create additional data transitions and improve timing recovery. This "filling" sequence must be recognized and replaced by the original string of zeros at the receiver. Two commonly used bipolar coding schemes for eliminating strings of zeros are described below and illustrated in Figure 5.5 [2, 5]:

Bipolar N-Zero Substitution (BNZS)

The most popular of the zero substitution codes is **Bipolar N-Zero Substitution (BNZS),** which replaces all strings of N 0's with a special N-bit sequence containing at least one bipolar violation. All BNZS formats are dc free and retain the balanced feature of AMI, which is achieved by forcing the substitution patterns to have an equal number of positive and negative pulses. For $N \geq 4$, this balanced property of BNZS can be achieved by using a **nonmodal code,** defined as a code whose substitution patterns do not depend on previous substitutions. In this case, two substitution patterns are used with the choice between the two sequences based solely on the polarity of the pulse immediately preceding the string of 0's to be replaced. Substitution sequences for nonmodal codes contain an equal number of positive and negative pulses, contain an even number of pulses, and force the last pulse to be the same polarity as the pulse immediately preceding the sequence. For $N < 4$, a **modal code** must be used, defined as a code wherein the choice of sequence is based on the previous substitution sequence as well as the polarity of the immediately preceding pulse. Modal codes are arranged to meet the balanced property by use of more than two substitution sequences and appropriate alternation of sequences. To further illustrate, using popular choices of these codes, BNZS substitution rules are given in Table 5.1 for B3ZS, a modal code, and for B6ZS and B8ZS, both nonmodal codes. These substitution rules can also be described by use of the following notation: B represents a normal bipolar pulse that conforms to the alternating polarity rule, V represents a bipolar violation, and 0 represents no pulse. Thus, in the B3ZS code, each block of three consecutive 0's is replaced by B0V or 00V; the choice of B0V or 00V is made so that the number of B pulses (that is, 1's) between consecutive V pulses is odd. B6ZS replaces six consecutive 0's with the sequence 0VB0VB. B8ZS replaces each block of eight consecutive 0's with 000VB0VB, such that bipolar violations occur in the fourth and seventh bit positions of the substitution. The B3ZS and B6ZS codes are specified for the DS-3 and DS-2 North American standard rates, respectively, and B8ZS is specified as an alternative to AMI for the DS-1 and DS-1C rates.

TABLE 5.1   Substitution Rules for BNZS and HDB3

(a) B3ZS

	Number of B Pulses Since Last Substitution	
Polarity of Preceding Pulse	Odd	Even
−	00−	+0+
+	00+	−0−

(b) B6ZS

Polarity of Preceding Pulse	Substitution
−	0−+0+−
+	0+−0−+

(c) B8ZS

Polarity of Preceding Pulse	Substitution
−	000−+0+−
+	000+−0−+

(d) HDB3

	Number of B Pulses Since Last Substitution	
Polarity of Preceding Pulse	Odd	Even
−	000−	+00+
+	000+	−00−

High-Density Bipolar N (HDBN)

High-density bipolar N (HDBN) limits the number of allowed consecutive 0's to N by replacing the $(N+1)$th 0 with a bipolar violation. Furthermore, in order to eliminate any possible dc offset due to the filling sequence, the coder forces the number of B pulses between two consecutive V pulses to be always odd. Then the polarity of V pulses will always alternate and a dc component is avoided. There are thus two possible $N+1$ sequences—B00 . . . V or 000 . . . V—where the first bit location is used to ensure an odd number of B pulses between V pulses, the last bit position is always a violation, and all other bit positions are 0's. Two commonly used HDB codes are HDB2, which is identical to B3ZS, and HDB3, which is used for coding of 2.048-Mb/s, 8.448-Mb/s, and 34.368-Mb/s multiplex within the European digital hierarchy [4]. HDB3 substitution rules are shown in Table 5.1(d).

5.2.5   Coded Mark Inversion (CMI)

Coded mark inversion (CMI) is another two-level coding scheme. In this case 1's are coded as either level for the full bit duration T, where the levels alternate with each occurrence of a 1, as in bipolar coding. Zeros are coded so that both levels are attained within the bit duration, each level for $T/2$ using the same phase relationship for each 0. CMI can also be viewed as representing each binary symbol by two code bits, with $0 \rightarrow 01$ and $1 \rightarrow 00$ or 11 alternatively. An example of CMI is illustrated in Figure 5.1. This example indicates that CMI significantly improves the transition density over NRZ. As with any redundant code, CMI provides an error detection feature through the monitoring of coding rule violations. Decoding is accomplished by comparing the first and second half of each interval. This decoder is therefore insensitive to polarity reversals. The CMI code has the disadvantage that the transmission rate is twice that of the binary input signal, since each input bit effectively generates two output bits. Coded mark inversion is specified for the coding of the 139.264-Mb/s multiplex within the European digital hierarchy [4].

5.3   Power Spectral Density of Binary Codes

The power spectral density of a baseband code describes two important transmission characteristics: required bandwidth and spectrum shaping. The bandwidth available in a transmission channel is described by its frequency response, which typically indicates limits at the high or low end. Further bandwidth restriction may be imposed by a need to stack additional signals into a given channel. Spectrum shaping can help minimize interference from other signals or noise. Conversely, shaping of the signal spectrum can allow other signals to be added above or below the signal bandwidth.

Power Spectral Density

Derivation of the power spectral density is facilitated by starting with the autocorrelation function, which is defined for a random process $x(t)$ as

$$R_x(\tau) = E\big[x(t)x(t + \tau)\big] \tag{5.3}$$

where $E[\cdot]$ represents the expected value or mean. The power spectral density describes the distribution of power versus frequency and is given by the Fourier transform of the autocorrelation function

$$S_x(f) = \mathscr{F}\big[R_x(\tau)\big] = \int_{-\infty}^{\infty} R_x(\tau)e^{-j2\pi f\tau}\, d\tau \tag{5.4}$$

In some cases $S_x(f)$ may be more conveniently derived directly from the Fourier transform of the signal. Assuming the signal $x(t)$ to be zero outside the range $-T/2$ to $T/2$, then the corresponding frequency domain signal is given by

$$
\begin{aligned}
X(f) &= \int_{-\infty}^{\infty} x(t) e^{-j2\pi f t}\, dt \\
&= \int_{-T/2}^{T/2} x(t) e^{-j2\pi f t}\, dt
\end{aligned}
\tag{5.5}
$$

By Parseval's theorem the average power across a 1-ohm load over the time interval $-T/2$ to $T/2$ is given by

$$
\begin{aligned}
P &= \frac{1}{T} \int_{-T/2}^{T/2} x^2(t)\, dt \\
&= \frac{1}{T} \int_{-\infty}^{\infty} |X(f)|^2\, df
\end{aligned}
\tag{5.6}
$$

But from the definition of power spectral density, the average power can also be expressed as

$$
P = \int_{-\infty}^{\infty} S_x(f)\, df
\tag{5.7}
$$

where $S_x(f)$ is in units of watts per hertz and P is in units of watts. From (5.6) and (5.7) we obtain

$$
S_x(f) = \frac{|X(f)|^2}{T}
\tag{5.8}
$$

In the following derivations we assume that the digital source produces 0's and 1's with equal probability. Each bit interval is $(0, T)$ and each corresponding waveform is limited to an interval of time equal to T.

5.3.1   Nonreturn-to-Zero (NRZ)

With polar NRZ transmission, we assume that the two possible signal levels are V and $-V$ and that the presence of V or $-V$ in any 1-bit interval is statistically independent of that in any other bit interval. The autocorrelation function for NRZ is well known and is given by

$$R_Q(\tau) = \begin{cases} V^2\left(1 - |\tau|/T\right) & |\tau| < T \\ 0 & |\tau| > T \end{cases} \tag{5.9}$$

Using the Fourier transform, we find the corresponding power spectral density to be

$$S_Q(f) = V^2 T\left(\frac{\sin \pi f T}{\pi f T}\right)^2 \tag{5.10}$$

This is shown in Figure 5.7a. With most of its energy in the lower frequencies, NRZ is often a poor choice for baseband transmission due to the presence of interfering signals around dc. Moreover, clock recovery at the receiver is complicated by the fact that random NRZ data do not exhibit discrete spectral components.

5.3.2   Return-to-Zero (RZ)
In the RZ coding format, the signal values are

$$f_1(t) = \begin{cases} V & 0 < t \le T/2 \\ 0 & T/2 < t \le T \end{cases} \tag{5.11}$$

$$f_2(t) = 0$$

The power spectral density can be found by first dividing the RZ signal into two components. The first component is a periodic pulse train of amplitude $V/2$ and duration $T/2$, defined by

$$p_1(t) = p_2(t) = \begin{cases} V/2 & 0 < t \le T/2 \\ 0 & T/2 < t \le T \end{cases} \tag{5.12}$$

The second component is a corresponding random pulse train defined by

$$r_1(t) = \begin{cases} V/2 & 0 < t \le T/2 \\ 0 & T/2 < t \le T \end{cases}$$

$$r_2(t) = \begin{cases} -V/2 & 0 < t \le T/2 \\ 0 & T/2 < t \le T \end{cases} \tag{5.13}$$

where now

$$f_1(t) = p_1(t) + r_1(t)$$
$$f_2(t) = p_2(t) + r_2(t) \tag{5.14}$$

The random component may be handled in a manner similar to the NRZ case, yielding a continuous power spectral density:

$$S_r(f) = \frac{V^2 T}{16} \left[\frac{\sin(\pi f T/2)}{\pi f T/2} \right]^2 \tag{5.15}$$

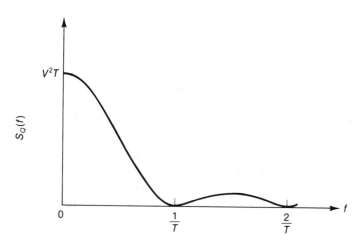

(a) Power spectral density of NRZ

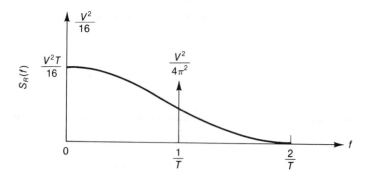

(b) Power spectral density of RZ

FIGURE 5.7   Power Spectral Density of NRZ and RZ

The periodic component may be evaluated by standard Fourier series analysis as

$$S_p(f) = \sum_{n=-\infty}^{\infty} \frac{V^2}{16} \left[\frac{\sin(n\pi/2)}{n\pi/2} \right]^2 \delta\left(f - \frac{n}{T}\right)$$ (5.16)

The power spectral density for RZ is simply the sum of (5.15) and (5.16). This is shown in Figure 5.7b. Note that the first zero crossing of RZ is at twice the frequency as that for NRZ, indicating that RZ requires twice the bandwidth of NRZ. However, the discrete spectral lines of RZ permit simpler bit synchronization.

5.3.3 Diphase

In the diphase format, the signal values are given by (5.2). The power spectral density may be found from (5.8) by first determining the Fourier transform of the two signal pulses as follows:

$$F_1(f) = j\frac{2V}{\pi f} \sin^2 \left(\frac{\pi f T}{2} \right)$$ (5.17)

$$F_2(f) = -F_1(f)$$

As determined by W. R. Bennett [6], the power spectral density depends on the difference between the Fourier transforms of the two pulses:

$$S_D(f) = \frac{1}{T} \left| F_1(f) - F_2(f) \right|^2$$ (5.18)

or from (5.17)

$$S_D(f) = V^2 T \frac{\sin^4(\pi f T/2)}{(\pi f T/2)^2}$$ (5.19)

As shown in Figure 5.8, the power spectral density of diphase has relatively low power at low frequencies and zero power at dc. The power spectral density is maximum at $0.743/T$ and has its first null at $2/T$. Hence the bandwidth occupancy is similar to that for RZ, although there are no discrete lines in the diphase power spectral density.

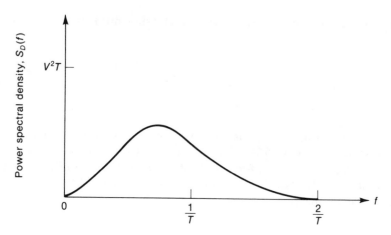

FIGURE 5.8   Power Spectral Density of Diphase (Manchester) Coding

5.3.4   Bipolar

The bipolar coding format uses three amplitudes: V, 0, $-V$. The bipolar waveforms are given by

$$f_1(t) = 0$$

$$f_2(t) = \begin{cases} \begin{array}{ll} V & 0 < t \le \alpha T \\ 0 & \alpha T < t \le T \end{array} & \text{first 1} \\ \begin{array}{ll} -V & 0 < t \le \alpha T \\ 0 & \alpha T < t \le T \end{array} & \text{next 1} \end{cases} \tag{5.20}$$

where $\alpha = 1$ represents nonreturn-to-zero (100 percent duty cycle) and $\alpha = \frac{1}{2}$ represents return-to-zero (50 percent duty cycle). The power spectral density of bipolar signaling is identical to that of *twinned binary* [1], which can be generated from NRZ by delaying it αT, subtracting it from the original, and dividing by 2 to provide proper amplitude scaling—that is,

$$B(t) = \frac{Q(t) - Q(t - \alpha T)}{2} \tag{5.21}$$

Taking the Fourier transform of both sides of (5.21), we have

$$B(f) = \frac{Q(f)}{2} (1 - e^{-j2\pi f \alpha T}) \tag{5.22}$$

From the definition of power spectral density (5.8),

$$S_B(f) = \frac{|Q(f)|^2}{4T} \left| 1 - e^{-j2\pi f \alpha T} \right|^2$$

$$= S_Q(f)\sin^2 \pi f \alpha T$$

(5.23)

The resulting power spectral density is shown in Figure 5.9 for both bipolar return-to-zero and bipolar nonreturn-to-zero. The class of bipolar codes that use filling sequences to substitute for strings of zeros have power spectral densities that are similar to simple bipolar, but each member of this class differs slightly as shown in Figure 5.9. Note the lack of discrete lines in any of the bipolar power spectral densities. The bandwidth occupancy of bipolar NRZ and bipolar RZ is similar to NRZ and RZ, respectively, although bipolar has relatively low power at low frequencies and zero power at dc—another advantage of bipolar over NRZ coding.

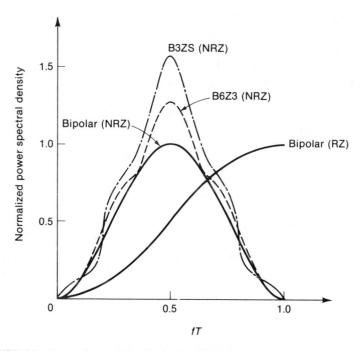

FIGURE 5.9   Power Spectral Density for Bipolar Codes

5.4   Error Performance of Binary Codes

Another important characteristic of binary codes is their error performance in the presence of noise and other sources of degradation. This performance characteristic is dependent on the signaling waveform and the
nature of the noise. Here we will assume the noise to be gaussian. This is a
reasonable assumption for linear baseband channels, where gaussian noise
is the predominant (and sometimes only) source of signal corruption. The
following analysis provides probability of error expressions for the binary
codes already considered. This analysis also builds the foundation for subsequent analysis of more complex modulation schemes and transmission
channels.

We begin with the simplest case by considering a polar NRZ signal having amplitudes $\pm V$ representing binary digits 0 and 1. At the receiver the
two signals are observed to be

$$
\begin{aligned}
y_1 &= V + n \\
y_0 &= -V + n
\end{aligned}
\tag{5.24}
$$

where n represents additive noise. We assume that this noise n is gaussian-
distributed with zero mean and variance σ^2. Since the noise is random and
has some probability of exceeding the signal level, there exists a probability
of error in the receiver decision process. To calculate this probability of
error, the receiver decision rule must be stated. Intuitively, if the noise is
symmetrically distributed about $\pm V$, the decision threshold should be set at
zero. Then the receiver would choose 1 if $y > 0$ and choose 0 if $y < 0$. For a
transmitted 1, an error occurs if at the decision time the noise is more negative than $-V$, with probability

$$
P(e\,|\,1) = P(n < -V)
\tag{5.25a}
$$

and similarly

$$
P(e\,|\,0) = P(n > V)
\tag{5.25b}
$$

Now let $P(0)$ and $P(1)$ be the probability of 0 and 1 at the source, where
$P(0) + P(1) = 1$. The total probability of error may then be expressed as

$$
P(e) = P(e\,|\,0)\,P(0) + P(e\,|\,1)\,P(1)
\tag{5.26}
$$

Since n is gaussian with probability density function

$$P(n) = \frac{1}{\sigma\sqrt{2\pi}}\, e^{-n^2/2\sigma^2} \tag{5.27}$$

it follows that y_1 and y_0 are also gaussian with variance σ^2 but with mean $\overline{y_1} = V$ and $\overline{y_0} = -V$. The probabilities of error $P(e\,|\,1)$ and $P(e\,|\,0)$ are simply the shaded areas under the curves $p(y_1)$ and $p(y_0)$ in Figure 5.10. In equation form these probabilities are

$$P(e\,|\,1) = \int_{-\infty}^{0} p(y_1)\,dy_1 = \frac{1}{\sigma\sqrt{2\pi}} \int_{-\infty}^{0} e^{-(y_1 - V)^2/2\sigma^2}\, dy_1$$

$$\tag{5.28}$$

$$P(e\,|\,0) = \int_{0}^{\infty} p(y_0)\,dy_0 = \frac{1}{\sigma\sqrt{2\pi}} \int_{0}^{\infty} e^{-(y_0 + V)^2/2\sigma^2}\, dy_0$$

By setting the decision threshold equal to zero, we observe from Figure 5.10 that the shaded areas are equal and further that the sum of the two areas is at a minimum. If we further assume that the transmitted binary digits are equiprobable (that is, $P(0) = P(1) = \frac{1}{2}$), then placing the decision threshold at zero also results in a minimum total probability of error, as given by (5.25). Thus $P(e)$ can be obtained directly from (5.28), where a change of variable $Z = (y_1 - V)/\sigma = (y_0 + V)/\sigma$ results in

$$P(e) = \frac{1}{\sqrt{2\pi}} \int_{V/\sigma}^{\infty} e^{-Z^2/2}\, dZ \tag{5.29a}$$

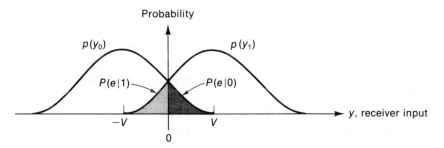

FIGURE 5.10   Probability Density Functions for Binary Transmission (Polar NRZ) over Additive Gaussian Noise Channel

We recognize this as the complementary error function,* which can be written in general form as

$$\text{erfc}(x) = \frac{1}{\sqrt{2\pi}} \int_x^{\infty} e^{-z^2/2} \, dz \tag{5.29b}$$

and approximated by

$$\text{erfc}(x) \simeq \frac{1}{x\sqrt{2\pi}} e^{-x^2/2} \quad \text{for } x > 3 \tag{5.29c}$$

Now (5.29a) can be rewritten as

$$P(e) = \text{erfc}\left(\frac{V}{\sigma}\right) \tag{5.30}$$

which is plotted in Figure 5.11.

These results can be related to signal-to-noise ratio by noting that the average signal power S is equal to V^2 and that the average noise power N is equal to σ^2. Hence (5.30) can be rewritten

$$P(e) = \text{erfc}\sqrt{\frac{S}{N}} \quad \text{(polar NRZ)} \tag{5.31}$$

The unipolar $(0,V)$ NRZ case follows in the same manner except that $S = V^2/2$ so that

$$P(e) = \text{erfc}\sqrt{\frac{S}{2N}} \quad \text{(unipolar NRZ)} \tag{5.32}$$

This derivation of probability of error applies only to those cases where the binary decision for each digit is made independently of all other bit decisions. Hence NRZ(L), RZ(L), and diphase are all characterized by the expressions derived to this point. However, certain binary codes make use of both the present bit value and previous bit values in forming the coded signal. The receiver must first make each bit decision followed by decoding using the present and previous bit values. As an example consider NRZ(I),

---

*There exist several definitions of erfc(x) in the literature. These are all essentially equivalent except for minor differences in the choice of constants.

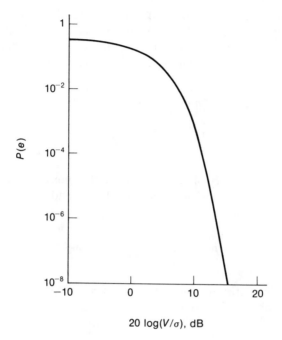

FIGURE 5.11   Probability of Error versus V/σ for Polar NRZ

which falls into this class of binary codes. The probability of error of the decoded signal depends on the binary decision made for both the present and previous bit; that is,

$$
\begin{aligned}
P(e) &= P[\text{present bit correct and previous bit incorrect}] \\
&\quad + P[\text{present bit incorrect and previous bit correct}] \\
&= (1 - p)\,p + p\,(1 - p) \\
&\approx 2p \quad \text{(for small } p\text{)}
\end{aligned}
\tag{5.33}
$$

where p is the probability of error in the binary decision as given in (5.30). In addition to NRZ(I), conditioned diphase has a probability of error given by (5.33).

For bipolar coding with levels $\pm V$ and 0, the probability of error in the receiver can be written

$$
P(e) = P(e\,|\,V)P(V) + P(e\,|\,{-}V)P(-V) + P(e\,|\,0)P(0)
\tag{5.34}
$$

As shown in Figure 5.12, the two decision thresholds are set midway between the three signal levels. The receiver decision rule is identical to the two-level case previously described: Select the signal level that is closest to the received signal. Expression (5.34) can then be rewritten

$$P(e) = P\left(n < \frac{-V}{2}\right)P(V) + P\left(n > \frac{V}{2}\right)P(-V)$$
$$+ \left[P\left(n < \frac{-V}{2}\right) + P\left(n > \frac{V}{2}\right)\right]P(0)$$

(5.35)

Because of the symmetry of gaussian-distributed noise, all the noise terms in (5.35) are equal. If we assume equiprobable 1's and 0's at the transmitter, then (5.35) reduces to

$$P(e) = \frac{3}{2}\,\text{erfc}\left(\frac{V}{2\sigma}\right) \quad \text{(bipolar)}$$

(5.36)

which simply indicates that bipolar coding has a probability of error that is 3/2 times that of unipolar NRZ.

5.5 Block Line Codes

The binary codes discussed to this point all can be viewed as bit-by-bit coders, in which each input bit is translated one at a time into a output symbol. A **block code** groups input bits into blocks and maps each block into another block of symbols for transmission. In general, block codes are used to introduce redundancy in order to meet one or more of the desired attributes of a line code. The two basic techniques used in block coding are

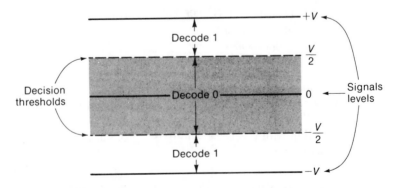

FIGURE 5.12   Bipolar Decoding Rules

(1) insertion of additional binary pulses to create a block of n binary symbols that is longer than the number of information bits m, or (2) translation of a block of input bits to a code that uses more than two levels per symbol. The motivation for the first approach lies in optical transmission, which is limited to two-level (on-off) modulation of the transmitter but is relatively insensitive to a small increase in transmission rate. The second approach applies to the case where bandwidth is limited but multilevel transmission is possible, such as wire lines used for digital subscriber loops. In the following sections two classes of block codes that are examples of these two basic techniques will be described.

5.5.1   mBnB Codes

The $mBnB$ code converts a block of m input bits to a block of n code bits. Coding efficiency is determined by the ratio m/n, but generally n is chosen as $m + 1$. One motivation behind $mBnB$ codes is elimination of the dc component that may exist in the data source. Hence code words are selected to preserve balance. A second motivation is the insertion of transitions to ensure adequate timing recovery. Hence the selection of code words also attempts to provide at least one transition in each code and to minimize the distance (number of bits) between transitions.

Those binary codes described earlier that have a transition in the middle of the symbol can be considered as part of the $1B2B$ class of codes. Diphase, for example, can be classified as a $1B2B$ code in which a binary input 0 is translated to a 10 code word, a binary input 1 is translated to 01, and code words 00 and 11 are disallowed. An example of the $2B3B$ family of codes, which converts two input bits into three code bits, is shown in Table 5.2. This format is seen to be modal, in which two codes alternate for a particular binary input. The first code is generated by simply repeating the binary input and adding a zero to it, while the second code is just the complement of the first. The alternating complementary codes ensure elimination of a dc component. A nonmodal $2B3B$ code would be undesirable, since dc balance can be achieved only by alternating code words. For the $3B4B$ format, some of the code words can be balanced since n is an even number, as indicated by the example shown in Table 5.2. Except for the 000 and 111 binary inputs, all inputs can be translated to a balanced code word by adding a 0 or 1 as appropriate. For the 000 and 111 blocks, a number of alternating codes are possible that preserve dc balance, but those shown provide at least one transition and ensure that there is a distance of not more than $n = 4$ bits between any two transitions. Using these same principles, it is possible to generate many other $mBnB$ formats that have greater efficiency (see Problem 5.24). Drawbacks to $mBnB$ codes are the complexity of the coder and decoder, the delay introduced by coding, and the increased bandwidth.

TABLE 5.2   Examples of mBnB Coding Rules

(a) 2B3B Code

	2B3B Code	
Input Bits	Mode 1	Mode 2
00	000	111
01	010	101
10	100	011
11	110	001

(b) 3B4B Code

	3B4B Code	
Input Bits	Mode 1	Mode 2
000	0010	1101
001	0011	
010	0101	
011	0110	
100	1001	
101	1010	
110	1100	
111	1011	0100

Variations on $mB(m + 1)B$ codes, also called bit insertion codes, have been popular in high speed fiber optics applications and are therefore worth describing here. The $mB1P$ code simply adds an odd parity bit to the m input bits, which ensures at least a single one in the $m + 1$ bits of the code and provides easy error detection. Diphase, for example, is a $1B1P$ code. The first transpacific submarine fiber cable (TPC-3) uses a $24B1P$ code [7]. Similar to the $mB1P$ code, the $mB1C$ code adds a complementary bit that is the opposite of the preceding information bit. With the $mB1C$ code the maximum number of consecutive identical symbols is $m + 1$, which occurs when the insertion bit and m succeeding bits are identical. In-service error detection is possible by using an EXCLUSIVE-OR applied to the last information bit and complementary (C) bit. This code has been adopted in a Japanese 400-Mb/s optical transmission system in the form of $10B1C$ [8]. Another similar code used with very high-speed optical transmission is $DmB1M$, or differential m binary with 1 mark insertion [9]. This code is generated by inserting a mark at the $(m + 1)$th bit and then applying differential coding as done with dicode (see Problem 5.21). A C bit (complementary bit) is automatically generated in the coder at the $(m + 1)$th bit, as in *mB1C,* but without the need for a special circuit to generate *C* bits. As with *mB1C, DmB1M* limits the maximum length of consecutive identical symbols to *m* + 1 and provides error detection through monitoring of the inserted bit every (*m* + 1)th bit. Unlike *mB1C,* however, *DmB1M* also ensures a mark rate of ½ regardless of the mark rate of the input signal, and has no discrete component except the dc component in the power spectrum, which indicates good jitter suppression. Its chief drawback is the error propagation inherent in any differential decoder. The continuous component of the power spectrum for both *mB1C* and *DmB1M* has a shape similar to AMI for small values of *m,* which ensures that high-frequency and low-frequency components are suppressed; as the block length *m* increases, however, the spectrum flattens and assumes the shape of a random signal. The problem of spectrum control in *mB(m* + *1)B* codes can be solved by adding further coding [10] or using scrambling techniques [11].

Example 5.1 _____

The physical layer of the Fiber Distributed Data Interface (FDDI), a standard for a high-bit-rate local area network [12], is based on a combination of NRZ(*I*) and *4B5B* coding. The first stage of encoding is the conversion of each group of four data bits to a block of five code bits. This *4B5B* code is then encoded into a sequence of five NRZ(*I*) bits. The decoder converts the NRZ(*I*) code bits back to NRZ bits, which are then decoded to the original hexadecimal symbols. Of the 32 possible code words, 16 are used for the hexadecimal data symbols, 3 are used as line state symbols, 5 are used as control symbols, and the remaining 8 are violation symbols, which are prohibited. The 8 invalid codes are not used because they cause an unacceptable dc component or violate limits on the number of consecutive 0's allowed. The allowed codes guarantee at least one pulse in every three bits, while the codes used for data symbols have a minimum pulse density of 40 percent.

5.5.2 *kBnT Codes*

The use of three levels as done in AMI can be generalized to a class of ternary codes known as *kBnT,* where *k* is the number of information bits and *n* is the number of ternary symbols per block. Like *mBnB* codes, *kBnT* codes are not saturated, that is, not all possible combinations of the *n* ternary symbols are used, so that the remaining redundancy may be used to create desirable code properties. Because these three-level codes are used to represent binary information, with a resulting loss in coding efficiency, these formats are also known as **pseudoternary codes.** In Table 5.3 we list

TABLE 5.3   *kBnT* Code Characteristics

Code	Input(k)	Output(n)	Redundancy	Bits/Symbol	Efficiency
1B1T	1	1	1/3	1	0.63
3B2T	3	2	1/9	3/2	0.95
4B3T	4	3	11/27	4/3	0.84
6B4T	6	4	17/81	3/2	0.95

several candidate $kBnT$ codes. Here redundancy is defined as the ratio of unused codes to total possible codes, and efficiency is given by the ratio $(m \log 2)/(n \log 3)$. Notice that the largest possible k for a particular n is given by $n \log_2 3$ rounded down to the nearest integer. AMI is a $1B1T$ code that has three possible symbols $(+,-,0)$ but for a given input bit can use only two of them, either $(+,0)$ or $(-,0)$, since AMI alternates the polarity for 1's. Thus one important feature of $kBnT$ codes is the use of different mappings, or **modes,** between the binary blocks and ternary blocks. With AMI there are two modes used to balance the number of positive and negative pulses, which also provides built-in error detection.

With $3B2T$ there are 8 possible combinations of 3 bits that are mapped into 9 possible combinations of 2 ternary signals. By the usual convention of eliminating 0's to maintain a high transition density, it is the 00 code that is not used. The $4B3T$ block code has greater redundancy than $3B2T$ and therefore fewer bits per symbol. In $4B3T$ there are 16 possible combinations of the 4 input bits, which are translated into 27 possible combinations of 3 ternary signals, leaving 11 unused codes. The objective is then to pair the binary and ternary blocks so as to provide some desired characteristic. The usual approach is to alternate modes of ternary alphabets in such a way that the running digital sum is maintained as close to 0 as possible. This approach has the effect of minimizing low-frequency spectral energy. As an example, we will considered $MMS43$, a four-mode version of $4B3T$ that has been proposed for digital subscriber loops [11]. The coding rules of $MMS43$, shown in Table 5.4, show that the three ternary pulses to be transmitted are a function of the four incoming bits and the "present mode" of the encoder. For example, if the incoming bits are 1111 and the present mode of the encoder is $M4$, then the output ternary pulses will be 00- and the next present mode will be $M3$. The present mode represents the past history of the signal in terms of the cumulative running digital sum. The next present mode is selected so that its associated code words will bias the RDS toward 0. Because each block of three ternary symbols is uniquely associated with one block of four bits, the decoder does not need to identify the present mode and can directly map the ternary sym-

TABLE 5.4   Coding Rules of *MMS43*

	Present Mode			
Input Bits	*M1*	*M2*	*M3*	*M4*
0001	0–+, M1	0–+, M2	0–+, M3	0–+, M4
0111	–0+, M1	–0+, M2	–0+, M3	–0+, M4
0100	–+0, M1	–+0, M2	–+0, M3	–+0, M4
0010	+–0, M1	+–0, M2	+–0, M3	+–0, M4
1011	+0–, M1	+0–, M2	+0–, M3	+0–, M4
1110	0+–, M1	0+–, M2	0+–, M3	0+–, M4
1001	+–+, M2	+–+, M3	+–+, M4	–––, M1
0011	00+, M2	00+, M3	00+, M4	––0, M1
1101	0+0, M2	0+0, M3	0+0, M4	–0–, M2
1000	+00, M2	+00, M2	+00, M4	0––, M2
0110	–++, M2	–++, M3	––+, M2	––+, M3
1010	++–, M2	++–, M3	+––, M2	+––, M3
1111	++0, M3	00–, M1	00–, M2	00–, M3
0000	+0+, M3	0–0, M1	0–0, M2	0–0, M3
0101	0++, M3	–00, M1	–00, M2	–00, M3
1100	+++, M4	–+–, M1	–+–, M2	–+–, M4

bols back to information bits by table look-up. Since the decoder is independent of the state of the encoder or the RDS, error propagation does not occur.

5.6   Pulse Shaping and Intersymbol Interference

In discussing binary codes thus far, we have assumed the transmission channel to be linear and distortionless. In practice, however, channels have a limited bandwidth, and hence transmitted pulses tend to be spread during transmission. This pulse spreading or **dispersion** causes overlap of adjacent pulses, giving rise to a form of distortion known as **intersymbol interference** (ISI). Unless this interference is compensated, the effect at the receiver may be errored decisions.

One method of controlling ISI is to shape the transmitted pulses properly. To understand this problem, consider a binary sequence $a_1a_2\cdots a_n$ transmitted at intervals of T seconds. These digits are shaped by the channel whose impulse response is $x(t)$. The received signal can then be written

$$y(t) = \sum_{n=-\infty}^{\infty} a_n x(t - nT)$$ (5.37)

One obvious means of restricting ISI is to force the shaping filter $x(t)$ to be zero at all sampling instants nT except $n = 0$. This condition, known as the **Nyquist criterion,** can be seen mathematically by rearranging (5.37) for a given time $t = nT$:

$$y_n = \sum_k a_k x_{n-k} \qquad (5.38)$$

where $y_n = y(nT)$ and $x_{n-k} = x[(n-k)T]$. Then, for time $t = 0$, Equation (5.38) can be written

$$y_0 = a_0 x_0 + \sum_{k \neq 0} a_k x_{-k} \qquad (5.39)$$

The first term of (5.39) represents the desired signal sampled at time $t = 0$; the second term arises from the overlap of other pulses contributing to the desired pulse at the zeroth sampling time. Clearly if the impulse function is zero for all sampling instants nT except $n = 0$, the second term of (5.39) disappears, thus eliminating intersymbol interference.

Example 5.2

In the following channel impulse response, let $x_{-1} = \frac{1}{4}$, $x_0 = 1$, $x_1 = -\frac{1}{2}$, and $x_i = 0$ for $i \neq -1,0,1$:

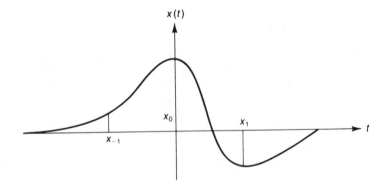

The sample x_{-1} interferes with the previous pulse, while the sample x_1 interferes with the succeeding pulse. The interference with past as well as future symbols is possible because of the time delay through a channel. The received pulse at time 0 is then the accumulation of the precursor and tail of two other transmitted pulses plus the value of the pulse transmitted at time 0:

$$y_0 = a_1 x_{-1} + a_0 x_0 + a_{-1} x_1$$
$$= \tfrac{1}{4} a_1 + a_0 - \tfrac{1}{2} a_{-1}$$

One pulse shape that produces zero ISI is the function

$$x(t) = \frac{\sin \pi t/T}{\pi t/T} \tag{5.40}$$

This is the impulse response of an ideal low-pass filter as shown in Figure 5.13. Note that $x(t)$ goes through zero at equally spaced intervals that are multiples of T, the sampling interval. Moreover, the bandwidth W is observed to be π/T rad/s or $1/2T$ hertz. If $1/2W$ is selected as the sampling interval T, pulses will not interfere with each other. This rate of $2W$ pulses per second transmitted over a channel with bandwidth W hertz is called the **Nyquist rate** [13]. In practice, however, there are difficulties with this filter shape. First, an ideal low-pass filter as pictured in Figure 5.13 is not physically realizable. Second, this waveform depends critically on timing precision. Variation of timing at the receiver would result in excessive ISI because of the slowly decreasing tails of the $\sin x/x$ pulse.

This ideal low-pass filter can be modified to provide a class of waveforms described by Nyquist that meet the zero ISI requirement but are simpler to attain in practice. Nyquist suggested a more gradual frequency cutoff with a filter characteristic designed to have odd symmetry about the ideal low-pass

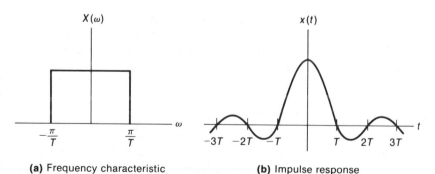

(a) Frequency characteristic   (b) Impulse response

FIGURE 5.13 $\sin x/x$ Filter Characteristic and Pulse Shape

cutoff point. Satisfaction of this characteristic gives rise to a set of Nyquist pulse shapes that obey the property of having 0's at uniformly spaced intervals. This set of pulse shapes is said to be *equivalent* since their filter characteristics all lead to the same sample sequence $\{x_n\}$ [14]. To maintain a rate of $2W$ pulses per second, this set of filter characteristics requires additional bandwidth over the so-called **Nyquist bandwidth,** which is defined by the ideal low-pass filter of Figure 5.13.

An example of a commonly used filter that meets this Nyquist criterion is the **raised-cosine** characteristic. This spectrum consists of a flat magnitude at low frequencies and a rolloff portion that has a sinusoidal form. The raised-cosine characteristic is given by

$$X(\omega) = \begin{cases} T & 0 \le |\omega| \le \frac{\pi}{T}(1-\alpha) \\ \frac{T}{2}\left\{1 - \sin\left[\frac{T}{2\alpha}\left(|\omega| - \frac{\pi}{T}\right)\right]\right\} & \frac{\pi}{T}(1-\alpha) \le |\omega| \le \frac{\pi}{T}(1+\alpha) \\ 0 & \omega > \frac{\pi}{T}(1+\alpha) \end{cases} \quad (5.41)$$

The parameter α is defined as the amount of bandwidth used in excess of the Nyquist bandwidth divided by the Nyquist bandwidth itself. The corresponding impulse response is

$$x(t) = \left(\frac{\sin \pi t/T}{\pi t/T}\right)\left[\frac{\cos \alpha \pi t/T}{1 - (4\alpha^2 t^2/T^2)}\right] \quad (5.42)$$

Plots of $x(t)$ and $X(\omega)$ are shown in Figure 5.14 for three values of α. Note that the case for $\alpha = 0$ coincides with the ideal low-pass filter depicted earlier in Figure 5.13. The case for $\alpha = 1$ is referred to as full (100 percent) raised cosine and doubles the bandwidth required over the Nyquist bandwidth. Larger values of α lead to faster decay of the leading and trailing oscillations, however, indicating that synchronization is less critical than with $\sin x/x$ ($\alpha = 0$) pulses.

5.7 Multilevel Baseband Transmission

With binary transmission, each symbol contains 1 bit of information that is transmitted every T seconds. The symbol rate, or **baud,** is given by $1/T$ and is equivalent to the bit rate for binary transmission. According to Nyquist theory, a data rate of $2W$ bits per second (b/s) can be achieved for the ideal

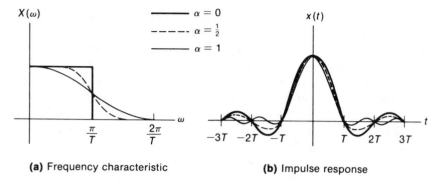

(a) Frequency characteristic   **(b)** Impulse response

FIGURE 5.14   Raised-Cosine Filter Characteristic and Pulse Shape

channel of bandwidth W hertz. The achievable data rate may be increased by using additional levels with the input symbols a_k. Multilevel baseband signaling may be viewed as a form of **pulse amplitude modulation (PAM),** in which the amplitude of a train of constant-width pulses is modulated by the data signal. Usually the number of levels M is a power of 2 so that the number of bits transmitted per symbol can be expressed as $m = \log_2 M$. The data rate R can then be stated as

$$R = 2W \log_2 M = \frac{1}{T} \log_2 M \quad \text{b/s} \tag{5.43}$$

Multilevel transmission thus increases the data rate packing or spectral efficiency by the factor m. To characterize the spectral efficiency, we can normalize the data rate to obtain

$$\frac{R}{W} = 2 \log_2 M \quad \text{bps/Hz} \tag{5.44}$$

This is a commonly used characterization of digital baseband and RF systems. As an example, for $M = 8$ level transmission through the theoretical Nyquist channel, the spectral efficiency is 6 bps/Hz. In practice, the signal-to-noise ratio required for acceptable error performance forces a limit on the number of levels that can be used. Moreover, since the theoretical Nyquist channel is unattainable the spectral efficiency given by (5.44) is an upper bound. As a more practical example, consider a raised-cosine system with $\alpha = 1$ and $M = 8$; here the spectral efficiency drops to 3 bps/Hz.

The error performance of M-ary baseband transmission can be found by extending the case for binary transmission, given by (5.30). Assume that the

M transmission levels are uniformly spaced with values of $\pm d$, $\pm 3d$, ..., $\pm(M-1)d$, that the M levels are equally likely, and that the transmitted symbols form an independent sequence $\{a_k\}$. Let the pulse-shaping filter be sin x/x and assume that no ISI is contributed by the channel. At the receiver, slicers are placed at the thresholds 0, $\pm 2d$, ..., $\pm(M-2)d$. Errors occur when the noise level at a sampling instant exceeds d, the distance to the nearest slicer. Note, however, that the two outside signals, $\pm(M-1)d$, are in error only if the noise component exceeds d and has polarity opposite that of the signal. The probability of error is then given by

$$
\begin{aligned}
P(e) = &P\big[e \mid \pm(M-1)d\big]P\big[\pm(M-1)d\big] \\
&+ P\big[e \mid \pm d, \pm 3d, \ldots, \pm(M-3)d\big] \\
&P\big[\pm d, \pm 3d, \ldots, \pm(M-3)d\big] \\
= &2\Big(1 - \frac{1}{M}\Big)P\,(n > d)
\end{aligned}
\tag{5.45}
$$

For zero mean gaussian noise with variance σ^2, this probability of error can be expressed by

$$
P(e) = 2\Big(1 - \frac{1}{M}\Big)\mathrm{erfc}\Big(\frac{d}{\sigma}\Big)
\tag{5.46}
$$

This result can be expressed in terms of signal-to-noise (S/N) ratio by first noting that the average transmitted symbol power is given by

$$
S = \overline{a^2} = \frac{d^2(M^2 - 1)}{3}
\tag{5.47}
$$

Assuming additive gaussian noise to have average power $N = \sigma^2$, the probability of error can now be written as

$$
P(e) = 2\Big(1 - \frac{1}{M}\Big)\mathrm{erfc}\left[\Big(\frac{3}{M^2 - 1}\frac{S}{N}\Big)^{1/2}\right]
\tag{5.48}
$$

Some important conclusions and comparisons can be made from the probability of error expression in (5.48). A comparison of the binary $P(e)$ expression derived earlier (Equation 5.31) with (5.48) for $M = 2$ reveals that the two expressions are identical, as expected. The effect of increased M, however, is that the S/N ratio is modified by $3/(M^2 - 1)$. Thus an $M = 4$ transmission system requires about 7 dB more power than the $M = 2$ case.

Another important comparison is that of bandwidth and bit rate. This comparison is facilitated by use of the normalized data rate given in (5.44) and by writing the argument within the erfc of (5.48) as

$$\Gamma = \frac{3}{M^2 - 1} \frac{S}{N} \tag{5.49}$$

After solving for M^2 in terms of Γ and S/N, the normalized data rate can be expressed as

$$\frac{R}{W} = \log_2 \left[\left(\frac{3}{\Gamma} \right) \left(\frac{S}{N} \right) + 1 \right] \text{ bps/Hz} \tag{5.50}$$

This expression can now be compared with Shannon's theoretical limit on channel capacity C over a channel limited in bandwidth to W hertz with average noise power N and average signal power S [15, 16]:

$$\frac{C}{W} = \log_2 \left(\frac{S}{N} + 1 \right) \text{ bps/Hz} \tag{5.51}$$

which represents the minimum S/N required for nearly error-free performance. A comparison between (5.50) and (5.51) reveals the difference to be the factor $3/\Gamma$. For a probability of error of 10^{-5}, $\Gamma \approx 20$, where the $2(1 - 1/M)$ factor has been ignored. Thus an additional 8 dB of S/N ratio is required to attain Shannon's theoretical channel capacity. The conclusion that can be made is that M-ary transmission may be spectrally efficient but suffers from effective loss of S/N ratio relative to theoretical channel capacity.

5.8   Partial Response Coding

To this point we have maintained that the Nyquist rate of $2W$ symbols per second for a channel of bandwidth W hertz can be attained only with a nonrealizable, ideal low-pass filter. Even raised-cosine systems require excessive bandwidth, so that practical symbol rate packing appears to be on the order of 1 symbol/Hz even though the theoretical limit is 2 symbols/Hz. This limit applies only to zero-memory systems, however—that is, where transmitted pulse amplitudes are selected independently.

In the following paragraphs we will discuss a class of techniques that introduce correlation between the amplitudes to permit practical attainment of the Nyquist rate. A. Lender first described this technique in a scheme termed **duobinary** [17, 18], which combines two successive pulses together to form a multilevel signal. This combining process introduces prescribed amounts of intersymbol interference, resulting in a signal that has three levels at the sampling instants. Kretzmer has classified and tabulated these schemes, which are more generally called **partial response** [19] or **correlative coding** [20]. Partial response coding provides the capability of increasing the transmission rate above W symbols per second but at the expense of additional transmitted power. Appropriate choice of precoder or filter makes available a variety of multilevel formats with different spectral properties. Furthermore, certain constraints on level transitions in the received signal make possible some error detection.

The basic ideas behind partial response schemes can be illustrated by considering the example of duobinary (or *class 1 partial response*). Consider an input sequence $\{a_k\}$ of binary symbols spaced T seconds apart. The transmission rate is $2W$ symbols per second over an ideal rectangular low-pass channel of bandwidth W hertz and magnitude T. These symbols are passed through the digital filter shown in Figure 5.15a, which simply sums two successive symbols to yield at the transmitter output

$$y_k = a_k + a_{k-1} \tag{5.52}$$

The cascade of the digital filter $H_1(\omega)$ and the ideal rectangular filter $H_2(\omega)$ is equivalent to

$$|H(\omega)| = |H_1(\omega)||H_2(\omega)| = 2T\cos\frac{\omega T}{2} \qquad |\omega| \leq \frac{\pi}{T} \tag{5.53}$$

since

$$\begin{aligned} H_1(\omega) &= 1 + e^{-j\omega T} \\ &= 2\cos\frac{\omega T}{2}\,e^{-j\omega T/2} \end{aligned} \tag{5.54}$$

Thus $H(\omega)$ can be realized with a cos-shaped low-pass filter, eliminating the need for a separate digital filter. The filter characteristic $H(\omega)$ and its corresponding impulse response

$$h(t) = \frac{4}{\pi}\left(\frac{\cos \pi t/T}{1 - 4t^2/T^2}\right) \tag{5.55}$$

are sketched in Figure 5.15. If the impulse response sampling time is taken at $t = 0$, then

$$h(nT) = \begin{cases} 1 & n = 0, 1 \\ 0 & \text{otherwise} \end{cases} \tag{5.56}$$

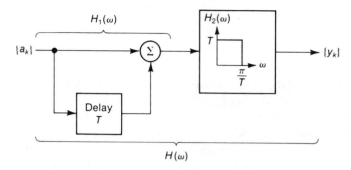

(a) Duobinary signal generation

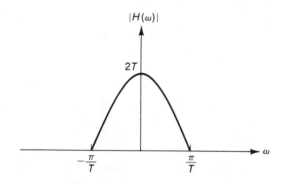

(b) Amplitude spectrum, composite filter

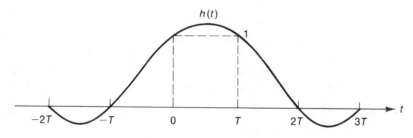

(c) Duobinary impulse response

FIGURE 5.15   Duobinary Signaling

The received signal can then be written as the sum of all pulse contributions

$$y_k = \sum_n a_n h_{k-n} = a_k + a_{k-1} \qquad (5.57)$$

This derivation of the duobinary signal duplicates (5.52), of course, but here we see more directly the beneficial role played by intersymbol interference. The duobinary response to consecutive binary 1's is sketched in Figure 5.16a, which illustrates the role of ISI. Figure 5.16b depicts the output y_k of the duobinary filter for a random binary input $\{a_k\}$. If the binary input $\{a_k\} = \pm d$, then y_k will have three values at the sampling instants: $-2d$, 0, and $2d$. At the receiver, the three-level waveform is sliced at $\frac{1}{4}$ and $\frac{3}{4}$ of its peak-to-peak value $4d$. The slicer outputs are then decoded to recover the binary data $\hat{a}_k$, using the following three rules:

- $y_k = +d, \hat{a}_k = +d$
- $y_k = -d, \hat{a}_k = -d$ $\qquad (5.58)$
- $y_k = 0, \hat{a}_k = -\hat{a}_{k-1}$

Owing to the correlation property of partial response systems, only certain types of transitions are allowed in the waveform. These level constraints can be monitored at the receiver, such that any violation of these rules results in an error indication. This feature of partial response, like error detection in bipolar coding, has proved to be valuable in performance monitoring [21]. For duobinary the following level constraints apply:

- A positive (negative) level at one sampling time may not be followed by a negative (positive) level at the next sampling time.
- If a positive (negative) peak is followed by a negative (positive) peak, they must be separated by an odd number of center samples.
- If a positive (negative) peak is followed by another positive (negative) peak, they must be separated by an even number of center samples.

Since the symbol in error cannot be determined, error correction is not possible without additional coding.

5.8.1 Duobinary Precoding

A problem with the decoder indicated in (5.58) is that errors tend to propagate. This occurs because correct decoding of $\hat{a}_k$ depends on correct decoding of $\hat{a}_{k-1}$. A method proposed by Lender eliminates this error propagation by precoding the input binary data prior to duobinary filtering. This logic

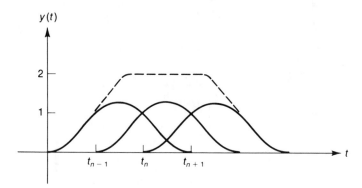

(a) Duobinary response to three consecutive 1's

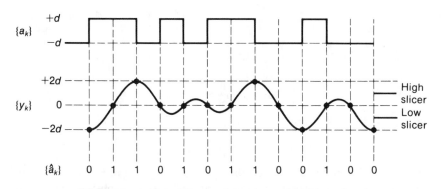

(b) Response of duobinary scheme to random binary input

FIGURE 5.16   Response of Duobinary Scheme

operation is shown in the block diagram of Figure 5.17a. The binary input sequence $\{a_k\}$ is first converted to another binary sequence $\{b_k\}$ according to the rule

$$b_k = a_k \oplus b_{k-1} \qquad (5.59)$$

where the symbol $\oplus$ represents modulo 2 addition or the logic operator EXCLUSIVE OR. The sequence $\{b_k\}$ is then applied to the input of the duobinary filter according to (5.52), which yields

$$\begin{aligned} y_k &= b_k + b_{k-1} \\ &= (a_k \oplus b_{k-1}) + b_{k-1} \end{aligned} \qquad (5.60)$$

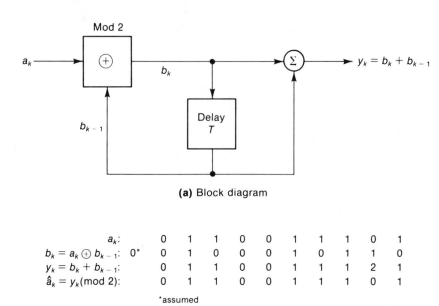

(a) Block diagram

a_k:		0	1	1	0	0	1	1	1	0	1
$b_k = a_k \oplus b_{k-1}$:	0*	0	1	0	0	0	1	0	1	1	0
$y_k = b_k + b_{k-1}$:		0	1	1	0	0	1	1	1	2	1
$\hat{a}_k = y_k(\text{mod } 2)$:		0	1	1	0	0	1	1	1	0	1

*assumed

(b) Example of signals

FIGURE 5.17   Precoded Duobinary Scheme

An example of the precoded duobinary signal is given in Figure 5.17b for a random binary input sequence. Examination of this example together with (5.60) indicates that $y_k = 1$ results only from $a_k = 1$ and that $y_k = 0$ or 2 corresponds only to $a_k = 0$. These values indicate that y_k can be decoded according to the simple rule

$$\hat{a}_k = y_k \,(\text{mod } 2) \tag{5.61}$$

Since each binary decision requires knowledge of only the current sample y_k, there is no error propagation.

5.8.2   Generalized Partial Response Coding Systems

As indicated earlier, duobinary is just one example of the classes of partial response signaling schemes. As shown in Figure 5.18, the design of a partial response coder is based on the generalized digital filter, $H_1(\omega)$. This so-called **transversal filter** provides a weighted superposition of N digits. Appropriate choice of the weighting coefficients h_n allows the spectrum to be shaped for a particular application. From application of the Fourier

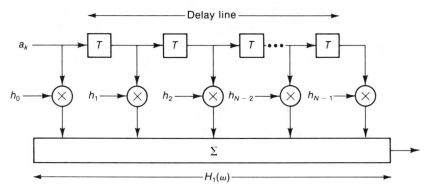

FIGURE 5.18   Generalized Partial Response Filter

series representation, the transfer function of this digital filter is known to be

$$H_1(\omega) = \sum_{n=0}^{N-1} h_n e^{-jn\omega T} \qquad (5.62)$$

Example 5.3

Consider the class of partial response signaling generated by subtracting pulses spaced by two sample intervals, yielding

$$y_k = a_k - a_{k-2}$$

The weighting coefficients of the generalized transversal filter are then $h_0 = 1$, $h_2 = -1$, with all other coefficients equal to zero. From (5.62), the transfer function is thus

$$H_1(\omega) = 1 - e^{-j2\omega T}$$

The amplitude characteristic is therefore

$$|H_1(\omega)| = 2 \sin \omega T$$

This scheme is known both as **modified duobinary** and as class 4 partial response. It has its advantage in the spectral shape that has no dc component. This spectrum characteristic is desirable for those transmission systems, such as the telephone plant, that cannot pass dc signals because of the use of transformers. Except for the spectral shape, modified duobinary has characteristics similar to duobinary. Both schemes result in three-level waveforms and in error propagation that can be eliminated with precoding. For transmitted symbols $\{a_k\} = \pm d$, modified duobinary (without precoding) is decoded according to the following rules:

- $y_k = +d, \hat{a}_k = +d$
- $y_k = -d, \hat{a}_k = -d$
- $y_k = 0, \hat{a}_k = \hat{a}_{k-2}$

Table 5.5 lists the classes of partial response signaling. The two classes that are most commonly used because of their simplicity, performance, and desirable spectral shape are *duobinary* (class 1) and *modified duobinary* (class 4). Partial response signaling is used in a variety of applications, including baseband repeaters for T carrier [22], voice-channel modems [23], and digital radio. To enhance data rate packing, partial response schemes are also used in conjunction with other modulation schemes such as single sideband [22], FM [24], and PSK [25]. Finally, partial response has been used with existing analog microwave radios to add data above or below the analog signal spectrum [26].

5.8.3 Partial Response Techniques for M-ary Signals

Partial response techniques can in general be utilized with M-ary input signals to increase the data rate possible over a fixed bandwidth. Instead of a binary $(M = 2)$ input to the partial response filter, the input is allowed to take on M levels. In the case of duobinary, we observed that a binary input

TABLE 5.5   Classes of Partial Response Signaling

Class	Generating Function	Transfer Magnitude Function $\|H(\omega)\|$	Number of Received Levels
1	$y_k = a_k + a_{k-1}$	$2\cos(\omega T/2)$	3
2	$y_k = a_k + 2a_{k-1} + a_{k-2}$	$4\cos^2(\omega T/2)$	5
3	$y_k = 2a_k + a_{k-1} - a_{k-2}$	$[(2 + \cos\omega T - \cos2\omega T)^2 + (\sin\omega T - \sin2\omega T)^2]^{1/2}$	5
4	$y_k = a_k - a_{k-2}$	$2\sin\omega T$	3
5	$y_k = a_k - 2a_{k-2} + a_{k-4}$	$4\sin^2\omega T$	5

results in a three-level received signal. The application of an M-ary input to a duobinary system results in $2M - 1$ received levels. The duobinary coder and decoder block diagrams for this M-ary application are similar to that of duobinary for binary input. Consider, for example, the case of precoded duobinary as shown in Figure 5.17. For M-ary input, the mod 2 operator must be replaced with a mod M operator. The precoder logic can then be described by the following rule:

$$b_k = a_k - b_{k-1} \quad (\text{mod } M) \tag{5.63}$$

which is analogous to the case for binary input (Equation 5.59). The sequence $\{b_k\}$ is then applied to the input of the duobinary filter, which yields

$$y_k = b_k + b_{k-1} \quad (\text{algebraic}) \tag{5.64}$$

The decoding rule is analogous to that of precoded duobinary for binary input, but here the operator is mod M; that is,

$$\hat{a}_k = y_k \quad (\text{mod } M) \tag{5.65}$$

Example 5.4

Suppose the number of levels M is equal to eight. The input symbols are assumed to be $\{0,1,2,3,4,5,6,7\}$. For the following random input sequence $\{a_k\}$, the corresponding precoder sequence $\{b_k\}$ and partial response sequence $\{y_k\}$ are shown in accordance with (5.63) to (5.65):

a_k:		7	5	5	0	4	6	3	6	1	4	2	2	0
b_k:	0	7	6	7	1	3	3	0	6	3	1	1	1	7
y_k:		7	13	13	8	4	6	3	6	9	4	2	2	8
$\hat{a}_k$:		7	5	5	0	4	6	3	6	1	4	2	2	0

The sequence $\{b_k\}$ is found by subtracting b_{k-1} from a_k, modulo M, so that starting from the first operation we have $7 - 0 = 7$; next, $5 - 7 = -2$ (mod 8) $= 6$; next, $5 - 6 = -1$ (mod 8) $= 7$; and so on. The sequence $\{y_k\}$ is simply the algebraic sum of b_{k-1} and b_k, so that the 15 levels $\{0,1,\ldots,14\}$ will result. Finally, $\hat{a}_k$ is obtained by taking the modulo 8 value of y_k.

The advantage of M-ary partial response signaling over zero-memory M-ary signaling is that fewer levels are needed for the same data rate. Fewer levels mean greater protection against noise and hence a better error rate performance. To show this advantage, first consider a zero-memory system based on a 100 percent raised-cosine filter of bandwidth R hertz. A binary sequence ($M = 2$) input results in a data rate packing of 1 b/s per hertz of bandwidth. Therefore an M-ary sequence ($M = 2^m$) input will result in an m bps/Hz data rate packing. Now consider a duobinary scheme used on a channel with the same bandwidth of R hertz. A binary input results in 2 bps/Hz. In general, to obtain a packing of k bps/Hz, a $2^{k/2}$-level input sequence is required that results in only $2(2^{k/2}) - 1$ output levels. For $k = 4$ bps/Hz, for example, 16 levels would be required with 100 percent raised-cosine transmission but only 7 levels with duobinary. For 6 bps/Hz, 100 percent raised-cosine transmission requires 64 levels whereas duobinary requires only 15 levels.

5.8.4   Error Performance for Partial Response Coding

For duobinary, the received signal in the absence of noise has three values, $+2d$, 0, and $-2d$, and the probabilities of receiving these levels are $\frac{1}{4}$, $\frac{1}{2}$, and $\frac{1}{4}$, respectively. For the gaussian noise channel, the probability of error is then

$$P(e) = \frac{3}{2} \operatorname{erfc}\left(\frac{d}{\sigma}\right) \tag{5.66}$$

where σ^2 is the variance of the noise. In terms of S/N ratio, the probability of error becomes [14]

$$P(e) = \frac{3}{2} \operatorname{erfc}\left[\frac{\pi}{4}\left(\frac{S}{N}\right)^{1/2}\right] \tag{5.67}$$

where S is the average signal power, N is the noise power in the Nyquist bandwidth, and the duobinary filter is split equally between the transmitter and receiver.

Comparing duobinary error performance with that of binary ($M = 2$) transmission (Equation 5.48), we find duobinary performance to be poorer by $\pi/4$, or 2.1 dB. The binary error performance is based on the theoretical Nyquist channel, however, while duobinary performance is based on a practical channel. A fairer comparison with duobinary is to use a raised-cosine channel with $\alpha = 1$. Then for the data rate packing (R/W) to be equal, we must choose $M = 4$. For this case duobinary requires 4.9 dB less S/N ratio to attain the same error performance as the raised-cosine channel.

For M-ary partial response, the probability of error is given by [14]

$$P(e) = 2\left(1 - \frac{1}{M^2}\right)\text{erfc}\left[\frac{\pi}{4}\left(\frac{3}{M^2 - 1}\frac{S}{N}\right)^{1/2}\right] \tag{5.68}$$

Note the familiar factor $3/(M^2 - 1)$ in the S/N term, which indicates the degradation in error performance due to additional signal levels.

5.9   Eye Patterns

A convenient way of assessing the performance of a baseband signaling system is by means of the **eye pattern**. This pattern is obtained by displaying the received signal on an oscilloscope. The time base of the oscilloscope is set to trigger at a rate $1/T$. The resulting pattern derives its name from its resemblance to the human eye for binary data. The persistence of the cathode ray-tube blends together all allowed signal waveforms. As an example based on bipolar signaling, consider the 3-bit sequences shown in Figure 5.19a, where the first 1 always has positive polarity. (A corresponding set of sequences exists in which the first 1 always has negative polarity.) Figure 5.19b gives the eye pattern for the sequences of Figure 5.19a by superimposing these waveforms in the same signaling interval. Together with the other allowed sequences, Figure 5.19c plots the complete bipolar eye pattern. Note that this pattern traces all allowed waveforms in a given sampling interval with a resulting three-level signal at the prescribed sampling time. Moreover, the sampling time is shown at the point where the opening of the eye is greatest.

The eye opening may be used as an indication of system "health." As noise, interference, or jitter is added by the channel, the eye opening will close. The height of this opening indicates what margin exists over noise. For example, an eye closing of 50 percent indicates an equivalent S/N degradation of $-20\log(1 - 0.5) = 6$ dB. The three-level partial response eye patterns of Figure 5.20 illustrate the observable degradation between a zero error rate and a 10^{-3} error rate.

5.10   Equalization

Real channels introduce distortion in the digital transmission process, resulting in a received signal that is corrupted by intersymbol interference. The concept of equalization is based on the use of adjustable filters that

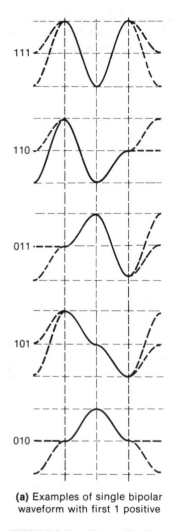

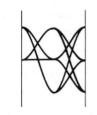

(b) Summation of waveforms
in part (a)

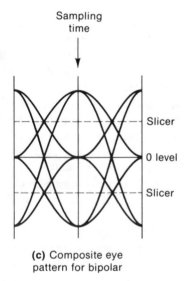

(a) Examples of single bipolar
waveform with first 1 positive

(c) Composite eye
pattern for bipolar

FIGURE 5.19   Bipolar Eye Patterns

compensate for amplitude and phase distortion. High-speed transmission
over telephone circuits or radio channels often requires equalization to
overcome the effects of intersymbol interference caused by dispersion. In
most practical cases, the channel characteristics are unknown and may vary
with time. Hence the equalizer must be updated with each new channel connection, as in the case of a voice-band modem operating in a switched net-

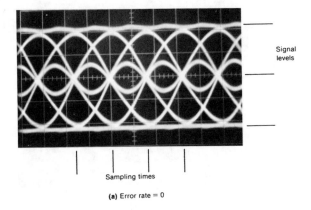

Signal
levels

Sampling times

(a) Error rate = 0

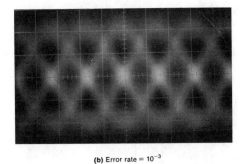

(b) Error rate = 10^{-3}

FIGURE 5.20   Eye Patterns of Three-Level Partial Response Signal

work, and must also adapt the filter settings automatically to track changes in the channel with time. Today automatic adaptive equalization is used nearly universally in digital radio applications [27] and in voice-band modems for speeds higher than 2400 b/s [28].

Because of its versatility and simplicity, the transversal filter is a common choice for equalizer design. This filter consists of a delay line tapped at T-second intervals, where T is the symbol width. Each tap along the delay line is connected through an amplifier to a summing device that provides the output. The summed output of the equalizer is sampled at the symbol rate and then fed to a decision device. The tap gains, or coefficients, are set to subtract the effects of interference from symbols that are adjacent in time to the desired symbol. We assume there are $(2N + 1)$ taps with coefficients $c_{-N}, c_{-N+1}, \ldots, c_N$, as indicated in Figure 5.21. Samples of the equalizer output

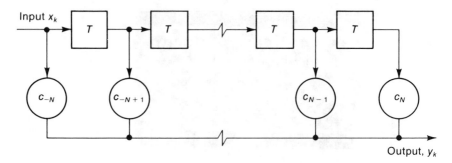

FIGURE 5.21   Transversal Filter

can then be expressed in terms of the input x_k and tap coefficients c_n as

$$y_k = \sum_{n=-N}^{N} c_n x_{k-n} \qquad k = -2N, \ldots, +2N \qquad (5.69)$$

Computation of the output sequence $\{y_k\}$ is simplified by use of matrices. If we let

$$\mathbf{y} = \begin{bmatrix} y_{-2N} \\ \vdots \\ y_0 \\ \vdots \\ y_{2N} \end{bmatrix} \qquad \mathbf{c} = \begin{bmatrix} c_{-N} \\ \vdots \\ c_0 \\ \vdots \\ c_N \end{bmatrix}$$

$$\mathbf{x} = \begin{bmatrix} x_{-N} & 0 & 0 & \cdots & 0 & 0 \\ x_{-N+1} & x_{-N} & 0 & \cdots & 0 & 0 \\ \vdots & \vdots & \vdots & & \vdots & \vdots \\ x_N & x_{N-1} & x_{N-2} & \cdots & x_{-N+1} & x_{-N} \\ \vdots & \vdots & \vdots & & \vdots & \vdots \\ 0 & 0 & 0 & & x_N & x_{N-1} \\ 0 & 0 & 0 & \cdots & 0 & x_N \end{bmatrix} \qquad (5.70)$$

then we can write (5.69) as

$$\mathbf{y} = \mathbf{x}\mathbf{c} \qquad (5.71)$$

Example 5.5 _____

For the channel impulse response of Example 5.2, assume that a three-tap transversal filter is used as the equalizer with tap coefficients $c_{-1} = -\frac{1}{4}$,

$c_0 = 1$, and $c_1 = \frac{1}{2}$. The output sequence is found by matrix multiplication:

$$\mathbf{y} = \mathbf{xc}$$

$$= \begin{bmatrix} \frac{1}{4} & 0 & 0 \\ 1 & \frac{1}{4} & 0 \\ -\frac{1}{2} & 1 & \frac{1}{4} \\ 0 & -\frac{1}{2} & 1 \\ 0 & 0 & -\frac{1}{2} \end{bmatrix} \begin{bmatrix} -\frac{1}{4} \\ 1 \\ \frac{1}{2} \end{bmatrix}$$

$$= \left(-\frac{1}{16} \quad 0 \quad \frac{5}{4} \quad 0 \quad -\frac{1}{4} \right)$$

Although adjacent symbol interference has been forced to zero, some ISI has been added at sample points further removed from the center pulse. In general, a finite transversal filter cannot completely eliminate ISI.

The criterion for selection of tap coefficients is generally based on the minimization of either peak distortion or mean square distortion. The so-called **zero-forcing equalizer** has been shown to minimize peak distortion [14]. Here the tap coefficients are selected to force the equalizer output to zero at N sample points on either side of the desired pulse. We may write this in mathematical form as $2N + 1$ constraint equations:

$$y_k = \begin{cases} 1 & k = 0 \\ 0 & k = \pm 1, \pm 2, \ldots, \pm N \end{cases} \tag{5.72}$$

We can then solve for the tap coefficients c_n by combining (5.69) and (5.72), which amounts to solving $2N + 1$ simultaneous linear equations.

The mean square minimization criterion also leads to $2N + 1$ linear simultaneous equations, whose solution minimizes the sum of squares of the ISI terms. The equalizer error signal at sample point k is simply the difference between the transmitted pulse δ_k and equalized received signal y_k, or

$$e_k = y_k - \delta_k \tag{5.73}$$

The mean square error is

$$\overline{e^2} = \frac{1}{2N + 1} \sum_{k=-N}^{N} (y_k - \delta_k)^2 \tag{5.74}$$

For minimum mean square error it is necessary that the gradient

$$\frac{\partial \overline{e^2}}{\partial c_j} = 0 \quad j = -N, \ldots, N \tag{5.75}$$

From (5.74) we obtain

$$\frac{\partial \overline{e^2}}{\partial c_j} = \frac{2}{2N+1} \sum_{k=-N}^{N} (y_k - \delta_k) \frac{\partial y_k}{\partial c_j} \tag{5.76}$$

Using (5.69) and (5.73) in (5.76) we obtain

$$\frac{\partial \overline{e^2}}{\partial c_j} = \frac{2}{2N+1} \sum_{k=-N}^{N} e_k x_{k-j} \quad j = -N, \ldots, N \tag{5.77}$$

Therefore the tap coefficients are optimum when the cross-correlation between the error signal and input pulse is forced to zero at each sample point within the equalizer range. In practice, the optimum tap settings are found by an iterative procedure. The tap coefficient vector $\mathbf{c}$ is updated during every symbol interval. Letting c^m be the mth such tap coefficient vector, we can write

$$\mathbf{c}^{m+1} = \mathbf{c}^m - \alpha \mathbf{e}^m \tag{5.78}$$

where α is the equalizer step size and $\mathbf{e}^m$ is the error vector at time m. If the sequence converges after some time m, that is,

$$\mathbf{c}^{m+1} = \mathbf{c}^m = \mathbf{c} \tag{5.79}$$

then

$$\frac{\partial \overline{e^2}}{\partial \mathbf{c}} = 0 \tag{5.80}$$

and the tap coefficients are optimized in a mean square error sense. Because of the presence of noise in the measurement of the error vector, in practice the gradient method is based on a noisy estimate and not the ideal estimate assumed above. The minimum mean square equalizer is more popularly used than the zero-forcing equalizer because it is more robust in the presence of noise or ISI and is superior in convergence performance.

Two general types of automatic equalization exist. In **preset equalization,** a training sequence is transmitted and compared at the receiver with a locally generated sequence. The resulting error voltages are used to adjust the tap gains to optimum settings (minimum distortion). The training period may consist of a periodic sequence of isolated pulses or a pseudorandom sequence. After this training period the tap gains remain fixed and normal transmission ensues. Two limitations of this form of automatic equalization are immediately apparent. First, training is required initially and must be repeated with breaks in transmission; second, a time-varying channel will change intersymbol interference characteristics and hence degrade performance since tap gains are fixed.

For these reasons **adaptive equalization** techniques have been developed that are capable of deriving necessary tap gain adjustments directly from the transmitted bits. In preset equalization, since a known training sequence is transmitted, the error voltages can be directly measured with tap gains set accordingly. During normal transmission, the transmitted symbols are unknown. However, a good estimate of this sequence is produced at the receiver output. In the case of adaptive equalization the output at the receiver is used as if it were the transmitted sequence to obtain an estimate of error voltages. Such a learning procedure has been called *decision-directed* since the receiver attempts to learn by employing its own decisions. If the bit decisions are good enough initially, the equalizer can obtain a sufficient estimate of the errors to iteratively improve tap gain settings. A potential drawback to the adaptive equalizer, then, is the settling, or convergence time for poor channels. A common solution to this problem is the use of a hybrid system that uses preset equalization to establish good error performance and then switches to adaptive equalization when normal transmission commences, as illustrated in Figure 5.22.

The adaptive equalizer techniques discussed here are based on the use of linear filters. To improve performance further, the introduction of nonlinear filter techniques has led to the development of the **decision feedback**

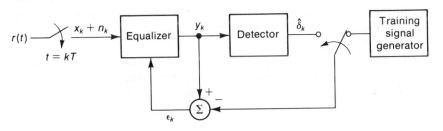

FIGURE 5.22 Adaptive Equalization

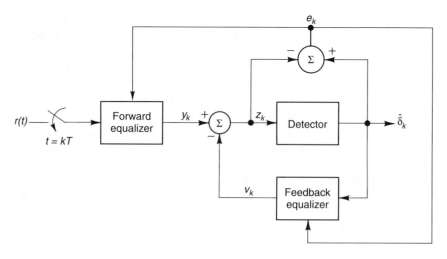

FIGURE 5.23   Adaptive Decision Feedback Equalizer

equalizer [29]. Figure 5.23 shows the basic structure of the decision feedback equalizer (DFE). A linear forward filter processes the received signal using the transversal equalizer described earlier. This filter attempts to eliminate ISI due to future symbols. Decisions made on the forward equalized signal are fed back to a second transversal filter. If the previous decisions are assumed to be correct, then the feedback equalizer cancels ISI due to previous symbols. The tap weights of the feedback equalizer are determined by the tail of the channel impulse response, including the effects of the forward equalizer. The decision feedback receiver that minimizes mean square error consists of a filter matched to the received signal followed by forward and backward transversal filters [30]. The equalizer is made adaptive through the use of the estimated gradient algorithm, in which the forward and feedback coefficients are continually adjusted to minimize mean square error. A necessary condition for convergence is that the probability of error be much less than one-half to avoid both error propagation in the feedback path and adaptation errors in the decision-directed process.

In the DFE of Figure 5.23, the forward equalizer is designed to eliminate precursor ISI only and therefore has a form that is anti-causal, that is, cannot have any positive delay terms, so that its output can be given as

$$y_k = \sum_{n=-N}^{0} c_n x_{k-n} \qquad (5.81)$$

Conversely, the feedback equalizer is designed to cancel the tails of ISI and is therefore causal, so that we may write its output as

$$v_k = \sum_{n=1}^{M} c_n \, \hat{\delta}_{k-n} \qquad (5.82)$$

where $\hat{\delta}_k$ is an estimate of the kth data symbol. The number of feedforward coefficients $N + 1$ and the number of feedback coefficients M can be different. The error signal between the detector input and output

$$e_k = \hat{\delta}_k - z_k \qquad (5.83)$$

is used to adapt the forward and feedback equalizers. Relative to the linear equalizer, which has no feedback equalizer, the main advantage of the DFE is that it reduces noise enhancement. The detector (slicer) eliminates any noise present at its input so that the feedback equalizer will have no noise at its output. The use of the feedback equalizer also allows more freedom in the choice of tap weights in the forward equalizer. Error propagation does occur, however, due to the presence of the feedback equalizer since any error made by the detector will affect future decisions so long as the errored symbol remains in the feedback delay line. (See Problem 5.46, which deals with the tail canceler and its error propagation effect.)

In practice, several conditions may limit the performance of adaptive equalizers:

1. Time-varying channels may exhibit a rate of change that exceeds the equalizer response time. This limitation can be offset by use of a larger step size in the equalizer, but at the expense of greater mean square error. For data transmission over telephone circuits, the step size is made larger for fast convergence during the training period and then reduced for operation during data transmission.
2. Practical equalizers must be based on a finite number of taps. The choice of equalizer range is based on the ratio of channel dispersion measured or estimated to data symbol width. Thus if the channel dispersion exceeds the equalizer range, performance is degraded due to the presence of ISI not cancelled by the equalizer.
3. Conventional adaptive equalizers with tap spacings of a symbol interval T may suffer from aliasing effects for channels with severe distortion. This problem has led to the use of **fractionally spaced equalizers** [31], whose taps are spaced closer than the inverse of twice the highest frequency component in the baseband signal, so that the sampling rate sat-

isfies the Nyquist rate requirement. Advantages of fractionally spaced over symbol-spaced equalizers include signal-to-noise performance, convergence time, and insensitivity to timing phase [32].

4. The problem of equalizer convergence time discussed earlier is not only affected by system probability of error and equalizer step size; it is also dependent on the transmission of random data. Many data terminals transmit repetitive patterns such as all 1's during idle periods. During these periods the adaptive equalizer settings may drift due to lack of sufficient data transitions. A simple solution to this problem is afforded by scrambling of the transmitted data to eliminate periods of data idling at the receiver.

5.11   Data Scrambling Techniques

Impairments in digital transmission systems often vary with the statistics of the digital source. Timing and equalization performance usually depend on the source statistics. For example, a long string of 0's or 1's can cause a bit synchronizer to degrade or even lose synchronization. Likewise, periodic bit patterns can create discrete spectral lines that cause crosstalk in cable transmission and cochannel or adjacent channel interference in radio transmission.

Scrambling the data can minimize long strings of 0's and 1's and suppress discrete spectral components. Scrambling devices randomize or "whiten" the data by producing digits that appear to be independent and equiprobable. There are two basic classes of scramblers: techniques that scramble via logic addition (such as modulo 2) of a pseudorandom sequence with the input bit sequence and those that scramble by performing a logic addition on delayed values of the input sequence itself. In the following discussion we will restrict our attention to applications of scramblers to binary transmission, but the techniques can be generalized to M-ary transmission by use of modulo M addition.

5.11.1   Pseudorandom Sequences

Known, fixed binary sequences that exhibit properties of a random signal can be used as the basis for data scrambling. These sequences are generated by using shift registers with certain feedback connections to modulo 2 adders. A shift register is composed of a number of flip-flops cascaded in series. When the shift register receives a clock pulse, the binary state of each flip-flop is transferred to the next flip-flop. The feedback connections consist of taps at certain stages; the tapped signals are added modulo 2 and fed back to the first flip-flop. To illustrate, consider the three-stage sequence generator shown in Figure 5.24. Suppose we initialize this generator by loading all 1's into the flip-flops. With each clock pulse, the contents of flip-flops 1 and 3 are mod 2 added, the contents of flip-flops 1 and 2 are shifted to flip-flops 2 and 3, and the mod 2 output is fed back to flip-flop 1. The contents of the shift register will cycle through seven different states and then repeat itself. The output of this three-stage sequence generator is taken from the last flip-flop of the shift register to yield the sequence 1110100. . . . This 7-bit sequence appears random to the outside observer. Because the sequence can be predicted from knowledge of the shift register length and taps, however, these sequences are known as **pseudorandom.**

The length of a pseudorandom sequence is determined by the choice of shift register length, feedback taps, and initial states of the flip-flops. As can be observed in Figure 5.24, an initial state of all 0's remains unchanged and thus is not useful for sequence generation. The *maximum-length sequence* produced by an n-stage shift register is therefore $2^n - 1$. Examples of feedback connections for maximum-length sequences are given in Table 5.6 for various shift register lengths. Each connection actually specifies two feedback configurations because the inverse arrangement provides a sequence of the same length, but reversed. Thus feedback from flip-flops 2 and 3 work as well as 1 and 3 to produce a sequence of length 7. Likewise, 3 and 4 work along with 1 and 4 for length 15. Other properties of maximum-length sequences are:

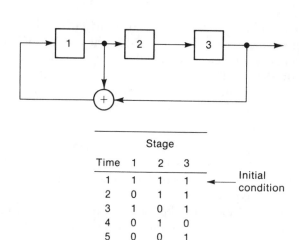

	Stage		
Time	1	2	3
1	1	1	1
2	0	1	1
3	1	0	1
4	0	1	0
5	0	0	1
6	1	0	0
7	1	1	0
8	1	1	1

Initial condition (pointing to Time 1)

FIGURE 5.24   Three-Stage Shift Register with Feedback for Generation of a 7-Bit Pseudorandom Sequence

TABLE 5.6   Examples of Maximum-Length Pseudorandom Sequences

Length of Shift Register	Feedback Taps	Period of Sequence
3	1, 3	7
4	1, 4	15
5	2, 5	31
6	1, 6	63
7	1, 7	127
8	1, 6, 7, 8	255
9	4, 9	511
10	3, 10	1,023
11	2, 11	2,047
12	2, 10, 11, 12	4,095
13	1, 11, 12, 13	8,191
14	2, 12, 13, 14	16,383
15	14, 15	32,767
16	11, 13, 14, 16	65,535
17	14, 17	131,071
18	11, 18	262,143
19	14, 17, 18, 19	524,287
20	17, 20	1,048,575

- The number of 1's in one cycle of the output sequence is 1 greater than the number of 0's.
- The number of runs of consecutive 0's or 1's of length n is twice the number of runs of length $n + 1$. That is, one-half the runs are of length 1, one-fourth of length 2, one-eighth of length 3, and so on.
- The autocorrelation of the sequence has a peak equal to the sequence length $2^n - 1$ at zero shift and at multiples of the sequence length. At all other shifts, the correlation is -1. Autocorrelation for a 7-bit pseudorandom sequence is shown in Figure 5.25.

The correlation property of pseudorandom sequences results in a flat (or white) power spectral density as the sequence length increases. Because these pseudorandom sequences simulate white noise, they are sometimes called *pseudonoise* sequences.

Pseudorandom sequences can be used to scramble data by mod 2 addition of the data with the pseudorandom sequence. Figure 5.26 is a block diagram of a scrambler and descrambler that uses a prescribed pseudorandom (PR) sequence. Note that the same PR sequence generator is required at both the scrambler and descrambler. Further, a state of synchronism must exist between the encoder and decoder; that is, the same set of pseudorandom and data bits must coincide at the input to the encoder and decoder. Initial synchronization can be accomplished by a handshaking exchange of

Sequence: +1 +1 +1 −1 −1 +1 −1

For $\tau = 0$

$$
\begin{array}{rrrrrrrr}
 & +1 & +1 & +1 & -1 & -1 & +1 & -1 \\
\times & +1 & +1 & +1 & -1 & -1 & +1 & -1 \\
\hline
 & +1 & +1 & +1 & +1 & +1 & +1 & +1 & = 7 = R(\tau = 0)
\end{array}
$$

For $\tau = 1$

$$
\begin{array}{rrrrrrrr}
 & +1 & +1 & +1 & -1 & -1 & +1 & -1 \\
\times & +1 & +1 & -1 & -1 & +1 & -1 & +1 \\
\hline
 & +1 & +1 & -1 & +1 & -1 & -1 & -1 & = -1 = R(\tau = 1)
\end{array}
$$

For $\tau = 2$

$$
\begin{array}{rrrrrrrr}
 & +1 & +1 & +1 & -1 & -1 & +1 & -1 \\
\times & +1 & -1 & -1 & +1 & -1 & +1 & +1 \\
\hline
 & +1 & -1 & -1 & -1 & +1 & +1 & -1 & = -1 = R(\tau = 2)
\end{array}
$$

For $\tau = -1$

$$
\begin{array}{rrrrrrrr}
 & +1 & +1 & +1 & -1 & -1 & +1 & -1 \\
\times & -1 & +1 & +1 & +1 & -1 & -1 & +1 \\
\hline
 & -1 & +1 & +1 & -1 & +1 & -1 & -1 & = -1 = R(\tau = -1)
\end{array}
$$

(a) Calculation of autocorrelation $R(\tau)$

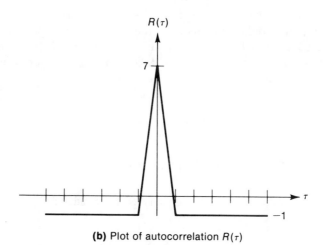

(b) Plot of autocorrelation $R(\tau)$

FIGURE 5.25   Example of Autocorrelation for a 7-Bit Pseudorandom Sequence

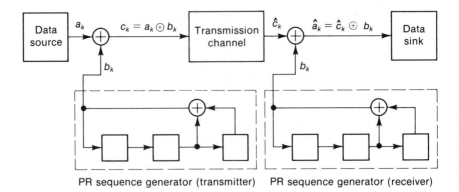

PR sequence generator (transmitter)   PR sequence generator (receiver)

FIGURE 5.26   Block Diagram of Pseudorandom Scrambler and Descrambler

a preamble to initiate a prestored sequence or by forward-acting operation where the initial shift register states are sent by transmitter and used to initialize the receiver shift register.

5.11.2   Self-Synchronizing Scrambler

The self-synchronizing scrambler derives its randomizing capability from a logic addition of delayed digits from the data source. Figure 5.27 shows a generalized self-synchronizing scrambler and descrambler. Each stage of the shift register represents a unit delay. For a scrambler containing M stages, the output may be written as

$$b_k = a_k \oplus \sum_{p=1}^{M} b_{k-p} \delta_p \qquad (5.84)$$

where $\oplus$ and $\sum$ denote modulo addition and δ_p is defined as

$$\delta_p = \begin{cases} 1 & \text{if stage } p \text{ is fed back and added} \\ 0 & \text{otherwise} \end{cases}$$

Decoding of the scrambler output b_k is possible after M error-free scrambler bits have been received, so that the shift register stages are identical at the transmitter and receiver. The equation for the decoder is then similar to that of the encoder:

$$\hat{a}_k = \sum_{p=0}^{M} \hat{b}_{k-p} \delta_p \qquad (5.85)$$

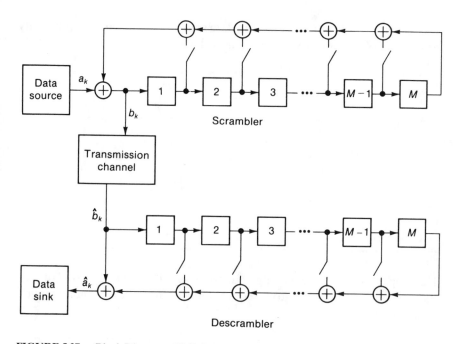

FIGURE 5.27   Block Diagram of Self-Synchronizing Scrambler and Descrambler

The scrambled and descrambled digits are estimates, as indicated in (5.85), due to the assumption of degradation in the channel and the resulting possibility of error in digit detection.

Example 5.6 _____

For a six-stage self-synchronizing scrambler, let δ_5 and $\delta_6 = 1$ and all other feedback connections $\delta_i = 0$ for $i = 1$ to 4. The scrambled data sequence is then

$$b_k = a_k \oplus b_{k-5} \oplus b_{k-6}$$

Assuming error-free transmission, the descrambled data sequence is

$$\hat{a}_k = b_k \oplus b_{k-5} \oplus b_{k-6}$$
$$= (a_k \oplus b_{k-5} \oplus b_{k-6}) \oplus b_{k-5} \oplus b_{k-6}$$
$$= a_k$$

which shows that the descrambled sequence is identical to the original data sequence. Now assume an initial loading of the binary sequence 110101 in the six stages of the scrambler. If the data source then produces an infinitely long string of 0's, the six stages will show the following sequence:

Time	Input	Stage 1	2	3	4	5	6
1	0	1	1	0	1	0	1
2	0	1	1	1	0	1	0
3	0	1	1	1	1	0	1
4	0	1	1	1	1	1	0
5	0	1	1	1	1	1	1
6	0	0	1	1	1	1	1
.	.	.	.	.	.	.	.
.	.	.	.	.	.	.	.
.	.	.	.	.	.	.	.

Development of the remaining sequences is left to the reader. For this example, the scrambled output sequence turns out to be maximum length, or $2^6 - 1 = 63$.

A potential drawback to the use of self-synchronizing scramblers is the inherent property of error extension. To observe this property, let us expand (5.85):

$$\hat{a}_k = \hat{b}_k\, \delta_0 + \hat{b}_{k-1}\, \delta_1 + \hat{b}_{k-2}\, \delta_2 + \cdots + \hat{b}_{k-M}\, \delta_M \qquad (5.86)$$

By inspection we see that for each scrambled bit received in error, n other data estimates are also in error, where n is the number of feedback paths that are closed. Thus the total probability of error is

$$P_T(e) = (1 + n)P(e) \qquad (5.87)$$

where $P(e)$ is the digital probability of error at the detector. In practice, the number of feedback connections required to realize a maximum-length sequence is on the order of 2 to 4. Hence the additional degradation due to error extension is usually considered negligible.

The length of the shift register in a data scrambler is selected according to the desired degree of randomness. The spectrum of the scrambler output is the appropriate measure of randomness. The power spectral density is composed of a number of discrete components within a $(\sin x/x)^2$ envelope;

the discrete components are spaced by $1/T_0$, where T_0 is the period of the data sequence. Although such a demonstration is beyond the scope of this book, it can be shown that a 20-stage scrambler, arranged to provide a maximum-length sequence by choice of appropriate feedback paths, provides the desired randomness for most applications [33].

5.12   Spread Spectrum

Certain advantages arise if the transmission bandwidth is made to greatly exceed the information bandwidth of the signal. The technique for spreading the bandwidth, known as **spread spectrum,** is accomplished by multiplying the data with a spreading signal that is independent of the data. The original signal is recovered at the receiver by using the same spreading signal, synchronized with the received spread signal, to despread the signal. There are several applications of spread spectrum, including: protection against interference, either hostile (jamming) or unintentional; separation of users sharing a common medium; and reduction in energy density to meet frequency allocation regulations, minimize detectability, or provide privacy. Spread spectrum has long been used by the military to provide jamming protection [34], but more recently its application has been extended to commercial communications [35]. Here we will discuss fundamental concepts, examine the two basic forms of spread spectrum, and look at two popular applications.

5.12.1   Fundamental Concepts

Consider the block diagram of Figure 5.28, which shows one way of achieving spread spectrum. The data signal $d(t)$ is multiplied by the spreading signal $p(t)$ and transmitted over some channel that adds an interfering signal $i(t)$. At the receiver the incoming signal $d(t)p(t) + i(t)$ is multiplied by $p(t)$ to yield $d(t) + p(t)i(t)$. The first term may be recovered using a filter of bandwidth $1/T$, while the second term indicates that the interfering signal has been spread over the bandwidth of the spreading signal. Thus even if the interfering signal is in the middle of the band of interest, only a fraction of its power can pass through the filter. Moreover, the bandwidth spreading effect on the interfering signal is independent of the characteristics of the interfering signal. Figure 5.29 compares the power spectrum of the spread versus nonspread signals at both the input and output of the receiver. These power spectra are easily found by recognizing that multiplication of two signals in the time domain is equivalent to convolution of their spectra in the frequency domain. Double multiplication of the desired signal is seen to protect it from the undesired signal, which is multiplied only once by the spreading signal.

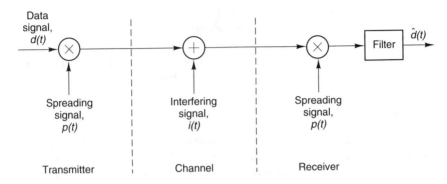

FIGURE 5.28   Spread Spectrum System

A commonly used figure of merit for spread spectrum systems is the processing gain, defined as the ratio of the spread spectrum signal bandwidth B_s to the data signal bandwidth B_d, or

$$G_p = B_s/B_d = f_c T \tag{5.88}$$

The processing gain indicates the degree of interference reduction realized by use of spread spectrum. A second advantage to spread spectrum systems is observed by considering the effect that spreading has on power. By spreading the desired signal over a large bandwidth, the transmitted power is spread over this same bandwidth, thus minimizing the power per unit bandwidth. The advantage here lies in the ability of spread spectrum to coexist on a noninterfering basis with other signals occupying the same band.

5.12.2   Direct Sequence and Frequency Hopping Techniques

There are two basic spread spectrum techniques, direct sequence and frequency hopping, which are typically used separately but can be combined in so-called hybrid systems. To illustrate the direct sequence technique, suppose that the $p(t)$ in Figure 5.28 is a pseudorandom sequence operating at a rate greater than the data rate itself. A pulse or bit of the pseudorandom sequence is known as a **chip.** Note that Figure 5.26 is a more detailed block diagram of direct sequence spread spectrum if the PR sequence generator is assumed to operate at a chip rate greater than the data rate. Further suppose that we use the 7-bit pseudorandom sequence shown in Figure 5.24 as the spreading sequence. In this case each data bit is modulo 2 added to the 7-bit pseudorandom sequence, so that there are seven chips per data bit.

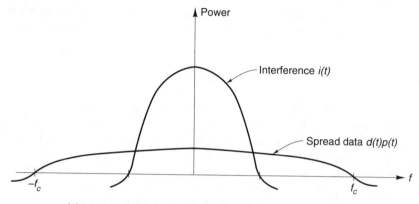

(a) Interference and spread data signals at receiver input

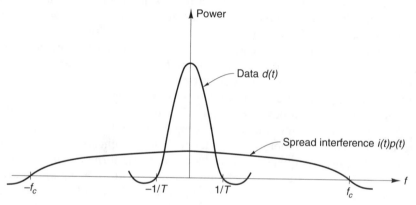

(b) Data and spread interference signals at receiver output

FIGURE 5.29   Power Spectra of Spread Spectrum Signals

The bandwidth expansion due to spreading is a factor of seven as is the processing gain. A more general definition for processing gain with the direct sequence technique is

$$G_p = 10 \log \text{[chip rate/bit rate] dB} \qquad (5.89)$$

The receiver uses the same pseudorandom sequence in a shift register to despread and recover the original signal.

The idea behind frequency hopping is to spread the spectrum sequentially rather than instantaneously, by pseudorandomly changing (hopping) the frequency of the carrier over the spread spectrum signal bandwidth B_s. The number of chips (n) required to specify the transmit frequency is given by

$$n = \log_2 (B_s / f_d) \qquad (5.90)$$

where f_d is the minimum frequency spacing between consecutive hops. In its simplest form, frequency hopping is done on a bit-by-bit basis. For M-ary multilevel transmission systems, however, the carrier frequency is hopped once per symbol. From (5.43) the symbol rate ($1/T$) is seen to be $R/\log_2 M$, which is also the hopping rate. The implementation of frequency hopping requires a frequency synthesizer whose output is changed every symbol according to the n-chip sequence from a pseudorandom sequence generator. At the receiver, the frequencies generated by the local synthesizer must be synchronized with the frequency pattern of the received signal in order to produce the dehopped output. The rate at which frequencies are hopped will determine the degree to which a spread spectrum signal is achieved. Frequency hopping may be classified as fast or slow. Fast frequency hopping is characterized as having one or more hops for each transmitted symbol, such that the hopping rate equals or exceeds the data rate. With slow frequency hopping there are several symbols per hop. The fast frequency hopping has greater robustness (say, against the smart jammer), while the slow frequency hopping approach is simpler to implement and effective against the simple jammer or in frequency sharing applications.

Both the direct sequence and frequency hopping forms of spread spectrum usually generate the spreading sequence with a maximum-length pseudorandom sequence, which has the properties described in Section 5.11.1. Depending on the application, additional properties may be desirable, such as protection against the smart jammer who might try to reconstruct the sequence. The linear feedback shift register described in Section 5.11.1 produces a maximum-length pseudorandom sequence and can therefore be used for spread spectrum, but it does not provide protection against the adversary trying to recover the code. Several techniques have been proposed to avoid this deficiency, including the use of nonlinear feedback shift registers [36]. Another example where nonmaximal pseudorandom sequences are necessary is code division multiple access (CDMA), which requires low cross-correlation among the user code sequences in order to minimize interference between pairs of users in the shared spectrum. Certain codes have been identified that are useful for CDMA because of their cross-correlation properties [37]. One such code is the set of Gold sequences, which are generated by the modulo 2 addition of certain maximum-length pseudorandom sequences [38].

At the receiver, a synchronized version of the spreading code is necessary to decode the signal. The synchronization process is accomplished in two steps, called acquisition and tracking. In the acquisition mode the locally generated spreading code is brought into coarse alignment with the received signal, after which the tracking mode provides finer synchronization. Code acquisition must account for several timing uncertainties, such as the distance between transmitter and receiver, clock instabilities, and Doppler shift. Acquisition can be done via parallel search, in which a bank of correlators is used to simultaneously compare all possible code positions, or via serial search, in which a single correlator is used to find the correct spreading code. With parallel search, correlator outputs are compared after a search dwell time and the code corresponding to the largest output is chosen. With serial search, the correlator output for each phase of the locally generated code is sequentially compared to a threshold, and when that threshold is exceeded, acquisition is assumed. Serial search implementation offers significant savings in complexity but at the expense of increased acquisition time. The choice of serial versus parallel search depends on the intended application, specifically on whether or not frequent acquisition and rapid acquisition times are necessary. A typical tracking loop contains a local code generator that is offset from the incoming sequence $p(t)$ by a time τ that is less than one-half the chip time. To provide finer synchronization, the local code generator outputs two sequences, delayed from each other by one chip. These two sequences are multiplied by the incoming signal $p(t)d(t)$, averaged in a feedback loop, and used to control the phase of the local code generator. When the tracking error τ converges to zero, the output of the local code generator becomes $p(t+\tau) = p(t)$, allowing the received signal to be despread without error.

5.12.3   Application of Spread Spectrum to Anti-Jamming and CDMA

The object of the jammer is to deny communications by placing interfering signals in the frequency band of interest. Spread spectrum counteracts the jammer by use of pseudorandom codes unknown to the jammer, making it difficult and costly for the jammer to effectively disrupt communications. Spread spectrum used in anti-jamming has a figure of merit known as the jamming-to-signal margin J/S defined by

$$(J/S)_{dB} = \left[G_p - (S/N)_{min} \right]_{dB} \tag{5.91}$$

where $(S/N)_{min}$ is the minimum signal-to-noise ratio required to meet the objective bit error rate. The parameter J/S is to be interpreted as the ratio of jamming signal to desired signal that can be tolerated at the receiver input. For example, if the processing gain is 30 dB and the minimum signal-to-noise ratio is 10 dB, then the (J/S) is 20 dB and the receiver can reliably detect the desired signal even when the jamming signal is 20 dB greater than the desired signal. Another figure of merit is the anti-jam (AJ) margin, defined as the difference between the (J/S) that is required against a particular jammer and the (J/S) that is actually received, or

$$M_{AJ} = (J/S)_{reqd} - (J/S)_{recd} \text{ dB} \tag{5.92}$$

Following the same example just cited, if $(J/S)_{reqd} = 20$ dB and the $(J/S)_{recd} = 15$ dB, then an AJ margin of 5 dB is provided. For more information on spread spectrum use in AJ applications, the reader is referred to [39].

Code Division Multiple Access (CDMA) is a spread spectrum system that allows multiple users to communicate simultaneously over the same frequency band. Each user is given a distinct sequence, $p_i(t)$, $i = 1$ to N for N users. At the input to any one receiver will be a linear combination of the energy from each of the users, $p_1(t)d_1(t) + p_2(t)d_2(t) + \ldots + p_N(t)d_N(t)$. Because of the spreading of the user bandwidth over the entire frequency band, a receiver "tuned" to a particular user's sequence will see all the energy of that particular user but only a fraction of the energy from other users. Thus the output of the receiver will consist of the following terms, assuming that the receiver is listening to user 1:

$$p^2_1(t)d_1(t) + p_1(t) \sum_{i=2}^{N} p_i(t)d_i(t) \tag{5.93}$$

The first term is of course the desired signal (since $p_i^2(t) = 1$) and the second term is the composite of all other undesired terms. Use of orthogonal codes, such as Gold codes, will ensure that the desired signal can be extracted and that undesired signals will be rejected. Even so, the receiver performance will be degraded when transmitters sharing the channel are not synchronized. The problem here is that two (or more) users operating asynchronously will lead to a partial correlation of the pseudorandom sequences. Any nonzero partial correlation will lead to interference between the two users. Careful selection of the code, such as use of Gold sequences, will ensure small partial cross-correlation. Although code design is crucial to CDMA system performance, a more serious limitation is the so-called **near-far** problem. Here an interfering signal from a nearby transmitter arrives with more power than the desired signal from a more distant location. This constraint on CDMA performance can be overcome by use of adaptive power control such that received power from each user is the same, or by use of frequency hopping instead of direct sequence CDMA.

CDMA is often compared with other multiple access techniques that have the same objective, that is, to share a communications media among several users. Time Division Multiple Access (TDMA) works by separating users in time using time slots, one per user, and using centalized control to synchronize time slots among all users. Time slots can be pre-assigned or demand-assigned. Frequency Division Multiple Access (FDMA) divides the frequency band into smaller bands, one per user. In comparison with TDMA, FDMA requires more complex implementation and is susceptible to nonlinearities as found in high-power amplifiers (see Chapter 9), but it eliminates the synchronization requirements of TDMA. Of interest here is a comparison of performance and capacity of CDMA versus TDMA or FDMA. Ignoring real-world constraints, all three multiple-access systems have the same theoretical capacity. CDMA has the advantages of not requiring synchronous operation and being able to easily add new users. Of course, the use of spread spectrum provides additional protection against interference such as multipath or external users. Studies of CDMA applied to mobile cellular communications and personal communications networks have concluded that CDMA has greater capacity potential primarily from better reuse of the frequencies in adjacent cells [35, 40, 41]. Recent interest in CDMA stems from the fact that there are no unallocated frequencies in the United States for wireless telephones, and spread spectrum offers the potential for sharing of the radio spectrum. The concept here is that the spread spectrum signals would overlay a frequency allocation already in use, perhaps a fixed microwave system, with negligible interference between the two systems.

5.13 Summary

Direct transmission of a signal without frequency translation is called baseband transmission. Certain codes facilitate transmission of data over baseband channels. By appropriate choice of code, certain desirable properties can be obtained, such as (1) increased data transition density to allow extraction of clock from the data signal, (2) shaping of the spectrum to minimize interference between channels, (3) reduction of bandwidth for increased spectral efficiency, and (4) error detection for on-line performance monitoring. Nonreturn-to-zero (NRZ) coding is a common choice for binary coding but lacks most of these desirable properties. Both diphase and bipolar coding result in increased data transitions and a desired reshaping of the data spectrum. Bipolar also provides inherent error detection, although the three levels used for bipolar result in a signal-to-noise (S/N) penalty compared to other two-level codes.

Block codes, which convert a group of bits into a block of symbols, are used to add desired characteristics to the baseband signal. For example, *mBnB* codes add one or more overhead bits to the information bits to provide parity, dc balance, or improved timing. Another class of block codes is the *kBnT* set of codes, which use three-level symbols to represent binary information. The added redundancy from the three levels is used to gain certain advantages over binary signaling, such as built-in error detection or reduction of low-frequency energy.

Because of bandwidth limitations in real channels, digital signals tend to be distorted in the transmission process. Unless compensated, this distortion causes intersymbol interference (ISI). However, certain pulse shapes called Nyquist waveforms can be used that contribute zero energy at adjacent symbol times. A sin x/x form provides the required shape but is not physically realizable; a raised-cosine shape provides zero ISI and also leads to physically realizable filters.

For a channel with bandwidth W hertz, the basic limitation in bit rate R is $2W$ bits per second for binary transmission. Hence the limitation in bandwidth efficiency (R/W) is 2 bps/Hz. With M-ary transmission, each symbol takes on one of M levels and hence represents $\log_2 M$ bits of information. M-ary transmission increases the spectral efficiency by a factor of $m = \log_2 M$ over binary transmission. However, comparison of probability of error performance indicates an S/N degradation of $3/(M^2 - 1)$ for M-ary transmission in comparison to binary $(M = 2)$ transmission.

Partial response coding uses prescribed amounts of intersymbol interference to reduce the transmission bandwidth further. There are several classes of partial response codes that permit choice of spectral shaping and number of input and output levels. The error performance is seen to be superior to that for binary and M-ary transmission.

Eye patterns are displays of random data on an oscilloscope. All allowed data transitions blend together to form a composite picture of the signal. At the receiver, the eye pattern viewed prior to data detection shows the effects of channel perturbations and is therefore an indication of system health.

Equalization of a signal distorted by a transmission channel is done by means of adjustable filters. The usual choice for equalizer design is the transversal filter, which consists of a tapped delay line. The preset equalizer uses a training sequence to initialize the tap settings. For time-varying channels, the equalizer must also be adaptive. This adaptation of the tap settings can be accomplished by using data decisions during periods of data transmission.

Data scrambling is required when performance of the communication system is dependent on the randomness of the data source. Pseudorandom sequences can be used for data scrambling by employing simple digital logic to add the data to the pseudorandom sequence. An alternative technique is the self-synchronizing scrambler, which performs a logic addition on delayed values of the data input sequence. Either technique can be used to improve performance in timing and equalization and minimize interference.

Spread spectrum is a technique used to increase the bandwidth of a transmitted signal to protect the information signal from interference. Generally, a pseudorandom sequence generator is used as the source of the spreading signal, which operates at a rate much greater than the information rate. The degree of protection from interference is determined by the ratio of the spreading rate to the information rate. The spreading signal can operate directly on the data sequence (much like a data scrambler) or can be used to select frequencies which are hopped according to the state of the pseudorandom sequence. Applications of spread spectrum are common in anti-jamming for military communications, but are increasing used in commercial communications such as code division multiple access for sharing of frequency bands among multiple users.

References

1. M. R. Aaron, "PCM Transmission in the Exchange Plant," *Bell System Technical Journal* 41(January 1962):99–141.
2. V. I. Johannes, A. G. Kain, and T. Walzman, "Bipolar Pulse Transmission with Zero Extraction," *IEEE Trans. on Comm. Tech.*, vol. COM-17, no. 2, April 1969, pp. 303–310.
3. J. H. Davis, "T2: A 6.3 Mb/s Digital Repeatered Line," *1969 International Conference on Communications,* Boulder pp. 34-9–34-16.
4. CCITT Yellow Book, vol. III.3, *Digital Networks—Transmission Systems and Multiplexing Equipment* (Geneva: ITU, 1981).
5. A. Croisier, "Introduction to Pseudoternary Transmission Codes," *IBM J. Res. Develop.*, July 1970, pp. 354–367.
6. W. R. Bennett, "Statistics of Regenerative Digital Transmission," *Bell System Technical Journal* 37(November 1958):1501–1542.
7. Y. Niro, Y. Ejiri, and H. Yamamoto, "The First Transpacific Optical Fiber Submarine Cable System," *1989 International Conference on Communications,* June 1989, pp. 50.1.1–50.1.5.
8. N. Yoshikai, K. Katagiri, and T. Ito, "mBIC Code and Its Performance in an Optical Communication System," *IEEE Trans. on Comm.*, vol. COM-32, no. 2, February 1984, pp. 163–168.
9. S. Kawanishi and others, "DmB1M Code and its Performance in a Very High-Speed Optical Transmission System," *IEEE Trans. on Comm.*, vol. COM-36, no. 8, August 1988, pp. 951–956.
10. W. Krzymien, "Transmission Performance Analysis of a New Class of Line Codes for Optical Fiber Systems," *IEEE Trans. on Comm.*, vol. COM-37, no. 4, April 1989, pp. 402–404.

11. I. J. Fair and others, "Guided Scrambling: A New Line Coding Technique for High Bit Rate Fiber Optic Transmission Systems," *IEEE Trans. on Comm.*, vol. COM-39, no. 2, February 1991, pp. 289–297.

12. ANSI Standard X3.148-1988, "Fiber Distributed Data Interface (FDDI)-Token Ring Physical Layer Protocol (PHY)," June 30, 1988.

13. H. Nyquist, "Certain Topics in Telegraph Transmission Theory," *Trans. AIEE* 47(April 1928):617–644.

14. R. W. Lucky, J. Salz, and E. J. Weldon, *Principles of Data Communication* (New York: McGraw-Hill, 1968).

15. C. E. Shannon, "A Mathematical Theory of Communication," *Bell System Technical Journal* 27(July 1948):379–423.

16. C. E. Shannon, "Communications in the Presence of Noise," *Proc. IRE* 37(January 1949):10–21.

17. A. Lender, "The Duobinary Technique for High Speed Data Transmission," *IEEE Trans. Comm. and Elect.* 82(May 1963):214–218.

18. A. Lender, "Correlative Digital Communication Techniques," *IEEE Trans. Comm. Tech.*, vol. COM-12, December 1964, pp. 128–135.

19. E. R. Kretzmer, "Generalization of a Technique for Binary Data Communication," *IEEE Trans. Comm. Tech.*, vol. COM-14, February 1966, pp. 67–68.

20. A. Lender, "Correlative Level Coding for Binary Data Transmission," *IEEE Spectrum* 3(February 1966):104–115.

21. D. R. Smith, "A Performance Monitoring Technique for Partial Response Transmission Systems," *1973 International Conference on Communications*, Seattle pp. 40-14–40-19.

22. S. Pasupathy, "Correlative Coding: A Bandwidth-Efficient Signaling Scheme," *IEEE Comm. Mag.*, July 1977, pp. 4–11.

23. F. K. Becker, E. R. Kretzmer, and J. R. Sheehan, "A New Signal Format for Efficient Data Transmission," *Bell System Technical Journal* 45(May/June 1966):755–758.

24. T. L. Swartz, "Performance Analysis of a Three-Level Modified Duobinary Digital FM Microwave Radio System," *1974 International Conference on Communications*, pp. 5D-1–5D-4.

25. C. W. Anderson and S. G. Barber, "Modulation Considerations for a 91 Mbit/s Digital Radio," *IEEE Trans. on Comm.*, vol. COM-26, no. 5, May 1978, pp. 523–528.

26. K. L. Seastrand and L. L. Sheets, "Digital Transmission over Analog Microwave Systems," *1972 International Conference on Communications*, pp. 29-1–29-5.

27. P. R. Hartman, "Digital Radio Technology: Present and Future," *IEEE Comm. Mag.*, July 1981, pp. 10–15.

28. S. Qureshi, "Adaptive Equalization," *IEEE Comm. Mag.*, March 1982, pp. 9–16.

29. C. A. Belfiore and J. H. Parks, Jr., "Decision Feedback Equalization," *Proc. IEEE* 67(August 1979):1143–1156.

30. P. Monsen, "Feedback Equalization for Fading Dispersive Channels," *IEEE Trans. on Information Theory*, vol. IT-17, January 1971, pp. 56–64.

31. R. D. Gitlin and S. B. Weinstein, "Fractionally Spaced Equalization: An Improved Digital Transversal Filter," *Bell System Technical Journal*, 60 (February 1981): 275–296.
32. S. U. Qureshi and G. D. Forney, "Performance and Properties of a T/2 Equalizer," *1977 National Telecommunications Conference*, pp. 11.1.1–11.1.9.
33. D. G. Leeper, "A Universal Digital Data Scrambler," *Bell System Technical Journal* 52(December 1973):1851–1866.
34. R. A. Scholtz, "The Origins of Spread Spectrum Communications," *IEEE Trans. on Comm.*, vol. COM-30, no. 5, May 1982, pp. 822–854.
35. D. L. Schilling and others, "Spread Spectrum for Commercial Communications," *IEEE Comm. Mag.*, April 1991, pp. 66–79.
36. R. L. Pickholtz, D. L. Schilling, and L. B. Milstein, "Theory of Spread-Spectrum Communications," *IEEE Trans. on Comm.*, vol. COM-30, no. 5, May 1982, pp. 855–884.
37. D. V. Sarwate and M. B. Pursley, "Cross Correlation Properties of Pseudo-Random and Related Sequences," *Proc. IEEE* 68(May 1980): 598–619.
38. R. Gold, "Optimum Binary Sequences for Spread Spectrum Multiplexing," *IEEE Trans. on Information Theory*, vol. IT-13, no. 5, September 1967, pp. 619–621.
39. D. J. Torrieri, *Principles of Secure Communication Systems* (Dedham MA: Artech House, 1985).
40. D. G. Smith, "Spread Spectrum for Wireless Phone Systems: The Subtle Interplay between Technology and Regulation," *IEEE Comm. Mag.*, February 1991, pp. 44–46.
41. J. T. Taylor and J. K. Omura, "Spread Spread Technology: A Solution to the Personal Communications Services Frequency Allocation Dilemma," *IEEE Comm. Mag.*, February 1991, pp. 48–51.
42. J. G. Proakis, *Digital Communications* (New York: McGraw-Hill, 1989).

Problems

5.1 Using the symbols +, −, and 0 to represent a positive pulse, negative pulse, and no pulse, respectively, determine the following transmission code sequences of the binary data sequence (assume 100 percent duty cycle):

0101 1100 0010 0011 0000 0000

(a) Bipolar with the most recent pulse being positive.
(b) Bipolar with the most recent pulse being negative.
(c) *B3ZS* with a +0+ substitution having just been made.
(d) *HDB3* with a +00+ substitution having just been made.
(e) *B6ZS* with the most recent pulse being positive.
(f) *B8ZS* with the most recent pulse being negative.

5.2 Using waveforms, determine the following transmission code sequences for the same binary sequence as in Problem 5.1: (a) NRZ(L); (b) NRZ(I); (c) Diphase; and (d) Conditioned Diphase.

5.3 Two binary codes, one using polar NRZ and the other using unipolar NRZ, are to be designed to operate at the same bit rate and to yield the same error rate when transmitted over identical additive noise channels. Determine the relative amplitudes of the two codes.

5.4 A ±5 volt, polar NRZ coded, binary signal is to be used over a channel with additive gaussian noise. If a probability of error equal to 1×10^{-6} is desired, what noise power can be tolerated?

5.5 Using Equation (5.3), derive Equation (5.9).

5.6 For a unipolar NRZ signal with levels $(0,V)$, find the (a) Autocorrelation; (b) Power spectral density.

5.7 The bipolar sequence $\{0,+V,-V\}$ is transmitted over a channel in which errors may occur for any or all of the symbols in the sequence. For each possible error pattern, determine which errors are detectable at the receiver. Assume that the receiver knows that the symbol immediately preceding this sequence is a $-V$.

5.8 Using the waveforms corresponding to the binary sequence shown in Figure 5.1, decode NRZ(M), diphase, and conditioned diphase for the following conditions: (a) The polarity of the transmitted signal is reversed during transmission. (b) A slip of one symbol interval occurs during transmission. (c) Now compare the performance of each of the three decoders under these two conditions.

5.9 Consider a unipolar RZ signal in which the pulse width is $T/4$ rather than $T/2$. (a) Find an expression for the autocorrelation and plot it. (b) Find an expression for the power spectral density and plot it. (c) Comment on the desirability of this code format from the standpoint of clock recovery.

5.10 A polar NRZ signal is transmitted over a noisy channel in which the noise has a laplacian probability density function. The signal amplitude is ±1 volt and the noise amplitude is 0.2 volts. Assuming equally likely binary signals, what is the probability of error?

5.11 An NRZ(L) transmission system operating at an error rate of 10^{-6} is to have its transmitted power increased in order to achieve an error rate of 10^{-8}. Assuming that the same data rate is used in both systems, what percentage of transmitter power increase is required?

5.12 Compare bipolar (with 100 percent duty cycle) and diphase codes as follows:

(a) Using waveforms, draw the code sequences corresponding to the following binary data sequence:

$$110 \ 000 \ 111 \ 101 \ 001$$

(b) Briefly compare the two with respect to decoder logic, probability of error, bandwidth, error detection capability, and clock recovery performance.

5.13 Miller code, also called **delay modulation,** is a baseband coding scheme whereby a level transition occurs at the midpoint of the symbol interval for each occurrence of a 1. No transition occurs for a 0 unless followed by a 0, in which case a transition occurs at the end of the symbol interval corresponding to the first 0. The advantage of this code is a very compact spectrum, relative to NRZ or diphase, concentrated at $f = 0.4/T$, and an insensitivity to 180° phase ambiguity.

(a) Express in equation form the allowed signal values. Assume $+V$ and $-V$ as the two signal levels.

(b) Starting with the voltage level $(-V)$, draw the transmitted sequence for the binary sequence 10110001101.

(c) Using digital logic, show the coder and decoder.

(d) Derive the expression for the probability of bit error.

5.14 Consider the conditioned diphase encoding scheme.

(a) Describe the encoder by giving two sets of equations which correspond to the operations of NRZ(I) and diphase.

(b) For the binary sequence 1100011, sketch the corresponding waveform.

(c) Describe the characteristics of this data encoding scheme as compared to other techniques discussed in this chapter.

5.15 Consider the decoding errors created when two different forms of bipolar are used at the encoder and decoder. For the information sequence {10000000011000000} what errors occur at the decoder output when: (a) The encoder uses AMI and the decoder uses $B3ZS$? (b) The encoder uses $B6ZS$ and the decoder uses AMI? (c) The encoder uses $B8ZS$ and the decoder uses $B3ZS$?

5.16 For coded mark inversion (CMI):

(a) Find the maximum possible length of consecutive identical symbols.

(b) Compare the probability of a mark versus space.

(c) Show that the probability of detecting a single error is given by

$$p \simeq \frac{8 - 2p_e}{8(1 + 2p_e)}$$

where $p_e = $ BER.

5.17 Find the running digital sum (RDS) for: (a) NRZ; (b) AMI; (c) *B6ZS*; and (d) *HDB3*.

5.18 Compatible High Density Bipolar of order N (*CHDBN*) is a bipolar code in which every sequence of $(N+1)$ 0's is replaced by

$$00 \ldots BOV \text{ or } 00 \ldots 00V$$

where the selection of substitution pattern is done in such a way that the number of B pulses between any two consecutive V pulses is odd. Like *HDBN*, the order N is the limit on the number of 0's allowed before a substitution pattern is inserted at the $(N+1)$th 0.

(a) Using B, *0*, and V, write out the *CHDB2* and *CHDB3* codes for the following binary data:

$$10011000011100010000011$$

(b) Compare the substitution pattern of *CHDB2* versus *HDB2*, and *CHDBN* versus *HDBN* for $N \geq 3$.

(c) Describe the decoder for *CHDBN*. How does the decoder change if the value of N is changed? What advantage does *CHDBN* decoding have over *HDBN* decoding?

(d) What is the digital running sum of *CHDBN*? How does that compare with the RDS of *HDBN*?

5.19 The WAL-2 is a two-level transmission code that has been proposed for digital subscriber loops. WAL-2 has two patterns corresponding to binary 0 and 1, as follows:

(a) Comment on the dc balance property of WAL-2.

(b) How does the bandwidth of WAL-2 compare with that of NRZ? Sketch the spectrum of WAL-2.

(c) Derive the probability of error.

5.20 Using the rules for forming *BNZS* codes: (a) Propose a substitution pattern for *B4ZS*. (b) Propose a substitution pattern for *B2ZS*. (c) Comment on the balance property of your *B4ZS* versus *B2ZS*.

5.21 A ternary transmission code known as **dicode,** or **twinned binary,** has transmitted symbols given by

$$a_k = b_k - b_{k-1}$$

where $\{b_k\}$ are the information bits with value 0 and 1.

(a) Draw a block diagram of the coder and decoder.

(b) Find an expression for the power spectrum.

(c) Give the coder output, using $\{-1,0,+1\}$, corresponding to the input $\{00011110110\}$.

(d) Describe the error extension problem with this code. Comment on the effect of transmission errors on decoding when the input to the coder is a long sequence of 1's followed by 0's.

(e) Describe the constraints on transitions (which is used to monitor errors).

(f) What is the bound on the running digital sum and what is the allowed digital sum variation?

5.22 In the *DmB1M* line code, the coder increases the speed of the input signal by $(m + 1)/m$ and then inserts a mark every $(m + 1)$ bits. The mark-inserted signal (Q) is converted to the *DmB1M* code (S) by the following logic:

$$S_k = S_{k-1} \oplus Q_k$$

(a) Draw a block diagram of the coder and decoder.

(b) Show that a C bit (complementary bit) is generated automatically by the coder at the $(m + 1)$th bit.

(c) Show that the error detection probability is approximately equal to $2p_e/(m + 1)$, where p_e is the probability of bit error in the transmission line.

(d) Show that the decoder has an error propagation factor of two.

5.23 **Paired select ternary** (PST) is a three-level transmission code in which two bits are encoded together into two ternary symbols per the following table:

Input data	Coded output
00	− +
01	−0 alternated with 0−
10	+0 alternated with 0+
11	+ −

(a) Using +, −, and 0, write out the PST code for the same binary data sequence as Figure 5.1.

(b) Comment on the dc balance and error-monitoring properties of PST.

(c) Since the PST is a block code, the receiver must be synchronized with the transmitter before decoding can be accomplished. How does the choice of allowed code pairs facilitate receiver synchronization? (Hint: examine invalid symbol patterns.)

(d) What is the running digital sum of PST?

5.24 A *5B6B* code is to be designed with certain characteristics.

(a) Using a table such as Table 5.2, show the coding rules. Where alternating patterns are required to achieve dc balance, your patterns can have no more than three adjacent 1's. Furthermore, your code set should have a maximum distance of six bits without a level transition.

(b) Draw a block diagram of the encoder and decoder logic.

(c) What coding delay is introduced by your encoder? Give your answer in terms of T, the pulse width of the data source.

(d) What bandwidth is required (in terms of T)?

5.25 Compute the ratio of peak positive amplitude (at $t = 0$) to peak negative amplitude (at first sidelobe) for: (a) ideal Nyquist pulse; and (b) raised cosine pulse with $\alpha = 1$.

5.26 Consider a binary transmission system in the absence of noise. The pulse shape used for transmission is

$$g(t) = \frac{\sin(\pi B t)}{\pi B t} \cos 2\pi f_c t \quad (f_c > 0)$$

(a) The transmission rate is to be $R = 1/T = 2B$. What should be the relation between f_c and B such that the above $g(t)$ satisfies the Nyquist criterion of zero ISI?

(b) What is the system bandwidth? Is Nyquist's relation between minimum bandwidth and signaling rate violated in this case? Explain.

(c) What is the minimum value of f_c for which part (a) is satisfied and what is the name of the resulting system?

5.27 The transfer function is shown below for a raised-cosine channel, which is to transmit at a rate $1/T$ symbols/sec.

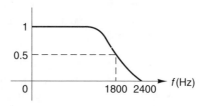

(a) What is the equivalent Nyquist bandwidth? (b) What is the rolloff factor α? (c) If the number of levels (M) per symbol is 2, what is the resulting bit rate?

5.28 The binary sequence {011001000011} is to be transmitted as a sequence of impulses $\delta(t - T/2)$ over a channel that has an ideal low-pass transfer function with cutoff frequency of $1/2T$. Sketch the channel output. What effect will decreasing the cutoff frequency of the low-pass filter have on performance?

5.29 A 16-kb/s modem uses 64-ary modulation and raised-cosine filtering to operate over a channel with 2.4-kHz bandwidth. What is the required rolloff factor α of the filter characteristic?

5.30 Using Equation (5.48), plot the probability of error versus S/N (in dB) for M-ary baseband transmission for $M = 2, 4, 8, 16$, and 32. Compare the signal bandwidths for $M = 4, 8, 16$, and 32 relative to $M = 2$.

5.31 A 16-level baseband transmission system transmits signals with levels ±1 V, ±3 V, ±5 V, ..., ±15 V over an additive noise channel with noise power equal to 5×10^{-2} watts. (a) Find the average transmitted symbol power. (b) Find the probability of error.

5.32 Consider the following partial response system:

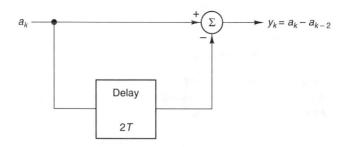

(a) For the binary input sequence $a_k = \{0111000101100\}$, determine the sequence of output signal levels, y_k. For this output sequence, indicate the decoding rule and apply this rule to obtain the original input sequence.

(b) For the given partial response system, indicate by block diagram a precoder (as used in duobinary) that eliminates error propagation. Using this form of precoded partial response, repeat part (a) for the binary sequence shown there.

5.33 For the following partial response systems, each with the four-level input sequence $\{a_k\} = \{1, -3, 1, -1, 3, 3, -3\}$, determine the sequence of output signal levels and indicate the decoding rules. Also, indicate the number of allowed output signal levels and the probability of occurrence for each level, assuming equally likely input levels.

(a) $y_k = a_k + a_{k-1}$

(b) $y_k = a_k - a_{k-1}$

(c) $y_k = a_k - a_{k-2}$

(d) Repeat part (c) but with a precoder (as used with duobinary) that eliminates error propagation. (Hint: Assume level conversion as follows $\{-3 \rightarrow 0, -1 \rightarrow 1, 1 \rightarrow 2, 3 \rightarrow 3\}$ and use a mod 4 adder as the first element in the precoder.)

5.34 Consider the partial response system given by

$$y_k = a_k + 2a_{k-1} + a_{k-2}$$

with equally likely binary inputs.

(a) Determine the number of output levels, their values, and their probability of occurrence.

(b) Find an expression for the power spectrum.

5.35 Using the sequence of binary digits shown in Figure 5.1:

(a) Determine the corresponding output sequence for duobinary and precoded duobinary.

(b) Decode the sequences found in part (a). State any necessary assumptions about previous input bits.

(c) Repeat part (b) but with the condition that the fourth bit has been received in error. Comment on the effect that this error has on decoding of duobinary versus precoded duobinary.

5.36 Like duobinary, modified duobinary has certain constraints that can be used for error detection. Describe these constraints and how they can be used for error detection.

5.37 A trellis can be used to represent the states and transitions between states of a code such as partial response. States are represented by a set of nodes arranged vertically and replicated horizontally, and the transitions are depicted as lines connecting the states at successive symbol times. Changes in the symbol value with time corresponding to changing input values can be traced by following a path through the trellis. (See Chapter 6 for several examples.) Draw the trellis and trace the output corresponding to the sequence of binary digits {11000101001} for: (a) Duobinary; and (b) Modified duobinary.

5.38 A binary digital transmission system uses pulses with equiprobable levels $\pm d$ and operates over an additive noise channel. The impulse response of the channel is $x(t)$, and the additive noise has power σ^2.

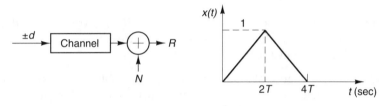

(a) For a symbol rate of $1/2T$ symbols per second, find an expression for the probability of error.

(b) Repeat part (a) but with the symbol rate increased to $1/T$ symbols per second.

(c) Carefully draw the eye pattern of part (a).

(d) Carefully draw the eye pattern of part (b).

5.39 A four-level baseband transmission system uses pulses with equally likely levels $\pm d$, $\pm 3d$, and operates over an additive noise channel with noise power σ^2. The impulse response $x(t)$ of the channel is identical to that shown in Problem 5.38. (a) Find the probability of error for a data rate of $1/2T$ b/s. (b) Repeat part (a) with the data rate increased to $1/T$. (c) Sketch the eye pattern for both parts (a) and (b).

5.40 Consider a channel that is driven by a binary sequence $\sum_k a_k \delta(t - T)$, where $T = 1$ ms, and that has an impulse response $x(t)$ shown below:

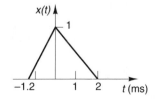

(a) Sketch the eye diagram.

(b) Find the probability of error in terms of d and σ, where $\pm d$ are the signal levels and σ^2 is the noise power of an additive channel.

5.41 Consider a three-level transmission system with levels ($\pm 2d$,0) and data rate $\log_2 3/T$, operating over a channel with impulse response given by $x(t)$:

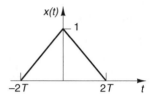

(a) Find the probability of symbol error if the symbols are detected independently by setting appropriate thresholds.

(b) Sketch the eye pattern.

5.42 For the sequence of binary digits given in Problem 5.37, sketch the output waveform (as done in Figure 5.16) for the following coders: (a) Duobinary; and (b) Modified duobinary.

5.43 For the sequence of binary digits given in Problem 5.37, sketch the output waveform (as in Figure 5.16) if the pulses are shaped with a raised-cosine filter with $\alpha = 1$.

5.44 Consider a channel with the following impulse response

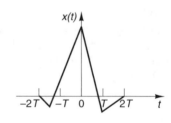

where
$$x_{-1} = 0.1$$
$$x_0 = 1$$
$$x_1 = -0.2$$
$$x_i = 0 \text{ for all other } i$$

(a) Determine the tap weights of a three-tap, zero-forcing equalizer.

(b) For the tap weights calculated in part (a), compute the values of the equalizer pulses y_k for $k = -3, -2, -1, 0, 1, 2, 3$.

5.45 Repeat Problem 5.44, parts (a) and (b), but with

$$x_{-1} = -0.2$$
$$x_0 = 1$$
$$x_1 = -0.3$$
$$x_i = 0 \text{ for all other } i$$

5.46 A transmitted sequence (a_n) of equally likely bits is received as

$$(y_n) = (a_n) + (a_{n-1})/2 + (n_n)$$

where the n_n are independent, identically distributed gaussian noise samples with 0 mean and with variance σ^2. A "tail cancellation" scheme is to be used whereby decisions $\{\hat{a}_n\}$ are used to cancel the intersymbol interference $(a_{n-1})/2$ by subtraction, so that the decision signal sequence (z_n) is

$$(z_n) = (y_n) - \{\hat{a}_{n-1}\}/2$$

(a) Draw a block diagram and describe the scheme further.
(b) Compute the probability of error if (a_n) takes on the values $\pm d$.
(c) Comment on stability, propagation of errors, and other features of this scheme.
(d) If an error occurs for a given bit, show that the probability of error in the next bit decision is approximately equal to $\frac{1}{2}$, that is, $P(e_n/e_{n-1}) = \frac{1}{2}$.
(e) Show that the probability of two successive errors is given by:

$$P(2e) = 1/2 \ P(e) \left[erfc(3d/\sigma) + 1 - erfc(d/\sigma) \right]$$

Hint: Use the basic definition of conditional probability.

5.47 Consider a channel that is noise free and has an impulse response with three nonzero samples $\{x_{-1} = 0.25, x_0 = 1, x_1 = 0.25\}$.
(a) Find the coefficients of a three-tap transversal equalizer that minimizes the mean square error in the resultant pulse response.
(b) Repeat part (a) for a zero-forcing three-tap transversal equalizer.

5.48 Consider a transmission system consisting of a raised-cosine signal with rolloff factor α transmitted over a dispersive channel. At the receiver, a fractionally spaced equalizer is to be used, with a sampling rate equal to the Nyquist rate [42].

(a) Express the sampling rate and tap spacing in terms of T, α, and f_{max}, the maximum frequency in the baseband signal.

(b) What tap spacing results if $α = 1$? $α = 0.5$?

5.49 The following free running mod 3 linear feedback shift register is initially loaded with 001:

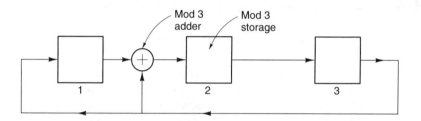

(a) Write out all subsequent states.

(b) If the sequence emanating from stage 3 is considered as a code word corresponding to 001, what is the code word corresponding to 121?

(c) Find the autocorrelation function of the sequence for shifts of 0, 1, and 2 if the following transformation is used

$$
\begin{aligned}
0 &\rightarrow -1 \\
1 &\rightarrow 0 \\
2 &\rightarrow +1
\end{aligned}
$$

5.50 The following self-synchronizing scrambler is initially loaded with the binary sequence 11111:

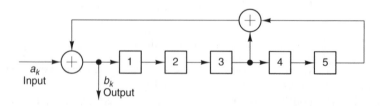

(a) Write out all subsequent states of the register and the output b_k, assuming the data source a_k produces an infinitely long string of 0's.

(b) What is the resulting length of the output sequence? Is it maximum length?

(c) Find the autocorrelation function of the output for shifts of 0, 1, 2, 3, and 4 using the transformation $\{0 ----> -1$ and $1 ----> +1\}$

5.51 Consider the same self-synchronizing scrambler shown in Problem 5.50.

(a) For a data source input $a_k = \{101010100000111\}$, find the scrambler output b_k assuming the initial content of the registers to be 0.

(b) Show the descrambler corresponding to the scrambler. Verify that when the sequence b_k of part (a) is applied to the input of the descrambler, the output is the original sequence a_k.

5.52 The following two self-synchronizing scramblers are connected in series, with each scrambler initially loaded with all 1's.

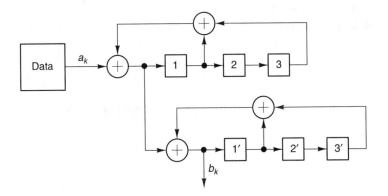

(a) Write out all subsequent states of each register element and the output b_k, assuming the data source produces an infinitely long string of 0's.

(b) What is the resulting length of the output sequence? Find an expression for the period of two equal-length scramblers connected in series, assuming each scrambler is itself maximum length.

(c) Comment on the balance, run, and correlation properties of this scrambler configuration.

(d) Show the descrambler corresponding to the scrambler configuration. Verify that when the sequence b_k of part (a) is applied to the input of the descrambler, the output is the original sequence a_k (all 0's).

5.53 Consider a 15-bit maximal-length PN code generated by feeding back certain stages of a four-stage shift register.

(a) Show by block diagrams two ways that a maximal-length PN code can be generated.

(b) Select one of the configurations from part (a), then assuming an initial state of all 1's, find all the other possible shift-register states in order of their occurrence.

(c) Determine and plot the correlation of the periodic PN sequence with a one-period replica.

5.54 Repeat problem 5.53 for a 7-bit maximal-length PN code using a three-stage shift register.

5.55 Consider a satellite communications system that is subjected to jamming. The jammer, ground station, and satellite have power levels of 10 kilowatts, 1 kilowatt, and 100 watts, respectively. The ground station user has a bit rate of 75 b/s and a required J/S of 20 dB.

(a) Determine the spreading bandwidth B_s required to provide an AJ margin of 20 dB against an up-link (ground station up to satellite) jammer. Assume that the jammer and ground station are the same distance from the satellite.

(b) Determine the B_s required to provide an AJ margin of 5 dB against a down-link (satellite down to ground station) jammer. Assume that the path loss from the satellite to the receiving ground station is 200 dB, and the path loss from the jammer to the receiving ground station is 170 dB. (The path losses are different because the jammer is assumed to be closer to the ground station than the satellite.)

6

Digital Modulation Techniques

OBJECTIVES

- Considers the three basic forms of digital modulation: amplitude-shift keying, frequency-shift keying, and phase-shift keying
- Compares binary modulation systems in terms of power efficiency and spectrum efficiency
- Explains the principle of quadrature modulation used with M-ary PSK
- Describes two modulation techniques that offer certain advantages for band-limited, nonlinear channels: offset quadrature phase-shift keying and minimum-shift keying
- Explains how the technique of quadrature partial response increases the bandwidth efficiency of QAM signaling
- Describes trellis coded modulation as a combination of convolutional coding and digital modulation, and shows examples in detail
- Compares digital modulation techniques in terms of error performance, complexity of implementation, and above all bandwidth efficiency

6.1 Introduction

In this chapter we consider modulation techniques used in digital transmission. Here we define **modulation** as the process of varying certain characteristics of a carrier in accordance with a message signal. The three basic forms of modulation are amplitude modulation (AM), frequency modulation (FM), and phase modulation (PM). Their digital representations are known as amplitude-shift keying (ASK), frequency-shift keying (FSK), and phase-shift keying (PSK). These digital modulation techniques can be characterized by their transmitted symbols, which have a discrete set of values

M and occur at regularly spaced intervals T. The choice of digital modulation technique for a specific application depends in general on the error performance, bandwidth efficiency, and implementation complexity. Binary modulation schemes use two-level symbols and are therefore simple to implement, provide good error performance, but are bandwidth inefficient. M-ary modulation schemes transmit messages of length $m = \log_2 M$ bits with each symbol and are therefore appropriate for higher transmission rates and more efficient bandwidth utilization. Here we will consider binary and M-ary modulation techniques by comparing various performance and design characteristics such as waveform shape, transmitter and receiver block diagram, probability of error, and bandwidth requirements.

6.1.1   Error Performance

In comparing error performance for each modulation scheme, we will use the classic additive white gaussian noise (AWGN) channel. The AWGN channel is not always representative of the performance for real channels, particularly for radio transmission where fading and nonlinearities tend to dominate. Nevertheless, error performance for the AWGN channel is easy to derive and serves as a useful benchmark in performance comparisons.

We know that error performance is a function of the signal-to-noise (S/N) ratio. Out of the many definitions possible for S/N, the one most suitable for comparison of modulation techniques is the ratio of energy per bit (E_b) to noise density (N_0). To arrive at this definition and to obtain an appreciation for its usefulness, let us relate definitions for S/N, given earlier in Chapter 5, to E_b/N_0. Earlier we defined the average power S for a signal $s(t)$ of duration T as

$$S = \frac{1}{T} \int_{-T/2}^{T/2} s^2(t) \, dt \tag{5.5}$$

This formula can be rewritten as

$$S = \frac{1}{T} \int_{-\infty}^{\infty} s^2(t) \, dt = \frac{E_s}{T} \tag{6.1}$$

where the integral of (6.1) is defined as the energy per signal or symbol, E_s. For a transmission rate R, given by (5.43), the average energy available per bit is therefore

$$E_b = \frac{S}{R} = \frac{E_s}{TR} = \frac{E_s}{\log_2 M} \tag{6.2}$$

The noise source is assumed to be gaussian-distributed per (5.27) and to have a flat spectrum as in Figure 6.1. In many instances, the power spectral density of gaussian noise, $S_N(f)$, is indeed flat over the frequency range of the desired signal. This characteristic leads to the concept of **white noise,** which has a (two-sided) power spectral density $N_0/2$, where N_0 has dimensions of watts per hertz. For positive frequencies only, the (one-sided) power spectral density is N_0 watts/Hz. The total noise power (N) in watts is given as the product of the bandwidth (B) and power spectral density (N_0), so that $N = N_0 B$.

With these relationships, we can now express the signal-to-noise ratio as a function of E_b/N_0:

$$\frac{S}{N} = \frac{E_s/T}{N_0 B} \tag{6.3}$$

Since $B = 1/2T$ is the minimum Nyquist bandwidth of the signal, we have, using (6.2),

$$\frac{S}{N} = \frac{(E_b \log_2 M)/T}{N_0/2T}$$
$$= 2 \log_2 M \left(\frac{E_b}{N_0} \right) \tag{6.4}$$

Caution is advised in converting S/N measures to E_b/N_0, however, since in most practical systems the receiver noise bandwidth is larger than the Nyquist bandwidth. A more practical comparison is provided by

$$\frac{S}{N} = \frac{E_b}{N_0} = \frac{R}{B_N} \tag{6.5}$$

where B_N is the noise bandwidth of the receiver.

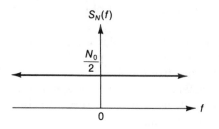

(a) Two-sided power spectral density

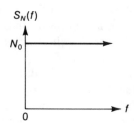

(b) One-sided power spectral density

FIGURE 6.1   White Noise Spectrum

The actual calculation of probability of error will be difficult or impossible for some M-ary modulation cases, making it necessary to use approximations. One useful approach is the union bound, which provides a tight upper bound by considering the union of all possible error events. To derive the union bound, first assume that the symbol s_i is transmitted from an M-ary alphabet. The probability of error is then the union

$$P(e\,|\,s_i) = P\left(\begin{array}{c} M \\ U\,e_{ij} \\ j = 1 \\ j \neq i \end{array}\right) \tag{6.6}$$

where e_{ij} denotes an error event in which the received signal is closer to symbol s_j than to s_i. But by the law of probability in which $P(A\ U\ B) \leq P(A) + P(B)$, (6.6) may be written as the union bound

$$P(e\,|\,s_i) \leq \sum_{\substack{j = 1 \\ j \neq i}}^{M} P(e_{ij}) \tag{6.7}$$

6.1.2   Bandwidth

Generally we will make use of the power spectral density in describing bandwidth characteristics. For each digital modulation technique, there exists a power spectral density from which the bandwidth can be stated according to any given definition of bandwidth. The signal spectrum depends not only on the modulation technique but also on the baseband signal. The usual practice for specifying power spectral density is based on assumptions of random data and long averaging times. Scramblers are usually employed to guarantee a certain degree of data randomness, which tends to produce a smooth spectrum. Baseband shaping also plays a role in the modulation spectrum. For many common modulation techniques, including ASK and PSK, the spectrum is identical to that of the baseband signal translated upward in frequency.

No universal definition of bandwidth exists for modulated signals. The classic channel used by Nyquist and Shannon assumes no power outside a well-defined band. Modulated signals are not band-limited, however, thus introducing the necessity for a more practical definition of bandwidth. The definitions discussed here are those commonly used by regulatory agencies, manufacturers, and telephone and telegraph companies:

- *Null-to-null bandwidth:* The lobed nature of digital modulation spectra suggests the definition of bandwidth as null-to-null, which is equal to the width of the main lobe. This definition lacks generality, however, since not all modulation schemes have well-defined nulls. Further, side lobes may contain significant amounts of power.

- *Half-power bandwidth:* This is simply the band between frequencies at which the power spectral density has dropped to half power below the peak value. Hence the half-power bandwidth is also referred to as the 3-dB bandwidth.

- *Percentage power bandwidth:* The fraction of power located within a bandwidth B is defined as

$$P_B = \frac{\int_{-B}^{B} S(f)\,df}{\int_{-\infty}^{\infty} S(f)\,df} \tag{6.8}$$

where $S(f)$ is the signal power spectral density. This definition requires that a certain percentage, say 99 percent, of the power be inside the allocated band B. Hence a certain percentage, say 1 percent, of the power is allowed outside the band. The 99 percent power bandwidth is somewhat larger than either the null-to-null or 3-dB bandwidth for most modulation techniques. This definition is used by many European Post, Telegraph, and Telephone (PTT) administrations.

- *Spectrum mask:* Another definition of bandwidth is to state that the power spectral density outside a spectrum mask must be attenuated to certain specified levels. The U.S. Federal Communications Commission (FCC) has defined bandwidth allocations for digital microwave radio using such a spectrum mask (see the accompanying box). For most modulation techniques, the FCC mask of Figure 6.2 results in less out-of-band power than the other bandwidth definitions considered here.

6.1.3   Receiver Types

The receiver performs demodulation to recover the original digital signal. This process includes detection of the signal in the presence of channel degradation and a decision circuit to transform the detected signal back to a digital (for example, binary) signal. There are two common methods of demodulation or detection. **Synchronous** or **coherent detection** requires a reference waveform to be generated at the receiver that is matched in frequency and phase to the transmitted signal. When a phase reference cannot be maintained or phase control is uneconomical, **noncoherent detection** is used. For ASK and FSK modulation, where the binary signals

---

FCC Part 21 Spectrum Occupancy Constraints

FCC Part 21 establishes policies and procedures for common carriers including the use of digital modulation techniques in microwave radio. To specify bandwidth, allowed out-of-band emissions are stated by use of a spectrum mask. For operating frequencies below 15 GHz, Part 21 specifies a spectrum mask as determined by the power measured in any 4-kHz band, the center frequency (f) of which is removed from the assigned carrier frequency (f_c) by 50 percent or more of the authorized transmission bandwidth. The following equation is used to define the mask:

$$A = 35 + 0.8(P - 50) + 10 \log_{10}(B) \qquad (6.9)$$

where A = attenuation (dB) below mean transmitted spectrum power output level
  P = percentage of autorized bandwidth of which center frequency of 4-kHz power measuremant bandwidth is removed from carrier frequency
  B = authorized transmitted bandwidth (MHz)

A minimum attenuation (A) of 50 dB is required, and attenuation (A) of greater than 80 dB is not required. A diagram of the mask for 20, 30, and 40-MHz bandwidth allocations is shown in Figure 6.2.

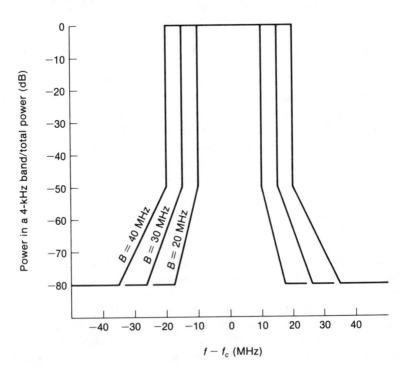

FIGURE 6.2   FCC Part 21 Spectrum Mask for 20, 30, and 40-MHz Bandwidth Allocations

are distinguished by a varying amplitude or frequency characteristic, non-coherent detection is accomplished by **envelope detection.** As the name indicates, detection is based on the presence or absence of the signal envelope. The simplest form of envelope detection is a half-wave rectifier (such as a diode) in series with an RC low-pass filter. Because PSK signals have a constant envelope, envelope detection cannot be used for PSK demodulation and some form of synchronous detection is therefore required.

For the AWGN channel the optimum detector is the matched filter, which maximizes the signal-to-noise ratio at its output [1]. For binary signals $s_1(t)$ and $s_0(t)$, the optimum detector consists of a pair of matched filters and a decision circuit. By definition, the impulse response of the matched filter is the mirror image of the transmitted signal $s(t)$ delayed by a signaling interval T. For the binary case, the matched filter impulse responses are

$$h_1(t) = s_1(T - t)$$
$$h_0(t) = s_0(T - t)$$

$$(6.10)$$

The output signal $y(t)$ of the matched filter is simply the convolution of the received signal $r(t)$ with the filter response. In the absence of noise, the received signal is $s_i(t)$, $i = 0$ or 1, and

$$y_i(t) = \int_{-\infty}^{\infty} s_i(\alpha) h_i(t - \alpha)\, d\alpha \qquad (6.11)$$

At $t = T$, the peak output of the matched filter is

$$y_i(T) = \int_{-\infty}^{\infty} s_i^2(\alpha)\, d\alpha = E_i \qquad (6.12)$$

where E_i is by definition the energy of the signal $s_i(t)$. Each filter produces a maximum output only in the presence of its matched input. Therefore the filter outputs can be subtracted to produce $\pm E_i$ at the input to the appropriate decision threshold circuit.

Later in this chapter we will see that matched filter (coherent) detection yields superior error performance to noncoherent techniques. This performance advantage is theoretically quite small, however, particularly at high S/N. Hence the simplicity of the noncoherent receiver usually makes it a more popular choice over coherent methods, except for applications such as satellite transmission where the higher S/N required for noncoherent detection may be expensive to attain.

6.2   Binary Amplitude-Shift Keying (ASK)

In **amplitude-shift keying** (ASK) the amplitude of the carrier is varied in accordance with the binary source. In its simplest form, the carrier is turned on and off every T seconds to represent 1's and 0's, and this form of ASK is known as on-off keying (OOK), as shown in Figure 6.3. The most general form of amplitude-shift keying is double sideband (DSB), represented by

$$s(t) = \frac{A}{2}\big[1 + m(t)\big]\cos\omega_c t \tag{6.13}$$

where $m(t)$ is the modulating signal (–1 and 1 for OOK) and ω_c is the carrier frequency. Since the carrier conveys no information, power efficiency is improved by suppressing the carrier and transmitting only the sidebands. The general form of the double sideband–suppressed carrier (DSB–SC) signal is

$$s(t) = Am(t)\cos\omega_c t \tag{6.14}$$

Double sideband signals contain an upper and lower sideband that are symmetrically distributed about the carrier frequency ω_c (Figure 6.4). For applications in which spectral efficiency is important, the required bandwidth

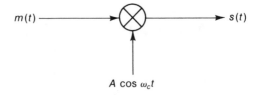

(a) Amplitude modulation

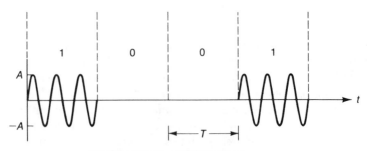

(b) On-off keying (OOK) signal

FIGURE 6.3   Amplitude-Shift Keying (ASK)

may be halved by use of single sideband (SSB) modulation. The unwanted sideband is removed by a bandpass filter. The sharp cutoff required for the bandpass filter has led to vestigial sideband (VSB) in which a portion of the unwanted sideband is transmitted along with the complete other sideband. Use of vestigial sideband allows a smooth rolloff filter at the expense of only slightly more bandwidth than single sideband.

The spectrum of ASK signals is found by application of the frequency-shifting property of the Fourier transform to Equations (6.13) or (6.14). For the case of DSB–SC, the effect of multiplication by cos $\omega_c t$ is to shift the spectrum of the original binary source up to and centered about the carrier frequency ω_c. The shaping and bandwidth of the modulated signal are determined by the baseband signal $m(t)$. If raised-cosine shaping with bandwidth B hertz is used, for example, the baseband spectrum $M(f)$ and modulated spectrum $F(f)$ for DSB–SC are as shown in Figure 6.4. Note that the bandwidth has been doubled by the modulation process to a total transmission bandwidth of $2\,B$ hertz.

Demodulation schemes for ASK fall into two categories, coherent and noncoherent, as illustrated in Figure 6.5. For coherent detection, matched filter detection is optimum; each filter has an impulse response that is

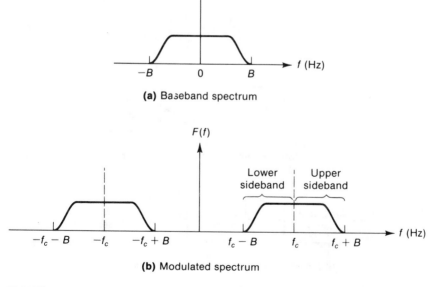

(a) Baseband spectrum

(b) Modulated spectrum

FIGURE 6.4   Example of ASK Spectrum for DSB-SC with Raised-Cosine Shaping

matched to the signal being correlated. For binary transmission, assuming equally likely 1's and 0's and assuming equal energy for both signals, the matched filter detector is as shown in Figure 6.5a. For OOK, only a single matched filter is required in order to detect the presence or absence of signal energy. For OOK, with $s_0 = 0$ and $s_1 = A\cos\omega_c t$, the output of the single matched filter at the sampling time T is

$$y_1(t) = \int_0^T s_1^2(t)\, dt = \frac{A^2 T}{2} \tag{6.15}$$

or

$$y_0(t) = 0 \tag{6.16}$$

The integral in (6.15) is recognized as the energy E_s of the signal $s_1(t)$. In the presence of additive noise $N(t)$, the matched filter detector makes a decision at $t = T$ based on the two signals $y_1(T) = E_s + N(T)$ and $y_0(T) = N(T)$. For the source that produces equally likely 1's and 0's and noise that has a gaussian distribution, the optimum receiver threshold is $E_s/2$. To determine the probability of error for the matched filter, we note that this case is identical to that of unipolar NRZ given by Equation (5.32), which is repeated here for convenience:

$$P(e) = \text{erfc}\sqrt{\frac{S}{2N}} \tag{5.32}$$

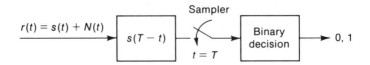

(a) Coherent detection by matched filter

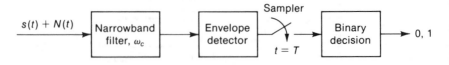

(b) Noncoherent detection by envelope detector

FIGURE 6.5   Receiver Structures for OOK

Here the average signal power $S = (\frac{1}{2})(A^2/2) = E_s/2T$. For white noise with (one-sided) power spectral density N_0, the noise power $N = N_0B$, where the Nyquist bandwidth $B = 1/2T$. For comparison with other modulation techniques, it is convenient to express $P(e)$ as a function of E_s and N_0. From (5.32) we have

$$P(e) = \text{erfc}\sqrt{\frac{E_s}{2N_0}} \qquad (6.17)$$

The result given in (6.17) pertains to the case of on-off keying, where all of the energy E_s is used to convey one binary state. This result holds for the more general case of ASK if E_s is interpreted as the total energy used for the two binary signals. Note that $E_s/2$ is then the average energy per binary signal, E_b, so that

$$P(e) = \text{erfc}\sqrt{\frac{E_b}{N_0}} \qquad (6.18)$$

Comparing (6.17) and (6.18) we note a 3-dB difference in performance when signal energy is referenced to average value (E_b) rather than peak value (E_s).

The noncoherent demodulator for OOK consists of a narrowband filter centered at ω_c, an envelope detector, and a decision device (Figure 6.5b). For equiprobable 1's and 0's at the source, the probability of error is [2]

$$P(e) \approx \frac{1}{2}\exp\left(\frac{-E_s}{4N_0}\right) = \frac{1}{2}\exp\left(-\frac{E_b}{2N_0}\right) \qquad (6.19)$$

A comparison of the $P(e)$ expressions in (6.19) and (6.17) for OOK reveals that for high signal-to-noise ratios the noncoherent detector requires less than a 1-dB increase in S/N to obtain the same $P(e)$ as a coherent detector. This small performance penalty usually allows the designer to select noncoherent detection because of its simpler implementation.

6.3   Binary Frequency-Shift Keying (FSK)

In frequency modulation, the frequency of the carrier varies in accordance with the source signal. For binary transmission the carrier assumes one frequency for a 1 and another frequency for a 0 as represented in Figure 6.6.

This type of on-off modulation is called **frequency-shift keying.** The modulated signal for binary frequency-shift keying may be written

$$s_1(t) = A\cos\omega_1 t \quad \text{for binary 1}$$
$$s_0(t) = A\cos\omega_0 t \quad \text{for binary 0} \tag{6.20}$$

An alternative representation of the FSK waveform is obtained by letting $f_1 = f_c - \Delta f$ and $f_o = f_c + \Delta f$, so that

$$s_1(t) = A\cos(\omega_c - \Delta\omega)t$$
$$s_0(t) = A\cos(\omega_c + \Delta\omega)t \tag{6.21}$$

where Δf is called the **frequency deviation.**

The frequency spectrum of the FSK signal is in general difficult to obtain [3]. However, transmission bandwidth (B_T) requirements can be estimated for two cases of special interest:

$$B_T \approx \begin{cases} 2\Delta f & \Delta f \gg B \tag{6.22a} \\ 2B & \Delta f \ll B \tag{6.22b} \end{cases}$$

where B is the bandwidth of the baseband modulation signal (such as NRZ). The ratio of the frequency deviation Δf to the baseband bandwidth B is called the **modulation index** m, defined as

$$m = \frac{\Delta f}{B} \tag{6.23}$$

A general relationship defining FM transmission bandwidth was established by Carson [4] as

$$B_T = 2B(1 + m) \tag{6.24}$$

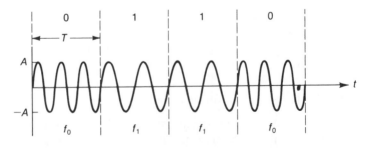

FIGURE 6.6   Frequency-Shift Keying (FSK) Waveform

Carson's rule approaches the limits given in (6.22) for both $m \gg 1$ and $m \ll 1$. These two cases are referred to as **wideband FSK** where $m \gg 1$ and **narrowband FSK** where $m \ll 1$. For $m > 1$, FSK requires more transmission bandwidth than ASK.

Coherent detection of FSK is accomplished by comparing the outputs of two matched filters, as shown in Figure 6.7a. The output signals of the matched filters are

$$y_1(T) = y_0(T) = \int_0^T A^2\cos^2\omega t \, dt = \frac{A^2T}{2} \tag{6.25}$$

The integral in (6.25) is of course the energy E_s of the signal. The summing device in Figure 6.7a thus produces a signal output of $+E_s$ if a 1 has been transmitted and $-E_s$ if a 0 has been transmitted (assuming no noise and

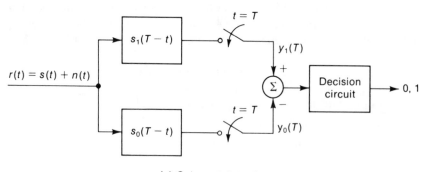

(a) Coherent detection

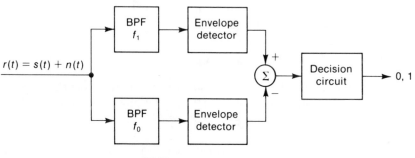

(b) Noncoherent detection

FIGURE 6.7   Receiver Structures for FSK

orthogonal FSK signals).* The probability of error for coherent FSK would appear to be analogous to that of polar NRZ except for the fact that the noise samples are subtracted by the summing device. For orthogonal FSK signals, these (white) noise samples are independent and therefore add on a power basis [5]. The total noise power is therefore doubled to $N = 2N_0B$. The probability of error can now be written, following the case for polar signaling but with twice the noise power, so that

$$P(e) = \text{erfc}\sqrt{\frac{E_s}{N_0}} = \text{erfc}\sqrt{\frac{E_b}{N_0}} \qquad (6.26)$$

Thus we can conclude that FSK has a 3-dB advantage over ASK on a peak power basis but is equivalent to ASK on an average power basis. The FSK waveform given in (6.20) and the receiver shown in Figure 6.7 imply other advantages of FSK over ASK. Frequency-shift keying has a constant envelope property that has merit when transmitting through a nonlinear device or channel. Further, the FSK receiver threshold is fixed at zero, independent of the carrier amplitude. For ASK, the threshold must be continually adjusted if the received signal varies in time, as happens with fading in radio transmission.

Conventional noncoherent detection for binary FSK employs a pair of bandpass filters and envelope detectors as shown in Figure 6.7b. The probability of error for additive white gaussian noise is [2]

$$P(e) = \frac{1}{2}\exp\left(-\frac{E_s}{2N_0}\right) = \frac{1}{2}\exp\left(-\frac{E_b}{2N_0}\right) \qquad (6.27)$$

A comparison of (6.27) and (6.26) for low error rates reveals that a choice of noncoherent detection results in an S/N penalty of less than 1 dB. Therefore, because of the complexity of coherent detection, noncoherent detection is more commonly used in practice.

6.4 Binary Phase-Shift Keying (BPSK)

In phase modulation the phase of the carrier is varied according to the source signal. For binary transmission the carrier phase is shifted by 180° to

---

*For binary FSK, $s_1(t)$ and $s_0(t)$ are orthogonal over $(0,T)$ if the signals have zero correlation; that is,

$$\int_0^T s_1(t)s_0(t)\, dt = 0$$

represent 1's and 0's and is called **binary phase-shift keying** (BPSK). The signal waveform for BPSK is given as

$$s(t) = \pm A \cos \omega_c t$$
$$= A \cos(\omega_c t + \phi_j) \qquad \phi_j = 0 \text{ or } \pi \tag{6.28}$$

A BPSK waveform example is shown in Figure 6.8, in which the data clock and carrier frequencies are exact multiples of one another.

The form of (6.28) suggests that the BPSK signal corresponds to a polar NRZ signal translated upward in frequency. The power spectral density of PSK has a double sideband characteristic identical to that of OOK. Thus the spectrum is centered at ω_c with a bandwidth twice that of the baseband signal.

It follows that coherent detection of a PSK signal is similar to the case for ASK. Detector outputs are $\pm E_s + N(T)$, and the threshold is set at 0 (assuming equiprobable 1's and 0's). The noise power $N = N_0 B$. For the case of ASK, we noted earlier that the detector outputs are $E_s + N(T)$ or $N(T)$, the noise power is $N = N_0 B$, and the optimum threshold is $E_s/2$. By analogy with the ASK case, we can directly write the probability of error for coherent PSK as

$$P(e) = \text{erfc}\sqrt{\frac{2E_s}{N_0}} = \text{erfc}\sqrt{\frac{2E_b}{N_0}} \tag{6.29}$$

Comparison of (6.29) with (6.26), (6.17) and (6.18) reveals that on a peak power basis PSK outperforms FSK by 3 dB and ASK by 6 dB, while on an average power basis PSK has a 3-dB advantage over both FSK and ASK.

For coherent detection, the receiver must have a phase reference available. For BPSK with the two phases separated by 180°, however, the carrier component is zero. Thus the phase reference must be generated by some other means, such as the frequency doubling phase-locked loop or the Costas loop. The frequency doubling loop shown in Figure 6.9a uses a square-law

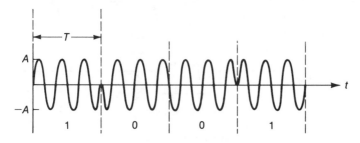

FIGURE 6.8   Phase-Shift Keying (PSK) Waveform

device to remove the modulation and create a line component in the spectrum at double the carrier frequency. A phase-locked loop (PLL) is used as a narrowband loop centered at twice the carrier frequency. Frequency division by 2 of the PLL output recovers the carrier reference. This frequency division causes a 180° phase ambiguity that can be accommodated by proper coding of the baseband signal, as in NRZ(I), which is transparent to the polarity of the carrier. The Costas loop multiplies the incoming signal by $\cos(\omega_c t + \phi)$ in one channel and $\sin(\omega_c t + \phi)$ in the other channel to generate a coherent phase reference from the suppressed carrier signal. As indicated in Figure 6.9b, the two

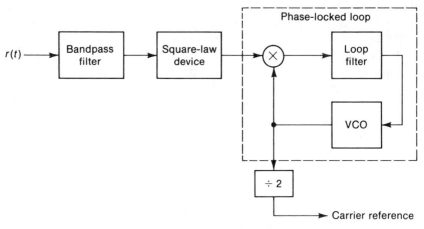

(a) Frequency doubling loop

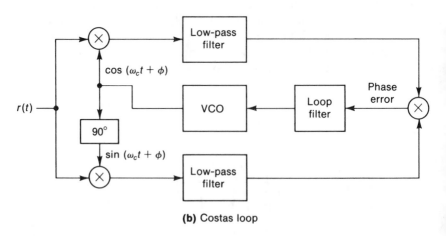

(b) Costas loop

FIGURE 6.9   Carrier Recovery Techniques for PSK

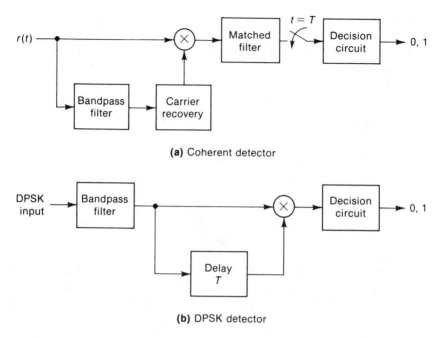

(a) Coherent detector

(b) DPSK detector

FIGURE 6.10   Receiver Types for PSK

phase detectors (multipliers) for sin and cos channels are used to control the phase and frequency of a voltage-controlled oscillator (VCO). The Costas loop is usually preferred over the simpler squaring loop for PSK coherent detection (Figure 6.10a) because of greater tolerance to shifts in the carrier frequency and the capability of wider bandwidth operation [6].

The added complexity of establishing a phase reference for coherent detection can be avoided by use of a technique known as **differential PSK** (DPSK). With DPSK, the data are encoded by means of changes in phase rather than by absolute value of phase in the carrier. For binary PSK, a 1 can be encoded as no change in the phase and a 0 encoded as a 180° change in the phase. (Note that this encoding scheme is identical to NRZ(S) shown in Figure 5.1.) The detector, shown in Figure 6.10b, uses the phase of the previous symbol as a reference phase to permit decoding of the current symbol. If the phases are the same, a 1 is decoded; if they differ, a 0 is decoded. The probability of error of DPSK is given by [2]

$$P(e) = \frac{1}{2}\exp\left(\frac{-E_s}{N_0}\right) = \frac{1}{2}\exp\left(\frac{-E_b}{N_0}\right) \tag{6.30}$$

Compared to coherent PSK, DPSK results in an S/N penalty of 1 dB or less for error rates of interest. However, differential encoding also causes errors to occur in pairs. Even so, DPSK is a popular alternative to coherent PSK, except for satellite applications where power limitations make coherent PSK preferable despite the added complexity of maintaining a phase reference.

6.5   Comparison of Binary Modulation Systems

In selecting a modulation system, the two primary factors for comparison are the transmitted power and channel bandwidth that are required to achieve a specified (error rate) performance. To measure power efficiency, the parameter E_b/N_0 is most commonly used. For spectrum efficiency, the ratio of transmission rate to transmission bandwidth is used. Other factors that may influence the choice of modulation technique include effects of fading, interference, channel or equipment nonlinearities, and implementation complexity and cost.

The probability of error for ASK, FSK, and PSK binary modulation is plotted in Figure 6.11 for both coherent and noncoherent detection. As noted earlier, PSK has a 3-dB performance advantage over both FSK and ASK, on an average power basis, for both coherent and noncoherent detection. Moreover, the difference in performance between coherent and noncoherent detection for a particular choice of modulation scheme is on the order of 1 dB for error rates of interest, that is, where $P(e) \leq 10^{-5}$.

None of the binary modulation schemes described so far is particularly efficient in bandwidth utilization. We noted earlier that ASK and PSK require the same bandwidth, whereas FSK usually (depending on choice of Δf) requires more bandwidth than ASK or PSK. The theoretical bandwidth efficiency of binary ASK and PSK is 1 bps/Hz. To obtain greater bandwidth efficiency, M-ary modulation schemes are required, but these schemes suffer an S/N penalty compared to the binary cases, as we will see later in the chapter.

In terms of implementation, the complexity and cost depend primarily on the choice of coherent versus noncoherent detection. Typically the added complexity of coherent detection is not justified by the slight improvement in error performance. Because satellite links tend to be limited in power, however, coherent detection is often used in lieu of noncoherent detection. Although ASK implementation is simple, it is not a popular choice because of its relatively poor error performance and susceptibility to fading and nonlinearities. Noncoherent FSK is commonly used for low data rates, while both coherent and differential PSK are preferred for higher data rate applications.

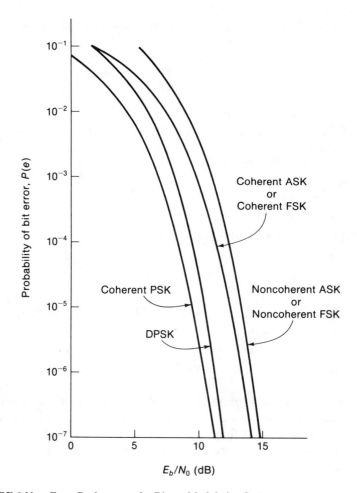

FIGURE 6.11   Error Performance for Binary Modulation Systems

6.6   *M*-ary FSK

In *M*-ary FSK, the *M*-ary symbols are represented by *M*-spaced frequencies selected from a set of equal energy waveforms

$$s_n(t) = A \cos \omega_n t \qquad 0 < t \leq T \qquad (6.31)$$

where $n = 1, 2, \ldots, M$ and where T is the symbol length. Here we will assume that the set of M-ary signals satisfies the orthogonality conditions

$$\int_0^T s_n(t)s_m(t)\, dt = \begin{cases} 0 & n \neq m \\ E_s & n = m \end{cases} \qquad (6.32)$$

where E_s is the energy per symbol. Orthogonality ensures no overlap among the detector outputs at the receiver. The minimum frequency separation $\Delta\omega$ required to satisfy (6.32) is

$$\Delta\omega = \omega_m - \omega_n = \frac{\pi}{T} \qquad (6.33)$$

Thus the minimum bandwidth of the signal set is

$$B \approx \frac{M}{2T} \text{ hertz} \qquad (6.34)$$

Correspondingly the minimum transmission bandwidth is given by (6.22b), so that for M-ary FSK,

$$B_T \approx \frac{M}{T} \qquad (6.35)$$

The optimum receiver for orthogonal M-ary FSK consists of a bank of M matched filters [7]. At the sampling times, $t = kT$, the receiver makes decisions based on the largest filter output. Symbol s_j is selected when the jth matched filter output is the largest. The probability of correct detection can be written

$$P(C\,|\,s_j) = P(\text{all } K_i < K_j,\ i \neq j) \qquad (6.36)$$

where K_i is the output of the ith matched filter at time T. Given that $s_j(t)$ is transmitted over an additive noise channel,

$$\begin{aligned} K_j &= S_j + N_j \\ &= \sqrt{E_s} + N_j \end{aligned} \qquad (6.37)$$

where N_j is a gaussian, zero-mean random variable with variance $N_0/2$. Now we can rewrite (6.36) as

$$P(C\,|\,s_j) = P\left(\text{all } N_i < \sqrt{E_s} + N_j\right),\ i \neq j \qquad (6.38)$$

Since the N_i's are independent random variables, each with the same probability density function $f(N)$, (6.38) may be averaged over the noise as follows

$$P(C \mid s_j) = \int_{-\infty}^{\infty} \prod_{\substack{i=1 \\ i \neq j}}^{M} P\left(N_i < \sqrt{E_s} + N_j\right) f(N) \, dN \qquad (6.39)$$

Since the pdf's of all N_i are identical, the product in (6.39) can be expressed as

$$\prod_{\substack{i=1 \\ i \neq j}}^{M} P\left(N_i < \sqrt{E_s} + N_j\right) = \left[\int_{-\infty}^{\sqrt{E_s} + N_j} \frac{\exp(-x^2/N_0)}{\sqrt{\pi N_0}} \, dx \right]^{M-1} \qquad (6.40)$$

$$= \left[1 - \mathrm{erfc}\left(u + \sqrt{2E_s/N_0}\right) \right]^{M-1}$$

where we have used the substitution $u = N_j \sqrt{2/N_0}$. Now, recognizing that $P(C \mid s_j)$ is independent of j, the average probability of symbol error $P(e)$ is $1 - P(C \mid s_j)$, so that from (6.39) and (6.40)

$$P(e) = 1 - \frac{1}{\sqrt{2\pi}} \int_{-\infty}^{\infty} e^{-u^2/2} \left[1 - \mathrm{erfc}\left(u + \sqrt{\frac{2E_s}{N_0}}\right) \right]^{M-1} du \qquad (6.41)$$

A plot of this symbol error probability is given in Figure 6.12 as a function of E_b/N_0 for several values of M. (Recall that $E_s = E_b \log_2 M$.) From Figure 6.12 it is apparent that for a fixed error probability, the required E_b/N_0 may be reduced by increasing M. The required bandwidth increases as M, however, and transmitter and receiver complexity also increase with large M.

Because the integral of (6.41) cannot be evaluated exactly but rather requires numerical integration techniques, it is useful to seek an approximation of $P(e)$. The union bound is given by

$$P(e) \leq (M - 1)\mathrm{erfc}\left(\sqrt{\frac{E_b \log_2 M}{N_0}}\right) \qquad (6.42)$$

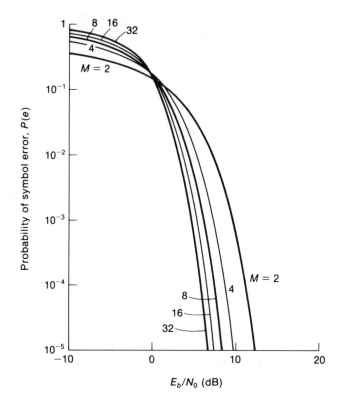

FIGURE 6.12   Error Performance for Coherent Detection of Orthogonal M-ary FSK

This bound becomes increasingly tight for fixed M as the E_b/N_0 is increased. For values of $P(e) \leq 10^{-3}$, the upper bound becomes a good approximation to $P(e)$. For binary FSK ($M = 2$), the bound of (6.42) becomes an equality.

Coherent FSK detection requires exact phase references at the receiver, which can be difficult to maintain. A simpler implementation, although slightly inferior performance, results when noncoherent detection is used. An optimum noncoherent detection scheme can be realized with a bank of M filters, each centered on one of M frequencies. Each filter is followed by an envelope detector, and the decision is based on the largest envelope output. For the AWGN channel, the probability of symbol error for orthogonal signaling is given by [8]

$$P(e) = \sum_{k=1}^{M-1} \frac{(-1)^{k+1}}{k+1} \binom{M-1}{k} \exp\left[\frac{-kE_s}{(k+1)N_0}\right] \quad (6.43)$$

A plot of this symbol error probability is given in Figure 6.13 as a function of E_b/N_0 for several values of M. The set of curves in Figure 6.13 exhibit the same behavior as the case for coherent detection shown in Figure 6.12— that is, for a fixed symbol error probability, the required E_b/N_0 decreases with increasing M while the required bandwidth increases. A comparison of Figure 6.13 with Figure 6.12 indicates that for $P(e) \leq 10^{-4}$, the E_b/N_0 penalty for choice of noncoherent detection is less than 1 dB for any choice of M. Thus, for orthogonal codes, noncoherent detection is generally preferred because of its simpler implementation.

An upper bound on the probability of symbol error for noncoherent detection is provided by the leading term of (6.43).

$$P(e) \leq \frac{M-1}{2} \exp\left(\frac{-E_s}{2N_0}\right) \qquad (6.44)$$

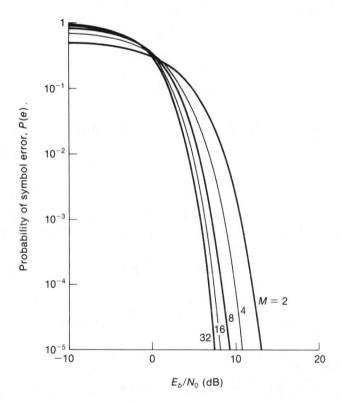

FIGURE 6.13   Error Performance for Noncoherent Detection of Orthogonal *M*-ary FSK

As the E_b/N_0 is increased, this bound becomes increasingly close to the true value of error probability. For binary FSK, the bound of (6.44) becomes an equality.

The error probability expressions presented to this point have been in terms of symbol error. This measure of error performance is appropriate where messages of length $m = \log_2 M$ bits are transmitted, such as alphanumeric characters or PCM code words. For binary transmission, however, the bit error probability is the measure of interest. For orthogonal signals, an error in detection is equally likely to be made in favor of any one of the $M - 1$ incorrect signals. For the symbol in error, the probability that exactly k of the m bits are in error is

$$P(k \text{ errors in } m \text{ bits} \,|\, \text{symbol error}) = \frac{\binom{m}{k}}{\displaystyle\sum_{k=1}^{m} \binom{m}{k}} \tag{6.45}$$

Hence the expected number of bits in error is

$$\frac{\displaystyle\sum_{k=1}^{m} k\binom{m}{k}}{\displaystyle\sum_{k=1}^{m} \binom{m}{k}} = \frac{m2^{m-1}}{2^m - 1} \tag{6.46}$$

The conditional probability that a given bit is in error given that the symbol is in error is then

$$P(e_b \,|\, e) = \frac{2^{m-1}}{2^m - 1} \tag{6.47}$$

In terms of the symbol error probability $P(e)$, the bit error probability, more commonly known as the bit error rate (BER), can be written

$$\text{BER} = \frac{2^{m-1}}{2^m - 1} P(e) \tag{6.48}$$

where $P(e)$ is given by (6.41) and (6.43) for coherent and noncoherent detection. Note that for large m, the BER approaches $P(e)/2$.

6.7 M-ary PSK

An M-ary PSK signal may be represented by the set of signals

$$s_n(t) = A\cos(\omega_c t + \theta_n) \qquad 0 \le t \le T \tag{6.49}$$

where the *M* symbols are expressed as the set of uniformly spaced phase angles

$$\theta_n = \frac{2(n-1)\pi}{M} \qquad n = 1, 2, \ldots, M \tag{6.50}$$

The separation between adjacent phases of the carrier is $2\pi/M$. For BPSK, the separation is π; for 4-PSK, the separation is $\pi/2$; and for 8-PSK, the separation is $\pi/4$. A more convenient means of representing the *M*-ary PSK signals of (6.49) is by a phasor diagram, which displays the signal magnitude and phase angle. Examples of BPSK, 4-PSK, and 8-PSK phasors are shown in Figure 6.14. The signal points of the phasor diagram represent the **signal constellation.** As indicated in Figure 6.14, each signal has equal amplitude *A*, resulting in signal points that lie on a circle of radius *A*. Decision thresholds in the receiver are centered between the allowed phases θ_n, so that correct decisions are made if the received phase is within $\pm\pi/M$ of the transmitted phase.

Another convenient representation of *M*-ary PSK (and other bandwidth-efficient modulation techniques) is provided by *quadrature signal representation*. The signals of (6.49) may be expressed, by use of trigonometric expansion, as a linear combination of the carrier signals $\cos \omega_c t$ and $\sin \omega_c t$:

$$s_n(t) = A\left[p_n \cos\omega_c t + q_n \sin\omega_c t \right] \qquad 0 \le t \le T \tag{6.51}$$

where

$$\begin{aligned} p_n &= \cos\theta_n \\ q_n &= \sin\theta_n \end{aligned} \tag{6.52}$$

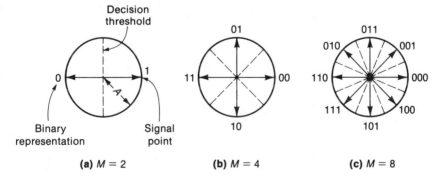

FIGURE 6.14   Phasor Diagrams of *M*-ary PSK Signals

For the case of BPSK as shown in Figure 6.14a, the signal phase angles of 0 and π are represented in quadrature by the coefficients

$$(p_n, q_n) = (1, 0), (-1, 0) \qquad \text{for } 0, \pi$$

Similarly, the corresponding sets of (p_n, q_n) for the 4-PSK and 8-PSK constellations shown in Figure 6.14 are given in Tables 6.1 and 6.2. Since $\cos \omega_c t$ and $\sin \omega_c t$ are orthogonal in a phasor diagram, they are said to be in quadrature. The cosine coefficient p_n is represented on the horizontal axis and is called the *in-phase* or I signal. The sine coefficient q_n is represented on the vertical axis and is called the *quadrature* or Q signal.

Quadrature representation leads to one common method of generating M-ary PSK signals as a linear combination of quadrature signals. A block diagram of a generalized PSK modulator based on quadrature signal structure is shown in Figure 6.15. Binary data at the input to the modulator are serial-to-parallel converted to create $m(= \log_2 M)$ parallel bit streams, each having a bit rate of R/m. The I and Q signal generator converts each m-bit word to a pair of numbers (p_n, q_n), which are the I and Q signal coefficients. The I and Q signals are multiplied by $\cos \omega_c t$ and $\sin \omega_c t$, respectively, and the multiplier outputs are summed to create the PSK modulated signal.

TABLE 6.1   Quadrature Signal Coefficients for 4-PSK Modulation

Bit Values	Quadrature Coefficients		Phase Angle θ_n
	p_n	q_n	
00	1	0	0
01	0	1	$\pi/2$
11	-1	0	π
10	0	-1	$-\pi/2$

TABLE 6.2   Quadrature Signal Coefficients for 8-PSK Modulation

Bit Values	Quadrature Coefficients		Phase Angle θ_n
	p_n	q_n	
000	1	0	0
001	$1/\sqrt{2}$	$1/\sqrt{2}$	$\pi/4$
011	0	1	$\pi/2$
010	$-1/\sqrt{2}$	$1/\sqrt{2}$	$3\pi/4$
110	-1	0	π
111	$-1/\sqrt{2}$	$-1/\sqrt{2}$	$-3\pi/4$
101	0	-1	$-\pi/2$
100	$1/\sqrt{2}$	$-1/\sqrt{2}$	$-\pi/4$

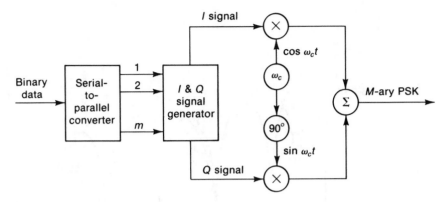

FIGURE 6.15   Generalized *M*-ary PSK Modulator

As observed earlier, the spectrum of a BPSK signal corresponds to that of an NRZ signal translated upward in frequency. This relationship also holds for conventional *M*-ary PSK, where symbols are independent and the baseband signals applied to the quadrature channels are orthogonal, multilevel waveforms. The individual *I* and *Q* signals produce the common sin *x*/*x* spectrum, and the composite spectrum is simply the sum of the individual spectra. The resulting one-sided power spectrum for a random binary output is

$$S(f) = A^2 T \left\{ \frac{\sin\left[(f - f_c)\pi T\right]}{(f - f_c)\pi T} \right\}^2 \tag{6.53}$$

Figure 6.16 plots the power spectrum for 2-, 4-, and 8-PSK systems operating at the same data rate.

Coherent detection of *M*-ary PSK requires the generation of a local phase reference. For BPSK, only a single reference, cos $\omega_c t$, is required to detect signals separated by 180°. For higher-level PSK, additional phase references are needed. One approach is to use two references that are in quadrature—that is, cos $\omega_c t$ and sin $\omega_c t$—along with a logic circuit to determine the linear combinations of the quadrature signals. A block diagram of a generalized PSK demodulator using two phase references is shown in Figure 6.17. A second method is to employ a set of *M* coherent detectors, using cos($\omega_c t + \theta_n$) as the set of reference signals. In the absence of noise, the ideal phase detector measures θ_n at the sampling times. Errors due to noise will occur if the measured phase ϕ falls outside the region

$$\theta_n - \frac{\pi}{M} \leq \phi < \theta_n + \frac{\pi}{M} \tag{6.54}$$

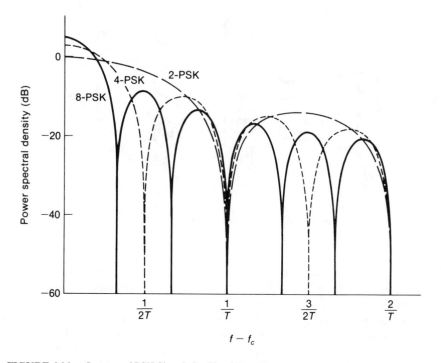

FIGURE 6.16   Spectra of PSK Signals for Fixed Data Rate

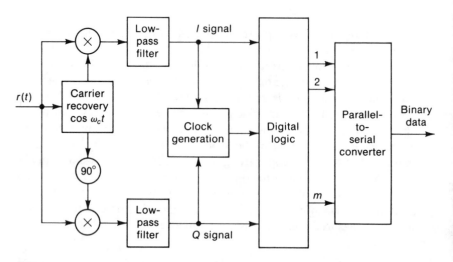

FIGURE 6.17   Generalized M-ary PSK Demodulator

Calculation of the probability of symbol error results in a closed-form expression only when $M = 2$ and $M = 4$. For $M = 2$, the probability of error $P(e)$ was given earlier by (6.29); for $M = 4$,

$$P(e) = 2\,\text{erfc}\sqrt{\frac{E_s}{N_0}}\left(1 - \frac{1}{4}\,\text{erfc}\sqrt{\frac{E_s}{N_0}}\right) \tag{6.55}$$

For $M > 4$, an approximation may be found by overbounding the error region, defined as the complement of the decision region given by (6.54). Consider two semicircles defined by

$$\begin{aligned} C_1 &= (\angle\theta_n - \pi/M \text{ to } \angle\theta_n - \pi/M - \pi) \\ C_2 &= (\angle\theta_n + \pi/M \text{ to } \angle\theta_n + \pi/M + \pi) \end{aligned} \tag{6.56}$$

which overlap by exactly $2\pi/M$, the angle of the decision region. An upper bound is obtained by noting that the true error region is exceeded by the sum of the areas in C_1 and C_2, so that

$$P(e) \le P(REC_1) + P(REC_2) \tag{6.57}$$

where R is the received signal. The two probabilities in the righthand side of (6.57) are equal due to the symmetry of the PSK signal constellation and of the noise distribution. The calculation of (6.57) is further simplified by resolving the noise into $N_\parallel$ and $N_\perp$, components parallel and perpendicular to the boundary of C_1 or C_2. Only $N_\perp$ is needed for the calculation, since only it can cause R to be in C_1 or C_2. Therefore, since $\sqrt{E_s}\sin(\pi/M)$ is the distance from the signal point to the boundary of C_1 or C_2

$$\begin{aligned} P(REC_1) = P(REC_2) &= P\left(N_\perp > \sqrt{E_s}\sin\,\pi/M\right) \\ &= \frac{1}{\sigma\sqrt{2\pi}}\int_{\sqrt{E_s}\sin(\pi/M)}^{\infty} \exp(-x^2/2\sigma^2)\,dx \\ &= \text{erfc}\left[\sqrt{2E_s/N_0}\,\sin^2\pi/M\right] \end{aligned} \tag{6.58}$$

and the upper bound becomes

$$P(e) \approx 2\,\text{erfc}\sqrt{\frac{2E_s}{N_0}\sin^2\frac{\pi}{M}} \qquad M > 2 \tag{6.59}$$

This approximation becomes increasingly tight for fixed M as E_s/N_0 increases. Results for an exact calculation of symbol error probability are plotted in Figure 6.18 as a function of $E_b/N_0 = E_s/mN_0$. The curves of Figure 6.18 show that for fixed symbol error probability, the E_b/N_0 must be increased with increasing M, while according to Figure 6.16 the required signal bandwidth is decreasing.

Coherent detection requires not only the use of multiple phase detectors but also the recovery of the carrier for phase coherence. Carrier recovery techniques were described in Section 6.4 for BPSK, and these can easily be generalized for M-ary PSK. An alternative to recovering a coherent reference is to use the phase difference between successive symbols. If the data are encoded by a phase shift rather than by absolute phase, then the receiver detects the signal by comparing the phase of one symbol with the phase of the previous symbol. This is the same technique described for BPSK in Section 6.4 and is known as differential PSK (DPSK) or differentially coherent PSK (DCPSK). Symbol errors occur when the measured

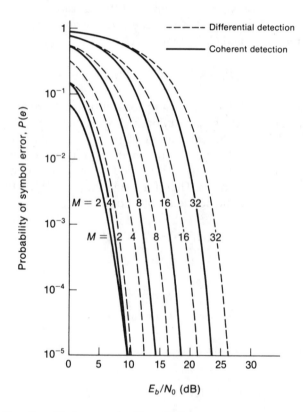

FIGURE 6.18   Error Performance for M-ary PSK

phase difference differs by more than π/M in absolute phase from the phase difference of the transmitted signal. As is typical with these calculations, a closed-form expression for probability of symbol error cannot be derived for *M*-ary differential PSK. For large E_b/N_0, however, a good approximation to $P(e)$ is given by [9]

$$P(e) \approx 2\,\mathrm{erfc}\sqrt{\frac{2E_s}{N_0}\sin^2\frac{\pi}{\sqrt{2}M}} \qquad (6.60)$$

A comparison of error probabilities for coherent and differentially coherent detection is given in Figure 6.18. Note that DPSK involves a tradeoff between signal power and data rate similar to that observed for coherent PSK. For BPSK, the performance of differential detection is within 1 dB of coherent detection for $P(e) < 10^{-3}$. This degradation increases with *M*, however, approaching 3 dB for $M \geq 8$.

The error probability expressions shown here for PSK are based on symbol error. If we assume that when a symbol error occurs the decoder randomly chooses one of the remaining $M - 1$ symbols, then the bit error probability is related to the symbol error probability by Equation (6.48). For PSK errors, however, the selected phase is more likely to be one of those adjacent to the transmitted phase. Further, Gray coding is usually employed with PSK symbols in which adjacent symbols differ in only one bit position (see Figure 6.14). Then each symbol error is most likely to cause only a single bit error. For high S/N,

$$\mathrm{BER} \approx \frac{P(e)}{\log_2 M} \qquad (6.61)$$

A comparison of BER for $M = 4$ using (6.59) and (6.61), and for $M = 2$ using (6.29), reveals that 4-PSK and BPSK have identical bit error performance.

Example 6.1 _____

In the U.S. 2-GHz common carrier band, digital radio systems are required to be able to carry 96 PCM voice channels within a 3.5-MHz bandwidth. For a choice of 4-PSK modulation with raised-cosine filtering, determine the filter rolloff factor α and the S/N required to provide BER $= 10^{-6}$.

Solution   Four PCM DS-1 (1.544 Mb/s) bit streams are required for 96 voice channels, which can be combined with a second-level multiplex into a DS-2 (6.312 Mb/s) bit stream. With a 3.5-MHz bandwidth allocation, the required bandwidth efficiency is 6.312 Mbps/3.5 MHz = 1.8 bps/Hz. For 4-PSK with raised-cosine filtering, the bandwidth efficiency is given by

$$\frac{R}{B} = \frac{(\log_2 M)/T}{(1+\alpha)/T} = 1.8 \text{ bps/Hz}$$

where α is the fraction of excess Nyquist bandwidth as defined in Chapter 5. Solving for α, we find that $\alpha = 0.1$ is required. Using (6.59) and (6.61), we obtain the following for BER = 10^{-6}:

$$10^{-6} = \text{erfc}\sqrt{\frac{2E_s}{N_0}\sin^2\frac{\pi}{4}} = \text{erfc}\sqrt{\frac{E_s}{N_0}} = \text{erfc}\sqrt{\frac{2E_b}{N_0}}$$

Solving for E_b/N_0, we find

$$\left(\frac{E_b}{N_0}\right)_{dB} = 11.5 \text{ dB}$$

From Equation (6.5),

$$\left(\frac{S}{N}\right)_{dB} = \left(\frac{E_b}{N_0}\right)_{dB} + 10\log\frac{R}{B_N} = 11.5 \text{ dB} + 2.6 \text{ dB} = 14.1 \text{ dB}$$

6.8 Quadrature Amplitude Modulation (QAM)

The principle of quadrature modulation used with M-ary PSK can be generalized to include amplitude as well as phase modulation. With PSK, the in-phase and quadrature components are not independent. Their values are constrained in order to produce a constant envelope signal, which is a fundamental characteristic of PSK. If this constraint is removed so that the quadrature channels may be independent, the result is known as **quadrature amplitude modulation** (QAM). For the special case of two levels on each quadrature channel, QAM is identical to 4-PSK. Such systems are more popularly known as **quadrature PSK** (QPSK). The signal constellations for higher-level QAM systems are rectangular, however, and therefore distinctly different from the circular signal sets of higher-level PSK systems. Figure 6.19 compares PSK versus QAM signal constellations for $M = 16$.

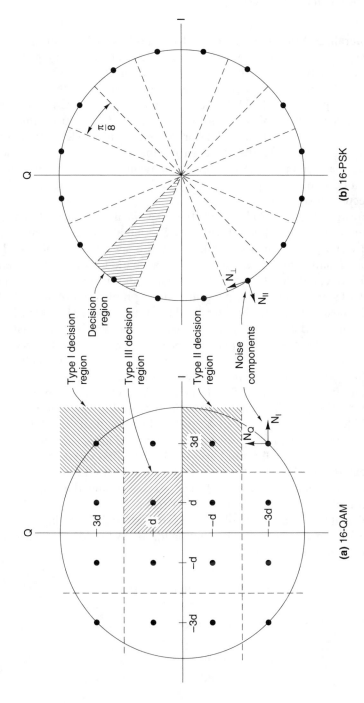

FIGURE 6.19 Comparison of 16-QAM and 16-PSK Signal Constellations with the Same Peak Power

A generalized block diagram of a QAM modulator is shown in Figure 6.20. The serial-to-parallel converter accepts a bit stream operating at a rate R and produces two parallel bit streams each at an $R/2$ rate. The 2-to-L ($L = \sqrt{M}$) level converters generate L-level signals from each of the two quadrature input channels. Multiplication by the appropriate phase of the carrier and summation of the I and Q channels then produces an M-ary QAM signal. The QAM demodulator resembles the PSK demodulator shown in Figure 6.17. To decode each baseband channel, digital logic is employed that compares the L-level signal against $L - 1$ decision thresholds. Finally, a combiner performs the parallel-to-serial conversion of the two detected bit streams.

The form of Figure 6.20 indicates that QAM can be interpreted as the linear addition of two quadrature DSB–SC signals. Thus it is apparent that the QAM spectrum is determined by the spectrum of the I and Q baseband signals. This same observation was made for M-ary PSK, leading us to conclude that QAM and PSK have identical spectra for equal numbers of signal points. The general expression for the QAM (or PSK) signal spectrum is given by (6.53). The transmission bandwidth B_T required for QAM is $2B$, the same as DSB–SC, where B is the (one-sided) bandwidth of the baseband spectrum (see Figure 6.4). For the Nyquist channel, the channel capacity for M-ary signaling was shown in Section 5.7 to be $2B\log_2 M$ b/s. Thus the transmission bandwidth efficiency for M-ary QAM (and PSK) is $\log_2 M$ bps/Hz. For example, 16-QAM (or 16-PSK) has a maximum transmission bandwidth efficiency of 4 bps/Hz and 64-QAM (or 64-PSK) has an efficiency of 6 bps/Hz. For practical channels, pulse shaping is employed by using specially constructed low-pass filters whose composite shape is usually split between the modulator and demodulator as indicated by Figure 6.20. For a choice of raised-cosine filtering, described in Chapter 5, the rolloff factor α determines the transmission bandwidth efficiency, given by $\log_2 M/(1 + \alpha)$ bps/Hz.

To calculate the probability of symbol error, we begin by giving a general characterization of the signal constellation for M-ary QAM. The signal constellation points may be represented by

$$s_i(t) = A_i \cos \omega_c t + B_i \sin \omega_c t \qquad 0 \leq t \leq T \qquad (6.62)$$

where A_i and B_i take on the amplitudes

$$A_i = \pm d, \pm 3d, \ldots, \pm(L-1)d$$
$$B_i = \pm d, \pm 3d, \ldots, \pm(L-1)d$$

$$(6.63)$$

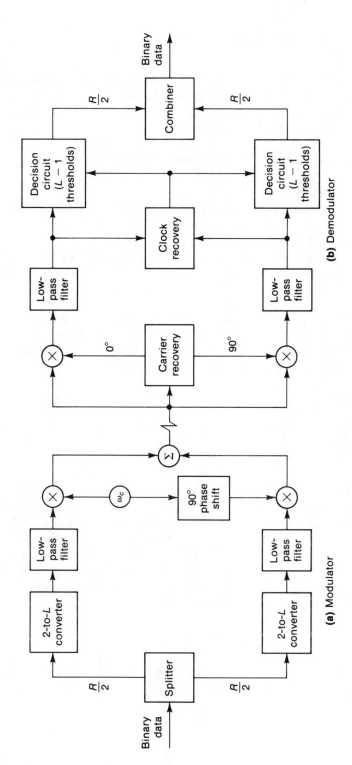

FIGURE 6.20 QAM Modulator and Demodulator

(a) Modulator

(b) Demodulator

and where $2d$ is the separation between nearest neighbors in the signal constellation. By considering the general structure of the QAM constellation and also studying the specific example of 16-QAM in Figure 6.19(a), we see that there are three types of signal locations:

Type I: Corner points that have two nearest neighbors and two decision boundaries. There are always four corner points in any square M-ary QAM constellation.

Type II: Outside points that have three nearest neighbors and three decision boundaries. There are eight outside points in 16-QAM and $4(L - 2)$ in square M-ary QAM.

Type III: Inside points that are surrounded by four nearest neighbors and by four decision boundaries. There are four inside points in 16-QAM and $M - 4(L - 1)$ in square M-ary QAM.

Each signal point of a given type will have an identical probability of error. The probability of symbol error can now be conveniently expressed as a function of the conditional probability of correct detection for each of the three types of points, assuming equally likely signals:

$$
\begin{aligned}
P(e) &= 1 - \Big[P(C \,|\, I)P(I) + P(C \,|\, II)P(II) + P(C \,|\, III)P(III) \Big] \\
&= 1 - \Big[P(C \,|\, I)(4/M) + P(C \,|\, II)(4L - 8)/M \\
&\quad + P(C \,|\, III)(M - 4L + 4)/M \Big]
\end{aligned}
\tag{6.64}
$$

To illustrate with one example, let us consider the case of 16-QAM. Figure 6.19(a) indicates one signal point and its associated decision region for each of the three signal types. For the calculations that follow, we will use those three points, with the understanding that all other points of the same type will have the same error probability by symmetry of the QAM constellation. The I and Q noise components, N_I and N_Q, are assumed to be zero-mean gaussian with variance $N_0/2$. Now consider the Type I point whose decision region is shaded in Figure 6.19(a), for which

$$
\begin{aligned}
P(C \,|\, I) &= P(N_I > -d)P(N_Q > -d) \\
&= \left[\frac{1}{\sqrt{\pi N_0}} \int_{-d}^{\infty} \exp(-x^2/N_0)\, dx \right]^2 \\
&= \left[1 - \mathrm{erfc}\!\left(\sqrt{2d^2/N_0} \right) \right]^2
\end{aligned}
\tag{6.65}
$$

For the Type II points,

$$P(C \mid II) = P(-d < N_Q < d)P(N_I > -d)$$
$$= \left[1 - 2\,\mathrm{erfc}\!\left(\sqrt{2d^2/N_0}\,\right)\right]\!\left[1 - \mathrm{erfc}\!\left(\sqrt{2d^2/N_0}\,\right)\right] \qquad (6.66)$$

and for the Type III points,

$$P(C \mid III) = P(-d < N_Q < d)P(-d < N_I < d)$$
$$= \left[1 - 2\,\mathrm{erfc}\!\left(\sqrt{2d^2/N_0}\,\right)\right]^2 \qquad (6.67)$$

Substituting (6.65), (6.66), and (6.67) into (6.64) yields

$$P(e) = 3\,\mathrm{erfc}\!\left(\sqrt{2d^2/N_0}\,\right) - 2.25\,\mathrm{erfc}^2\!\left(\sqrt{2d^2/N_0}\,\right) \qquad (6.68)$$

The probability of symbol error can be found as a function of E_b/N_0 by first finding the average signal energy in terms of d. If we let D_m equal the Euclidean distance of the mth signal type and P_m equal its probability, then the average signal energy is

$$\bar{E}_s = \sum_{m=1}^{3} |D_m|^2 P(m)$$
$$= \left(3\sqrt{2}\,d\right)^2 (4/16) + \left(\sqrt{d^2 + 9d^2}\,\right)^2 (8/16) + \left(\sqrt{2}\,d\right)^2 (4/16) \qquad (6.69)$$
$$= 10d^2$$

Then, we can express

$$\frac{\bar{E}_b}{N_0} = \frac{1}{\log_2 M}\frac{\bar{E}_s}{N_0} = \frac{5d^2}{2N_0} \qquad (6.70)$$

Substituting (6.70) into (6.68), we obtain

$$P(e) = 3\,\mathrm{erfc}\!\left(\sqrt{4\bar{E}_b/5N_0}\,\right) - 2.25\,\mathrm{erfc}^2\!\left(\sqrt{4\bar{E}_b/5N_0}\,\right) \qquad (6.71)$$

Figure 6.21 plots the symbol error rate, given by (6.64), for M-ary QAM with $M(= L^2) = 4, 16, 32$, and 64. To relate this symbol error rate to bit error rate, we assume that the multilevel signal is Gray coded so that symbol errors are most likely to produce only single bit errors in each quadrature channel. Hence

$$\text{BER} \approx \frac{P(e)}{\log_2 L} \tag{6.72}$$

Note that for $M = 4(L = 2)$, the BER equation for QAM given by (6.72) is identical to the BER equation for PSK given by (6.61), indicating the equiv-

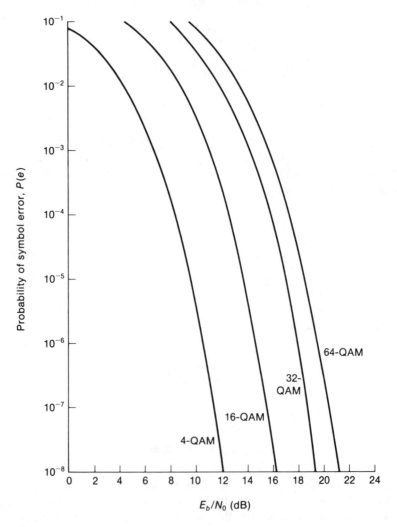

FIGURE 6.21   Error Performance for QAM

alence of 4-PSK and 4-QAM. For higher-level systems, however, the error performance for QAM systems is superior to that of PSK systems (compare Figure 6.21 with Figure 6.18). A comparison of the signal constellations for 16-QAM and 16-PSK, as shown in Figure 6.19, reveals the reason for this difference in error performance. The distance between signal points in the PSK constellation is smaller than the distance between points in the QAM constellation.

QAM is seen to have the same spectrum and bandwidth efficiency as M-ary PSK, but it outperforms PSK in error performance for at least the AWGN channel. These properties of QAM have made it a popular choice in current applications of high-speed digital radio [10, 11]. However, QAM is sensitive to system nonlinearities such as those found in the use of traveling wave tube (TWT) amplifiers. This effect requires that TWT amplifiers be operated below saturation for QAM systems. (See Chapter 9 for a discussion of the effects of nonlinear amplifiers in digital radios.)

6.9   Offset QPSK (OQPSK) and Minimum-Shift Keying (MSK)

For band-limited, nonlinear channels, two modulation techniques known as **offset quadrature phase-shift keying** (OQPSK) and **minimum-shift keying** (MSK) offer certain advantages over conventional QPSK [12, 13]. In a nonlinear channel, the spectral side lobes of a filtered QPSK signal tend to be restored to their initial characteristics prior to filtering. With OQPSK and MSK, the signal envelope is constant, which makes these modulation techniques impervious to channel nonlinearities. A choice of MSK further facilitates band limiting because its power spectral density decreases more rapidly than QPSK beyond the minimum bandwidth. These advantages are achieved with the same $P(e)$ performance and bandwidth efficiency (2 bps/Hz) as QPSK and with only modest increase in modulator and detector complexity.

Offset QPSK has the same phasor diagram as QPSK and thus can be represented by Equation (6.51). The difference between the two modulation techniques is in the alignment of the in-phase and quadrature bit streams, as illustrated in Figure 6.22. With OQPSK (also referred to as staggered QPSK), the I and Q bit streams are offset in time by one bit period T_b. With QPSK, the transitions of the I and Q streams coincide. This difference in quadrature signal alignment results in different characteristics in the phase changes of the carrier. The phase change in QPSK may be 0, $\pm 90°$, or $180°$ with each symbol interval $(2T_b)$, depending on the quadrature coefficients (p_n, q_n). In OQPSK, the quadrature coefficients cannot change state simultaneously. This characteristic eliminates $180°$ phase changes and results in only $0°$ and $\pm 90°$ phase changes.

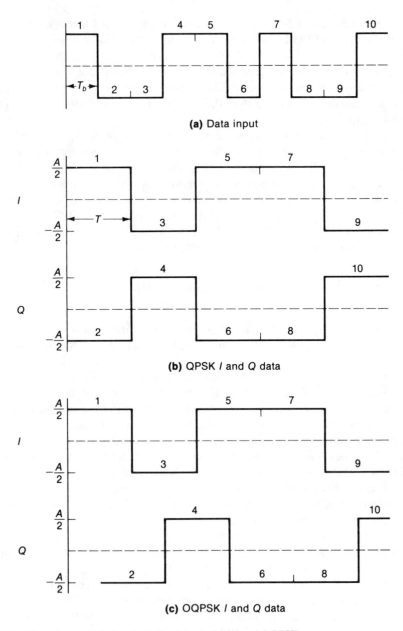

(a) Data input

(b) QPSK *I* and *Q* data

(c) OQPSK *I* and *Q* data

FIGURE 6.22   *I* and *Q* Data Relationships in QPSK and OQPSK

The usefulness of OQPSK is found in its application to the band-limited, nonlinear channel. In practical applications, band limiting is necessary to meet spectrum occupancy allocations. Bandpass filtering of QPSK causes the envelope of the signal to fluctuate and go to zero at each 180° phase reversal. If a nonlinear amplifier is used as the output power stage of the transmitter, envelope fluctuations are reduced and the spectrum sidebands are restored to their original level prior to filtering. These sidebands introduce out-of-band emissions that interfere with other signals in adjacent frequency bands. OQPSK when band-limited has no envelope zeros. Thus the elimination of large phase changes with OQPSK facilitates the use of nonlinear amplification, since the undesired sidebands removed by filtering are not regenerated by the amplifier.

The advantages of OQPSK result from eliminating the largest phase change (180°) associated with QPSK. This feature suggests that further suppression of out-of-band emissions is possible if phase transitions are eliminated altogether. One modulation technique that provides such a constant-envelope, continuous-phase signal is minimum-shift keying (MSK). MSK is a special case of FSK in which continuous phase is maintained at symbol transitions using a minimum difference in signaling frequencies. The general class of continuous-phase FSK (CPFSK) has a waveform defined by

$$s_i(t) = A \cos \left[\omega_i t + \phi(t) \right] \tag{6.73}$$

where the phase $\phi(t)$ is made continuous at the signal transitions. The simplest way to ensure continuous phase is to use waveforms that consist of an integer number of cycles, such that

$$\omega_i - \omega_j = (m_i - m_j) \, 2\pi/T_b \tag{6.74}$$

where m_i and m_j are integers. The choice of frequencies is made to satisfy (6.74) and the orthogonality conditions of (6.32), and usually to conserve bandwidth by minimizing the separation between frequencies. For binary FSK, the minimum frequency separation that also satisfies (6.74) and (6.32) would be one cycle, which means that $|m_i - m_j| = 1$. Figure 6.23 illustrates binary CPFSK with minimum frequency separation. For the general case of M-ary CPFSK, with frequencies $\omega_1 < \omega_2 < \ldots < \omega_M$, minimum separation is given by

$$\omega_i - \omega_{i-1} = 2\pi/T_b, \quad 2 \le i \le M \tag{6.75}$$

We shall now see that it is possible to further reduce the frequency separation by a factor of two and still meet the orthogonality and phase conti-

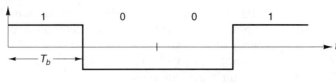

(a) Binary data signal

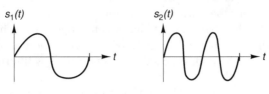

(b) Binary CPFSK waveforms

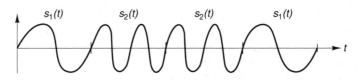

(c) CPFSK signal corresponding to data signal

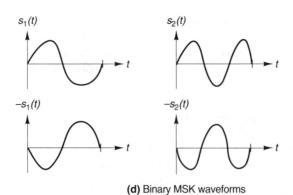

(d) Binary MSK waveforms

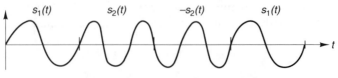

(e) MSK signal corresponding to data signal

FIGURE 6.23   Comparison of CPFSK and MSK Waveforms

nuity requirements. Consider the binary case with

$$s_1(t) = \pm\cos\,\omega_1 t \qquad 0 \le t \le T_b$$
$$s_2(t) = \pm\cos\,\omega_2 t \qquad 0 \le t \le T_b \tag{6.76}$$

The sign + or − is picked to maintain phase continuity at $t = kT_b$. We can now let

$$\omega_1 - \omega_2 = (m_1 - m_2)\,\pi/T_b \tag{6.77}$$

indicating that the frequency separation has been reduced to a half-cycle between waveforms. At the same time, (6.77) provides the minimum allowable separation to maintain orthogonality. This technique, known as **minimum shift keying (MSK)**, is illustrated in Figure 6.23 for the binary case. For M-ary MSK with frequencies $\omega_1 < \omega_2 < \ldots < \omega_M$,

$$\omega_i - \omega_{i-1} = \pi/T_b \qquad 2 \le i \le M \tag{6.78}$$

MSK may now be viewed as a special case of CPFSK having the waveform

$$s_i(t) = A\,\cos\!\left[\omega_c t + (b_i \pi t)/2T_b + \phi_i\right] \tag{6.79}$$

where ω_c is the carrier frequency, $b_i = \pm 1$ represents the binary digits 0 and 1, and $\phi_i = 0$ or π as necessary to maintain phase continuity at $t = kT_b$. The requirement for phase continuity results in the following recursive relationship for ϕ_i:

$$\phi_i = \left[\phi_{i-1} + \pi i(b_{i-1} - b_i)/2\right] \qquad \text{modulo } 2\pi \tag{6.80}$$

Thus with each bit interval T_b the phase of the carrier changes by $\pm\pi/2$, according to the value of the binary input, 0 or 1. Assuming an initial phase $\phi(0) = 0$, the phase value of $\phi(t)$ advances in time according to the phase trellis shown in Figure 6.24. Each path represents a particular combination of 1's and 0's. Since the phase change with each bit interval is exactly $\pm\pi/2$, the accumulated phase $\phi(t)$ is restricted to integral multiples of $\pi/2$ at the end of each bit interval. Further, as seen in Figure 6.24, the phase is an odd multiple of $\pi/2$ at odd multiples of T_b and an even multiple of $\pi/2$ at even multiples of T_b.

Minimum-shift keying can also be viewed as a special case of OQPSK with sinusoidal pulse shaping used in place of rectangular pulses. Using the

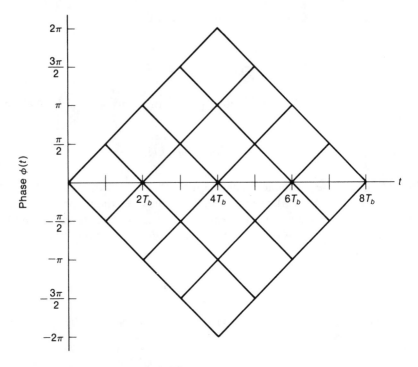

FIGURE 6.24   Phase Trellis for MSK

quadrature representation of QPSK, given in (6.51), and adding sinusoidal pulse shaping, we obtain

$$s_n(t) = A \left[p_n \cos\left(\frac{\pi t}{2T_b}\right) \cos \omega_c t + q_n \sin\left(\frac{\pi t}{2T_b}\right) \sin \omega_c t \right] \qquad (6.81)$$

where $\cos(\pi t/2T_b)$ is the sinusoidal shaping for the in-phase signal and $\sin(\pi t/2T_b)$ is the sinusoidal shaping for the offset quadrature signal. Applying trigonometric identities, Expression (6.81) can be rewritten as

$$s_n(t) = \begin{cases} A p_n \cos\left(\omega_c t + \dfrac{\pi t}{2T_b}\right) & p_n \neq q_n \\[4mm] A p_n \cos\left(\omega_c t - \dfrac{\pi t}{2T_b}\right) & p_n = q_n \end{cases} \qquad (6.82)$$

By comparing (6.82) with (6.79), we see that MSK is identical to this special form of OQPSK, with

$$b_n = -p_n q_n \tag{6.83}$$

Thus MSK can be viewed as either an OQPSK signal with sinusoidal pulse shaping or as a CPFSK signal with a frequency separation (Δf) equal to one-half the bit rate.

The power spectral density for OQPSK is the same as that for QPSK, given by (6.53) with $T = 2T_b$:

$$S(\omega) = 2A^2 T_b \left\{ \frac{\sin\left[2(f - f_c)\pi T_b\right]}{2(f - f_c)\pi T_b} \right\}^2 \tag{6.84}$$

The power spectral density for MSK is

$$S(\omega) = \frac{16A^2 T_b}{\pi^2} \left(\frac{\cos\left[2\pi(f - f_c)T_b\right]}{1 - 16(f - f_c)^2 T_b^2} \right)^2 \tag{6.85}$$

Plots of (6.84) and (6.85) are shown in Figure 6.25. The MSK spectrum rolls off at a rate proportional to f^{-4} for large values of f whereas the OQPSK spectrum rolls off at a rate proportional to only f^{-2}. This relative difference in spectrum rolloff is expected, since MSK inherently maintains phase continuity from bit to bit. Notice that the MSK spectrum has a wider main lobe than OQPSK; the first nulls fall at $3/4T_b$ and $1/2T_b$, respectively. A better comparison of the compactness of these two modulation spectra is the 99 percent power bandwidth, which for MSK is approximately $1.2/T_b$ while for QPSK and OQPSK it is approximately $10.3/T_b$. This spectral characteristic suggests that MSK has application where a constant envelope signal and little or no filtering are desired.

The error performance of OQPSK and QPSK is identical, since the offsetting of bit streams does not change the orthogonality of the I and Q signals. Furthermore, since MSK is seen to be equivalent to OQPSK except for pulse shaping, MSK also has the same error performance as QPSK and OQPSK. Coherent detection of the QPSK, OQPSK, or MSK signals is based on an observation period of $2T_b$. If MSK is detected as an FSK signal with bit decisions made over a T_b observation period, however, MSK would be 3 dB poorer than QPSK. Hence MSK detected as two orthogonal signals has a 3-dB advantage over orthogonal FSK. MSK can also be noncoherently detected, which permits a simpler demodulator for situations where the E_b/N_0 is adequate.

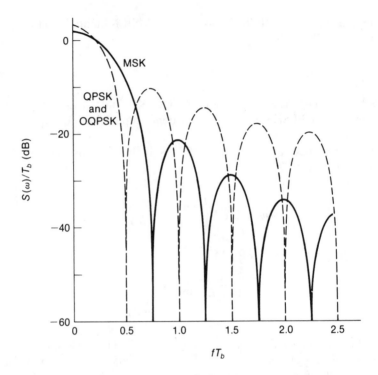

FIGURE 6.25   Power Spectral Density of QPSK, OQPSK, and MSK

6.10 Quadrature Partial Response (QPR)

To further increase the bandwidth efficiency of QAM signaling, the in-phase and quadrature channels can be modulated with partial response coders. This technique is termed **quadrature partial response** (QPR). The quadrature addition of two partial response signals can be expressed mathematically as

$$s(t) = y_I \sin \omega_c t + y_Q \cos \omega_c t \tag{6.86}$$

As an example, assume the use of class 1 partial response coders in the I and Q channels of a 4-QAM (QPSK) modulator. Recall from Section 5.8 that the effect of the partial response coder is to produce three levels from a binary input. As shown in Figure 6.26a, the resulting QPR signal constellation is a 3×3 rectangle with nine signal states. Similarly, 16-QAM that has four levels on each I and Q channel produces seven levels after partial response coding, a 7×7 signal constellation, and 49 signal states, as shown in Figure 6.26b.

A block diagram of the QPR modulator and demodulator is shown in Figure 6.27. This QPR implementation is very similar to that of QAM

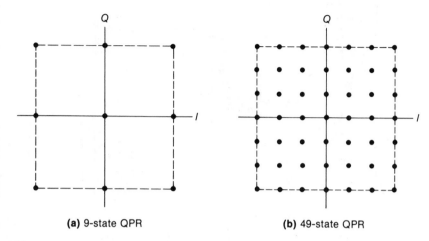

(a) 9-state QPR   **(b)** 49-state QPR

FIGURE 6.26   Phasor Diagrams of QPR

shown in Figure 6.20. The essential differences are in the filtering and detection. Partial response shaping can be accomplished at baseband with low-pass filters or after quadrature modulation with bandpass filters. In either case, the filter characteristic is usually split between the transmitter and receiver, such that the receiver filter not only finishes the required shaping but also rejects out-of-band noise. A choice of bandpass filtering allows the use of nonlinear power amplifiers prior to filtering when the signal still has a constant-envelope characteristic. After amplification, the bandpass filter converts the QAM signal to QPR. A disadvantage of bandpass filtering is the higher insertion loss compared to low-pass filtering. After final filtering in the QPR demodulator, the I and Q baseband signals are independently detected using the same partial response detection scheme described in Section 5.8.

Earlier we noted that the spectrum and error performance of M-ary QAM are determined by the corresponding characteristics of the I and Q baseband signals. Thus QPR has a spectrum defined by the partial response signal (see Table 5.5). Recall from Section 5.8 that partial response coding attains the Nyquist transmission rate. Therefore QPR permits practical realization of the transmission bandwidth efficiency theoretically possible with QAM. This efficiency is given by $\log_2 M$ bps/Hz. With class 1 partial response coding applied to the I and Q channels, for example, 4-QAM becomes 9-state QPR and achieves 2 bps/Hz whereas 16-QAM becomes 49-state QPR and achieves 4 bps/Hz efficiency. The error performance of QPR is given by the corresponding expression for partial response (Equation 5.68), but with M replaced by L, the number of levels on each quadrature

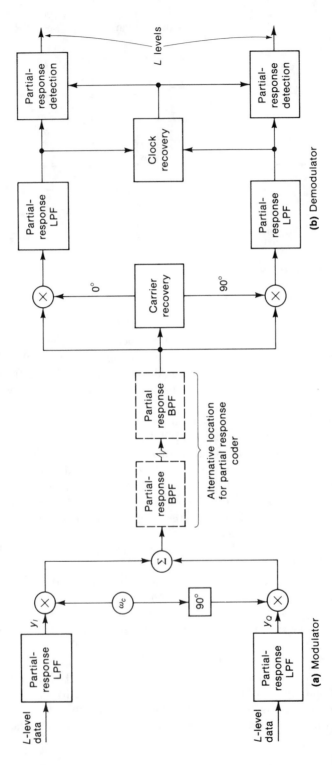

FIGURE 6.27 QPR Modulator and Demodulator

(a) Modulator

(b) Demodulator

channel prior to filtering. Using (6.34) to express S/N as a function of E_b/N_0, the QPR error probability can be rewritten from (5.68) as

$$P(e) = 2\left(1 - \frac{1}{L^2}\right)\mathrm{erfc}\left[\frac{\pi}{4}(\log_2 L)^{1/2}\left(\frac{6}{L^2-1}\right)^{1/2}\left(\frac{E_b}{N_0}\right)^{1/2}\right] \qquad (6.87)$$

Assuming the use of Gray coding on each baseband signal, the bit error probability is given by (6.87) divided by $\log_2 L$.

A comparison of error performance for 9-QPR versus 4-QAM, using Expressions (6.87) and (6.64), indicates a net loss of 2 dB in E_b/N_0 for QPR. However, QPR attains a 2 bps/Hz spectral efficiency with realizable filtering, while 4-QAM requires the theoretical Nyquist channel in order to attain the same efficiency. A fairer comparison with 9-QPR is 16-QAM with 100 percent ($\alpha = 1$) raised-cosine filtering. Both systems provide 2 bps/Hz spectral efficiency, but now 9-QPR requires 1.9 dB less E_b/N_0 to provide the same error performance as 16-QAM. Its error performance, spectral efficiency, and simplicity of implementation have made QPR a popular choice in current digital radio systems. Several QPR radio systems have been described in the literature. A digital radio designed for use in the Canadian 8-GHz frequency band uses nine-state QPR to transmit 91 Mb/s in a 40-MHz RF bandwidth—an efficiency of 2.25 bps/Hz [14, 15]. A digital radio designed for the U.S. military also uses nine-state QPR to achieve an RF bandwidth efficiency of 1.9 bps/Hz [16]. Implementation of the partial response coding has been done both at baseband [16] and at RF after power amplification [14].

Example 6.2

A 9-QPR radio system is designed to transmit 90 Mb/s in a 40-MHz bandwidth. The transmit signal power is 1 watt and net system loss is 111 dB. For transmission over an additive white gaussian noise channel with one-sided power spectral density equal to 4×10^{-21} watts/Hz, compute the BER for this system. How would this performance compare with a differentially coherent 8-PSK system, assuming that the bandwidth is reduced to 30 MHz?

Solution   From (5.67) the BER for QPR modulation is

$$\mathrm{BER} = \frac{3}{2}\mathrm{erfc}\left[\frac{\pi}{4}\left(\frac{S}{N}\right)^{1/2}\right]$$

$$= \frac{3}{2}\mathrm{erfc}\left\{\frac{\pi}{4}\left[\frac{7.9 \times 10^{-12}\ \mathrm{watts}}{(4 \times 10^{-21}\ \mathrm{watts/Hz})(40 \times 10^6\ \mathrm{Hz})}\right]^{1/2}\right\}$$

$$= \frac{3}{2}\mathrm{erfc}\ 5.53$$

$$= 2.4 \times 10^{-8}$$

For 8-PSK modulation, assuming the use of Gray coding, the BER is obtained from (6.60) and (6.61) as

$$\text{BER} = \frac{P(e)}{\log_2 8} = \frac{2}{3}\,\text{erfc}\left[\frac{2E_b\log_2 8}{N_0}\sin^2\frac{\pi}{8\sqrt{2}}\right]^{1/2}$$

From (6.5) the S/N is given by

$$\frac{S}{N} = \frac{R}{B_N}\frac{E_b}{N_0}$$

so that

$$\text{BER} = \frac{2}{3}\,\text{erfc}\left[(0.451)\left(\frac{30\times10^6}{90\times10^6}\right)\frac{7.9\times10^{-12}}{(4\times10^{-21})(30\times10^6)}\right]^{1/2}$$

$$= \frac{2}{3}\,\text{erfc}\,3.15$$

$$= 5.4\times10^{-4}$$

6.11 Trellis Coded Modulation

Coding has long been used for forward error correction, especially with power-limited channels such as deep space or satellite transmission. The combination of coding with modulation considered here does not perform error correction but rather provides a "coding gain" compared to conventional modulation techniques. Coding gain can be realized by either block coding or state-oriented trellis coding such as convolutional coding. A trellis code can be found that will always perform as well as block coding, however, with greater simplicity. Ungerboeck [17, 18] first proposed the technique known as trellis coded modulation, in which a coding gain of up to 6 dB can be achieved within the same bandwidth by doubling the signal constellation set from $M = 2^m$ to $M = 2^{m+1}$ and using a rate $k/(k+1)$ convolutional code. The redundant bit added by the convolutional coder introduces constraints in the allowed transitions between signal points in the constellation. As a result, the minimum distance between signal points in the coded system is greater than the minimum distance for the uncoded system. Use of a maximum likelihood detector, such as the Viterbi decoder, yields a coding gain over the uncoded system.

In this section we shall describe convolutional coding and then see how it is used with modulation to effect trellis coded modulation.

6.11.1   Convolutional Coding

A convolutional encoder is a finite memory system consisting of a shift register with feedback to exclusive OR's. The encoder is described by the number of input bits (k), the number of output bits (n), the constraint length (K), and the feedback connections. The *code rate* is given by the ratio k/n, and the constraint length K is equal to the length of the shift register. The placement of the feedback connections or taps determines possible state transitions where the number of possible states is defined as $2^{(K-1)}$. The impulse response or *generator function* of the encoder is the output of the coder when the input sequence is a 1 followed by all 0's.

To illustrate, consider the example convolutional encoder shown in Figure 6.28. The code rate is ½ and the constraint length is three; that is, two code bits are produced for every input bit, and the length of the shift register is three bits. There exist two generating functions (g_1, g_2), one for each output bit. For a sample input 100, the impulse response of the encoder is (assuming that the register starts in the all-zero state)

$$(g_1, g_2) = (1 \oplus D^2, 1 \oplus D \oplus D^2)$$
$$= (11 \ 01 \ 11) \tag{6.88}$$

where D^i represents a delay of i bits in the shift register and $\oplus$ represents exclusive OR. Thus, the code bits (c_1, c_2) depend on the current bit ($S1$) as well as the previous input bit pattern stored in stages $S2$ and $S3$ of the shift register. The four bit patterns corresponding to ($S2,S3$) are called states, which can be conveniently labeled $\{a,b,c,d\}$. The output bits and transitions among states can be traced by means of a tree diagram, trellis diagram, or state diagram. We will now consider each method using the encoder of Figure 6.28 as our example.

The tree diagram of Figure 6.29 shows all possible output sequences for any input sequence. For a given input sequence, a particular path is followed from left to right. Each new input bit causes a branching upward (for a 0) or

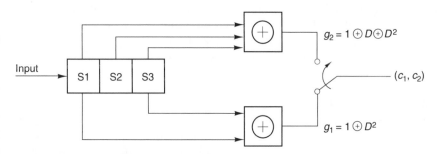

FIGURE 6.28   Example of ½ Rate Convolutional Encoder

downward (for a 1). Four branch points, or states, are observed for the example encoder and are labeled {a,b,c,d}. The state is determined by the two previous bits, $S2$ and $S3$. Shown with each branch are the shift register contents (in the ovals) and the output sequence. Now consider an input sequence (1101), and assume that the shift register is initially loaded with all zeros. The output sequence can be traced in Figure 6.29: The first input bit (1) results in an output sequence (11) and shift register contents (100). The second input bit (1) has an output sequence (10) and shift register contents (110). The third input bit (0) has an output sequence (10) and shift register contents (011). And the fourth input bit (1) has an output sequence (00) and shift register contents (101). The complete output sequence is then (11 10 10 00). Note that for our example encoder the corresponding tree diagram is repetitive after the third branching. From that point on, all branches emanating from the same state will generate identical sequences, so that the upper and lower halves of the tree are the same.

The problem with the tree diagram is that it quickly becomes unwieldly for a long sequence of input bits or for a more complex coder. By inspection of Figure 6.29, however, note that any two transitions leading to the same state at the same time t_i can be merged since all succeeding paths will be indistinguishable. This observation leads to a more useful technique, the trellis diagram shown in Figure 6.30. The trellis conveys the same information as the tree but takes advantage of the finite memory of the encoder, which causes the repetitive nature of the tree and the limited number of states. Examining Figure 6.30, we see that there again two branches corresponding to the current bit, upper branches (shown solid) and lower branches (shown dashed) for input bits of 0 and 1, respectively. The diagram starts in state a, the all-zero condition, and after the first two input bits, the encoder will be in one of its four states, {a,b,c,d}. For any input sequence, the corresponding output sequence is easily determined by taking the proper branch at each state and noting the corresponding output shown for each branch.

The notion of states and transitions suggests a third approach, a state diagram as shown in Figure 6.31 for our example encoder of Figure 6.28. The transitions among the states are once again denoted by a solid line for a 0 input bit and dashed line for a 1 input bit. Output bits are shown for each transition. For example, assume that the encoder is in state b and the current bit is a 1. Then the coder will generate an output (10) and will transition to state d.

6.11.2   Viterbi Decoding

There are several different methods for decoding convolutional codes, but only Viterbi decoding will be considered here since it provides maximum-likelihood decoding [19]. Refering to the trellis diagram of Figure 6.30, a maximum-likelihood decoder could be realized by computing the likeli-

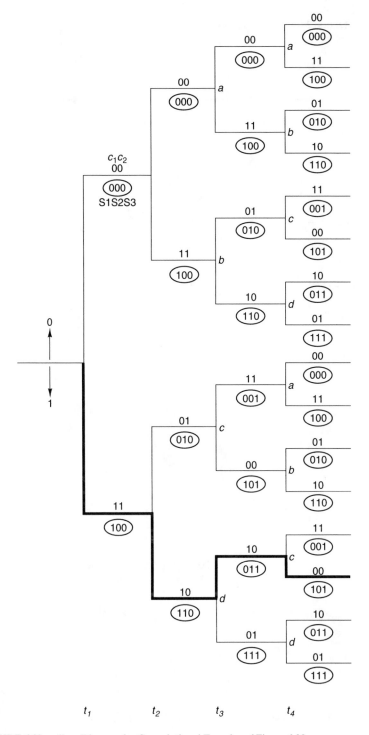

FIGURE 6.29   Tree Diagram for Convolutional Encoder of Figure 6.28

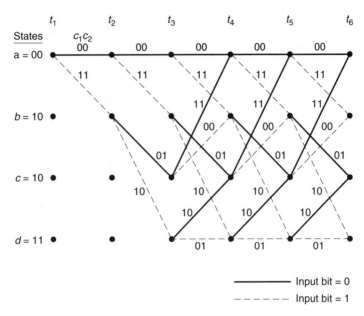

FIGURE 6.30   Trellis Diagram for Convolutional Encoder of Figure 6.28

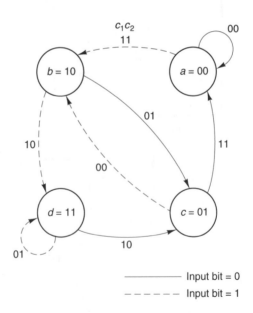

FIGURE 6.31   State Diagram for Convolutional Encoder of Figure 6.28

hood of the received data being any one of the paths through the trellis. The path with the largest likelihood would then be selected, and the corresponding code word would be decoded into the original data. However, the number of paths for an L bit sequence is 2^L so that this brute force decoder quickly becomes impractical as L increases. The Viterbi decoder reduces this computational complexity by calculating the likelihood of each of the paths entering a given state and eliminating from further consideration all but the most likely path that leads to that state. The most likely path is that which has the smallest distance between any of the allowed paths and the actual received sequence. The **Hamming distance** for two code words of the same length is defined as the number of bits that differ between the two sequences. The Hamming distance of a given path at time t_i is the sum of the branch Hamming distances along that path up to time t_i. Thus Viterbi decoding is accomplished by computing the Hamming distances for all paths entering each state and eliminating all but that with the smallest cumulative Hamming distance, which is called the *survivor*.

Consider the example of maximum-likelihood decoding shown in Figure 6.32 for the ½ rate convolutional coder of Figure 6.28. The two paths entering state a at time t_i are shown to have Hamming distances of 2 and 3. These distances are found as follows: the path entering state a from state a has a distance $2 + 0 = 2$, while the path entering state a from state c has a distance $1 + 2 = 3$. The first term in these sums is the cumulative distance up to $t = i - 1$ and the second term is the Hamming distance for the most recent branch from $t = i - 1$ to $t = i$. In this example the first term is arbitrary, but the second term is computed by comparing the received code at time t_i ($y_i = 00$) with the allowed code words (00 and 11) on the two branches merging into state a. Then the path with distance 2 is chosen as the survivor and the other path is eliminated from further consideration.

The Viterbi decoder can now be described as the following procedure. First draw the trellis (or tree) up to the point where all states occur after which the trellis assumes a fixed periodic structure (this occurs at a depth of three in the example of Figure 6.30 or a depth K in general). At this depth (or time) there will be more than one path merging into each next state. The Hamming distance is computed for each path entering each state by adding the Hamming distance of the survivor to the most recent branch distance. For each state, the path with the smallest distance is selected as the survivor. There will always be as many survivors as there are states (four survivors in the case of the ½ rate coder). In case of ties a random decision is made in picking one and only one as the survivor. These survivors are extended and their distances are stored along with the code words they represent. The depth of the trellis is increased by one and the procedure of summing distances and finding the survivors is repeated.

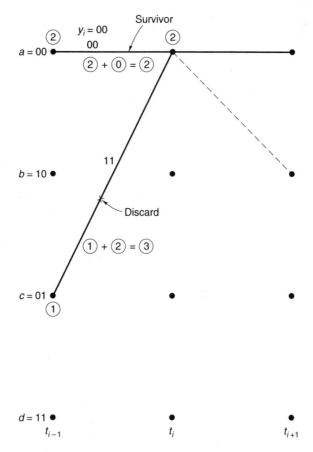

FIGURE 6.32   Example of Maximum-Likelihood Calculation in Viterbi Decoder

The decoder described thus far never actually decides upon one most likely path. There are always 2^{K-1} survivors and also paths left after each decoding step. Each surviving path is the most likely path to have entered a given encoder state. A common way of eliminating survivors and thus forcing a decoder decision is to append a $(K-1)$ fixed sequence to the end of the information sequence. This tail sequence forces the encoder into a known state, allowing the decoder to select that path leading to the known state at the end of the received sequence. In practice, another approach is more commonly used to avoid the need for periodic fixed sequences added to the transmitted data. It can be shown that all of the 2^{K-1} paths tend to have a common trunk that eventually branches off to the various states [20]. Thus if the decoder stores enough of the past history of the 2^{K-1} surviving paths, then the oldest bits will have a common trunk, that is, they will be identical. Assume that the decoder stores surviving paths up to some *truncation length h*. If the current depth of the decoder is i, the algorithm will make a decision on the state (and therefore bit value) at depth $i - h$. The decoder can thus output one more bit (the oldest) each time it steps deeper into the trellis. The use of truncation results in some performance loss compared to the maximum-likelihood detector, but proper choice of truncation length will make the error probability due to truncation small [21].

Example 6.3 _____

Suppose that the Viterbi decoder for the convolutional encoder of Figure 6.28 has the input sequence $y = (11\ 10\ 11\ 00\ 00\ 11)$. Assume that a tail of two 0's was appended to the end of the source sequence prior to coding at the transmitter. What is the most likely code word to have been transmitted?

Solution   Decoding will be accomplished through a series of trellis diagrams shown in Figure 6.33. As the first step, shown in Figure 6.33a, the trellis is drawn up to the depth where two (or more) paths merge into the same state. Consider state 00 at time t_4. The upper path merging into state 00 has a Hamming distance of five compared to the received code word. The lower path merging into state 00 has a Hamming distance of two. Therefore the upper path is eliminated and the lower path with distance two becomes the survivor. The same comparison is made with the two paths entering each of the three remaining states. Each path no longer to be considered is marked out with an X, and the distance between the received code word and surviving path is encircled above each state.

The four surviving paths are now extended to the next trellis depth shown in Figure 6.33b. Note that eliminated paths are not extended. Now consider the two paths entering state 00 at time t_5. The Hamming distance for the path from state 00 is found by summing the stored distance (two) to the distance over the new branch (zero), yielding a accumulated distance of two. Likewise, the Hamming distance for the path from state 01 is found as the sum of the old distance (one) and new distance (two) for a total of three. Hence the first path survives and the other is eliminated. Similar comparisons for the other three states result in the surviving paths and Hamming distances shown. As the survivors are extended to form Figure 6.33c, note that the oldest code bit pair can now be decoded as 11 since there is only one path left at time t_2.

The final two decoding steps shown in Figure 6.33c and 6.33d take advantage of the known tail, which forces the encoder and decoder into the 00 state. Hence at time t_6 in Figure 6.33c only two states are permissible, 00 and

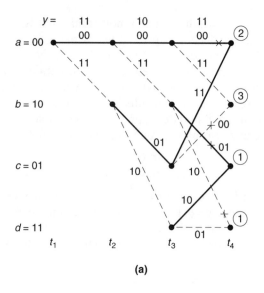

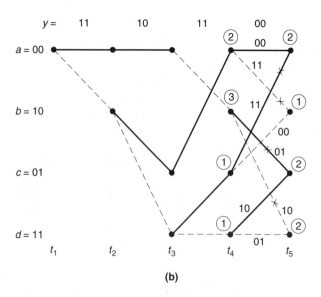

FIGURE 6.33   Survivor Paths for the Viterbi Decoder of Example 6.3

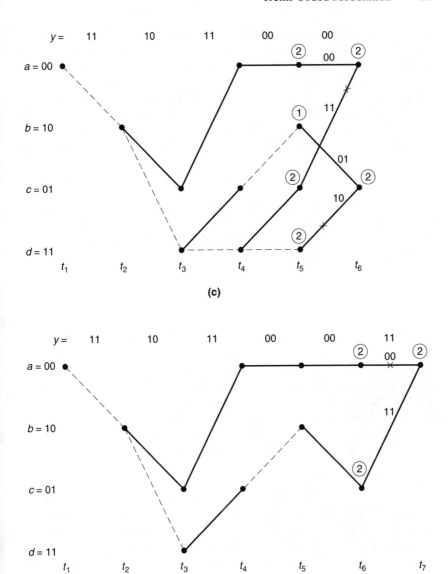

FIGURE 6.33   Survivor Paths for the Viterbi Decoder of Example 6.3

01, since only those two states will lead to state 00 in the final trellis depth. As before, the two paths leading to each state are compared for Hamming distance, with one path surviving each comparison. At the final depth only two paths remain, since it is known that the decoder must end in state 00. After this final comparison, only one path remains that has a corresponding code word of (11 10 10 00 01 11). Comparison of this decoded word with the received code word indicates two transmission errors at bit locations five and nine. As an exercise for the reader, determine the original source sequence.

The error-correcting property of convolutional codes depends on the minimum Hamming distance d_{min} in the set of all possible paths that diverge from and remerge into the same state. The number of code bit errors that can be corrected is then

$$t = \frac{d_{\min} - 1}{2} \tag{6.89}$$

If more than t errors occur, the trellis path corresponding to the actual transmitted sequence will not survive in the decoding process, leading to a decoding error. To determine d_{min}, first assume some arbitrary transmitted sequence (anyone will do) and then examine all paths that diverge from the starting state (for instance, all 0's) and then remerge at the ending state (for instance, all 0's) some time later, to find that path with minimum distance. The reader is left to verify that the convolutional decoder of Figure 6.33 has $d_{min} = 5$ and therefore can correct any two code bit errors.

6.11.3   Trellis Coding

Trellis coding applied to modulation achieves coding gain at the expense of coder and decoder complexity but without increase in power or bandwidth. In conventional M-ary modulation, the modulator maps an m bit sequence into one of $M = 2^m$ possible symbols, and the demodulator recovers each m-bit sequence by making decisions on each received symbol independent of other symbols. Bit errors occur if the received signal is closer in Euclidean distance to any symbol other than the one transmitted. An increase in bit rate requires a greater number of symbols packed into the constellation, implying a loss in noise margin or increase in transmit power to compensate for the increased symbol packing. In conventional coding applications, the bandwidth is expanded by code bits that are added to provide error correction. In trellis coded modulation, convolutional coding is used to increase the Euclidean distance between allowed symbols rather than perform error correction. To accommodate the added code bits, the modulation constellation must be increased, usually from 2^m to 2^{m+1} symbols. The resulting coding gain more than compensates for the loss in noise margin introduced by the increase in number of symbols. As we have seen in Section 6.11.1, convolutional coding introduces interdependencies between sequences of states and therefore limitations on the allowed transitions from one state (now read symbol) to another state (symbol). Hence, decoding is performed using the Viterbi algorithm but with a different metric, that of Euclidean distance. The intent of trellis coding here is to maximize the *Euclidean* distance among the code words, rather than to maximize Hamming distance as was accomplished with convolutional coding alone.

Because of the limitations in symbol transitions introduced by trellis coding, error performance is no longer determined by closest neighbors in the signal set. Rather, minimum distances between members of the set of allowable symbol sequences will determine error performance. Proper coding will ensure that the minimum distance between any two possible sequences will be greater than the minimum distance in the uncoded signal constellation. The distances referred to here are Euclidean distances, not Hamming distances considered in convolutional coding. Using Ungerboeck's notation, let $d(a_n, b_n)$ denote the Euclidean distance between two symbols a_n and b_n at time n. The coder is then designed to provide maximum free Euclidean distance

$$d_{free} = \min_{\{a_n\} \neq \{b_n\}} \left[\sum_n d^2(a_n, b_n) \right]^{1/2} \qquad (6.90)$$

between all pairs of symbol sequences $\{a_n\}$ and $\{b_n\}$ that the encoder can produce. The Viterbi algorithm can then be used to determine the coded symbol sequence $\{a_n\}$ closest to the received symbol sequence. The most likely errors will occur between those symbol sequences that are at the minimum free distance d_{free}.

To illustrate, we will compare uncoded 4-PSK with coded 8-PSK, while keeping the transmit bandwidth and power constant. In conventional 4-PSK, bits are grouped in pairs by the modulator and mapped into one of four phases. Now consider Figure 6.34, which shows one of these two bits (b_2) being convolutionally encoded and the other (b_1) left unmodified. Note that the convolutional encoder is identical to that shown earlier in Figure 6.28. The three code bits (c_1, c_2, c_3) are input to the an 8-PSK modulator, with code bits c_2 and c_3 assigned symbol values to provide maximum free distance. The mapping of code bits to symbols, called **set partitioning,** will be discussed later. The resulting trellis diagram shown in Figure 6.35a has four states as in Figure 6.30 but with eight sets of parallel transitions due to the presence of the uncoded bit c_1. The states are labeled from the alphabet

$\{a,c,b,d\}$ as in Figure 6.30, and each transition is labeled with its corresponding code word $\{c_1, c_2, c_3\}$. The two parallel transitions correspond to the two possible values of the first source bit (b_1), with the upper branch representing $b_1 = 0$ and the lower branch representing $b_1 = 1$. The diverging transitions from each state represent the second source bit, with a solid line indicating $b_2 = 0$ and a dashed line indicating $b_2 = 1$.

A look at possible decoding errors will suggest a heuristic method for the mapping of transitions into symbols. As observed before with the description of the Viterbi decoder, the most likely errors are those with the smallest distance from the correct path. For the trellis diagram of Figure 6.35a, the parallel transitions show that single error events can occur and are therefore the closest error events. To minimize this error probability, we will maximize the Euclidean distance between the symbols corresponding to parallel transitions. The next most likely errors occur for those transitions that diverge from or merge into the same state, so we assign the next largest Euclidean distance to those symbols. Ungerboeck first identified this general set of rules for the assignment of bits to symbols that provides maximum Euclidean distances:

1. First, maximize the distance between parallel transitions.
2. Next, maximize the distance between transitions diverging from or merging into the same state.
3. Use all symbols with equal frequency.

These rules are heuristic and do not necessarily lead to an optimum code.

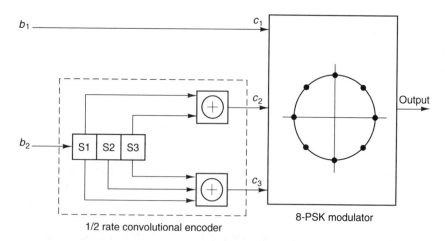

FIGURE 6.34   Example of Trellis Coded Modulation

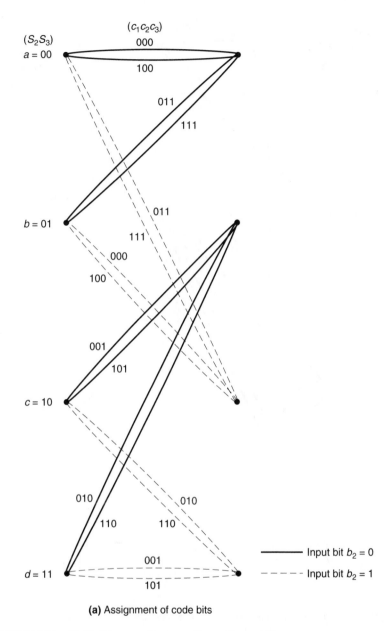

(a) Assignment of code bits

FIGURE 6.35   Trellis Diagrams Corresponding to Figure 6.34

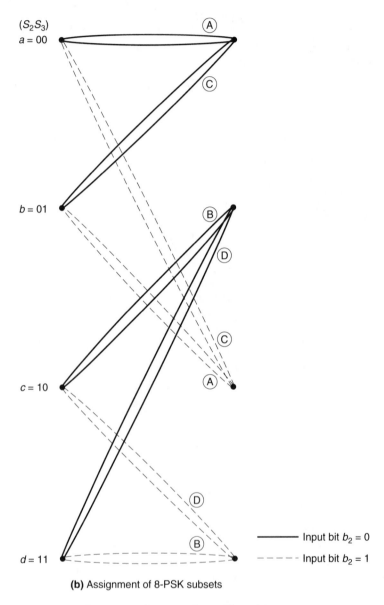

(b) Assignment of 8-PSK subsets

FIGURE 6.35   Trellis Diagrams Corresponding to Figure 6.34

Now we shall apply these rules to the example of trellis coding shown in Figure 6.34. We begin by examining the distances between 8-PSK symbols and numbering of these symbols, as shown in Figure 6.36b. (Note that Figure 6.36 also shows comparison of uncoded 4-PSK versus coded 8-PSK.) From smallest to largest, these distances are $D_0 = 2\sin(\pi/8)$, $D_1 = \sqrt{2}$, and $D_2 = 2$, assuming that the symbols all have unit amplitude. Referring back to Figure 6.35a, note that parallel transitions correspond to the following pairs of (decimal) numbers: (0,4), (1,5), (2,6), and (3,7); also note that merging and diverging transitions correspond to the following groups of numbers: (0,3,4,7) and (1,2,5,6). According to Ungerboeck's rules, parallel transitions are to have maximum distance D_2, so the pairs (0,4), (1,5), (2,6), and (3,7) are mapped into subsets (A,B,C,D), respectively, in the signal constellation to ensure a distance $D_2 = 2$. Transitions that diverge from or merge into one state are assigned symbols with distance D_1 between them, which means that groups (1,2,5,6) and (0,3,4,7) are to be mapped into subsets (B,D) and (A,C), respectively, which have distance $D_1 = \sqrt{2}$. Figure 6.36c indicates the resulting assignment of bits (in decimal equivalent) to signals in the 8-PSK constellation. The trellis diagram of Figure 6.35a can now be redrawn to show the assignment of code bits to subsets. As shown in Figure 6.35b, code bits c_2 and c_3 are assigned to subset A, B, C, or D, and code bit c_1 is mapped to the proper point within the subset. Inspection of Figure 6.35b reveals that any two paths that diverge from one state and remerge into that state after more than one transition will have a squared distance of at least $(D_1)^2 + (D_0)^2 + (D_1)^2 = (D_2)^2 + (D_0)^2$. For example, the two paths shown in Figure 6.37 have this distance, where the upper path (all-zero code) is assumed to be the correct path. The distance between such paths as found in Figure 6.37 is greater than the distance between symbols assigned to parallel transitions, $D_2 = 2$, which is thus the free distance for 8-PSK trellis coded modulation. The coding gain may now be found as the ratio of $(d_{free})^2$ for coded 8-PSK to $(d_{free})^2$ for uncoded 4-PSK. If the symbols of 4-PSK have unit amplitude, then their free distance is $\sqrt{2}$, and

$$\text{Coding Gain} = 10 \log \frac{2^2}{\sqrt{2^2}} = 3 \text{ dB} \tag{6.91}$$

It should be noted that there exist other mappings of coded bits to symbols for the 8-PSK example considered here that result in this same coding gain of 3 dB. However, all such mappings obey the three rules given earlier. Further gain is possible by increasing the number of states in the trellis code. We can obtain about 4-dB gain with 16 states, 5 dB with 64 states, and nearly 6 dB with 256 states [22, 23].

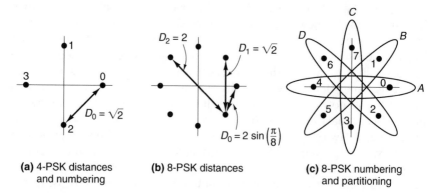

(a) 4-PSK distances and numbering

(b) 8-PSK distances

(c) 8-PSK numbering and partitioning

FIGURE 6.36   Uncoded 4-PSK versus Coded 8-PSK

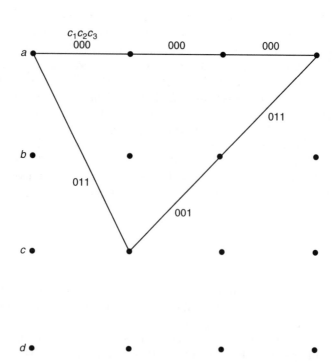

FIGURE 6.37   A Minimum-Error Event for TCM of Figure 6.34

Another way to view set partitioning is illustrated in Figure 6.38 for 8-PSK and 16-QAM. The signal constellation is successively partitioned into subsets with increasing minimum distance $D_0 < D_1 < D_2$ between the symbols in each of these subsets. The partitioning is repeated $k + 1$ times, where k is the number of bits input to the convolutional coder. Hence, if $k = 1$ as in Figure 6.34, two partitions are needed to yield four subsets; the uncoded bit c_1 selects a symbol from the subset, and the two coded bits c_2 and c_3 select the subset. To use the fourth row of the 16-QAM constellation shown in Figure 6.38b, a trellis coder with $k = 2$ is needed. Alternatively, a trellis coder with $k = 1$ and two uncoded bits would use the third row in Figure 6.38b. In this case, the two uncoded bits select one of four symbols within the subset, and the two coded bits select one of the four subsets. For two-dimensional rectangular constellations such as 16-QAM, each successive partition increases the minimum distance, d_{free}, by a factor of $\sqrt{2}$, that is, $D_{i+1} = \sqrt{2}D_i$. For one-dimensional PAM constellations, $D_{i+1} = 2D_i$. Nonlattice-type constellations such as PSK have unique sequences of distances.

The Viterbi decoder operates in essentially the same manner as shown in Example 6.3. The procedure may be summarized as follows. At each time period, every state in the trellis can have several paths merging into it, but

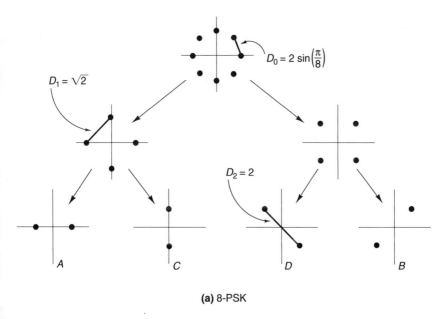

(a) 8-PSK

FIGURE 6.38   Partitioning of Signal Constellations into Subsets with Increasing Minimum Distances

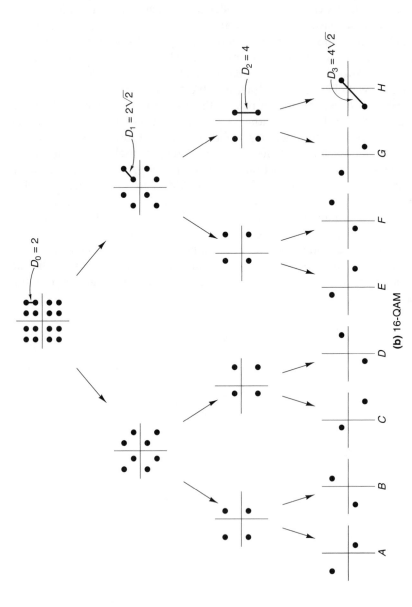

FIGURE 6.38 Partitioning of Signal Constellations into Subsets with Increasing Minimum Distances

(b) 16-QAM

only the path with minimum Euclidean distance will be kept and the others discarded. The signal and distance for the surviving path are stored for each state. Storing these data creates a history so that it is possible to trace back and find the correct output of the decoder.

A general structure of the encoder and modulator for trellis coded modulation is shown in Figure 6.39. A total of m bits are transmitted, using a constellation of 2^{m+1} symbols. The m bits are separated into k, which are coded, and $m-k$, which are uncoded. Coding is performed by a rate $k/(k+1)$ convolutional encoder, which produces $k+1$ coded bits. These coded bits are used to select one of 2^{k+1} subsets of the 2^{m+1}-ary symbol set. The remaining $m-k$ uncoded bits are used to select one of the 2^{m-k} symbols in this subset as the transmitted signal. A trellis coder with $m-k$ uncoded bits will have 2^{m-k} parallel transitions between states. For the special case of $m = k$, there are no parallel transitions and therefore the subsets contain only one symbol.

We can further generalize trellis coded modulation by first noting that the convolutional coder is not constrainted to a $k/(k+1)$ rate but may take on the rate $k/(k+p)$, where p is the number of parity bits. In this case the encoder will increase the required size of the signal set from 2^{m+1} to 2^{m+p}. Ungerboeck, in his research, however, found that little increase in channel capacity would occur if the signal set were expanded by more than a factor of two.

Another more practical variation of TCM is the choice of a nonlinear convolutional coder, used to solve the problem of phase offset in the receiver. To briefly describe the problem, a phase offset may rotate the received signal such that the distance between received signals and allowed signals will be less than d_{free}. Figure 6.40a shows the implementa-

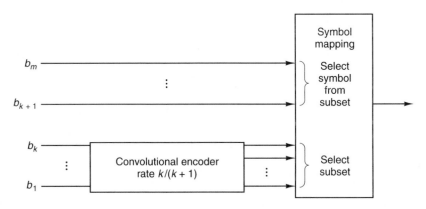

FIGURE 6.39   General Structure of Encoder and Modulator for Trellis Coded Modulation

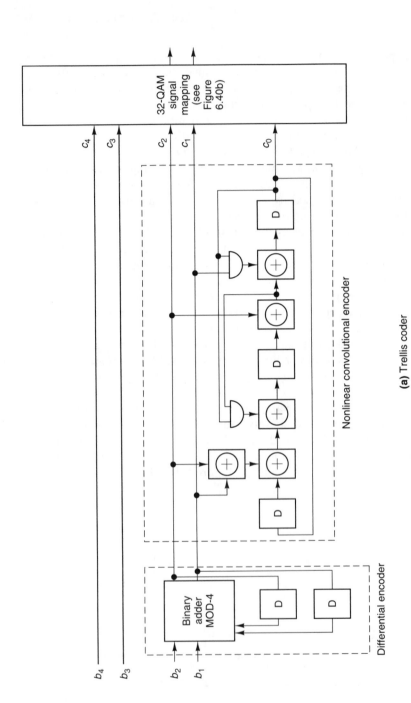

FIGURE 6.40 Trellis Coding for 9.6-kb/s Modems per CCITT Rec. V.32 (Courtesy CCITT [23])

(a) Trellis coder

422

tion of CCITT Rec. V.32 for 9.6-kb/s voiceband modems, which is based on a nonlinear convolutional encoder matched with 32-QAM. The signal constellation shown in Figure 6.40b is commonly referred to as a **cross constellation,** and is preferred over a square constellation because of its greater minimum Euclidean distance and smaller ratio of peak to average power [24]. This trellis coder solves the problem of phase ambiguity in the receiver by using a differential encoder to provide 90° of phase invariance. As a result, c_4 and c_3 are the same for all points on the 32-cross constellation that are 90° apart. The trellis diagram will have eight states because the convolutional encoder has three delay elements. The output of the encoder consists of three (coded) bits, c_0, c_1, and c_2. The five code bits c_0 through c_4 are then mapped into the transmitted symbol according to the 32-cross signal constellation shown in Figure 6.40b. The reader is referred to Problem 6.47 and CCITT Rec. V.32 [25] for more investigation of the V.32 modem.

Trellis coded modulation (TCM) is not limited to one- or two-dimensional signal space and has been implemented in more than two dimensions to gain

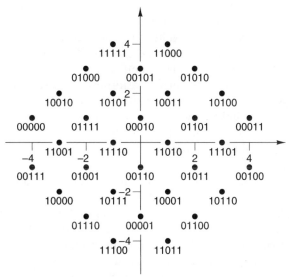

The binary numbers denote $c_0c_1c_2c_3c_4$

(b) Signal constellation

FIGURE 6.40   Trellis Coding for 9.6-kb/s Modems per CCITT Rec. V.32 (Courtesy CCITT [23])

certain advantages [23]. The motivation behind an increase in the number of dimensions is a savings in transmit power. For two-dimensional rectangular signal sets such as 16-QAM, the minimum distance D_0 must be reduced by $\sqrt{2}$ (3 dB) to maintain the same signal power as the uncoded case. This loss in signal separation must be more than compensated by the coder to achieve overall coding gain. Higher dimensionality schemes provide less than 3-dB loss. For example, four-dimensional TCM leads to only 1.5-dB loss and eight-dimensional TCM has only 0.75-dB loss. In practice, multidimensional signals are transmitted as sequences of two-dimensional signals. Thus, for example, a four-dimensional constellation can be realized by transmitting the first two bits as a symbol from a two-dimensional constellation and the last two bits as the next symbol.

6.12   Summary

The summary for this chapter takes the form of a comparison of digital modulation techniques. The choice of digital modulation technique is influenced by error performance, spectral characteristics, implementation complexity, and other factors peculiar to the specific application (such as digital radio or telephone channel modem). Earlier we observed that binary modulation schemes provide good error performance and are simple to implement, but they lack the bandwidth efficiency required for most practical applications. Hence our emphasis here is on M-ary modulation schemes that provide the necessary bandwidth efficiency.

The symbol probability of error performance for several PSK, QAM, and QPR systems is compared in Figure 6.41. Note that with increasing M, the QAM systems hold an advantage over PSK. For example, 16-QAM requires about 4 dB less power than 16-PSK systems. Moreover, QPR systems are in general more power efficient than PSK systems for equivalent bandwidth efficiency.

Spectral efficiency among candidate modulation techniques is compared in Figure 6.42 as a function of the S/N required to provide $P(e) = 10^{-6}$. Points on this graph are labeled with the number of signal states present in the signal constellation. Two sets of curves are given for both PSK and QAM, corresponding to $\alpha = 0$ and $\alpha = 1$, where α is the fractional bandwidth required in excess of the Nyquist bandwidth. Of course, the penalty in the bandwidth efficiency of PSK or QAM systems for a choice of $\alpha = 1$ is a factor of 2 as compared to $\alpha = 0$, the minimum Nyquist bandwidth. As an example of the use of Figure 6.42, consider a requirement for 2 bps/Hz. Here QPR appears to be the best choice, when one trades off the required S/N against equipment complexity, considering that systems with $\alpha = 0$ are impractical.

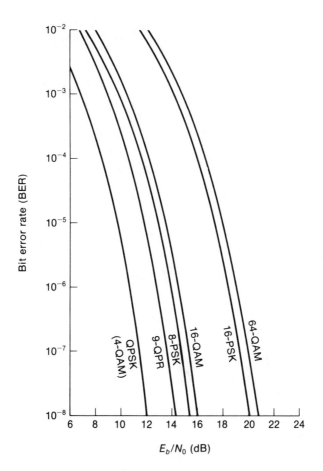

FIGURE 6.41   Bit Error Rate Performance for *M*-ary PSK, QAM, and QPR Coherent Systems

Trellis coded modulation combines convolutional coding with digital modulation to provide coding gain over uncoded modulation. Table 6.3 provides a summary of the coding gain achievable in trellis coded modulation. Each entry corresponds to a particular choice of the number of states 2^{K-1}, the number of coded bits (k), the number of information bits (m), and the modulation techniques. The code rate is given by $k/(k + 1)$ for all cases. Coding gains on the order of 3 to 6 dB are readily obtainable, but additional gains require considerably more complexity in the coder and decoder and are furthermore limited by channel capacity bounds.

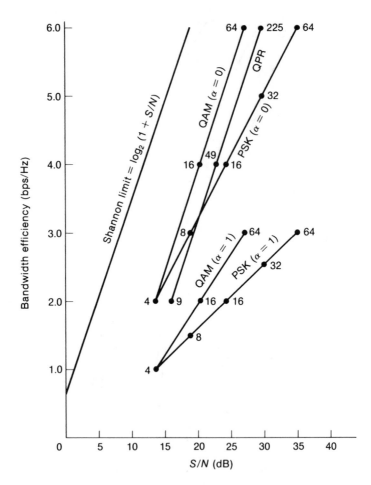

FIGURE 6.42   Bandwidth Efficiency for *M*-ary PSK, QAM, and QPR Coherent Systems on the Basis of *S/N* Required for $P(e) = 10^{-6}$

The actual bandwidth required in digital signal transmission is a function of not only the modulation technique but also the definition of bandwidth. Table 6.4 lists the required bandwidths corresponding to three different definitions for several modulation techniques. The bandwidths are in units of *R,* the bit rate. The high values of 99 percent bandwidth for PSK and QAM systems are due to a slow rate of spectral rolloff. Conversely, the low value of 99 percent bandwidth for MSK is due to a greater rate of spectral rolloff, which is proportional to f^{-4} compared to the f^{-2} characteristic for PSK.

TABLE 6.3   Coding Gains for Trellis Coded Modulation [23]

Number of States	k	Asymptotic Coding Gain (dB)			
		$G_{8\text{-PSK/4-PSK}}$ $m = 2$	$G_{16\text{-QAM/8-PSK}}$ $m = 3$	$G_{32\text{-QAM/16-QAM}}$ $m = 4$	$G_{64\text{-QAM/32-QAM}}$ $m = 5$
4	1	3.01	4.36	3.01	2.80
8	2	3.60	5.33	3.98	3.77
16	2	4.13	6.12	4.77	4.56
32	2	4.59	6.12	4.77	4.56
64	2	5.01	6.79	5.44	5.23
128	2	5.17	7.37	6.02	5.81
256	2	5.75	7.37	6.02	5.81

TABLE 6.4   Bandwidths for Digital Modulation in Units of Bit Rate R

Modulation Technique	Bandwidths		
	Null-to-Null	3 dB	99%
2-PSK	2	0.88	20.56
4-PSK (QAM)	1	0.44	10.28
8-PSK	2/3	0.29	6.85
16-PSK (16-QAM)	1/2	0.22	5.14
MSK	1.5	0.59	1.18

Other factors that may influence the choice of modulation technique include channel nonlinearities, radio fading, and interference from adjacent radio channels. Discussion of these topics is deferred to the treatment of digital radio system design in Chapter 9.

References

1. J. M. Wozencraft and I. M. Jacobs, *Principles of Communication Engineering* (New York: Wiley, 1967).
2. A. B. Carlson, *Communication Systems: An Introduction to Signals and Noise in Electrical Communication* (New York: McGraw-Hill, 1975).
3. R. W. Lucky, J. Salz, and E. J. Weldon, *Principles of Data Communication* (New York: McGraw-Hill, 1968).
4. J. R. Carson, "Notes on the Theory of Modulation," *Proc. IRE* 10 (February 1922):57–64.
5. M. Schwartz, W. R. Bennett, and S. Stein, *Communication Systems and Techniques* (New York: McGraw-Hill, 1966).
6. J. J. Spliker, *Digital Communications by Satellite* (Englewood Cliffs, NJ: Prentice Hall, 1977).
7. R. E. Ziemer and R. L. Peterson, *Digital Communications and Spread Spectrum ~   - (New York: Macmillan, 1985).

428 Digital Modulation Techniques

8. W. C. Lindsey and M. K. Simon, *Telecommunication Systems Engineering* (Englewood Cliffs, NJ: Prentice Hall, 1973).

9. E. Arthurs and H. Dym, "On the Optimum Detection of Digital Signals in the Presence of White Gaussian Noise—A Geometric Interpretation and a Study of Three Basic Data Transmission Systems," *IRE Trans. Comm. Systems,* vol. CS-10, no. 4, December 1962, pp. 336–372.

10. H. Kawahara and T. Murase, "First Implementation of an Operating 256 QAM 400 Mb/s Microwave Radio System," *GLOBECOM 1990,* pp. 504.1.1–504.1.6.

11. G. D. Richman and P. C. Smith, "Transmission of Synchronous Digital Hierarchy by Radio," *1990 International Communications Conference,* pp. 208.1.x–208.1.y.

12. S. A. Gronemeyer and A. L. McBride, "MSK and Offset QPSK Modulation," *IEEE Trans. on Comm.,* vol. COM-24, no. 8, August 1976, pp. 809–820.

13. S. Pasupathy, "Minimum Shift Keying: A Spectrally Efficient Modulation," *IEEE Comm. Mag.,* July 1979, pp. 14–22.

14. C. W. Anderson and S. G. Barber, "Modulation Considerations for a 91 Mbit/s Digital Radio," *IEEE Trans. on Comm.,* vol. COM-26, no. 5, May 1978, pp. 523–528.

15. I. Godier, "DRS-8: A Digital Radio for Long-Haul Transmission," *1977 International Conference on Communications,* June 1977, pp. 5.4-102–5.4-105.

16. C. M. Thomas, J. E. Alexander, and E. W. Rahneberg, "A New Generation of Digital Microwave Radios for U.S. Military Telephone Networks," *IEEE Trans. on Comm.,* vol. COM-27, no. 12, December 1979, pp. 1916–1928.

17. G. Ungerboeck, "Channel Coding with Multilevel/Phase Signals," *IEEE Trans. on Information Theory,* vol. IT-28, no. 1, January 1982, pp. 55–67.

18. G. Ungerboeck, "Trellis-Coded Modulation with Redundant Signal Sets Part I: Introduction," *IEEE Comm. Mag.,* February 1987, pp. 5–11.

19. G. D. Forney, "The Viterbi Algorithm," *Proc. IEEE* 61 (March 1973):268–278.

20. J. A. Heller and I. M. Jacobs, "Viterbi Decoding for Satellite and Space Communication," *IEEE Trans. on Comm. Tech.,* vol. COM-19, no. 5, October 1971, pp. 835–848.

21. F. Hemmati and D. J. Costello, "Truncation Error Probability in Viterbi Decoding," *IEEE Trans. on Comm.,* vol. COM-25, no. 5, May 1977, pp. 530–532.

22. G. D. Forney and others, "Efficient Modulation for Band-Limited Channels," *IEEE Journal on Sel. Areas in Comm.,* vol. SAC-2, no. 5, September 1984, pp. 632–647.

23. G. Ungerboeck, "Trellis-Coded Modulation with Redundant Signal Sets Part II: State of the Art," *IEEE Comm. Mag.,* February 1987, pp. 12–21.

24. L. F. Wei, "Rotationally Invariant Convolutional Channel Coding with Expanded Signal Space—Part II: Nonlinear Codes," *IEEE Journal on Sel. Areas in Comm.,* (September 1984):672–686.

25. CCITT Recommendation V.32, "A Family of 2-Wire, Duplex Modems Operating at Data Signalling Rates of up to 9600 bit/s for Use on the General Switched Telephone Network and on Leased Telephone-Type Circuits," Blue Book, vol. VIII.1, *Data Communication over the Telephone Network* (Melbourne: ITU, 1988).

Problems

6.1 In a jamming environment, the jamming power is generally much greater than the thermal noise power. Consequently, the figure of merit is taken to be E_b/J_0, where J_0 is the power spectral density of the jamming signal. Derive an expression for E_b/J_0 as a function of the processing gain G_p and $(J/S)_{reqd}$ defined in Chapter 5.

6.2 Calculate the E_b/N_0 required to yield a probability of error of 1×10^{-6} for: (a) coherently detected binary ASK; and (b) noncoherently detected binary ASK.

6.3 A coherent binary ASK system produces an E_b/N_0 of 13 dB, in which $N_0 = 1 \times 10^{-5}$ W/Hz. Find the transmitter peak amplitude for this system.

6.4 Consider a coherent ASK binary system in which the probabilities of 1 and 0 are given by P_1 and P_0, the transmitter amplitudes corresponding to binary 1 and 0 are given by V_1 and V_0, and the decision threshold is given by V_T. Using these system characteristics, find an expression for probability of error in terms of the complementary error function.

6.5 For an E_b/N_0 of 13 dB, calculate the probability of error for the following systems: (a) coherently detected binary FSK; and (b) noncoherently detected binary FSK.

6.6 Repeat Problem 6.3 but for a coherent FSK binary system.

6.7 Consider two orthogonal sinusoids used for binary FSK transmission, in which the peak voltage is 1.0 volts and the bit rate is 1200 bps. Calculate the probability of error if $N_0 = 6.7 \times 10^{-5}$ W/Hz and coherent detection is used.

6.8 Calculate the probability of error for a BPSK system that has a pulse duration of 0.0002 sec and a peak amplitude of 0.5 volts. Assume coherent detection and that $N_0 = 4 \times 10^{-6}$ W/Hz. Repeat the calculation for a differentially coherent detection scheme.

6.9 Consider a raised-cosine baseband signal with $\alpha = 0.2$ that is to transmit 1.544 Mb/s, use binary modulation with coherent detection, and provide a BER = 1×10^{-6}. Find the E_b/N_0 in dB and the transmission bandwidth for: (a) SSB ASK; (b) PSK; and (c) narrowband FSK.

6.10 Derive Equation (6.42).

6.11 Derive Equation (6.41) from Equations (6.39) and (6.40).

6.12 Consider a 4-FSK system with bit rate 2400 b/s and peak amplitude of 1.0 volts. Assuming equally likely signals, orthogonal waveforms, coherent detection, and $N_0 = 4 \times 10^{-8}$, calculate the probability of error.

6.13 Consider a 16-PSK digital radio that transmits 90 Mb/s, occupies 20 MHz of bandwidth, and uses coherent detection. The transmit signal power is 4 watts and the net system loss is 112 dB. For transmission over an additive white gaussian channel with one-sided power spectral density equal to 4×10^{-21} W/Hz, compute the BER for this radio system. Assume the use of Gray coding.

6.14 By comparing the $P(e)$ expressions for coherent versus differentially coherent detection of M-ary PSK, find an expression for the increased power requirements of differentially coherent detection. What is this increased power requirement for $M = 4$?

6.15 A given coherent PSK system is designed to transmit data at 45 Mb/s in the presence of additive white gaussian noise with one-sided power spectral density $N_0 = 1.67 \times 10^{-20}$ W/Hz. Assume Gray coding, a transmitter average power of 1 watt, and an expected net system loss of 110 dB. For this system, compute the bit error rate (BER) with a choice of: (a) BPSK; (b) 4-PSK; and (c) 8-PSK.

6.16 Repeat Problem 6.15 for differentially coherent PSK.

6.17 For 8-PSK:
   (a) Sketch the decision regions and boundary regions used in deriving the upper bound on probability of error given by Expression (6.59).
   (b) Suggest a reasonable expression for a lower bound on probability of error for 8-PSK.

6.18 For the following modulation schemes, show a trellis to represent the allowed states and transitions between those states. Using the trellis, sketch the path corresponding to the binary input sequence {00111010011}: (a) coherent 4-PSK; (b) differentially coherent 4-PSK; and (c) OQPSK.

6.19 Consider a coherently detected QPSK signal that is received with a phase error of ϕ degrees. (a) Derive an expression for probability of error. (b) What increase in S/N is required to compensate for this phase error ϕ?

6.20 An M-ary PSK system with Gray coding and a fixed data rate is to be designed such that the bit error probability remains the same for $M = 4, 8$, and 16 in the presence of the same additive white gaussian noise.

(a) Draw the resulting signal constellations for $M = 4$, 8, and 16 using a common scale.

(b) What change in energy per bit is required as M is increased?

6.21 A digital radio system is designed to transmit 135 Mb/s in the presence of additive white gaussian noise with a one-sided power spectral density of 6×10^{-20} W/Hz. Assume use of 16-QAM, Gray coding, a transmitter power of 10 watts, total system losses of 110 dB, and coherent detection.

(a) Calculate the BER.

(b) Now assume a transmitted signal bandwidth of 45 MHz, for a spectral efficiency of 3.0 bps/Hz. For a choice of raised-cosine filtering, what rolloff factor α is required to provide this spectral efficiency?

(c) Repeat part (a) for a choice of 16-PSK.

6.22 For a 64-QAM modulation scheme:

(a) Sketch the signal constellation with a spacing of $2d$ between closest symbols. For any four contiguous points, indicate a possible choice of bit values assigned to these points, using Gray coding.

(b) Find the average signal power as a function of d.

(c) Find the probability of symbol error as a function of d.

(d) Express the probability of error in part (c) in terms of E_b/N_0 and compute the approximate probability of error when the $E_b/N_0 = 10$ dB.

(e) Find the spectral efficiency and the transmission bandwidth for 64-QAM with raised cosine ($\alpha = \frac{1}{2}$) and data rate = 90 Mbps.

6.23 A convenient method of comparing signal-to-noise performance between PSK and QAM is through the differences in the Euclidean distance, which is defined as the minimum distance between points in the signal constellation.

(a) Calculate the Euclidean distance as a function of the peak amplitude A for each of the two modulation techniques. Which is more immune to noise and why? Which is more immune to phase jitter and why? Which is more immune to amplitude nonlinearities and why?

(b) Derive a general expression for the Euclidean distance for M-ary PSK and one for M-ary QAM, for peak amplitude A.

(c) What is the difference in signal-to-noise (in dB) between 16-QAM and 16-PSK for the same peak power?

6.24 For the signal constellation of 16-QAM:

(a) Determine the I and Q signal values to generate each of the 16 signals, assuming that the four allowed signal levels of the I or Q component are +3, +1, –1, and –3.

(b) Using Gray coding, assign binary values to the 4-bit words represented by the 16 signal points.

6.25 Consider the signal constellations for 4-QAM and 16-QAM.

(a) Find the exact probability of symbol error for both constellations in terms of the signal-to-noise ratio, defined as

$$SNR = E[a_k^2]/2\sigma^2$$

From the exact probability of error expressions, give expressions for the approximate probability of error.

(b) Determine the difference in SNR required to maintain the same probability of error for 4-QAM versus 16-QAM when the average signal power is the same.

(c) Repeat part (b) for the case where the peak signal power is the same.

6.26 Consider a square QAM signal constellation which by definition has 4^n points, $n = 1,2,3,\ldots$. Find a general expression that relates average signal energy (E_s) to the minimum separation between signals $(2d)$. Note that Equation (6.69) is the special case of $n = 2$.

6.27 Refer to the 32-point cross constellation of Figure 6.40b. (a) Find the probability of symbol error in terms of minimum separation $2d$ and N_0. (b) Find the average and peak signal power. (c) Express part (a) in terms of E_b/N_0. (d) Compare parts (a) and (b) to values for 32-PSK.

6.28 The dynamic range of a signal constellation is defined as the ratio of the maximum to minimum power found in the set of constellation points. Find the dynamic range in dB for: (a) 64-QAM; (b) 256-QAM; and (c) 4^n-QAM, $n = 1,2,3,\ldots$.

6.29 A QAM system is created with an I signal of ± 2 volts and a Q signal of ± 4 volts. (a) Sketch the signal constellation. (b) Determine the average signal power. (c) Determine the phases of the signal points. (d) Show a PSK constellation with the same average signal power.

6.30 A performance comparison of M-ary PSK versus M-ary QAM can be done by determining the relative increase in average signal power required to maintain the same error probability as M increases.

(a) Find an expression for the relative increase in average signal power required for PSK as M increases.

(b) Find an expression for the relative increase in average signal power required for QAM as M increases.

(c) Show that the ratio of the expression found in part (a) to that found in part (b) can be expressed as

$$\frac{3M^2}{2(M-1)\pi^2}$$

and determine the ratio in dB for $M = 8$, 16, and 64.

6.31 One approach to deriving the probability of error for QAM systems is to consider QAM as two independent multilevel baseband signals multiplied by two quadrature carriers. For QAM with square signal constellations, the probability of correct detection is

$$P(c) = \left[1 - P_{BB}(e)\right]^2$$

$P_{BB}(e)$ is the probability of error for a multilevel baseband system as given by Equation (5.48), but with M replaced with L, the number of levels on each of the two inputs to the QAM system. Using this approach, show that the probability of symbol error for QAM with square constellation is upper bounded by

$$P(e) = 2\ \text{erfc}\left[\sqrt{\frac{3}{L^2-1}\frac{S}{N}}\right]$$

6.32 An MSK modulation system is to be designed for a symbol interval of $T_b = 0.25$ µsec. What is the minimum required separation in MHz?

6.33 Starting with Equation (6.76), show that the frequency separation given by Equation (6.78) is the minimum separation that will provide orthogonality.

6.34 Using the same sequence as shown in Figure 6.22, sketch the phase for QPSK, OQPSK, and MSK.

6.35 For both 9-state QPR and 49-state QPR, calculate (a) the ratio of peak to average signal power; and (b) the probability distribution of the states.

6.36 Find the bandwidth for a DS-3 (45 Mb/s) signal using the following modulation techniques, assuming the bandwidth is defined as null-to-null: (a) BPSK; (b) QPSK; (c) 8-PSK; (d) 16-QAM; and (e) MSK.

6.37 For the following modulation systems, the symbol error probability is $P(e) = 1 \times 10^{-6}$. Assuming the use of Gray coding of the transmitted symbols and the use of coherent detection, what is the approximate bit error probability? (a) 16-QAM; (b) 16-PSK; and (c) 16-FSK.

6.38 For the following rate ½, $K = 7$ convolutional encoder, find the output sequence corresponding to an input of {10000000}.

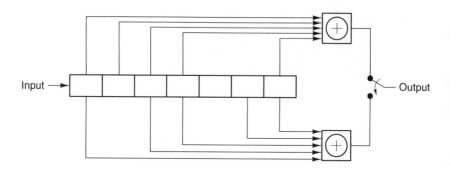

6.39 Consider a convolutional coder with rate k/n and constraint length K. For a tree diagram of the encoder: (a) How many branches emanate from each node? (b) How many states are there?

6.40 Consider the rate ⅔ convolutional coder shown in the figure below

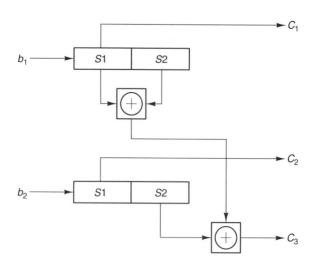

(a) Draw the tree diagram, trellis diagram, and state diagram with each transition labeled.

(b) For the all-zero path, find the Hamming distance of the minimum distance error event.

6.41 Repeat Problem 6.40, parts (a) and (b), for the rate ⅓ convolutional coder shown below.

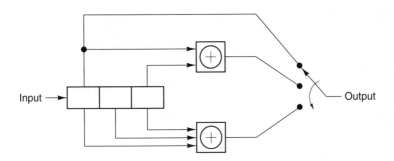

6.42 Draw the convolutional coder block diagram for the following cases:
   (a) Rate ½, $K = 5$, $(g_1,g_2) = 1 \oplus D^3 \oplus D^4, 1 \oplus D \oplus D^2 \oplus D^4$
   (b) Rate ⅔, $K = 4$, $(g_1,g_2,g_3) = 1 \oplus D^3 \oplus D^4 \oplus D^5 \oplus D^6, 1 \oplus D \oplus D^3 \oplus D^4$ $\oplus D^6, 1 \oplus D \oplus D^3 \oplus D^4 \oplus D^5 \oplus D^6 \oplus D^7$
   (c) Rate ⅓, $K = 4$, $(g_1,g_2,g_3) = 1 \oplus D \oplus D^3, 1 \oplus D \oplus D^2 \oplus D^3, 1 \oplus D$

6.43 The generators for convolutional coders are often expressed in binary or octal form. For example, the generators of Figure 6.28 may be expressed in binary form as $g_1 = [101]$ and $g_2 = [111]$ and in octal form as $g_1,g_2 = (5,7)$. Express the convolutional coders of Problem 6.42 in both binary and octal form.

6.44 Consider the rate ⅓, $K = 3$ convolutional coder shown in the figure below:
   (a) Draw the tree diagram, trellis diagram, and state diagram.

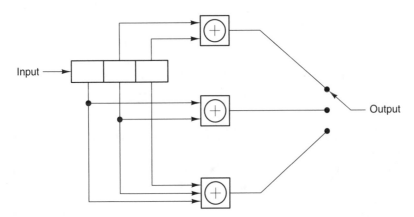

(b) What is the impulse response?

(c) Determine the output sequence when the input sequence is {111}.

6.45 Assume the use of the convolutional coder shown in Figure 6.28. At the receiver, the sequence {1110011011} is observed followed by a tail of all 0's.

(a) Find the maximum likelihood path through the trellis diagram and determine the first four decoded information bits.

(b) Identify any received (coded) bits that are in error.

6.46 Assume that the trellis coded modulator shown in Figure 6.34 is used with 8-PAM rather than 8-PSK. Let the eight signal levels be represented by $\pm1, \pm3, \pm5,$ and ±7.

(a) Using set partitioning rules, show the assignment of each three-bit code $c_1c_2c_3$ to a PAM symbol.

(b) Show an uncoded 4-PAM signal set which has the same average power as the coded 8-PAM signal set.

(c) By use of a trellis, sketch the minimum distance error event for the coder.

(d) Find the coding gain in dB of coded 8-PAM over uncoded 4-PAM.

6.47 Consider the trellis coded modulation used in CCITT Rec. V.32

(a) Using set partitioning, assign 5-bit codes to symbols within the 32-ary constellation.

(b) Draw the trellis diagram showing all states and transitions.

(c) Find the coding gain of coded 32-cross modulation over uncoded 16-QAM.

6.48 Consider the trellis coded modulation shown in the following figure.

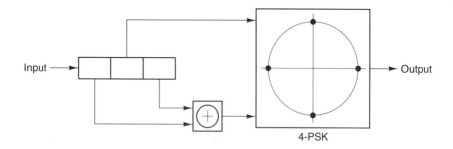

4-PSK

(a) What is the minimum Hamming distance of the convolutional coder alone?

(b) Make an assignment of code bits c_1c_2 to 4-PSK symbols, and draw the state diagram showing the code bits for each transition.

(c) Find the minimum distance error event assuming the correct trajectory is all 0's.

6.49 Consider the trellis coded system shown in the following figure.

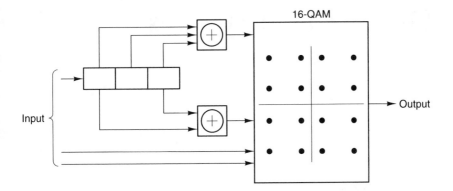

(a) Using set partitioning, design a mapping of each 4-bit code to a symbol in a 16-QAM constellation.

(b) Show the distance between adjacent signal points at each stage of partitioning.

6.50 Consider a trellis coder with the following block diagram:

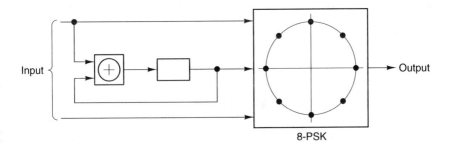

(a) Using set partitioning, design the mapping of code bits to symbols in the 8-PSK constellation.

(b) Draw the trellis diagram showing all states and transitions.

(c) Determine the coding gain of coded 8-PSK over uncoded 4-PSK.

7

Digital Cable Systems

OBJECTIVES

- Provides the theory and practices necessary for designing digital cable systems
- Describes twisted-pair and coaxial cable, including physical and propagation characteristics
- Explains the design and operation of regenerative repeaters in detail
- Assesses the performance of tandemed repeaters in terms of bit error rate, jitter, and crosstalk
- Presents methods for calculating repeater spacing for each cable type
- Discusses additional implementation factors including power feeding, fault location, automatic protection switching, and orderwires

7.1 Introduction

Digital transmission over cable employs the basic elements shown in Figure 7.1. At each end office the digital signal to be transmitted is converted to a form suitable for cable transmission using the baseband coding techniques described in Chapter 5. In the case of metallic cable systems, the digital signal can then be directly interfaced with the cable, whereas for optical cable systems the electrical signal must first be converted into light pulses by an optical transmitter. After propagation along the cable, the received signal is an attenuated and distorted version of the transmitted signal. Digital repeaters are stationed at regular intervals along the transmission path to provide reshaping, retiming, and regeneration of the received signal. In the case of optical cable systems a repeater includes an optical receiver and

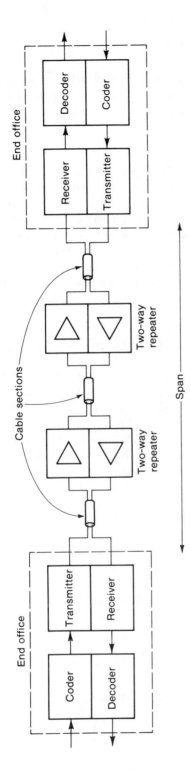

FIGURE 7.1 Digital Transmission Over Cable

439

transmitter. The receiver in each end office is identical to a one-way repeater, although for optical cable transmission the receiver must first provide optical-to-electrical signal conversion.

7.2   Cable Characteristics

Two cable types are considered here: twisted-pair and coaxial. Historically, twisted-pair was the first of these cable types to be used for digital transmission, dating back to the early 1960s with the first T-carrier systems. Limitations in bandwidths, repeater spacing, and crosstalk performance, however, have led to use of coaxial and more recently to optical cable systems. When compared to metallic cable systems, optical cable systems provide greater bandwidths, lower attenuation, freedom from crosstalk and electrical interference, and potentially lower costs. Optical cable systems are considered in Chapter 8.

7.2.1   Twisted-Pair Cable

Twisted-pair cable consists of two insulated conductors twisted together. The standard configuration uses copper conductors and plastic (such as polyethylene) or wood pulp insulation. For multipair cable assemblies, the individual cable pairs are grouped in units or concentric layers to form the core of the cable. An outer sheathing, of lead or plastic, is applied over the core for protection. Adjacent pairs are twisted at a different pitch to minimize interference (crosstalk) between pairs in multipair cable.

The four primary electrical characteristics of twisted-pair cable are its series resistance R, series inductance L, shunt capacitance C, and shunt conductance G per unit length of cable. Transmission characteristics, such as characteristic impedance Z and propagation constant γ, can be calculated from these primary electrical characteristics. The impedance is given by

$$Z = \sqrt{\frac{R + j\omega L}{G + j\omega C}} \tag{7.1}$$

For polyethylene cable, the conductance G is small, so that at low frequencies, where $\omega \ll R/L$,

$$Z \approx \sqrt{\frac{R}{j\omega C}} \tag{7.2}$$

while at high frequencies, where $\omega \gg R/L$,

$$Z \approx \sqrt{\frac{L}{C}} \tag{7.3}$$

The propagation constant γ describes the attenuation α and phase shift β as

$$\gamma = \alpha + j\beta = \sqrt{(R + j\omega L)(G + j\omega C)} \qquad (7.4)$$

At low frequencies, the phase and attenuation characteristics are proportional to ω, as seen by

$$\alpha = \beta \approx \sqrt{\frac{\omega RC}{2}} \qquad (7.5)$$

For high frequencies, the attenuation becomes a constant and is given by

$$\alpha = \frac{R}{2}\sqrt{\frac{C}{L}} \qquad (7.6)$$

while the phase has a linear dependence on frequency. Other factors that influence the attenuation characteristic are the skin effect, where attenuation exhibits a $\sqrt{\omega}$ dependence for very high frequencies, and temperature change, where an additional factor proportional to $\sqrt{R}$ appears in the expression for attenuation. Typical attenuation values for polyethylene-insulated cable (PIC) of various sizes are plotted in Figure 7.2 for typical carrier frequencies.

Twisted-pair cable has a long history as a major form of transmission for voice frequency (VF) and FDM signals. Twisted-pair cable has also been used for digital transmission, beginning in the early 1960s with T1 carrier in the Bell System. Engineering design for T carrier was based on the capabilities of standard VF cable [1]. Using standard 22-gauge cable, T1 repeaters were designed for 6,000-ft spacings, which coincided with the distance between loading coils (inductors) used in VF transmission to improve the frequency response. Because loading coils introduce a nonlinear phase response, they can cause intersymbol interference and must therefore be removed for T1 service. With the growing use of T carrier in the United States, cables have been designed specifically for T1 and T2 carrier. The use of low-capacitance PIC cable reduces loss at high frequencies, allowing greater distance between repeaters. To reduce crosstalk when both directions of transmission are in one sheath, *screened* cable has been introduced in which an insulated metallic screen separates the two directions of transmission within the cable core. In North America, T carrier systems operate over twisted-pair cable at rates of 1.544 Mb/s (T1), 3.152 Mb/s (T1C), and 6.312 Mb/s (T2). Similar applications exist in Europe and elsewhere at standard rates of 2.048 and 8.448 Mb/s.

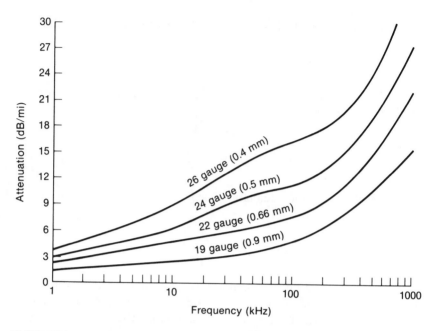

FIGURE 7.2   Typical Attenuation Characteristic for PIC Cable at 68°F

There are other short-haul digital transmission applications of twisted-pair cable worth mentioning as well. For two-wire subscriber loops, in-place twisted-pair will support the transmission of two 64-kb/s B-channels plus one 16-kb/s D-channel in a composite 144-kb/s channel. For ISDN applications in North America, the standard transmission method calls for a 2B1Q (two binary, one quaternary) line code over twisted-pairs for distances up to several kilometers [2]. The 2B1Q code is formed by grouping two bits into a pair that is then mapped into one of four PAM levels, or *quats*. Echo cancelers then provide full duplex operation over two-wire loops. Twisted-pair cable is also used with local area network (LAN) transmission for rates up to 100 Mb/s, including applications with Ethernet, token ring, and FDDI.

At transmission rates above 10 Mb/s, the use of twisted-pair cable for long-haul transmission becomes undesirable because of higher loss and greater crosstalk. As the next section explains, coaxial cable is preferred at such transmission rates.

7.2.2   Coaxial Cable

A **coaxial cable** consists of an inner conductor surrounded by a concentric outer conductor. The two conductors are held in place and insulated by a dielectric material, usually polyethylene disks that fit over the inner conductor and are spaced at approximately 1-in. intervals. The inner conductor is solid copper; the outer conductor consists of a thin copper tape covered by steel tape. A final covering of paper tape insulates coaxial pairs from each other. The standard cable size for long-haul trans-mission has an inner conductor diameter of 0.104 in. (2.6 mm) and an outer conductor diameter of 0.375 in. (9.5 mm), usually expressed as 2.6/9.5-mm cable. Other sizes commonly used are 1.2/4.4-mm and 0.7/2.9-mm coaxial cable.

Coaxial cables have low attenuation at carrier frequencies, where loss is directly proportional to resistance R and is relatively independent of the other primary characteristics. Attenuation due to resistive loss is found to be proportional to the square root of frequency and inversely proportional to the cable diameter, as seen in Figure 7.3, which is based on data extracted from CCITT recommendations [2]. The attenuation can be expressed as a function of frequency (in megahertz) by

$$\alpha = a + b\sqrt{f} + cf \quad \text{dB/km} \tag{7.7}$$

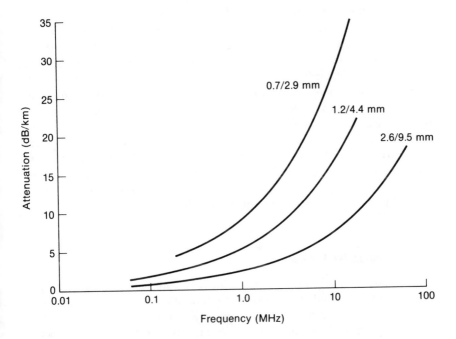

FIGURE 7.3   Attenuation Characteristic for Standard Coaxial Cable at 10°C

where the set of constants (a, b, c) are peculiar to a particular size of cable. With 1.2/4.4-mm coaxial cable, for example,

$$a = 0.07$$
$$b = 5.15$$
$$c = 0.005$$

The characteristic impedance of these cables is also a function of frequency and conductor diameter ratio and can be expressed as

$$Z = Z_\infty + \frac{A}{\sqrt{f}}(1 - j) \tag{7.8}$$

where Z_∞ is the impedance at infinite frequency and f is given in megahertz. The constant A is peculiar to each size of cable. For the cable sizes mentioned here, the characteristic impedance varies by only a few ohms over the frequency range of interest. A typical value for the real part of the impedance is 75 ohms at 1 MHz for 0.7/2.9-mm or 1.2/4.4-mm cable.

Coaxial cable is a standard transmission medium for many FDM and analog video systems. A good example of coaxial cable application is the AT&T L carrier, which provides capacities up to 10,800 voice channels (Chapter 12). Digital coaxial cable systems are not as widespread or standardized as analog coaxial cable. Nevertheless, there are digital coaxial cable systems in use today, particularly at standard rates of 8.448 Mb/s and above. The CCITT has identified the characteristics of digital line systems on coaxial cable for rates up to 564.992 Mb/s [3]. Most of the digital cable systems currently in use take advantage of the characteristics of existing analog cable plant and repeater spacing. For example, North American 274-Mb/s systems [4], European 565-Mb/s systems [5], and Japanese 400- and 800-Mb/s systems [6, 7] all are compatible with 60-MHz FDM systems, which operate on 2.6/9.5-mm coaxial cable with a nominal repeater spacing of 1.5 km. Table 7.1 lists the standard data rates and cable types used with twisted-pair and coaxial cable.

7.3   Regenerative Repeaters

The purpose of a regenerative repeater is to construct an accurate reproduction of the original digital waveform. Its application is described here for cable systems, but regenerative repeaters are common to all digital transmission systems. The functions performed in a digital repeater can be grouped into three essential parts—reshaping, retiming, and regeneration, as illustrated in Figure 7.4 and described in the following paragraphs.

TABLE 7.1   Transmission Rates and Cable Types for Digital Transmission Using Twisted-Pair and Coaxial Cable

Transmission Rate (Mb/s)	Application	Cable Type
1.544	North America (T1) and Japan	VF cable with paper or polyethylene insulation
2.048	Europe	
3.152	North America (T1C)	
6.312	North America (T2) and Japan	Low-capacitance PIC
8.448	Europe	0.7/2.9-mm coaxial cable
34.368	Europe	
8.448	Europe	1.2/4.4-mm or 2.6/9.5-mm coaxial cable
34.368	Europe	
139.264	Europe	
34.368	Europe	2.6/9.5-mm coaxial cable
44.736	North America	
97.728	Japan (PCM-100M)	
139.264	Europe	
274.176	North America (T4)	
400.352	Japan (PCM-400M)	
564.992	Europe	
800	Japan	

After transmission over a section of cable, each received digital pulse is flattened and spread over several symbol times due to the effects of attenuation and dispersion. Before these pulses can be sampled for detection and regeneration, reshaping is necessary; it is accomplished by equalization and amplification. The equalizer compensates for the frequency-dependent nature of attenuation in cable. Ideally the equalizer provides a characteristic that is inverse to the cable characteristic, so that the overall response is independent of frequency. In practice, however, perfect equalization is neither possible nor necessary. Equalized pulses still tend to appear "rounded off," but this shape is adequate for proper regeneration. The equalizer and amplifier combination is designed for a standard length of cable. Variations in cable length are compensated by a network that builds out the cable to the equivalent of a standard section of cable. This line build-out feature is automatic (ALBO) and typically covers a wide range of line loss. The ALBO network is controlled by a peak detector, whose output controls a variable resistance in the ALBO.

Retiming in a regenerative repeater is accomplished by deriving timing from the data. This self-timing method proves to be acceptable if the data maintain a sufficient density of transitions, or zero crossings. A discussion of suitable transmission codes and scramblers for proper clock recovery is given in Chapter 5. The alternatives to this retiming scheme include use of highly accurate oscillators at each repeater or separate transmission of a

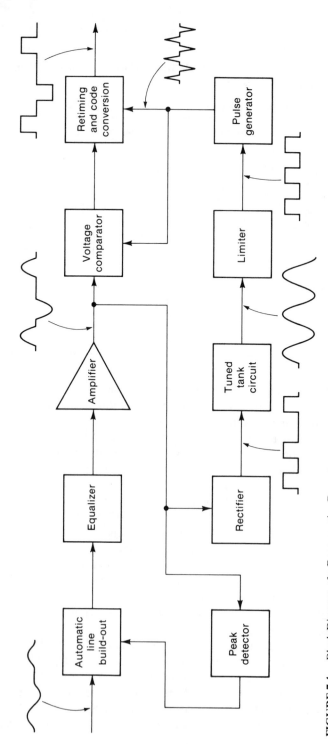

FIGURE 7.4   Block Diagram of a Regenerative Repeater

clock signal, but such schemes are costly and therefore seldom used. Typically the clock recovery circuit consists of a resonant tank circuit tuned to the timing frequency (Figure 7.4). This circuit is preceded by a rectifier that converts the reshaped pulses into unipolar marks. Each mark stimulates the tank circuit, which produces a sine wave at the desired timing frequency. This sine-wave signal decays exponentially until another mark restimulates the tank circuit. So long as these marks arrive often enough, the tank circuit can be forced to reproduce the correct frequency of the incoming data signal. The sinusoidal timing signal is then limited to produce a constant-amplitude square wave. A pulse generator yields positive and negative timing spikes at the zero crossings of the square wave.

The regenerator performs voltage comparison, clock sampling, and any necessary code conversion. Using the timing spikes generated by the clock recovery circuit, the reshaped signal is sampled near the center of each symbol interval. The result is compared with a threshold by a decision circuit. The width of each regenerated pulse is controlled by the recovered clock. Finally, code conversion is applied as would be required, for example, with a bipolar format.

7.4   Clock Recovery and Jitter

In a regenerative repeater, a form of short-term phase modulation known as **jitter** is introduced during the clock recovery process. The sampled and regenerated data signal then has its pulses displaced from their intended position, as illustrated by the photograph in Figure 7.5 for a binary signal. This display in Figure 7.5 is generated by use of an oscilloscope in which many transitions of the data signal are superimposed to indicate the magnitude of jitter. Jitter leads to sampling offsets and potential data errors in a repeater and also causes distortion in the reconstructed analog waveform for PCM systems.

Sources of jitter may be classified as systematic or random depending on whether or not they are pattern-dependent. Since each repeater in a long repeater chain operates on the same data pattern, any jitter source related to the data pattern produces the same effect at each repeater and hence is systematic. Random sources of jitter are uncorrelated at each repeater and should not cause a significant buildup of jitter in a repeater chain. Thus accumulation of jitter in a long repeater chain is caused primarily by systematic sources.

7.4.1   Sources of Jitter

Major sources of jitter include intersymbol interference, threshold misalignment, and tank circuit mistunings. These sources are difficult to analyze individually, which complicates complete jitter analysis for a typical repeater.

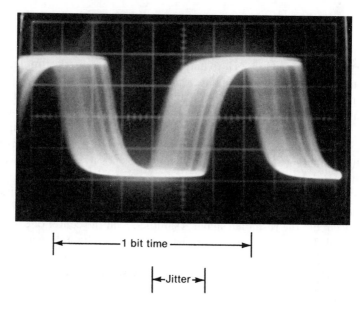

FIGURE 7.5   Photograph of Jitter for Binary Signal

Imperfect equalization during pulse reshaping causes a skewing of the pulses due to residual intersymbol interference. The effective center of each pulse varies from pulse to pulse depending on the surrounding data pattern. The resulting shifts in peak pulse position cause a corresponding phase shift in the timing circuit output.

In the repeater block diagram shown in Figure 7.4, the clock signal is generated at the zero crossings of the limiter output. If the threshold detector is offset from the ideal zero crossings, the timing spikes used for retiming will incur a phase shift. Variation in the timing signal amplitude leads to corresponding variation in the triggering time.

The frequency of the tank circuit, f_T, is tuned to the expected line frequency, f_0. Mistuning results from initial frequency offset plus changes in f_r and f_0 with time. Both static and dynamic phase shifts in the timing signal occur from mistuning. Static phase shifts, however, can be compensated by phase adjustment in the retiming portion of the regenerator. Dynamic phase shifts are caused by variation in the transition density of the data signal. The presence of a strong spectral component at f_0 causes the tank circuit to converge on f_0, while a weak spectral component allows the tank circuit to drift toward its natural frequency, f_T.

These major jitter components are all to some extent pattern-dependent and thus accumulate in a systematic fashion. Other random components of jitter, due for example to noise and crosstalk, also exist. These components are uncorrelated at successive repeaters and do not accumulate in a systematic manner.

7.4.2   Timing Jitter Accumulation

A model for the analysis of timing jitter accumulation in a chain of repeaters has been developed by Byrne and colleagues [8] and provides the basis of our discussion (see Figure 7.6). To make the model tractable, several assumptions must be made:

1. The same jitter is injected at each repeater.
2. All significant jitter sources can be represented by an equivalent jitter at the input to each clock recovery circuit.
3. Jitter adds linearly from repeater to repeater.
4. Clock recovery is performed by a tank circuit tuned to the pulse repetition frequency.
5. Since the rate of change of jitter is small compared to the recovered clock rate, the tank circuit is equivalent to a low-pass filter with a single pole corresponding to the half-bandwidth of the tank circuit.

The input and output jitter amplitude spectra, $\theta_i(f)$ and $\theta_o(f)$, are related by the jitter transfer function at each repeater $Y(f)$:

$$\theta_o(f) = \frac{1}{1 + j(f/B)}\theta_i(f) \tag{7.9}$$

$$= Y(f)\theta_i(f)$$

where

$$B = \frac{f_0}{2Q} = \text{half-bandwidth of tank circuit}$$

f_0 = timing frequency of data

Q = quality factor of tank circuit

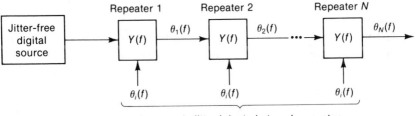

Systematic jitter injected at each repeater

FIGURE 7.6   Model of Jitter Accumulation Along a Chain of Repeaters

The result of (7.9) is that low-frequency jitter is passed by the tank circuit while higher-frequency jitter is attenuated and shifted in phase.

The jitter at the end of a chain of N repeaters is the sum of jitter introduced at each repeater and operated upon by all succeeding repeaters. Since the jitter introduced in each repeater is assumed to be identical, the accumulated jitter spectrum can be expressed as

$$\theta_N(f) = \theta_i(f)\left[Y(f) + Y^2(f) + \cdots + Y^N(f)\right] \tag{7.10}$$

The right-hand side of (7.10) is the sum of a geometric series that is given by

$$\theta_N(f) = \theta_i(f)\frac{B}{jf}\left[1 - \left(\frac{1}{1 + jf/B}\right)^N\right] \tag{7.11}$$

The power density of jitter at the Nth repeater output is obtained by squaring the magnitude of the transfer function:

$$S_N(f) = \left(\frac{B}{f}\right)^2 \left|1 - \left(\frac{1}{1 + jf/B}\right)^N\right|^2 S_i(f) \tag{7.12}$$

where $S_i(f)$ is the power density of the jitter injected in each repeater. The normalized jitter power spectrum, obtained from (7.12), has been plotted in Figure 7.7 for $N = 1, 10, 100$, and 1000 repeaters. At very low frequencies the normalized jitter power spectrum is proportional to the square of N. For higher frequencies the power gain falls off as the inverse square of frequency. As a result of this spectrum shape, the bulk of jitter gain for large N is at low frequency where jitter increases with an amplitude proportional to N.

Random Patterns
In normal operation, the data transmitted along a chain of repeaters have a nearly random bit pattern. For this case, Byrne and colleagues have determined that the low-frequency power density of jitter injected in each repeater is independent of frequency, such that $S_i(f)$ in Equation (7.12) is a constant.

The mean square value of jitter can be found by integrating the jitter spectrum (Equation 7.11) over all frequencies. The result is

$$\overline{J_N^2} = \frac{S_i B}{2}\left(N - \frac{1}{2}\left\{\frac{(2N - 1)!}{4^{N-1}[(N - 1)!]^2}\right\}\right) \tag{7.13}$$

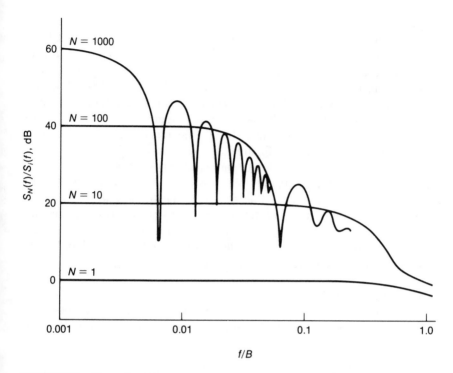

FIGURE 7.7   Normalized Jitter Power Spectrum for N Repeaters in Tandem

For $N > 100$, this expression reduces to the approximation

$$\overline{J_N^2} \approx \frac{1}{2} S_i BN \tag{7.14}$$

Thus the mean square jitter in a long repeater chain increases as N, and the rms amplitude increases as the square root of N. Moreover, Expression (7.14) indicates that the jitter power is directly proportional to the timing circuit bandwidth B and inversely proportional to the Q of the circuit.

Analysis of rms jitter amplitude due to uncorrelated sources indicates an accumulation according to only the fourth root of N [9]. For example, the use of scramblers with repeaters will result in uncorrelated jitter from repeater to repeater and accumulation that is proportional to the fourth root of N. Dejitterizers reduce the jitter amplitude from a single repeater and also result in jitter accumulation that is proportional to the fourth root of N (see CCITT Rec. G.823 [3]).

Repetitive Patterns

Any data bit pattern **m** can be shown to have a characteristic jitter, $\theta_m(t)$. A change from pattern **m** to pattern **n** causes a jump in jitter amplitude from θ_m to θ_n. This change in jitter stems from the response of the timing circuit to the new pattern. The rate of change is controlled by the bandwidth B of the timing circuit and is given by $B(\theta_m - \theta_n)$. The total change in jitter at the end of N repeaters accumulates in a linear fashion and is given by $N(\theta_m - \theta_n)$. These characteristics of accumulated jitter for N repeaters due to transition in the data pattern are illustrated in Figure 7.8.

The worst-case jitter for all data pattern changes is characterized by the maximum value of $(\theta_m - \theta_n)$ for all possible pairs of repetitive patterns. Experimental results indicate that the jitter magnitude due to transitions between two fixed patterns is affected by the 1's density (for bipolar) and intersymbol interference. In practice such repetitive data patterns are rare, except when the line is idle, test patterns are being transmitted, or a PCM system is being used to sample slowly changing analog signals.

7.4.3   Alignment Jitter

So far we have not related jitter to the error performance of a regenerative repeater. Here we will see that the misalignment caused by jitter between the

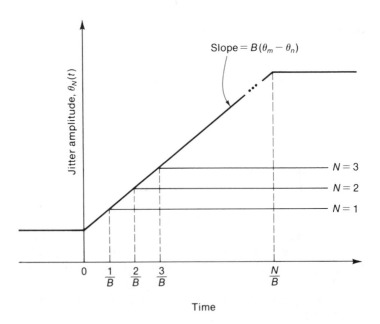

FIGURE 7.8   Jitter at End of N Repeaters Due to Change in Data Pattern

data signal and sampling times does not contribute significantly to bit errors. The alignment jitter at the Nth repeater, Δ_N, is defined as the difference between the input and output jitter, so that for the Nth repeater we have

$$\Delta_N(t) = \theta_N(t) - \theta_{N-1}(t) \tag{7.15}$$

From (7.11) we can rewrite (7.15) as

$$\Delta_N(f) = \theta_i(f)\left(\frac{1}{1+jf/B}\right)^N \tag{7.16}$$

Using the Fourier transform, Equation (7.16) can be rewritten in the time domain as

$$\begin{aligned}\Delta_N(t) &\le M_i(1 - e^{-Bt})^N \\ &\le M_i\end{aligned} \tag{7.17}$$

where M_i is the maximum value of $\theta_i(t)$, the jitter injected at each repeater. Thus the maximum alignment jitter does not increase with the length of the chain and is bounded by the maximum jitter injected at each repeater. This property can also be seen from Figure 7.8, where the constant slope of jitter change indicates that the alignment jitter is the same between any two successive repeaters.

7.5   Crosstalk

Crosstalk is defined as the coupling of energy from one circuit to another, resulting in a form of interference. In cable transmission, crosstalk takes the form of electrical coupling between twisted pairs in a multipair cable or between coaxials in a multicoaxial cable. As noted before, crosstalk is negligible in fiber optic cable. Depending on the cable arrangement, crosstalk may be the limiting factor in determining transmission distance between repeaters. Crosstalk is eliminated by a regenerative repeater, however, so that it does not accumulate in a chain of repeaters.

Figure 7.9 illustrates the two principal types of crosstalk. **Near-end crosstalk (NEXT)** refers to the case where the disturbing circuit travels in a direction opposite to that of the disturbed circuit. **Far-end crosstalk (FEXT)** is the case where the disturbed circuit travels in the same direction as the disturbing circuit. Of the two types, NEXT is the dominant source of interference because a high-level disturbing signal is coupled into the disturbed signal at the receiver input where signal level is normally low. For this reason, opposite directions of transmission are often isolated by using separate cables or a screened cable to avoid NEXT so that only FEXT is of concern.

Crosstalk is usually measured and specified by a ratio of the power in the disturbing circuit to the induced power in the disturbed circuit. Specifically, using the notation shown in Figure 7.9 the NEXT ratio is given by

$$\text{NEXT} = \frac{P_{2n}}{P_{1n}} \tag{7.18}$$

and the FEXT by

$$\text{FEXT} = \frac{P_{2f}}{P_{1f}} \tag{7.19}$$

where P_{1n} = transmit power of disturbing circuit

P_{1f} = receiver power of disturbing circuit

P_{2n} = near-end power in disturbed circuit due to coupling from disturbing circuit.

P_{2f} = far-end power in disturbed circuit due to coupling from disturbing circuit

For long transmission lines and high (carrier) frequencies, the NEXT ratio is given by

$$\frac{P_{2n}}{P_{1n}} \approx k_n f^{3/2} \tag{7.20}$$

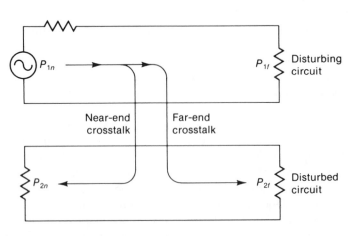

FIGURE 7.9   Types of Crosstalk

or, since NEXT is usually given in decibels,

$$\text{NEXT} = 10 \log k_n + 15 \log f \qquad (7.21)$$

where k_n is the NEXT loss per unit frequency and f is frequency in the same units as k_n. Note that NEXT increases with frequency at a rate of 15 dB per frequency decade and is independent of cable length. For like pairs or coaxials, the FEXT ratio is given by

$$\frac{P_{2f}}{P_{1f}} \approx k_f f^2 L \qquad (7.22)$$

or, in decibels,

$$\text{FEXT} = 10 \log k_f + 20 \log f + 10 \log L \qquad (7.23)$$

where k_f is the FEXT loss of a unit length cable per unit frequency, and L is the cable length. Note that FEXT increases with frequency at a rate of 20 dB per frequency decade and with length at a rate of 10 dB per length decade.

For multipair cable, the total crosstalk in a given pair is the power sum of individual contributions from the other pairs. Measurements of total crosstalk indicate that the distribution of crosstalk values, in decibels, is normal (gaussian). Additionally, the distribution of pair-to-pair crosstalk, in decibels, has been found to be normal [10].

The effect of crosstalk on error rate depends on its mean square value, analogous to the case of additive gaussian noise used in modeling thermal noise. The total interference at a repeater is then the sum of crosstalk interference σ_c^2 and thermal noise σ_n^2:

$$\sigma_T^2 = \sigma_c^2 + \sigma_n^2 \qquad (7.24)$$

If the noise is gaussian and the crosstalk is known to be gaussian, the total interference σ_T^2 must also be gaussian, thus allowing a straightforward calculation of error rate using the appropriate expressions of the next section.

7.6   Error Performance for Tandem Repeaters

Here we are interested in the behavior of the error rate after n tandem repeaters, where each repeater has probability of error p. The probability of error for various digital transmission codes has already been shown in Chapter 5. Each transmitted digit may undergo cumulative errors as it passes from repeater to repeater. If the total number of errors is even, they cancel out. A final error is made only if an odd number of errors is made in the repeater chain. For a given bit the probability of making k errors in n repeaters is given by the binomial distribution:

$$P_k = \binom{n}{k} p^k (1-p)^{n-k} \tag{7.25}$$

The net probability of error for the n repeater chain, $P_n(e)$, is obtained by summing over all odd values of k, yielding

$$
\begin{aligned}
P_n(e) &= \sum_{k=1}^{n} \binom{n}{k} p^k (1-p)^{n-k} \quad (k \text{ odd}) \\
&= np(1-p)^{n-1} + \frac{n(n-1)(n-2)}{3!} p^3 (1-p)^{n-3} + \cdots \\
&\approx np
\end{aligned}
\tag{7.26}
$$

where the approximation holds if $p \ll 1$ and $np \ll 1$. This approximation indicates that if the probability of a single error is small enough, the probability of multiple errors is negligible. In this case, the net error probability increases linearly with the number of repeaters.

It is interesting to note here the advantage of digital repeaters over analog repeaters. An analog repeater is nothing more than an amplifier that amplifies both signal and noise. Assuming the presence of additive gaussian noise in each repeater, the noise accumulates in a linear fashion for a chain of repeaters. Consequently, the signal-to-noise ratio progressively decreases with each repeater so that after n repeaters the final signal-to-noise ratio is

$$(S/N)_n = \frac{S/N}{n} \tag{7.27}$$

where S/N is the value after the first repeater. Choosing NRZ as the transmission code for binary data, the error probability for n digital repeaters in tandem is obtained from (7.26) and (5.31) as

$$P_n(e) \approx n \, \text{erfc} \sqrt{S/N} \tag{7.28}$$

For n analog repeaters in tandem,

$$P_n(e) = \text{erfc} \sqrt{\frac{S/N}{n}} \tag{7.29}$$

Figure 7.10 indicates the power savings in using digital (regenerative) repeaters for a fixed 10^{-5} probability of error.

7.7   Repeater Spacing in Multipair Cable Systems

Repeaters are designed to provide a certain error rate objective on the basis of each repeater section and for the end-to-end system. The error rate in multipair cable systems is affected by the transmitted signal level, cable loss, crosstalk, and the required receiver signal-to-noise ratio. The required spacing between repeaters can in general be written as

$$R_s = \frac{L_D}{L_c} \tag{7.30}$$

where R_s = repeater spacing (units of distance)

L_D = maximum section loss (dB)

L_c = cable attenuation (dB/unit distance)

The maximum section loss L_D is the loss a signal may undergo and still provide an adequate signal-to-noise ratio at the repeater input to meet the error rate objective.

Repeatered lines for multipair cable may be designed with both directions of transmission in the same cable or with each direction in a separate cable. In one-cable operation, near-end crosstalk is the limiting factor in determining L_D and hence repeater spacing. In two-cable (or one-cable T-

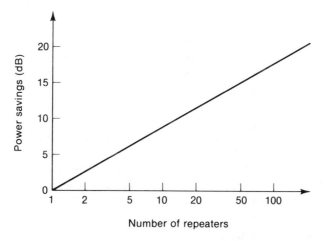

FIGURE 7.10   Power Savings Gained by Digital Repeaters Over Analog Repeaters for a Fixed $P(e) = 10^{-5}$

screen) operation, the physical separation of transmit and receive directions eliminates NEXT. The limiting factor in determining repeater spacing is then a combination of FEXT and thermal noise, which permits wider repeater spacing than in one-cable operation.

While crosstalk can dictate the repeater spacing, particularly in one-cable systems, there is a maximum spacing determined by cable attenuation alone. Cable attenuation characteristics depend on the gauge (size), construction, insulation, and operating temperature. Temperature changes in multipair cable affect the resistance and hence the attenuation characteristic. Therefore the maximum operating temperature must be known in determining the maximum section loss. Cable attenuation and section loss values are often given for a standard temperature, such as 55°F. Then a temperature correction factor, f_T, can be used to account for other operating temperatures, where we define

$$f_T = \frac{\text{loss at } T°F}{\text{loss at } 55°F} \qquad (7.31)$$

Cable attenuation for a given maximum operating temperature is given by

$$L_c(T°F) = L_c(55°F)f_T \qquad (7.32)$$

The maximum section loss corresponding to a given maximum operating temperature is then given by

$$L_D(T°F) = \frac{L_D(55°F)}{f_T} \qquad (7.33)$$

For the design of T1 carrier systems (1.544 Mb/s), values of f_T and L_c are listed in Table 7.2 for various types of cable. Based on U.S. climate, the maximum ambient temperature is assumed to be 100°F for buried cable and 140°F for aerial cable.

For T1 carrier systems, repeaters are spaced at a nominal distance of 6,000 ft, where the loss for 22-gauge cable is in the range of 27 to 33 dB. Hence T1 repeaters are designed to operate with a maximum transmission loss of typically 32 to 35 dB. The exact choice of maximum allowed loss is based on the expected maximum loss plus a margin to account for variation in loss. The range of repeater operation is extended to lower-loss sections by using automatic line build-out techniques.

7.7.1   One-Cable Systems

In one-cable systems, near-end crosstalk depends on the type of cable, the frequency, the physical separation of the pairs in the two directions of trans-

TABLE 7.2   Cable Losses at 772 kHz for T1 Carrier Design

Cable Gauge	Cable Type		L_c at 55°F	Buried Cable (100°F Max. Temp.)		Aerial Cable (140°F Max. Temp.)	
	Construction	Insulation	(dB/1000 ft)	f_T	L_c (dB/1000 ft)	f_T	L_c (dB/1000 ft)
19	Unit or layer	Paper	3.00	1.038	3.11	1.071	3.21
	Unit	PIC	3.18	1.043	3.31	1.080	3.43
	T-screen, unfilled	PIC	3.28	1.056	3.46	1.108	3.63
	T-screen, filled to be watertight.	PIC	2.94	1.056	3.10	1.108	3.26
22	Unit or layer	Paper	5.10	1.042	5.31	1.078	5.49
	Unit	PIC	4.39	1.044	4.58	1.083	4.76
	T-screen, unfilled	PIC	4.47	1.056	4.72	1.108	4.95
	T-screen, filled to be watertight	PIC	3.99	1.056	4.20	1.108	4.39
24	Unit or layer	Paper	6.80	1.044	7.09	1.083	7.36
	Unit	PIC	5.58	1.026	5.72	1.052	5.87
	T-screen, unfilled	PIC	5.52	1.056	5.83	1.108	6.12
	T-screen, filled to be watertight	PIC	4.92	1.056	5.19	1.108	5.40
26	Unit or layer	Pulp	6.79	1.054	7.16	1.101	7.48
	Unit	PIC	7.48	1.024	7.66	1.046	7.82
	T-screen, filled to be watertight	PIC	6.65	1.056	7.02	1.098	7.30

f_T = temperature correction factor
L_c = cable loss at 772 kHz

459

mission, and the number of carrier systems contained within the cable. Engineering of a one-cable system first requires detailed knowledge of the cable construction. Multipair cable is constructed with pairs divided into units, layers, or groups. To minimize NEXT, it is standard practice to use pairs in nonadjacent units (layers or groups). Greater separation decreases the interference from NEXT. For small cables, however, it may be necessary to use pairs in adjacent units, or even pairs in the same unit, for opposite directions of transmission. In smaller cables, therefore, crosstalk interference increases significantly with the number of carrier systems.

Table 7.3 presents typical figures for NEXT on 22-gauge plastic insulated cable at 772 kHz [11]. This choice of frequency applies to T1 systems using bipolar coding. The frequency correction factor of 15 dB per decade can be used to adjust these figures to other frequencies. Some differences in crosstalk values exist between plastic, pulp, and paper-insulated cable. For a highly filled system, however, it has been found that the values do not differ noticeably. Therefore, as a first approximation, the values presented in Table 7.3 can be used for any type of insulation [12]. NEXT values do vary with cable gauge, though, according to the following expression:

$$\text{NEXT}_1 = \text{NEXT}_2 + 10\log\frac{\alpha_1}{\alpha_2} \qquad (7.34)$$

where subscripts 1 and 2 denote the wire gauge being compared and α is attenuation per unit length. The values given in Table 7.3 are **NEXT coupling loss** defined as the ratio of power in the disturbing circuit to the induced power in the disturbed circuit. Because of variations in the basic physical and electrical parameters of individual cable pairs, crosstalk coupling loss varies from pair to pair. The distribution of crosstalk coupling loss values is known to be normal (gaussian) so that only the mean (m) and standard deviation (σ) are needed to characterize the distribution. As crosstalk becomes worse, the crosstalk coupling loss values get smaller. Note that the values in Table 7.3 are affected both by the pair count and by the cable layout. The smaller cable sizes, such as 12- and 25-pair cable, have inherently greater crosstalk because of the close proximity of pairs. With larger cable sizes, it is possible to select widely separated pairs, in separate units, and thus provide tolerable levels of crosstalk. Screened cable may be used to improve crosstalk performance further, especially if a small pair cable is involved.

The relationship between maximum section loss (L_D), near-end crosstalk coupling loss (m and σ), and number of carrier systems (n) can be written in the general form

$$L_D \le am - b\sigma - c\log n - d \qquad \text{(decibels)} \qquad (7.35)$$

TABLE 7.3   Typical Near-End Crosstalk Coupling Loss at 772 kHz for 22-Gauge Cable [11]

Cable Size	Location of Opposite-Direction Pairs	Mean (m), dB	Standard Deviation (σ), dB
12-Pair	Adjacent pairs (same layer)	67	4.0
cable	Alternate pairs (same layer)	78	6.5
25-Pair	Adjacent pairs (same layer)	68	7.2
cable	Alternate pairs (same layer)	74	6.5
50-Pair	Adjacent pairs (same layer)	67	9.8
cable	Alternate pairs (same layer)	70	6.4
	Adjacent units	83	8.2
	Alternate units	94	8.7
100-Pair	Adjacent pairs (same layer)	66	5.5
cable	Alternate pairs (same layer)	73	8.2
	Adjacent units	85	8.6
	Alternate units	101	6.7
200-Pair	Adjacent pairs (same layer)	65	5.8
cable	Alternate pairs (same layer)	73	7.1
	Adjacent units	84	9.0
	Alternate units	103	6.7
Screened			
cable			
25-Pair ⎞		103	7.5
50-Pair ⎪ Opposite halves		107	7.5
100-Pair ⎪		112	7.0
200-Pair ⎠		117	6.5

where a, b, c, and d are constants whose values depend on the system. The U.S. industry standard for T1 cable engineering, for example, is given as [1]

$$L_D \le (m - \sigma) - 10 \log n - 32 \qquad (7.36)$$

This expression for T1 repeater section engineering is based on a requirement that at least 99 percent of the lines in a repeater section should have an error rate below 1×10^{-7}. The constant term, $d = 32$, in Equation (7.36) includes a 6-dB margin of safety factor to allow for variations in cable manufacturing, additional interference from far-end crosstalk, and other sources of degradation such as jitter. For convenience, Expression (7.36) can be put in the form of a chart as in Figure 7.11. For selected values of $(m - \sigma)$ and n, the value of L_D can be read directly off this chart. This value of L_D may then be used in (7.30) to determine repeater spacing.

A value for L_D from (7.35) or (7.36) that would lead to a section loss greater than the basic loss due to transmission attenuation simply means

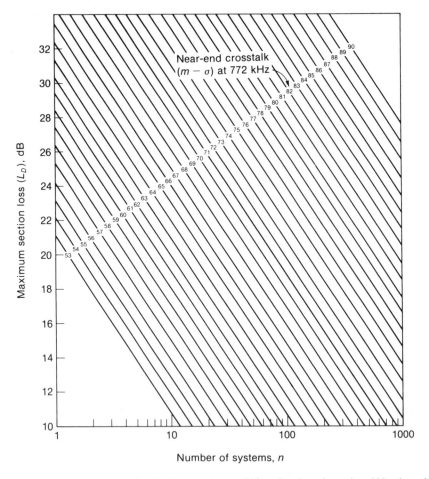

FIGURE 7.11   Maximum Section Loss vs. Crosstalk Coupling Loss ($m - \sigma$) and Number of Systems

that the allowed section loss is set equal to the attenuation loss. Thus for T1 carrier the allowed section loss cannot exceed 32 to 35 dB, the fundamental limit due to cable attenuation.

As noted earlier, the number of carrier systems in a cable affects crosstalk performance. Using Expression (7.36), this effect is illustrated for T1 carrier in Figure 7.12, which shows the crosstalk performance required to support a given number of carrier systems, assuming a certain allowed repeater section loss. Figure 7.12 also includes two plots of actual crosstalk performance values extracted from Table 7.3 for two cases: (1) transmit and receive cable pairs that are in adjacent units and (2) pairs that are separated by screening. A comparison of the required and actual NEXT performance indicates the improvement needed (case 1) or margin provided (case 2) by actual crosstalk performance to match the performance required for 100 percent cable fill. Conversely, use of Figure 7.12 and Table 7.3 can show the backoff from 100 percent cable fill required to allow use of existing cable. For 50-pair cable, for example, the near-end crosstalk coupling loss ($m - \sigma$) for adjacent units is 75 dB from Table 7.3. From Figure 7.12 the near-end crosstalk loss required to support 25 (two-way) systems at maximum allowed section loss is 80 dB. Therefore the improvement in crosstalk loss required for maximum fill is 5 dB. Such an improvement is not easily realized even by redesign of the cable. Faced with this predicament, the system designer must therefore reduce the number of carrier systems in the cable or use a screened or two-cable system in place of a one-cable system.

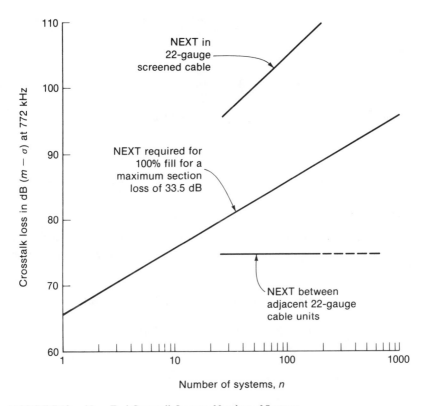

FIGURE 7.12   Near-End Crosstalk Loss vs. Number of Systems

Example 7.1 _____

A capacity for 50 T1 carrier systems is to be provided in a buried 200-pair cable of unit construction and 22-gauge plastic insulated conductors. The 50-pair unit used for one direction of transmission will be adjacent to the 50-pair unit used for the opposite direction. Calculate the required repeater spacing. Repeat the calculation but for only 10 carrier systems.

Solution   The maximum section loss due to NEXT is found from (7.36), where from Table 7.3 we have $m - \sigma = 75$. Therefore

$$L_D = 75 - 32 - 10 \log 50 = 26 \, dB$$

(Note that the L_D can also be read directly from Figure 7.11.) Repeater spacing is found by using (7.30), where from Table 7.2 we have $L_c(100°F) = 4.58 \, dB/1000$ ft. Therefore

$$R_s = \frac{L_D}{L_c} = \frac{26 \, dB}{4.58 \, dB/1000 \, ft} = 5677 \, ft$$

For only 10 carrier systems,

$$L_D = 75 - 32 - 10 \log 10 = 33 \, dB$$

so that the repeater spacing can be increased to

$$R_s = \frac{33 \, dB}{4.58 \, dB/1000 \, ft} = 7205 \, ft$$

Example 7.2 _____

Find the maximum number n of T1 carrier systems that can be accommodated by the cable type given in Example 7.1 if the repeater spacing is fixed at 6,000 ft.

Solution   The maximum section loss due to NEXT is found from (7.30) as

$$L_D = R_s L_c = (6000 \, ft)(4.58 \, dB/1000 \, ft)$$
$$= 27.5 \, dB$$

Then from (7.36), or using Figure 7.11

$$10 \log n = 75 - 32 - 27.5 = 15.5$$

Therefore the maximum number of T1 carrier systems is

$$n \le 35$$

7.7.2 Two-Cable Systems

For the engineering of two-cable systems, an expression for the maximum section loss (L_D) or maximum cable fill (n) can be obtained by using the general form given by (7.35). Here, however, the statistical parameters m and σ apply to FEXT performance. For many two-cable system applications, losses due to far-end coupling do not exceed the level established by cable losses alone. It follows, therefore, that the maximum section loss is usually determined by only cable attenuation and not crosstalk, even for a large fill (n) in the cable.

Example 7.3 _____

Calculate the required repeater spacing for a T1 carrier system, given the use of 19-gauge PIC cable constructed in units and a maximum section loss designed at 33.5 dB for:
(a) Aerial cable (140°F maximum temperature)
(b) Buried cable (100°F maximum temperature)

Solution   Using (7.30) and Table 7.2, we obtain

$$\textbf{(a)} \ R_s = \frac{33.5 \text{ dB}}{3.43 \text{ dB}/1000 \text{ ft}} = 9767 \text{ ft}$$

$$\textbf{(b)} \ R_s = \frac{33.5 \text{ dB}}{3.31 \text{ dB}/1000 \text{ ft}} = 10,121 \text{ ft}$$

7.8   Repeater Spacing in Coaxial Cable Systems

For coaxial cable systems, repeater spacing is determined by transmitter signal level, cable attenuation, and the required receiver signal-to-noise ratio. Whereas crosstalk is usually the major contributor to section loss in multipair cable, the use of coaxial cable virtually eliminates crosstalk as a factor in repeater spacing design. Thus repeater spacing is calculated by using Expression (7.30), which in the case of coaxial cable amounts to dividing the allowed attenuation between repeaters (L_D) by the attenuation of the cable per unit length (L_c).

The attenuation characteristic for several standard coaxial cables was given in Figure 7.3 for a fixed temperature, 10°C. However, attenuation increases with temperature because of the increase in conductor resistance with temperature. The attenuation characteristic at temperature T°C is referenced to 10°C by

$$\alpha(T°C) = \alpha(10°C)\left[1 + K(T°C - 10°C)\right] \tag{7.37}$$

The temperature coefficient, K, depends on both cable size and frequency. A typical value for frequencies above 1 MHz is $K = 2 \times 10^{-3}$ per degrees Celsius for any of the cable sizes shown in Figure 7.3.

Example 7.4 _____

A 34.368-Mb/s system is to be applied to a 2.6/9.5-mm coaxial cable system. Maximum operating temperature is anticipated as 40°C for buried cable. Assuming an allowed attenuation between repeaters of 75 dB, calculate the required repeater spacing. Repeat the calculation for 1.2/4.4-mm coaxial cable.

Solution   The attenuation of 2.6/9.5-mm cable is found from Figure 7.3 and Equation (7.37). Assuming the use of bipolar signaling and a temperature coefficient of $K = 2 \times 10^{-3}/°C$, the attenuation at 17.184 MHz and 40°C is

$$\alpha(40°C) = 10\left[1 + (0.002)(30)\right] = 10.6 \text{ dB/km}$$

The repeater spacing is found from (7.30) as

$$R_s = \frac{75 \text{ dB}}{10.6 \text{ dB/km}} = 7.1 \text{ km}$$

For 1.2/4.4-mm cable, the attenuation is given by

$$\alpha(40°C) = 22\left[1 + (0.002)(30)\right] = 23.3 \text{ dB/km}$$

and the repeater spacing is then $R_s = 3.2$ km.

CCITT Recs. G.953 and G.954 provide nominal repeater spacings for coaxial cable as a function of hierarchical bit rates. These repeater spacings for the three standard sizes of coaxial cable are shown here in Table 7.4.

7.9   Implementation Considerations

Figure 7.13 shows representative implementations of twisted-pair and coaxial PCM repeatered lines, both of which provide the same 672 voice channel

TABLE 7.4   Nominal Repeater Spacing for Standard Coaxial Cable [3]

	Nominal Repeater Spacing (km)		
Bit Rate (Mb/s)	0.7/2.9 mm	1.2/4.4 mm	2.6/9.5 mm
8.448	4.0		
34.368	2.0	3.0-4.0	
97.728			4.5
139.264		2.0	3.0-4.5
4 × 139.264			1.5

capacity. Figure 7.13 also illustrates the savings in equipment and increase in repeater spacings afforded by coaxial cable systems over conventional T-carrier cable. The coaxial cable system operates at the T3 rate, 44.736 Mb/s, with one operating line and one spare line (1:1 redundancy). The T1 wire-line system uses 28 operating and 4 spare lines (1:7 redundancy). At the end offices, transmitted and received signals are fed through a protection switch that contains monitoring and control logic and allows a failed line to be switched off-line and replaced with a spare line. Repeaters for regeneration are placed at regular intervals according to the repeater spacing calculations. These repeaters are housed in protective cases and may be above or below ground, depending on the application. Access to the repeater is then typically by telephone pole or manhole. Because repeater sites are frequently in remote locations, power is often unavailable locally and must be fed along the communication lines. For troubleshooting, fault isolation devices and an orderwire channel are also standard features in repeatered line equipment.

7.9.1   Power Feeding
In metallic cable systems, line repeaters are powered by using the cable itself for dc power transmission. In multipair cable, power from the end offices is passed through the pairs and decoupled from the signal through transformers. The required voltage V_M for a cable section is determined by

$$V_M = RV_R + LV_L \tag{7.38}$$

where $\quad V_R$ = voltage drop of one repeater
R = number of repeaters
V_L = voltage drop per unit length of cable
L = cable length

For a specified maximum voltage V_M and for given voltage drops V_R and V_L, Equation (7.38) determines the maximum number of repeaters allowed.

Typical values supplied by the end office power supply are 50 to 100 mA constant current and 48 to 260 volts maximum voltage.

In coaxial cable systems, standard practice is to feed dc power via the inner conductors of the two coaxial cables used to form a span.

7.9.2 Automatic Protection Switching

The use of spare lines with automatic switching provides increased availability and facilitates maintenance, especially at remote unattended locations. The ratio of spare lines to operating lines depends on the required availability, the failure rate of individual line components (repeaters, cable), and the failure rate of the protection switch itself. The usual practice is to drive spare lines with a dummy signal (from a pseudorandom sequence generator, for example) and monitor the received signal for errors. At the same time, operational lines are monitored for failure—for example, by counting bipolar violations or consecutive zeros and comparing the count to a threshold. Logic circuitry controls the transfer of operating lines to spare lines.

7.9.3 Fault Location

When a fault occurs, a means of locating the fault from a remote location or end office is desirable and usually standard practice in cable systems. Fault location techniques can be categorized as **in-service,** where the traffic is not interrupted during testing, or as **out-of-service,** in which case operational traffic is replaced by special test signals. A faulty repeater section can be identified by monitoring the error rate at each repeater. This may be done on an in-service basis by use of parity error or bipolar violation checking. Or it may be done out-of-service by transmitting a pseudorandom sequence from the end office, looping back this signal successively from one repeater to the next, and detecting errors in the sequence at the end office. In theory, the preferred method of fault location is in-service, since the techniques employed are usually simple to implement, provide continuous and automatic monitoring, and do not result in user outage. Out-of-service techniques are more accurate, however, and are more commonly used for fault location.

For 24- and 30-channel PCM transmission over multipair cable, fault location is performed out-of-service by use of a pseudorandom test sequence applied to the line under test. A particular repeater is accessed for test by use of an address transmitted as one of several audio frequencies or bit sequences. The addressed repeater responds by closing a loop from the transmit path to the receive path, so that the test signal returns to the fault location unit. A comparison of detected versus transmitted patterns indicates the exact bit error rate. Each repeater may be progressively tested by use of the appropriate address until the faulty section is discov-

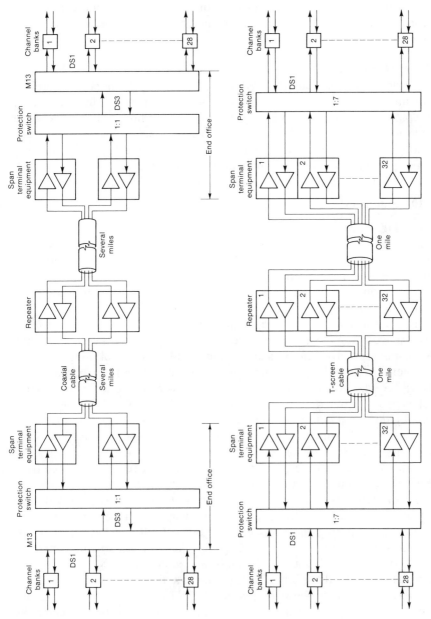

FIGURE 7.13 Coaxial Cable vs. T1 Carrier for PCM Repeatered Lines

469

ered. The number of addresses is typically 12 to 18 for a fault location unit in an end office. Since this fault location can be carried out at both end offices, a line span may contain a maximum of 24 to 36 repeaters for proper fault location.

7.9.4   Orderwire

Orderwires are voice circuits used by maintenance personnel for coordination and control. In cable transmission, the orderwire is usually provided by a separate pair in parallel with the communication lines. Alternatively, the orderwire can be modulated onto the same cable by using frequencies below or above the digital signal spectrum. Using the orderwire, maintenance personnel can dial any number from a repeater location.

7.10   Summary

Digital transmission via metallic cable has been developed for twisted-pair and coaxial cables. The performance of twisted-pair cable generally limits its application to below 10 Mb/s, but nevertheless twisted-pair cable has been used extensively for T carrier systems. Coaxial cable is a standard transmission medium for FDM and has been adapted for digital transmission at rates approaching 1 Gb/s (see Table 7.1).

Regenerative repeaters are required in a digital cable system at regularly spaced intervals to reshape, retime, and regenerate the digital signal. Reshaping involves amplification and equalization; retiming requires recovery of clock from the data signal; and regeneration performs sampling, voltage comparison, and code conversion. Each regenerative repeater contributes jitter and data errors in the process of regenerating the digital signal. Jitter amounts to a displacement of the intended data transition times, which leads to sampling offsets and data errors. Sources of jitter may be systematic or random. Jitter related to the data pattern is systematic and tends to accumulate along a chain of repeaters. Jitter from uncorrelated sources is random and does not accumulate as rapidly in a repeater chain. The error rate in a chain of n repeaters is approximately equal to np, where p is the probability of error for a single repeater.

Repeater spacing in multiple twisted-pair (multipair) cable is determined largely by attenuation and crosstalk. Crosstalk occurs when energy is coupled from one circuit to another. Near-end crosstalk (NEXT) refers to the case when the disturbing circuit travels in a direction opposite to that of the disturbed circuit. Far-end crosstalk (FEXT) is the case where the disturbed circuit travels in the same direction as the disturbing circuit. The degree of crosstalk depends on cable size and construction. When both directions of transmission are present in the same cable, NEXT is the limiting factor in determining repeater spacing. When two separate cables are used for the two directions, NEXT is eliminated and repeater spacing is determined by FEXT and noise.

For coaxial cable systems, crosstalk is virtually eliminated and repeater spacing becomes a function of only the attenuation characteristic. Repeater spacing is calculated by dividing the allowed attenuation between repeaters by the attenuation of the cable per unit length.

Implementation of metallic cable systems must also include consideration of power feeding, redundancy with automatic protection switching, fault location using in-service or out-of-service techniques, and voice orderwire for maintenance personnel.

References

1. H. Cravis and T. V. Crater, "Engineering of T1 Carrier System Repeatered Lines," *Bell System Technical Journal* 42(March 1963):431–486.
2. ANSI T1.601-1991, *Integrated Services Digital Network (ISDN)—Basic Access Interface for Use on Metallic Loops for Application on the Network Side of the NT (Layer 1 Specification)*, January 1991.
3. CCITT Blue Book, vol. III.3, *Transmission Media Characteristics* (Geneva: ITU, 1989).
4. CCITT Blue Book, vol. III.5, *Digital Networks, Digital Sections, and Digital Line Systems* (Geneva: ITU, 1989).
5. J. Legras and C. Paccaud, "Development of a 565 Mbit/s System on 2.6/9.5 mm Coaxial Cables," *1981 International Conference on Communications*, pp. 75.3.1–75.3.5.
6. N. Inoue, H. Kasai, T. Miki, and N. Sakurai, "PCM-400M Digital Repeatered Line," *1975 International Conference on Communications*, pp. 24.11–24.15.
7. H. Kasai, K. Ohue, and T. Hoshino, "An Experimental 800 Mbit/s Digital Transmission System Over Coaxial Cable," *1981 International Conference on Communications*, pp. 75.5.1–75.5.5.
8. C. J. Byrne, B. J. Karafin, and D. B. Robinson, "Systematic Jitter in a Chain of Digital Regenerators," *Bell System Technical Journal* 42(November 1963):2679–2714.
9. H. E. Rowe, "Timing in a Long Chain of Binary Regenerative Repeaters," *Bell System Technical Journal* 37(November 1958):1543–1598.
10. Bell Telephone Laboratories, *Transmission Systems for Communications* (Winston-Salem: Western Electric Company, 1970).
11. L. Jachimowicz, J. A. Olszewski, and I. Kolodny, "Transmission Properties of Filled Thermoplastic Insulated and Jacketed Telephone Cables at Voice and Carrier Frequencies," *IEEE Trans. on Comm.*, vol. COM-21, no. 3, March 1973, pp. 203–209.
12. T. C. Henneberger and M. D. Fagen, "Comparative Transmission Characteristics of Polyethylene Insulated and Paper Insulated Communication Cables," *Trans. AIEE Communications Electronics* 59(March 1962):27–33.

Problems

7.1. Consider a twisted-pair cable system used for transmission of a 1.544-Mb/s signal.

 (a) Determine the repeater spacing for a buried cable application if the maximum section loss due to crosstalk is 30 dB, the attenuation at 55°F is 5 dB/1000 ft, and the temperature correction factor f_T for buried cable (100°F) is 1.042.

 (b) Assume ten repeaters are required for the span between end offices. Further assume the probability of error to be 1×10^{-8} for nine of the repeaters and 5×10^{-8} for the tenth repeater. What is the expected probability of error from end office to end office?

 (c) If the number of repeaters is increased from 10 to 100, by what value (in dB) will the jitter power spectrum increase for low frequencies?

7.2 Binary NRZ signals are transmitted over a cable system with twenty repeaters.

 (a) Find the overall probability of error of the cable system if the probability of error at any repeater is 10^{-7}.

 (b) Find the signal-to-noise ratio required at each repeater.

 (c) Now assume the use of analog repeaters and find the S/N required to maintain the same end-to-end probability of error found in part (a).

7.3. Assume the jitter transfer function of a regenerative repeater is an ideal low-pass filter. Assume the jitter input to each repeater is given by $\theta_i(f) = \theta$, that is, a white power spectrum.

 (a) Determine the overall (total) jitter power spectrum for N repeaters in tandem.

 (b) How does the jitter power increase with the number of repeaters?

7.4. Consider a 100-repeater cable system which has a peak-to-peak jitter of five signaling intervals measured at the end of the cable system. Assuming each repeater has an identical timing circuit with bandwidth 1 kHz, find the jitter phase slope in radians per second for each repeater.

7.5 A rectangular pulse of amplitude V and width T is transmitted through a cable system with transfer function

$$Y(f) = \frac{1}{1 + jf/B}$$

 (a) Show an expression for the received pulse, $p(t)$.

 (b) Find the peak value of $p(t)$ for $B = 1/T$.

8

Fiber Optic
Transmission Systems

OBJECTIVES

- Compares multimode versus single-mode optical fiber, for both attenuation and dispersion characteristics
- Describes optical sources, including light emitting diodes (LED) and laser diodes
- Describes optical detectors, including PIN photodiodes and avalanche photodiodes
- Explains wavelength-division multiplexing as a way to expand the bandwidth or accommodate multiple signals in a single fiber.
- Presents methods for calculating the repeater spacing as a function of system parameters such as bandwidth, rise time, transmitter power, and receiver sensitivity.
- Discusses innovative approaches to fiber optic transmission, including coherent transmission, optical amplifiers, and soliton transmission
- Describes standards for fiber optic transmission developed in North America as the Synchronous Optical Network (SONET) and by the CCITT as the Synchronous Digital Hierarchy (SDH)

8.1 Introduction

A fiber optic transmission system has the same basic elements as a metallic cable system—namely a transmitter, cable, repeaters, and a receiver—but with an optical fiber system the electrical signal is converted into light

473

pulses. The principal application of fiber optics is with digital transmission, where rates in excess of 1 Gb/s are now in use. Fiber optic transmission now dominates long-haul national and international communications systems, and is expected to dominate short-haul systems as well. Today's fiber optic systems typically use binary transmission with on-off keying (OOK) of the optical source and direct detection of the received signal. Other technologies described in this chapter, such as coherent detection, wavelength-division multiplexing, optical amplifiers, and soliton transmission, are now available and will eventually extend existing transmission rates and repeater spacings by one or two orders of magnitude. Also of significance are national [1] and international [2] standards, which now provide the opportunity for common interfaces and interoperable systems among the world's fiber optic transmission systems.

The advantages of fiber optic cable over metallic cable and digital radio are illustrated here with a series of comparisons:

- *Size:* Since individual optic fibers are typically only 125 μm in diameter, a multiple-fiber cable can be made that is much smaller than corresponding metallic cables. For example, AT&T's AccuRibbon contains seventeen 12-fiber ribbons stacked to provide 204 fibers in a 0.6-in diameter cable. To equal that bandwidth capacity with coaxial cable would require a cable with several hundred times the fiber cable's cross-sectional area.
- *Weight:* The weight advantage of fiber cable over metallic cable is small for single-fiber, low-rate systems (such as T1) but increases dramatically for multiple-fiber, high-rate systems. As a result of this weight advantage, the transporting and installation of fiber optic cable is much easier than for other types of communication cable.
- *Bandwidth:* Fiber optic cables have bandwidths that can be orders of magnitude greater than metallic cable. Low-data-rate systems can be easily upgraded to higher-rate systems without the need to replace the fibers. Upgrading can be achieved by changing light sources (LED to laser), improving the modulation technique, improving the receiver, or using wavelength-division multiplexing.
- *Repeater spacing:* With low-loss fiber optic cable, the distance between repeaters can be significantly greater than in metallic cable systems. Moreover, losses in optical fibers are independent of bandwidth, whereas with coaxial or twisted-pair cable the losses increase with bandwidth. Thus this advantage in repeater spacing increases with the system's bandwidth.
- *Electrical isolation:* Fiber optic cable is electrically nonconducting, which eliminates all electrical problems that now beset metallic cable. Fiber

optic systems are immune to power surges, lightning-induced currents, ground loops, and short circuits. Fibers are not susceptible to electromagnetic interference from power lines, radio signals, adjacent cable systems, or other electromagnetic sources.

- *Crosstalk:* Because there is no optical coupling from one fiber to another within a cable, fiber optic systems are free from crosstalk. In metallic cable systems, by contrast, crosstalk is a common problem and is often the limiting factor in performance.

- *Radiation:* Unlike metallic cable systems, fiber optic cable does not radiate electromagnetic energy, which is important in applications involving military security. This means that the signal can be detected only if the cable is physically accessed.

- *Environment:* Properly designed fiber optic systems are relatively unaffected by adverse temperature and moisture conditions and therefore have application to underwater cable. For metallic cable, however, moisture is a constant problem particularly in underground (buried) applications, resulting in short circuits, increased attenuation, corrosion, and increased crosstalk.

- *Reliability:* The reliability of optical fibers, optical drivers, and optical receivers has reached the point where the limiting factor is usually the associated electronics circuitry.

- *Cost:* The numerous advantages listed here for fiber optic systems have resulted in dramatic growth in their application with attendant reductions in cost due to technological improvements and sales volume. Today fiber optic systems are more cost-effective than metallic cable for long-haul, high-bit-rate applications. Fiber optic cable is also expected eventually to overtake metallic cable in short-haul applications, including metro facilities and local networks. One final cost factor in favor of fiber optics is the choice of material—copper, which may someday be in short supply, versus silicon, one of the earth's most abundant elements.

- *Frequency allocations:* Fiber (and metallic) cable systems do not require frequency allocations from an already crowded frequency spectrum. Moreover, cable systems do not have the terrain clearance, multipath fading, and interference problems common to radio systems.

Fiber optic and metallic cable systems do have the disadvantage, however, of requiring right-of-way along the physical path. Cable systems are also susceptible to inadvertent cuts of the cable, jokingly referred to as a *backhoe fade.*

8.2 Fiber Optic Cable

An **optical fiber** is a cylindrical waveguide made of two transparent materials each with a different index of refraction. The two materials, usually high-quality glass, are arranged concentrically to form an inner core and an outer cladding, as shown in Figure 8.1. Fiber size is designated by two numbers: core diameter and cladding diameter. For example, 50/125 μm is a standard fiber size specified by CCITT Rec. G.651 [3]. Light transmitted through the core is partly reflected and partly refracted at the boundary with the cladding. This property of optics, illustrated in Figure 8.2, is known as Snell's law:

$$n_1 \sin\theta_1 = n_2 \sin\theta_2 \tag{8.1}$$

where θ_1 = angle of incidence = angle of reflection
θ_2 = angle of refraction
n_1 = refractive index of medium 1
n_2 = refractive index of medium 2

If $n_1 > n_2$ there is a critical angle θ_c in which $\sin \theta_2 \geq 1$ when

$$\theta_1 \geq \theta_c = \sin^{-1}\left(\frac{n_2}{n_1}\right) \tag{8.2}$$

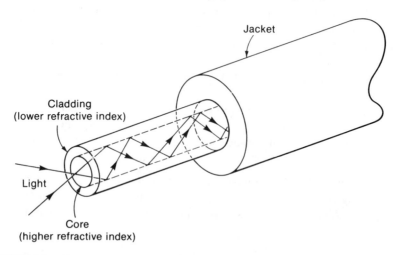

FIGURE 8.1   Geometry of an Optical Fiber

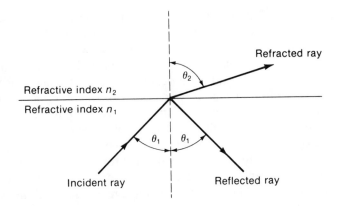

FIGURE 8.2   Optical Fiber Reflection and Refraction

For all $\theta_1 \geq \theta_c$ the incident ray is totally reflected with no refractive loss at the boundary of n_1 and n_2. Thus fiber optic cables are designed with the core's index of refraction higher than that of the cladding.

The maximum angle θ_m at which a ray may enter a fiber and be totally reflected is a function of the refractive indices of the core (n_1), cladding (n_2), and air (assumed to be $n_0 = 1$). Figure 8.3 traces the incoming rays at maximum angle, which become the rays at critical angle within the fiber. The angle θ_m measured with respect to the fiber axis is called the **acceptance angle,** while the sine of the acceptance angle is called the **numerical aperture (NA)** and is given by

$$NA = \sin\theta_m = \sqrt{n_1^2 - n_2^2} \qquad (8.3)$$

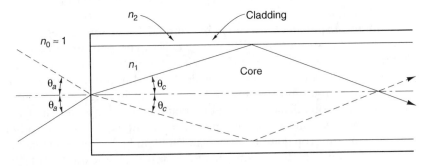

FIGURE 8.3   Acceptance Angle of Incident Light Ray

Numerical aperture indicates the fiber's ability to accept light, where the amount of light gathered by the fiber increases with θ_m. Different entry angles of the light source result in multiple modes of wave propagation. The total number of modes that can occur in an optical fiber is determined by the core size, wavelength, and numerical aperture. A distinction is usually made between high-order modes, which are propagation paths that have relatively large angles with respect to the fiber axis, and low-order modes, which are propagation paths that have relatively small angles with respect to the fiber axis. The high-order modes tend to be lost because of refractive loss at the boundary of the core and cladding. Propagation can be restricted to a single mode by using a small-diameter core, by reducing the numerical aperture, or both. The choice between **single-mode** and **multimode fiber** depends on the desired repeater spacing or transmission rate; single mode is the preferred choice for long-haul or high-data-rate systems. The range of NA for most multimode fiber is 0.18 to 0.24.

8.2.1 Dispersion

Certain propagation characteristics in fiber optic transmission systems give rise to dispersion that causes a spreading of the transmitted pulse. This dispersive effect in fiber optic cable arises because of two separate phenomena, one in which the various wavelengths of a single mode have different propagation velocities and the other in which the different modes or paths have different propagation velocities. Thus as the signal propagates along the optical fiber, components of the signal will arrive at the destination at different times, causing a spreading of the signal relative to its original shape. The total dispersion is then a function of the length of the cable. Furthermore, because bandwidth will be limited by the degree of dispersion, the total available bandwidth is known to be inversely proportional to the cable length. This relationship has led to the definition and manufacturers' use of the **bandwidth-distance product,** usually given in terms of MHz-km. Typical values are 10 MHz-km for step-index fibers and 400 to 1200 MHz-km for graded-index fibers.

Multimode dispersion (also called **intermodal dispersion**) arises in multimode cable because of the different velocities and path lengths of the different rays (modes). These differences result in different arrival times for rays launched into the fiber coincidentally. Multimode dispersion is predominant in multimode cable and nonexistent in single-mode cable. Values of multimode dispersion can be calculated by comparing the transit time of two rays, one having minimum propagation delay and the other having maximum propagation delay (see Problem 8.3). This direct calculation of multimode dispersion, however, requires knowledge of the refractive index profile of the fiber, information which is not readily available to the link designer. Manufacturers usually specify the coefficient of multimode dispersion, from which the total multimode dispersion can be found using

$$t_{mo} = c_{mo} \, D^{\gamma} \qquad (8.4)$$

where t_{mo} = multimode dispersion in ns

c_{mo} = coefficient of multimode dispersion in ns/km

D = distance in km

γ = length dependence factor (typically $0.5 \leq \gamma \leq 1$, where γ varies with fiber size, numerical aperture, and wavelength)

The factor γ accounts for the fact that multimode dispersion does not always accumulate linearly, due to scattering and attenuation of higher order modes.

Although multimode dispersion can be eliminated by use of single-mode fibers, **material dispersion** (also called **intramodal dispersion** or **chromatic dispersion**) may be produced by a variation in propagation velocity within the wavelengths of the source line spectrum. This phenomenum occurs because the index of refraction is a function of wavelength, and any light source, no matter how narrow in spectrum, will have a finite spectral width. Thus this type of dispersion is a function of both cable length and line spectrum, as given by

$$t_{ma} = c_{ma} \, \Delta\lambda \, D \qquad (8.5)$$

where t_{ma} = material dispersion in ns

c_{ma} = coefficient of material dispersion in ns/nm-km

$\Delta\lambda$ = line spectrum in nm

D = distance in km

Any given fiber has an associated coefficient of material dispersion, which is a function of the chemical content of the fiber and the operating wavelength. Typical values of the material dispersion coefficient are plotted in Figure 8.4 for glass fiber.

8.2.2 Attenuation

A second propagation characteristic of optical fiber is attenuation, measured in decibels per kilometer. Whereas the dispersion characteristic determines the achievable bandwidth, attenuation determines the available

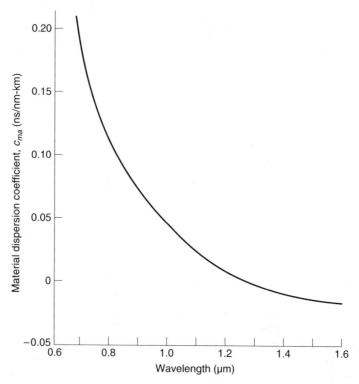

FIGURE 8.4   Typical Material Dispersion Coefficient for Fused Silica Glass Fiber [4]

signal-to-noise ratio. Attenuation has two sources, absorption and scattering, as illustrated in Figure 8.5 for a typical fiber. Absorption is caused by the natural presence of impurities in glass, such as OH ions at 945, 1240, and 1390 nm. The fiber-manufacturing process must control the concentration of such impurities to ensure that attenuation is kept low. Rayleigh scattering arises because of small variations in density that are introduced in the glass during the manufacturing process. Rayleigh scattering is known to be inversely proportional to the fourth power of wavelength and represents the fundamental limit of low-loss fibers.

Other sources of attenuation are deformations or microbends in the fiber, which cause refractive losses or scattering of the light. A change in the diameter of the fiber may cause the angle of the incident ray to be smaller than the critical angle, resulting in refractive losses. Microbends are sharp curvatures of the fiber involving axial displacements of a few micrometers and spatial wavelengths of a few millimeters. Although large radius bends in optical fibers have negligible effect on attenuation, microbends

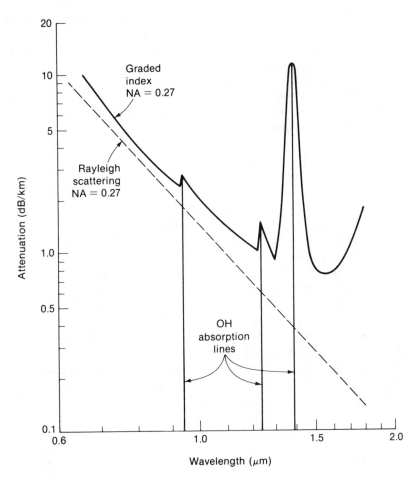

FIGURE 8.5   Spectral Attenuation for Low-Loss Fibers

can decrease the critical angle so that light normally reflected within the core will escape into the cladding. Microbending can also cause micro-cracks, which will increase attenuation. Proper manufacturing and installation of fiber optic cables minimize the possibility of microbending.

Examination of Figure 8.5 reveals three low-loss windows, at approximately 850, 1200, and 1600 nm. Early applications of fiber optics for communications applications were based on the short wavelength band of roughly 800 to 860 nm. Operation in the longer wavelength bands, particularly at 1300 and 1550 nm, is attractive because of improved attenuation and dispersion characteristics at these wavelengths. Typically today the shorter-wavelength band is used for short-haul, low-data-rate systems, and the longer-wavelength bands are applied to long-haul, high-data-rate systems. As shown in Figure 8.4, a choice of 1300 nm will minimize the effects of intramodal dispersion. Special fibers have been developed that shift the minimum dispersion to about 1550 nm to take advantage of lower attenuation as well as minimum dispersion. These fibers are called **dispersion-shifted fibers,** and are important to single-mode fiber applications where intermodal dispersion does not exist.

8.2.3   Types of Optical Fiber

There are three basic types of optical fiber, whose structures are sketched in Figure 8.6. These fiber types are differentiated by their refractive index profile and number of propagation modes. The earliest form of multimode fiber was the **step-index,** where the core has a uniform index of refraction and the concentric cladding also has a uniform but lower index. In this case the propagation velocity within the core is constant, so that rays traveling a longer path arrive behind rays traveling a shorter path, thus producing intermodal dispersion. These dispersive effects may be remedied by constructing a fiber whose refractive index increases toward the axis, with a resulting refractive index profile that is parabolic. With a **graded-index** fiber, rays that travel longer paths have greater velocity than rays traveling the shorter paths due to decreasing refractive index with radial distance. The various modes then tend to have the same arrival time, such that dispersion is minimized and greater bandwidths become possible for multimode fibers. Typical values for core diameter are 50, 62.5, and 85 microns (µm) with cladding diameter being 125 microns. Graded-index fibers frequently are designed to provide two wavelength regions of operation, at approximately 850 nm and at 1300 nm. Attenuation characteristics are typically 2–4 dB/km at 850 nm and 0.5–2.0 dB/km at 1300 nm. Bandwidths are determined mainly by the intermodal dispersion characteristics, with typical bandwidth-distance products of 200 to 1000 MHz-km for the shorter wavelength region, and 200 to 2000 MHz-km in the longer wavelength region [3]. One disadvantage of graded-index relative to step-index fiber is that its acceptance angle is smaller, making coupling of light into the fiber more difficult.

The core of a fiber can be reduced to a size (typically 2–10 µm) where only the axial mode (also called the zero mode) can propagate. Such fibers are known as **single mode.** Single-mode fibers are generally designed to provide operation at either or both of two wavelengths, 1300 and 1550 nm. Attenuation coefficients are in the range of 0.3–1.0 dB/km at the 1300-nm wavelength and 0.15–0.5 dB/km at the 1550-nm wavelength. A lower numerical aperture is used, which means a lower acceptance angle and greater coupling loss compared to graded-index. However, no intermodal dispersion exists in single-mode fibers since only one mode exists. Material

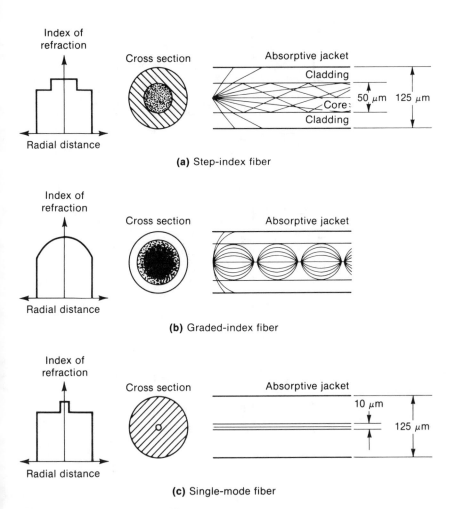

FIGURE 8.6   Types of Optical Fiber

dispersion characteristics will determine the available bandwidth in single-mode fibers. Typically, the zero dispersion wavelength is found at 1310 ± 10 nm, while at 1550 nm the dispersion is on the order of 20 ps/nm-km [3].

Conventional design of single-mode fibers employs a step-index refractive profile. Two such profiles are used, matched-cladding and depressed-cladding, as shown in Figure 8.7a and b. With matched-cladding, the entire cladding has a uniform refractive index approximately 0.3 to 0.4 percent lower than the core. In depressed-cladding fiber, the core and outer cladding are separated by a ring of inner cladding having a refractive index lower than the core or outer cladding. Depressed-cladding fiber provides greater resis-

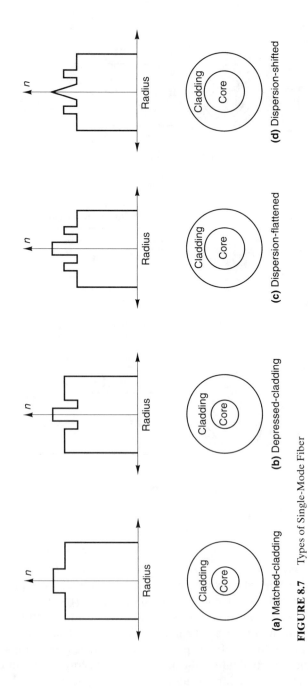

FIGURE 8.7 Types of Single-Mode Fiber

(a) Matched-cladding

(b) Depressed-cladding

(c) Dispersion-flattened

(d) Dispersion-shifted

484

tance to microbending. To affect these differences in refractive index, dopants such as germanium are added to the core to raise its refractive index above that of a pure silica cladding. Alternatively, the cladding can be doped with fluorine to lower its refractive index relative to a pure silica core.

To obtain even greater performance from single-mode fibers, the dispersion characteristic may be shifted to higher wavelengths (1550 nm) or flattened for a range of wavelengths, hence the names **dispersion-shifted** and **dispersion-flattened** fiber. This shift is achieved by varying the index of refraction across the fiber radius instead of using a step index profile. The original approach used to shift zero dispersion to 1550 nm was based on a segmented core. Here the core consisted of an inner region of high index, a middle region of lower index, and an outer region of high index, with the entire core surrounded by a cladding of low-index. Other approaches include reduction of the core diameter or different distribution of the core refractive index such as triangular, trapezoidal, or W-shaped. Two examples of a **W-index** fiber are shown in Figure 8.7c and d, corresponding to a dispersion shift and to a dispersion flattening. Typical values of dispersion for a unshifted, shifted, and flattened characteristic in single-mode fiber are shown in Figure 8.8.

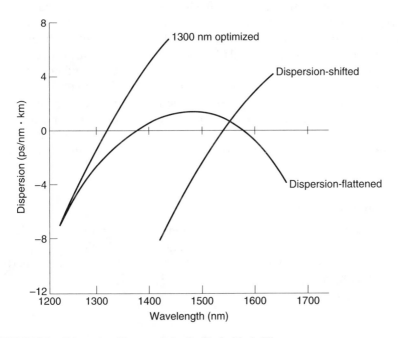

FIGURE 8.8   Dispersion Characteristics for Single-Mode Fiber

Other efforts to improve performance of optical fibers have focused on lowering attenuation through the use of new materials. Ultralow-loss fibers have been developed based on fluoride glasses, which consist of compounds of fluorine with zirconium, cadmium, beryllium, and other elements. These fibers attain losses that are two or three orders of magnitude less than silica, but shifted to wavelengths in the range of 2000 to 5000 nm as compared to the 800- to 1550-nm wavelengths used in conventional fiber systems. The manufacturing process for these ultralow-loss fibers is complex, however, making production impractical outside the laboratory. To render these new fibers practical, mechanical and chemical properties must be improved, the zero dispersion point must be moved to coincide with minimum loss, and new sources and detectors must be developed for operation at higher wavelengths.

Fiber cables designed for long-distance transmission must provide protection against various mechanical and environmental forms of stress. The method of installation—duct, aerial, trenched, or underwater—will also determine the design characteristics. Fibers are placed within the cable in a *loose buffer* or *tight buffer* configuration. A loose buffer or tube decouples the fiber(s) from the tube, permitting the fiber(s) to move independently of the tube. The inside diameter of the tube is much larger than the diameter of the fiber or group of fibers, which allows the fiber(s) to contract or expand with temperature changes or mechanical forces without stressing the fiber. A tight buffer encapsulates the fiber(s) such that the fiber moves with its tube. Tubes may be filled with a gel compound to protect against water entry and to provide impact and crush resistance. Fiber cables are typically designed with a central member made of steel or a dielectric material to add strength, and may also include a copper wire or tube to carry electric power. An outer jacket of plastic provides further protection against corrosion and penetration. Depending on the application, other sheath or armor layers may be required, as is the case for undersea cable as described in Section 8.6.1.

8.3 Optical Sources

Optical sources used in fiber optic systems operate on the principle that photons are emitted when an electrical signal is applied to a junction formed by two contiguous semiconductor materials. A photon is a particle of light with energy

$$E_p = hv \tag{8.6}$$

Here h is Planck's constant (6.626×10^{-34} Joule-sec) and ν is the optical frequency, which is given by

$$\nu = c/\lambda \tag{8.7}$$

where c is the speed of light (3×10^8 m/sec) and λ is the optical wavelength. The energy of the photons, and therefore their wavelength, depends on the energy band gap of the semiconductor materials. The optical wavelength can thus be controlled by choice of different materials, which inherently have different energy gaps.

The properties of optical sources considered important for communications applications include speed, efficiency, power, coupling loss, wavelength, spectral width, temperature range, reliability, and cost. Before proceeding with a comparison of these factors for various types of light emitting diode (LED) and laser diode (LD) optical sources, a few definitions are needed for those parameters not yet introduced. The speed of the source refers to its **response time,** the speed with which the device can be turned on and off as in the use of intensity modulation (for example, nonreturn to zero). **Quantum efficiency** of an optical source is defined as the number of photons generated per electron, or

$$\eta_q = \text{photons generated/electrons injected} \tag{8.8}$$

Typical values of η_q vary from 0.5 to 0.8, meaning that 50 to 80 percent of the injected electrons will produce photons. The **power efficiency** is defined by

$$\eta_p = P_{op}/P_{el} \tag{8.9}$$

where P_{op} is the optical power output and P_{el} is the electrical power input. **Coupling efficiency** η_c is a measure of the loss of signal power when the light from the source is coupled into the fiber. In decibels,

$$\eta_c = 10 \log (\text{Power into fiber/Power out of source}) \tag{8.10}$$

The coupling efficiency is a function of the mismatches between the source and fiber diameters and between the source radiation angle and the fiber numerical aperture. Typically, the source radiation angle is much larger than the fiber acceptance angle, resulting in considerable coupling loss.

Example 8.1 _____

Consider an LED whose diameter is perfectly matched to the fiber core area. The LED is assumed to radiate power as a Lambertian source, meaning that its radiation pattern has a peak at an angle $\theta = 0$ where θ is measured with respect to the direction perpendicular to the LED surface. At any other θ, the relative intensity falls off as cos θ. For this case the coupling loss is given by

$$\eta_c = 10 \log (NA)^2 \qquad (8.11)$$

For a fiber with NA = 0.20, the coupling loss is then $\eta_c = -14$ dB.

_____

In general, the light sources used in fiber optics are not monochromatic, that is, they do not produce just a single frequency or wavelength. Rather, these sources produce a narrow band of wavelengths whose bandwidth is referred to as the **linewidth** or **spectral width.** One common definition is the width in wavelengths, $\Delta\lambda$, at which the power has dropped to one half of the maximum (the 3-dB point), referred to as the full width at half maximum (FWHM). The linewidth plays an important part in determining the maximum bandwidth of a fiber optic system, since linewidth is directly proportional to dispersion. Typical linewidths of an LED and LD are shown in Figure 8.9, where the peak intensities have been normalized to the same peak value. In practice, the actual peak intensity of a laser diode is much greater than that of an LED.

Example 8.2 _____

Determine the frequency bandwidth of an optical source whose FWHM is 4 nm, with 3-dB points at 1298 and 1302 nm.

Solution   Using (8.7) we can express the bandwidth as the difference in frequencies between the 3-dB points:

$$
\begin{aligned}
v_2 - v_1 &= c/\lambda_2 - c/\lambda_1 \\
&= c/(1298 \times 10^{-9}) - c(1302 \times 10^{-9}) \\
&= 710 \text{ GHz}
\end{aligned}
$$

_____

The light emitting diode (LED) is a spontaneous emission device (as opposed to the laser, which depends on stimulated emission). Two types of LEDs are popular, the surface-emitting LED and the edge-emitting LED.

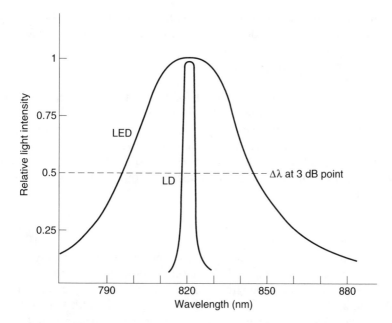

FIGURE 8.9   Typical Linewidths for Light Emitting Diode (LED) and Laser Diode (LD) [4]

As the name implies, the surface-emitting LED emits light from its surface, which leads to large radiation angles, up to 120°, and poor coupling efficiency. Because of this property, surface-emitting LEDs are used almost exclusively with large-core multimode fibers. Edge-emitting LEDs produce a more focused beam, about 30°, and greater power by confining the emission area to the edge. Spectral widths range from 40 to 170 nm for surface-emitting LEDs and 30 to 100 nm for edge-emitting LEDs. Edge-emitting LEDs, however, are more temperature-sensitive and more complex than surface-emitting versions. Relative to laser diodes, LEDs are slower speed devices with rise times in the nanosecond region, and have broad linewidths with correspondingly large material dispersion factors.

Laser diodes are distinguished from LEDs by the presence of a resonant cavity for coherent light amplification. The cavity confines the photons to a small area where photoelectron collisions occur, producing more photons and light amplification (*lasing*) in the process. For example, the Fabry Perot cavity has mirrors at both ends, which cause multiple reflections in the cavity and resonance from the standing waves. Two types of radiation are possible, *longitudinal* along the length of the cavity and *transverse* along its width. The wavelengths produced are determined by the length of the cavity, $L;$ to produce a wavelength λ, the cavity must be a multiple of $\lambda/2$. In general, a laser produces a spectrum consisting of multiple radiation peaks, or **modes,** with a main mode having largest amplitude at the center frequency and smaller modes with progressively smaller amplitudes away from the center frequency. Separation (in wavelengths) between modes is given by $c/2L$.

Of the many variations developed for lasers, the most important for communications applications is the distributed feedback (DFB) laser. The DFB laser introduces wavelength-dependent loss, provided by a grating inside the cavity, to select a particular lasing wavelength and suppress all other modes. The grating has a periodic corrugation that generates maximum reflections (distributed feedback) at the Bragg wavelength λ_B, given by [5]

$$\lambda_B = 2n\,\Upsilon \tag{8.12}$$

where n and Υ are the index of refraction and period, respectively, of the corrugated waveguide. The DFB laser thus produces single-mode operation corresponding to wavelength λ_B, since other modes are prevented from oscillation and lasing. The single mode has a spectral width that is determined by *laser chirp,* in which modulation of the injection current of a DFB laser produces shifts in the wavelength of its output. Typical linewidths of the DFB laser are less than 0.1 nm, with side modes attenuated 30 dB below the main mode. By comparison, the Fabry Perot laser can have linewidths two orders of magnitude larger than the DFB laser.

Lasers have many advantages including a narrow line spectrum, fast rise times (picosecond region) meaning high speed, and a wide operating temperature range. Compared to LEDs, lasers output higher power in the range of −10 to 0 dBm and can be coupled into a single-mode fiber with greater efficiency; however, lasers have the disadvantages of shorter life, greater temperature sensitivity, and higher costs.

8.4   Optical Detectors

Optical detectors are based on the same quantum mechanics principles used for optical sources. With the use of a **photodetector,** incoming photons are absorbed, photoelectrons (electron-hole pairs) are created, and electrons are emitted. The quantum efficiency of the photodetector is defined as

$$\eta_q = \frac{\text{average number of electrons generated}}{\text{average number of incident photons}} \tag{8.13}$$

which is similar to the quantum efficiency of an optical source as defined in (8.8). If the incident optical power in watts is P_i, then from (8.6), the number of incident photons per second is $P_i/h\nu$. The **photocurrent** I_p (current generated by absorption of photons) can then be given by

$$I_p = \eta_q \, q_e \, (P_i/h\nu) \tag{8.14}$$

where q_e is the electron charge (1.6×10^{-19} Coulombs). A more commonly used term to describe detector performance is its **responsivity** R, defined as

$$R = I_p/P_i \tag{8.15}$$
$$= q_e \, \eta_q / h\nu$$

According to (8.15), responsivity increases with wavelength, but in practice R will peak at a particular wavelength and taper off on either side of the peak response. The need for high quantum efficiency and responsivity over a particular wavelength region of interest will determine the choice of detector semiconductor material. Silicon devices are used in the 0.6- to 1.0-micron range, while detectors most suitable for 1.3 to 1.7 microns use germanium or InGaAs (indium-gallium-arsenide). A typical responsivity curve for a germanium PIN photodiode is shown in Figure 8.10.

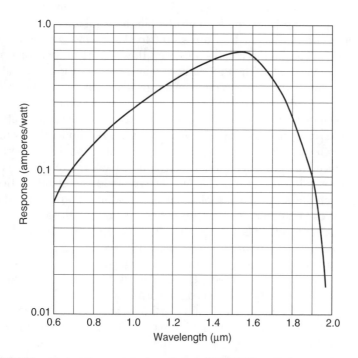

FIGURE 8.10   Typical Responsivity for a Germanium PIN Photodiode

Example 8.3 _____

A PIN (positive-intrinsic-negative) photodiode has an operating wavelength of 1300 nm and a quantum efficiency of 65 percent. Find its responsivity.

Solution   From (8.15) and (8.7), the responsivity can be rewritten as

$$R = \frac{\eta_q \lambda}{1.24} \tag{8.15a}$$

with λ in μm and η_q as a fraction. Then, substituting,

$$R = 0.68 \; A/W$$

Two types of detectors are used in fiber optics communications, the PIN photodiode and the avalanche photodiode (APD). By far the most popular detector, the PIN photodiode consists of intrinsic (nondoped) material sandwiched between a positively doped layer and a negatively doped layer. The diode is operated with a reverse bias (typically 5 to 10 volts), which fully depletes the intrinsic region. Therefore as incident photons pass through the thin *p*-layer and are absorbed in the depleted *i*-layer, electron-hole pairs are generated that are swept to the *n*- and *p*-doped regions, respectively. This action creates a current flow that is proportional to the incident optical power. The PIN diode is characterized by low internal noise but has no internal amplification mechanism. Therefore performance tends to be determined by the noise introduced in the external amplifier.

The avalanche photodiode, in contrast, utilizes an avalanche effect to obtain internal amplification. Compared to the PIN diode, the avalanche diode uses a much larger reverse bias voltage so that the diode operates in the avalanche breakdown region. Electron-hole pairs created by the incoming photons will accelerate in this high electric field and collide with other atoms to create an avalanche of other electron-hole pairs. Since it takes more time to build up these electron-hole pairs through the avalanche process, the APD is not as fast as the PIN diode, limiting the attainable bandwidth. Avalanche photodiodes are also more expensive, more temperature-sensitive, and, because of the high voltage levels, less reliable than PIN diodes. The main advantage of the avalanche photodiode is its sensitivity (responsivity), which is on the order of 100 times (20 dB) greater than that of a PIN diode. Its greater sensitivity makes the avalanche diode the favored choice for long-haul, high-data-rate systems, while lower cost, higher reliability, and greater temperature range of the PIN diode make it applicable to short-distance, low-bit-rate links.

Detector performance is determined by the number of photoelectrons received and type of noise present in a signaling interval of the received signal. At optical frequencies, a fundamental law of quantum theory known as the **quantum limit** establishes a bound on performance. To illustrate, consider a binary on-off (OOK) modulation scheme in which a binary 1 is represented by a burst of optical energy and a 0 by the absence of optical energy. Then, in the absence of all other forms of noise, the probability of error is determined by the presense or absence of photoelectrons at the detector. To detect a particular bit, the output of the photodiode is integrated over the signaling interval T, yielding an average number of photons arriving at the detector (when a 1 is sent) given by

$$N = E_b/h\nu \tag{8.16}$$

where E_b is the optical energy in the signaling interval. At the detector, photons have random arrival times assumed to have a Poisson distribution. Thus, the probability of detecting n photoelectrons is

$$p(n) = \frac{N^n e^{-N}}{n!} \tag{8.17}$$

To find the bound on error performance, assume that a pulse is transmitted but no photons are received. For this case, using (8.17) for the probability of observing 0 photoelectrons ($n = 0$), the probability of error becomes

$$P_e = \frac{e^{-N}}{2} \tag{8.18}$$

where we have assumed equally likely binary signals at the source. In terms of the average number of photons per bit, $P_b = N/2$, so (8.18) becomes

$$P_e = \frac{e^{-(2P_b)}}{2} \tag{8.19}$$

In practice, noise in the receiver makes this quantum limit unrealizable unless coherent detection is used (as discussed later).

At the receiver, two types of electrical noise can be distinguished: shot and thermal. **Shot noise** results from the random arrival times of photons and the resulting discrete nature of electrical charges generated at discrete points in time. Shot noise is therefore assumed to be a Poisson process. The current generated by a photodiode is a shot noise process with intensity that is proportional to the incident optical power. Even in the absence of incident light, a form of shot noise known as **dark current** will be generated in the detector due to thermal activity of electrical charges. With no incident photons present, the shot noise current is given by

$$i_s = \sqrt{2 q_e I_d W} \qquad (8.20)$$

in which I_d is the dark current and W is the bandwidth of the detector. With incident light, the photocurrent I_p will be much larger than the dark current, so that the expression for shot noise can ignore the effects of I_d, resulting in

$$i_s = \sqrt{2 q_e I_p W} \qquad (8.21)$$

Thermal noise is present in the receiver amplifier, and can be considered white noise at frequencies in the microwave band and below. Even though thermal noise becomes negligible at optical frequencies, it is significant at the frequencies (microwave and below) found in the receiver amplifier. When using a PIN diode detector, thermal noise becomes the dominant noise term. For the case of the avalanche photodiode, shot noise introduced by the avalanche gain effect is significant in addition to thermal noise.

Example 8.4 _____

For a detector with bandwidth 400 kHz, calculate the shot noise for a dark current $I_d = 50 \, nA$ and a photocurrent $I_p = 0.5 \, \mu A$.

Solution   For shot noise due to dark current, from (8.20)

$$i_s = \sqrt{(2) (1.6 \times 10^{-19}) (50 \times 10^{-9}) (400 \times 10^{3})}$$
$$= 80 \, pA$$

For shot noise due to photocurrent, from (8.21)

$$i_s = \sqrt{(2) (1.6 \times 10^{-19}) (0.5 \times 10^{-6}) (400 \times 10^{3})}$$
$$= 253 \, pA$$

_____

Detection schemes are based on direct or coherent techniques. Direct detection has been assumed so far, in which the source uses some form of intensity modulation, such as NRZ, and the detector converts incoming light directly into an electrical signal, as shown in Figure 8.11a. Coherent detection, as illustrated in Figure 8.11b, increases receiver sensitivity by use of heterodyne or homodyne detection, similar to techniques used in radio communications. A coherent receiver requires only 10 to 20 photons per bit to achieve a 10^{-9} BER, whereas direct detection via a APD requires on the order of 1000 photons per bit, an increase in sensitivity of about 20 dB that coherent detection provides. Homodyne detection requires coherent phase detection and approaches the quantum limit in performance. Heterodyne detection does not use the phase of the carrier and consequently is 3 dB less sensitive than homodyne detection. In either case, a locally generated optical signal is added to the received optical signal and the sum is detected by a PIN photodiode. The resulting signal is a replica of the original signal, translated from optical frequencies to microwave frequencies, where conventional electronics can be used to demodulate and amplify the signal. The transmitted signal and locally generated signal are both produced using single-frequency lasers.

In a homodyne receiver, both the incoming signal and the locally generated reference are at the same frequency and phase. We will consider two cases of the homodyne receiver, the first in which the local signal has the same magnitude, frequency, and phase as the received signal. Assume that the laser output is a BPSK-modulated signal given by

$$s(t) = \pm A \cos \omega_0 t \tag{8.22}$$

The idealized homodyne detector adds to the received signal a locally generated signal with amplitude A, yielding

$$r(t) = (A \pm A) \cos \omega_0 t \tag{8.23}$$

When the sum signal $r(t)$ is detected by a photodiode and the output is integrated over the signaling interval T, the average number of photoelectrons N (assuming perfect quantum efficiency) will be either $4A^2T$ or 0, in which case the probability of error is

$$P_e = \frac{e^{-(4A^2T)}}{2} \tag{8.24}$$

But since the average number of received photons per bit is $P_b = A^2T$,

$$P_e = \frac{e^{-(4P_b)}}{2} \tag{8.25}$$

This idealized homodyne receiver actually performs 3 dB better than the quantum limit and has been called the *super quantum limit* [6]. A more real-

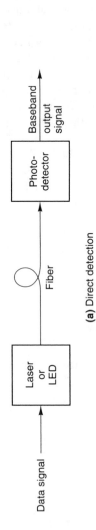

(a) Direct detection

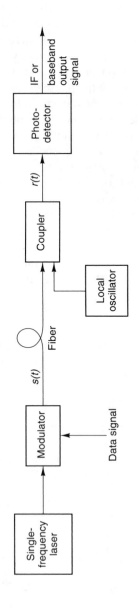

(b) Coherent detection

FIGURE 8.11 Block Diagram of Optical Transmission System

496

istic case supposes that the locally generated signal has a different magnitude B, so that the summed signal can be represented as

$$r(t) = (B \pm A) \cos \omega_0 t \qquad (8.26)$$

where we will assume $B \gg A$. The output of the integrator is then $(B \pm A)^2 T$, or expanding, $(B^2 + A^2 \pm 2AB)T$. The common term $(B^2 + A^2)T$ can be subtracted, leaving $\pm 2ABT$ corresponding to binary one and zero. For this case the probability of error becomes [6]

$$P_e = e^{-(2 P_b)} \qquad (8.27)$$

which asymptotically (large P_b) approaches the quantum limit given by (8.19).

In a heterodyne receiver, the incoming signal is translated to an intermediate frequency rather than to baseband, which is accomplished by offsetting the frequencies of the received and local signals by an amount equal to the desired IF. To determine error performance, again assume that the laser signal is a BPSK signal with magnitude $\pm A$. The locally generated signal is assumed to have magnitude B, with $B \gg A$. The addition of the two signals results in

$$r(t) = \pm A \cos \omega_0 t + B \cos \omega_1 t \qquad (8.28)$$

where ω_1 is the frequency of the local signal. Expressing $r(t)$ in terms of the envelope $E(t)$ and phase $\beta(t)$,

$$r(t) = E(t) \cos \left(\omega_1 t + \beta(t) \right) \qquad (8.29)$$

where the envelope of the signal is

$$E(t) = \sqrt{B^2 + A^2 \pm 2AB \cos (\omega_{IF} t)} \qquad (8.30)$$

and ω_{IF} is the intermediate frequency (IF) given by $\omega_0 - \omega_1$. The output of the photodiode will be a shot noise process with the mean rate of photoelectrons given by the square of the envelope, $E^2(t)$. As with the homodyne case, the common term in (8.30) can be subtracted out, leaving $\pm 2AB \cos (\omega_{IF} t)$ as the antipodal signal pair to be input to the slicer. The probability of error in this case is asymptotically

$$P_e = \text{erfc} \left(\sqrt{2A^2 T} \right) \simeq e^{-(A^2 T)} = e^{-(P_b)} \qquad (8.31)$$

where $P_b = A^2 T$ is as before the average number of photons per bit.

PSK provides the best sensitivity of any competing modulation techniques, but requires very narrow linewidth sources, on the order of 1 percent of the system bit rate. Lasers typically have a large linewidth, on the order of 5 to 50 MHz, that limits coherent detection. The combination of a laser with external optical feedback, however, has been effective in reducing the linewidths down to several kilohertz [7]. Phase noise is also a problem in lasers, manifested as a broadening of the linewidth of the laser output. As a result of these limitations in the application of PSK, other modulation techniques have also been implemented [8,9]. For example, FSK and ASK may be used but with a penalty of 3 and 6 dB, respectively, in the probability of error expressions relative to PSK. FSK is a popular choice because it has less stringent requirements on the laser linewidth and can be detected noncoherently.

The most obvious advantages to coherent detection are the greater distances allowed between repeaters due to increased sensitivity and greater bandwidth available when modulating an optical carrier. Heterodyne detection has other advantages as well. Heterodyne receivers hold a significant advantage over homodyne in that the local oscillator need not be as accurate or stable and noncoherent detection is possible. At a heterodyne receiver, IF equalization can be applied to compensate for dispersion in the optical signal [10]. But the biggest advantage lies in the ability to simultaneously transmit several carriers using a form of optical frequency-division multiplexing (FDM) known as wavelength-division multiplexing (WDM). With heterodyning, WDM channels can be selected at the receiver by simply adjusting the local oscillator, similar to dialing a channel as done in radiowave reception.

8.5   Wavelength-Division Multiplexing

The use of single-frequency optoelectronics in conventional fiber optic systems fails to take advantage of the enormous bandwidth available in the optical spectrum. The obvious solution is to stack optical channels by using multiple wavelengths, each corresponding to a separate, independent signal, much like frequency-division multiplexing was used in the days of analog carrier. The use of parallel, narrowband channels rather than a single wideband, serial channel has several advantages in fiber optics. First, since dispersion is directly proportional to bandwidth, the use of narrow rather than wideband channels mitigates the overall effects of dispersion. The use of separate channels also introduces the possibility of transporting and switching different signal types through the use of optical filtering and switching technology. This latter advantage is significant in the extension of fiber optics from conventional long-haul applications to local area network and fiber-to-the-home applications.

Design considerations for wavelength-division multiplexing (WDM) include the number of channels, separation between channels, insertion loss, and crosstalk. Where dynamic rather than fixed assignment of wavelengths is used, tuning range and tuning speed also become important parameters. The use of multiple wavelengths means that different dispersion and attenuation characteristics must be taken into account; the effect that dispersion and attenuation differences have on link design can be reduced by selecting wavelengths within the same performance window. The insertion loss is defined as the attenuation a particular wavelength λ_k suffers from its input power level, I_k, to its output power level, O_k, of the WDM device. In decibels, the insertion loss is

$$L(\lambda_k) = -10 \log \frac{I_k}{O_k} \tag{8.32}$$

Similar to crosstalk in metallic cable, crosstalk occurs in WDM because of power leakage from one wavelength to another. Using Figure 7.9 and the notation of (7.18) and (7.19), we can define P_{ij} as the amount of undesired power that is coupled from a disturbing channel λ_i into a disturbed channel λ_j. Crosstalk attenuation C_{ij} is then the difference in decibels between P_{ij} and I_i, or

$$C_{ij} = -10 \log \frac{P_{ij}}{I_i} \tag{8.33}$$

where I_i is the power in the disturbing channel. Both far-end and near-end crosstalk exists in WDM systems, just as defined for metallic cable in Figure 7.9. Proper channel separation and filtering are used to combat crosstalk in WDM systems.

One of the earliest implementations of wavelength-division multiplexing (WDM) with fixed wavelengths combined only two wavelengths, at 1300 and 1550 nm. These two signals can be multiplexed onto the same fiber, demultiplexed at each repeater and individually regenerated before being recoupled to the fiber, and separated again at the receiver for detection. Multiplexing and demultiplexing of multiple wavelengths is possible through the use of diffraction gratings in devices that can be used as either a multiplexer or demultiplexer. The basic structure of a Littrow grating multiplexer and demultiplexer [11] is shown in Figure 8.12. In the grating demultiplexer, light emerging from the fiber into the device is first collimated by a lens before striking the grating, which is oriented at an oblique angle to the optical axis. The diffractive grating will then reflect each wavelength at a different angle. Each wavelength will be spatially separated as it passes back through the lens, allowing the individual wavelengths to be separately detected or coupled to separate fibers. The multiplexing operation is identical, except that the beam directions are reversed. As an alternative to the lens, a concave mirror can be used to provide collimation and refocusing [12]. Actual implementations have accommodated 20 channels with channel spacings of 1 to 2 nm. Crosstalk attenuation between adjacent channels is typically better than 25 dB, while the minimum insertion loss is about 1 dB [13].

A second and more powerful technique possible for WDM is based on the use of tunable transmitters or receivers that can filter, select and change wavelengths, which permits dynamic connections as opposed to fixed connections. The most commonly proposed approach involves frequency-tunable receivers combined with single-frequency laser transmitters, as shown in Figure 8.13. The elements of this wavelength-tunable system are the N lasers all operating at fixed but different wavelengths carrying independent signals, an $N \times M$ passive star coupler that connects every one of the N wavelengths to each of the M lines, M wavelength-tunable filters, and M optical detectors. The laser sources must be both narrow in spectrum and stable with temperature and

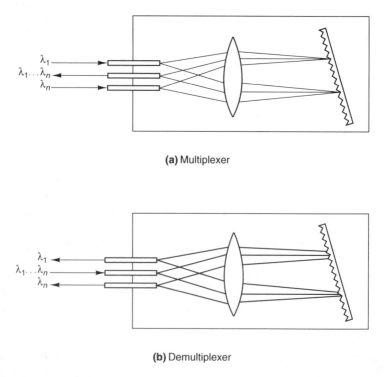

(a) Multiplexer

(b) Demultiplexer

FIGURE 8.12   Littrow Configuration of a Diffraction Grating Multiplexer and Demultiplexer (Adapted from [13])

aging, because of the close spacing between frequencies. The star coupler distributes the power of each of the N incoming signals evenly among the M outgoing fibers, but as a result suffers a loss due to signal splitting given by 10 log M dB, assuming $M > N$. Other limitations on the number of channels, N, are introduced by the choice of the tunable-filter technology. The number N is given by the ratio of the total tuning range to the channel spacing required to meet crosstalk objectives. Typical values are a 200-GHz tuning range, 1-GHz channel bandwidth, 5-GHz channel spacing, and 40-channel capacity [14].

Tunable receivers used with WDM are based on one of three basic types: interferometric filters, mode-coupling filters, and semiconductor laser diodes [14]. The first two types of tunable receivers use direct detection preceded by a tunable filter, while the third type is used with coherent detection and a tunable local oscillator. The Fabry Perot (FP) cavity described in Section 8.3 is an example of the interferometric filter. This filter has its resonant frequency tuned by finely changing the cavity length L. An important performance parameter of a resonator is its Q factor, or the *finesse* as it is called in optics. A narrow linewidth implies a high finesse, and vice versa, so that the FP filter's finesse is an indication of how many channels can be packed into the available spectrum. A couple of variations of the FP filter are worth mentioning. To reduce the size of the filter, the FP filters can be deposited directly on the fiber, by coating two fiber ends with reflective films. Two such fiber FP filters can be placed in tandem to provide greater filter finesse. Overall, FP filters have large tuning ranges (50 nm), but are constrained in switching speed to a few milliseconds due to limitations in mechanical tuning. The second filter type based on mode coupling uses acoustic-optic [15], electro-optic [16], or magneto-optic [17] effects. The acoustic-optic filter is particularly powerful because of its ability to simultaneously select multiple optical wavelengths, while also providing a very wide tuning range (400 nm) and fast (μs) switching times. Semiconductor lasers, the third type of filter, are used in coherent

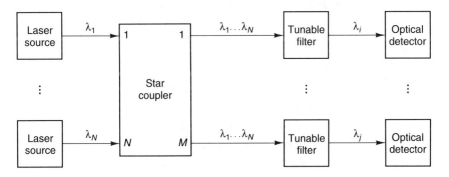

FIGURE 8.13   Wavelength-Division Multiplexing with Dynamic Connections

heterodyne detection as a tunable local oscillator. These semiconductor lasers are based on a modification of the single-frequency laser diode, such as the DFB laser, examined in Section 8.3. According to (8.12), the lasing wavelength can be tuned by changing the index of refraction of the waveguide. The resonant frequency may be tuned by changing temperature since the index of refraction rises with temperature, but this approach results in slow speed and small tuning range. A preferred strategy is to change the index of refraction by injecting current into the passive region (waveguide) of the device. Response times are on the order of one ns, so switching times are fast, but tuning ranges are limited to about 10 to 15 nm.

8.6   Repeater Spacing in Fiber Optic Systems

The maximum spacing of repeaters in a fiber optic system is determined by two transmission parameters, dispersion and attenuation, both of which increase with cable distance. Dispersion reduces the bandwidth available due to time spreading of digital signals, while attenuation reduces the signal-to-noise ratio available at the receiver. The distance between repeaters is then determined as the smaller of the distance limitation due to dispersion and that due to attenuation. In general, path length is limited by attenuation at lower data rates and limited by bandwidth at higher data rates.

To calculate repeater distance as determined by the attenuation characteristic, a power budget is required. Figure 8.14 illustrates the composition of a repeater section together with the concept of a power budget. Each component of the repeater section—transmitter, connectors, fiber, splices, and receiver—has a loss or gain associated with it. For repeater section design, it is convenient to divide the power budget into a system gain and loss. Each component has a characteristic gain or loss, usually expressed in decibels, which may be expressed in statistical terms (such as mean and standard deviation) or as a worst case, depending on the component supplier's specifications. The system gain G accounts for signal power characteristics in the transmitter and receiver of the repeater section and is given by

$$G = P_T - C_T - L_T - P_R - C_R - L_R \tag{8.34}$$

where P_T = average transmitter power

P_R = minimum input power to receiver required to achieve the error rate objective

C_T, C_R = connector loss of transmitter and receiver

L_T, L_R = allowance for degradation of transmitter and receiver with temperature variation and aging

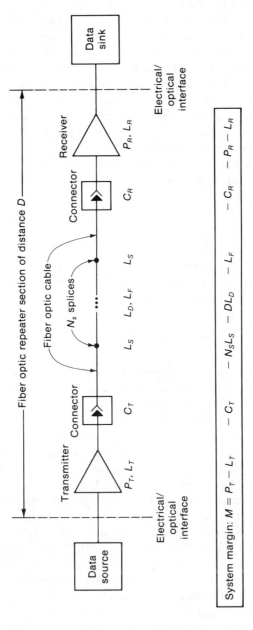

FIGURE 8.14 Fiber Optic Repeater Section Composition and Power Budget

System margin: $M = P_T - L_T - C_T - N_S L_S - DL_D - L_F - C_R - P_R - L_R$

503

The average transmitter power, P_T, depends on the peak output power of the light source, the duty cycle or modulation technique, and the loss in coupling the light source into the fiber. At both the transmitter and receiver, connector losses are included to account for joining of the fibers. Losses in a connector are due to misalignment of the two fiber ends or to mismatches in the characteristics of the two fibers, resulting in total losses of 0.5 to 1.5 dB. The receiver sensitivity and choice of modulation technique then determine P_R, the minimum required input power. Because source output power and receiver sensitivity may degrade with temperature excursions or with time, an allowance is made for these additional losses.

The loss L takes into account cable losses, including fiber attenuation and splice losses, and is given by

$$L = DL_D + L_F + N_S L_S \qquad (8.35)$$

where D = distance

L_D = average loss per unit length of fiber

L_F = allowance for degradation of fiber with temperature variation and aging

N_S = number of splices

L_S = average loss per splice

The loss in fiber optic cable consists of two components: intrinsic loss, which is due to absorption and scattering introduced in the fiber manufacturing process, and cabling loss caused by microbending. Total cable loss is found by multiplying cable distance (D) by the attenuation characteristic (L_D) and adding a loss factor (L_F) to account for the worst-case effects of temperature and age on cable loss. Losses due to splices are added to cable loss in order to yield the total loss L. Splices are necessary to join fibers when repeater spacing exceeds the standard manufacturing lengths (typically a few km). The two standard techniques are the fusion splice, accomplished by applying an electric arc to melt the glass, or the mechanical splice, consisting of a positioning device and optical cement. Fusion splices have lower losses, typically 0.01 to 0.2 dB, while mechanical splice losses are in the 0.15- to 0.5-dB range. Because splices are permanent junctions, the loss per splice can be made lower than that of a connector.

After calculation of system gain G and loss L, the system margin M is obtained as G—L. To assure that error rate objectives are met, the system margin must be a nonnegative value. Often the system designer sets a minimum acceptable margin such as 3 or 6 dB that then accounts for additional sources of degradation not included in the gain and loss calculations. If the required system margin is not met, a system redesign is required, involving the selection of an improved component (such as the transmitter, fiber, connectors, splices, or receiver) or a reduction in repeater spacing.

The calculation of fiber optic link parameters for the power budget is facilitated by the form shown in Table 8.1. Here the system designer may determine the required value for a component, such as fiber loss or receiver sensitivity, by fixing the values of all other parameters including the system margin and working back to arrive at the value of the component in question. Conversely, the system designer may fix all component values and work forward to calculate the system margin or allowed link distance. To illustrate the use of the form in Table 8.1, sample values are given for two hypothetical fiber optic systems:

1. Case 1 is a low-data-rate (OC-1 = 51.84 Mb/s), short-haul (2-km) system based on an LED transmitter, a PIN diode receiver, and 1.5-dB/km fiber attenuation.
2. Case 2 is a high-data-rate (OC-12 = 622.08 Mb/s), long-haul (120-km) system based on a laser diode transmitter, avalanche photodiode (APD) receiver, and 0.19-dB/km fiber attenuation.

Repeater spacing must also be determined as a function of dispersion, since dispersion increases with distance and limits the available bandwidth. The bandwidth required to support a given data rate depends on the baseband coding scheme. For NRZ, which is commonly used in fiber optic systems, the required bandwidth is equal to the data rate. (See Figure 5.7a for NRZ bandwidth requirements.) The dependence of bandwidth on length varies with fiber type and wavelength and in general is not linear. The relationship for calculating the bandwidth of an optical fiber section is

$$B_e = \frac{B_d}{D^\gamma} \qquad (8.36)$$

where B_e = end-to-end bandwidth in megahertz

B_d = bandwidth-distance product in megahertz-kilometers

D = distance in kilometers

γ = length dependence factor

Table 8.2 provides a convenient form for calculating the required bandwidth-distance product, B_d, according to (8.36). Sample values for the two cases described above are given in Table 8.2. This table may also be used to determine the maximum repeater spacing, D, for the required bandwidth B_e.

TABLE 8.1 Power Budget Calculation for Fiber Optic System

Parameter	Source	Sample Value for Low-Data-Rate, Short-Haul System	Sample Value for High-Data-Rate, Long-Haul System
Average transmitter power, P_T	Manufacturer's specifications	−20 dBm (LED)	0 dBm (laser diode)
Allowance for transmitter degradation with temperature variation and aging, L_T	Manufacturer's specifications	2 dB	2 dB
Transmitter connector loss, C_T	Manufacturer's specifications	1.5 dB	1.5 dB
Receiver connector loss, C_R	Manufacturer's specifications	1.5 dB	1.5 dB
Minimum receiver power input, P_R	Manufacturer's specifications	−40 dBm* (PIN diode)	−39 dBm* (APD)
Allowance for receiver degradation with temperature variation and aging, L_R	Manufacturer's specifications	2 dB	2 dB
System gain, G	$P_T - L_T - C_T - C_R - P_R - L_R$	13 dB	32 dB
Average loss per unit length of fiber, L_D	Manufacturer's specifications	1.5 dB/km	0.19 dB/km
Repeater section distance, D	System design	2 km	120 km
Allowance for fiber degradation with temperature variation and aging, L_F	Manufacturer's specifications	1 dB	3 dB
Total fiber loss	$DL_D + L_F$	4 dB	25.8 dB
Average loss per splice, L_S	Manufacturer's specifications	0.2 dB	0.1 dB
Number of splices, N_S	System design	0	20
Total splice loss	$N_S L_S$	0 dB	2 dB
Total loss, L	$DL_D + L_F + N_S L_S$	4 dB	27.8 dB
System margin, M	$G - L$	9 dB	4.2 dB

*For BER = 10^{-6}.

TABLE 8.2   Bandwidth Calculation for Fiber Optic System

Parameter	Source	Sample Value for Low-Data-Rate, Short-Haul System	Sample Value for High-Data-Rate, Long-Haul System
Data rate, R	System design	51.84 Mb/s	622 Mb/s
End-to-end bandwidth, B_e	Dependent on choice of base-band coding scheme	52 MHz for NRZ	622 MHz for NRZ
Repeater section distance, D	System design	2 km	120 km
Fiber size	System design	50/125 μm	10/125 μm
Numerical aperture, NA	Manufacturer's specifications	0.22	0.24
Wavelength	System design	1300 nm	1550 nm
Length dependence factor, γ	Manufacturer's specifications	0.88	0.90
Bandwidth-distance product, B_d	$B_e D^\gamma$	95.4 MHz-km	46.2 GHz-km

Another measure of an optical fiber system's dispersive properties and bandwidth is the **system rise time,** usually defined as the time for the voltage to rise from 0.1 to 0.9 of its final value. For detection of NRZ data, the system rise time is typically specified to be no more than 70 percent of the bit interval. Rise time is determined by the dispersive properties of each system component—transmitter, fiber, and receiver. The overall rise time is the square root of the sum of the squares of the rise times of the individual system components, multiplied by a constant k_t:

$$t_s = \text{system rise time} = k_t(t_t^2 + t_r^2 + t_{ma}^2 + t_{mo}^2)^{1/2} \qquad (8.37)$$

$$\text{where} \quad t_t = \text{transmitter rise time}$$
$$t_r = \text{receiver rise time}$$
$$t_{ma} = \text{material dispersion}$$
$$t_{mo} = \text{multimode dispersion}$$

The constant k_t is determined by the shape of the pulse at the detector. For a raised-cosine or gaussian pulse shape, k_t is approximately 1.1.

A budget for rise time is derived in much the same manner as the power budget. Once again a form such as Table 8.3 is handy in calculating the system rise time. Sample values are shown for the two cases: a low-data-rate, short-haul system and a high-data-rate, long-haul system.

TABLE 8.3 Rise Time Calculation for Fiber Optic System

Parameter	Source	Sample Value for Low-Data-Rate, Short-Haul System	Sample Value for High-Data-Rate, Long-Haul System
Data rate, R	System design	51.84 Mb/s	622 Mb/s
Maximum allowed rise time, t_s	$0.7/R$	19.2 ns	1.61 ns
Repeater section distance, D	System design	2 km	120 km
Transmitter rise time, t_t	Manufacturer's specifications	6 ns (LED)	0.3 ns (laser diode)
Coefficient of multimode dispersion, c_{mo}	Manufacturer's specifications	0.2 ns/km (graded-index)	NA (single-mode)
Length dependence factor, γ	Manufacturer's specifications	0.88	0.90
Total fiber rise time due to multimode dispersion, t_{mo}	$c_{mo}D^{\gamma}$	0.37 ns	0
Coefficient of material dispersion, c_{ma}	Manufacturer's specifications	0.1 ns/nm-km	0.02 ns/nm-km
Line spectrum, $\Delta\lambda$	Manufacturer's specifications	40 nm	0.1 nm
Total fiber rise time due to material dispersion, t_{ma}	$c_{ma}\Delta\lambda D$	8 ns	0.24 ns
Receiver rise time, t_r	Manufacturer's specifications	5 ns (PIN diode)	0.5 ns (APD)
System rise time, t_s	$1.1\,(t_t^2 + t_r^2 + t_{ma}^2 + t_{mo}^2)^{\frac{1}{2}}$	12.3 ns	0.7 ns

8.6.1   Submarine Cable Applications

Fiber optic cable has replaced metallic cable as the cable of choice for undersea applications, principally because of its cost, capacity, and performance advantages. Submarine cable systems fall into two categories, repeatered and repeaterless. Where distances are short enough, repeaterless operation is preferred, since costs are less and capacity upgrades are easier [18]. Repeaters are required for transatlantic and transpacific systems and other long distance applications. Because of the time and expense involved in recovering and repairing undersea cable, high reliability is imposed, which leads to some degree of redundancy and automatic protection switching in the design of repeaters and cable.

Undersea cable is designed to withstand the stesses associated with installation on the sea bottom and recovery for repairs. In addition, the cable is often armored for installation in shallow water to protect against trawlers, anchors, and even fish bites. Cables are also designed to protect the optical fibers from the submarine environment, including hydrostatic pressure, water ingress, and hydrogen. Repeaters likewise must be designed to withstand this environment, including depths of up to 7,500 m and temperatures of 0 to 35°C. A lightweight submarine cable typical of those used to protect against shark bites is shown in Figure 8.15. The center copper-clad steel *kingwire* is surrounded by several optical fibers. These fibers are laid straight and embedded in an elastomer, which protects the fibers and holds them in place. Two copper sections are then added along with an epoxy compound injected to protect against water seepage. Next a double layer of high tensile steel wires are added to provide strength. After another application of epoxy compound for water blocking, aluminium armor is placed between double polyethylene sheaths.

A repeatered submarine cable system is designed with a supervisory system that allows both in-service and out-of-service performance monitoring. Typical in-service measurements include received optical power, laser bias, and parity violations. For out-of-service measurements, any repeater may be interrogated and any section of the cable system may be tested via loopbacks located at each repeater. Should a failure occur in a laser, detector, or repeater electronics, or should a fiber break occur, the supervisory system can be used to isolate the problem to a single cable section. Starting at one end, repeaters can be looped back one at a time until the returned signal cannot be detected. In addition, redundancy is used, typically one fiber pair and its associated repeaters dedicated as a spare on a $N + 1$ basis, where N is the number of operational fiber pairs. A typical protection switching arrangement is shown in Figure 8.16 for a $3 + 1$ redundancy scheme. Loopback (symbolized by ‖) and switchover (symbolized by X) are arranged so that each operational fiber pair can either be switched to the spare pair or

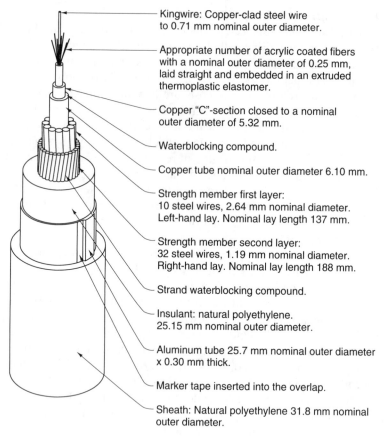

Kingwire: Copper-clad steel wire to 0.71 mm nominal outer diameter.

Appropriate number of acrylic coated fibers with a nominal outer diameter of 0.25 mm, laid straight and embedded in an extruded thermoplastic elastomer.

Copper "C"-section closed to a nominal outer diameter of 5.32 mm.

Waterblocking compound.

Copper tube nominal outer diameter 6.10 mm.

Strength member first layer: 10 steel wires, 2.64 mm nominal diameter. Left-hand lay. Nominal lay length 137 mm.

Strength member second layer: 32 steel wires, 1.19 mm nominal diameter. Right-hand lay. Nominal lay length 188 mm.

Strand waterblocking compound.

Insulant: natural polyethylene. 25.15 mm nominal outer diameter.

Aluminum tube 25.7 mm nominal outer diameter x 0.30 mm thick.

Marker tape inserted into the overlap.

Sheath: Natural polyethylene 31.8 mm nominal outer diameter.

FIGURE 8.15   Lightweight Optical Cable Used in Submarine Applications (Courtesy STC)

looped back at each repeater. In the example shown in Figure 8.16, an operational pair (pair 2) suffers a fiber break plus failed transmitter, is switched to the spare pair at repeater Q1, and switched back to pair 2 at repeater Q2; thus the failed portion of the system is bypassed, but the path is returned to pair 2 at the next available repeater location Q2. In this fashion, the spare pair can be utilized in other cable sections in the event of additional failures.

The first transatlantic fiber optic system was TAT-8, with landing stations located in the United States, United Kingdom, and France. TAT-8 employed 124 repeaters spaced an average of 53 km apart. Single-mode fibers with loss of 0.40 dB/km were used, each driven by 1.3 micron lasers. Total capacity is 560 Mb/s, supplied as 280 Mb/s on each of two operational fibers. Another transatlantic fiber optic cable, PTAT-1, has 138 repeaters

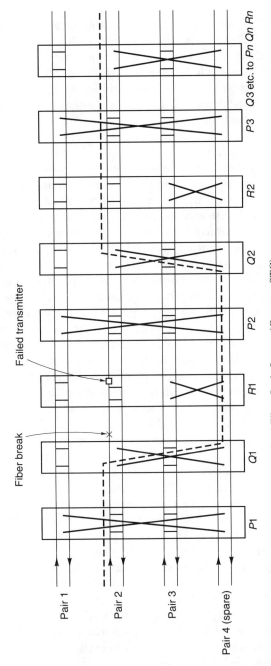

FIGURE 8.16 Protection Switching for a 3 + 1 Fiber Optic System (Courtesy STC)

and uses 420 Mb/s per fiber, with three operational fiber pairs plus one spare fiber pair. TAT-9 uses 560 Mb/s per fiber with 1.5 micron lasers, which permit increased repeater spacing (about 90 repeaters are required). Similar choices have been made for the various transpacific fiber optic cables. Future submarine cable systems will see significant improvements in optical components (sources, detectors, and fiber) along with a change from electronic to optical repeaters.

8.6.2   Optical Repeaters

Conventional repeaters make use of optoelectronics to convert the optical signal to an electrical signal for amplification and regeneration prior to conversion back to an optical signal. The electronic repeater is constrained to operate at a fixed bit rate and therefore must be replaced if the fiber optic system is to be upgraded to a higher speed. The concept of an optical amplifier, which has been recognized since the 1960s, is to replace the electronics in the repeater and create a repeater that is transparent to signal format and bit rate, allowing upgrade of a fiber optic system without replacement of the repeaters. Today the development of optical amplifiers has opened the way to less complex equipment, lower cost, higher reliability, increased repeater spacing, and increased capacity for long-haul fiber optic systems.

Optical amplifiers are based on either semiconductor laser devices or specially doped optical fiber. Lasers used as amplifiers are based on the same principle of stimulated emission used in the laser diode. In fabricating a laser amplifier, however, the cavity resonance normally found in a laser diode is suppressed, resulting in a laser that acts as an amplifier. The operation of a laser below the oscillation threshold is referred to as a traveling-wave amplifier, shown in Figure 8.17a. To create traveling waves in the laser, the most common approach is to apply an anti-reflection coating at the two ends of the structure, which then eliminates feedback and the possibility of oscillation. Gains on the order of 20 dB are possible [19], but several limitations exist with the laser amplifier. Of these problems, the most significant are low coupling efficiency between the amplifier and fiber and signal gain that is sensitive to polarization [20]. The advantage of the laser amplifier over the fiber amplifier is response time, which means that laser amplifiers are required for high-speed optical switches and modulators.

In a fiber amplifier, shown in Figure 8.17b, a suitable length (typically 5–20 m) of specially doped fiber is spliced to standard single-mode fiber. A semiconductor laser couples a high-power, continuous wave (CW) *pump* signal into the fiber amplifier. As the pump signal and the information signal travel along the fiber amplifier, energy from the pump signal is absorbed by the information signal, thus amplifying the information signal. Several rare earth elements have been used as the fiber dopant, but the most popu-

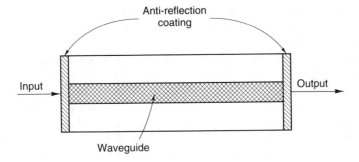

(a) Traveling-wave semiconductor laser amplifier

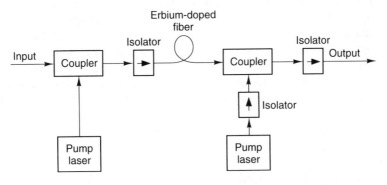

(b) Erbium-doped fiber amplifier

FIGURE 8.17   Optical Amplifiers

lar has been erbium, which produces a gain centered at 1550 nm and a 3-dB bandwidth of approximately 40 nm [20]. Other rare earth elements (neodymium and praesodymium) have been used for the 1300-nm band. The erbium-doped fiber amplifier (EDFA) takes advantage of the photoluminescent property of erbium: photons at any of several wavelengths are absorbed, which triggers the population of erbium atoms at higher energy states to release photons with a wavelength of 1550 nm. The absorbed photons are supplied by the pump laser at wavelengths of 980 nm and 1480 nm, which are absorption lines in the erbium dopant. This process is similar to that of a laser, but lasing is prevented by eliminating reflections of light in the fiber. Figure 8.17b indicates a bidirectional arrangement for pumping the amplifier at both input and output. Two other arrangements are possible: the pump signal may be fed only at the input of the amplifier, so that it travels in the same direction as the information signal (hence the term co-propagating pump); or the pump signal may be fed at the output of the amplifier so that it travels in the opposite direction of the information signal (counter-propagating pump).

The fiber amplifier has many advantages over the laser amplifier, including higher gain (30 dB is typical), higher output power (15 dBm is typical), polarization insensitivity, lower coupling loss, and lower noise figure. The gain of a fiber amplifier is slow to respond to changes in the pump or information signal, which means that the gain is independent of signal format, allowing any form of analog or digital modulation to be used. Gain of a fiber amplifier is also linear, even when operated in saturation. Thus, if several signals are present, such as a wavelength-division multiplexing system, crosstalk and intermodulation effects are negligible. The linear range of the fiber amplifier can be extended by use of automatic gain control, based on detection of power in the doped fiber and feedback control of the laser pump signal. The theoretical limitation on performance is due to noise generated by spontaneous emission and amplified along with the signal, resulting in a limit on the amplifier's signal-to-noise ratio. Fiber amplifiers have many potential applications, including both direct detection and coherent detection systems, submarine cable systems, and soliton transmission. For example, when used with coherent detection, experimental systems have demonstrated 560-Mb/s transmission over a 1028-km distance with 10 EDFAs [21] and 2.5-Gb/s transmission over a 2223-km distance with 25 EDFAs [22].

8.6.3   Soliton Transmission

A **soliton** can in general be defined as any pulse that propagates without dispersion. In the case of fiber optics, the soliton is free from chromatic dispersion, which spreads the pulse in time, and from nonlinear effects, which would broaden the frequency spectrum of the pulse. Mathematically, the soliton is a special solution to the nonlinear Schrödinger equation for a single mode fiber. This equation contains terms for chromatic dispersion, nonlinear effects, and loss. Since amplifiers properly spaced in a fiber optic system can make the average path loss zero, only the first two terms are significant. With the loss term set equal to zero, the dispersion and nonlinear terms cancel each other, and the Schrödinger equation has as its solution

$$u(z,t) = \text{sech}(t) \, \exp(iz/2) \qquad (8.38)$$

where $u(z,t)$ is known as the soliton, z is distance and t is time. The phase $\exp(iz/2)$ has no effect on the pulse shape, so that the soliton shape is independent of distance and therefore free of dispersive effects.

To force the soliton solution of Schrödinger's equation, the soliton power must be controlled to a specific value averaged over the path. Furthermore, the dispersion characteristic of the fiber, which tends to vary from one fiber section to another, must be constrained to have a constant average value over distances approximately equal to the repeater spacing. If these constraints are met, the nonlinear term associated with the average pulse power will cancel the dispersive term. The repeater spacing D must be significantly smaller than the soliton period to retain the shape of the soliton over long distances. The soliton period or unit distance z_0, the distance at which the pulse shape repeats, is given by [23]

$$z_0 = 0.322 \ \frac{\pi^2 c \tau^2}{\lambda^2 c_{ma}} \tag{8.39}$$

where τ is the soliton pulse width and c_{ma} is the chromatic dispersion averaged over the path length.

Example 8.5

Find the soliton period for a pulse width of 25 ps, a wavelength of 1560 nm, and an average chromatic dispersion of 1.0 ps/nm-km.

Solution   Substituting into (8.39), we find

$$z_0 = 0.322 \ \frac{\pi^2 (3 \times 10^8) \, (25 \times 10^{-12})^2}{(1560 \times 10^{-9})^2 (1 \times 10^{-6})}$$
$$= 245 \text{ km}$$

Soliton transmission has been realized experimentally with erbium-doped fiber amplifiers and dispersion-shifted single-mode fiber. Typical characteristics are a soliton pulse width of 30–50 ps, chromatic dispersion of a few ps/nm-km, and repeater spacings that are several tens of kilometers [24]. Ultralong distances may be spanned by solitons, making them suitable for transoceanic applications. Experiments have demonstrated error-free soliton transmission at distances over 10,000 km, sufficient to span any transoceanic application, for rates up to 10 Gb/s. Wavelength-division multiplexing is also possible over long distances, since solitons with different wavelengths are known to be transparent to each other [25].

8.7 SONET

A standard known as the **Synchronous Optical Network (SONET)** now defines a hierarchy of rates and formats to be used in optical systems, as well as other forms of high-speed digital transmission. Originally developed in the United States, the SONET standard was adopted by the CCITT but renamed the Synchronous Digital Hierarchy (SDH). These standards provide a complete set of specifications to allow national and international connections at various levels. Optical interfaces are defined that provide a universal fiber interface and permit mid-span interconnection of different vendor equipment. A standardized signal structure allows any existing hierarchical rates (for example, DS-1, DS-3, E-1, and E-3) to be accommodated. Overhead within the SONET signals facilitates synchronization, add and drop multiplexing, electronic switching, performance monitoring and network management of the composite and tributary signals. The SONET hierarchy is built on synchronous multiplexing of a basic SONET rate of 51.84 Mb/s, so that higher SONET rates are simply $N \times 51.84$ Mb/s. The basic signal structure provides sufficient flexibility to carry a variety of lower-level rates within the 51.84 Mb/s signal.

8.7.1   SONET Layer Structure

SONET uses an architecture that incorporates several layers: photonic, section, line, and path, as shown in Figure 8.18 and described below:

- The *photonic* layer prescribes such characteristics as optical pulse shape, power levels, and wavelength. Two configurations are identified corresponding to long-haul, high-performance and short-haul, low-cost applications. The long-haul version calls for optical rates up to $N = 48$, while the short-haul version is limited to rates up to OC-12 and distances up to 2 km. Single-mode fiber and wavelengths of 1300 or 1550 nm are specified as well [26].
- The *section* layer is that portion of the transmission facility—including the fiber optic cable and connectors—between an optical terminal and a repeater, between two adjacent repeaters, or between two adjacent optical terminals. The section terminating equipment (STE) consists of the network elements that perform functions such as facility performance monitoring and provides an orderwire.
- A *line* is defined as the transmission medium and associated equipment necessary to interconnect two network elements, known as line terminating equipment (LTE), which originate or terminate OC-N signals. The LTE performs line performance monitoring and automatic protection switching.
- A *path* provides the connection between the network elements, called path terminating equipment (PTE), that multiplex and demultiplex the SONET payload. The PTE provides path performance monitoring and labeling of the STS path within the SONET signal structure.

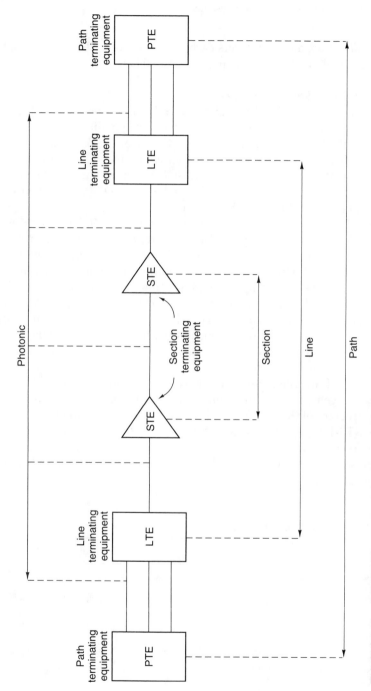

FIGURE 8.18 SONET Diagram Depicting Photonic Section, Line, and Path Definitions

The section, line, and path layers each have overhead allocated within the SONET signal to allow the functions described here.

8.7.2 SONET Signal Structure

The basic electrical signal within SONET is known as the Synchronous Transport Signal-Level 1 (STS-1), and the corresponding optical signal is known as the Optical Carrier–Level 1 (OC-1). Both have the rate of 51.84 Mb/s. Higher level signals, known as STS-N and OC-N, are at rates of N times 51.84 Mb/s. Table 8.4 lists those SONET rates that have been standardized [1], along with the equivalent number of DS-0, DS-1, and DS-3 channels carried within the SONET signal.

The STS-1 frame structure is shown in Figure 8.19. The frame is 90 columns by 9 rows of 8-bit bytes. The bytes are transmitted row by row, left to right. With a transmission rate of 8000 frames per second, the overall bit rate is then 51.84 Mb/s. The first 3 columns correspond to the *transport overhead,* and the remaining 87 columns contain the *synchronous payload envelope (SPE).* The transport overhead is further divided into *section overhead* and *line overhead.* The section overhead is contained in the first 3 rows (9 bytes) of the transport overhead, and the line overhead is found in the remaining 6 rows (18 bytes) of the transport overhead. The SPE is an 87-column-by-9-row matrix, for a total of 783 bytes. One of its columns is dedicated to path overhead and the rest is available for the payload. STS-N signals are generated by byte interleaving N frame-aligned STS-1s. The frame structure of the STS-N signal is identical to that shown in Figure 8.19 for the STS-1, except that the STS-N frame has $90 \times N$-byte columns. All of the section and line overhead channels are included in the STS-N signal, but only those associated with the first STS-1 are completely used, while others are partially used.

TABLE 8.4   SONET Digital Hierarchy

		Equivalent Number of:		
Level	Line Rate (Mb/s)	DS-3s	DS-1s	DS-0s
OC-1	51.84	1	28	672
OC-3	155.52	3	84	2,016
OC-9	466.56	9	252	6,048
OC-12	622.08	12	336	8,064
OC-18	933.12	18	504	12,096
OC-24	1244.16	24	672	16,128
OC-36	1866.24	36	1008	24,192
OC-48	2488.32	48	1344	32,256

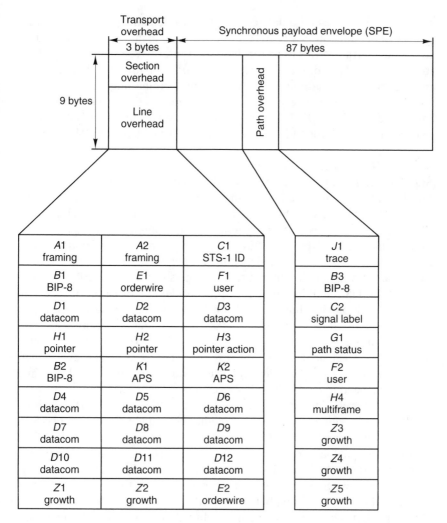

FIGURE 8.19   STS-1 Frame Structure

The use of section, line, and path overhead bytes (the photonic layer has no overhead) is shown in Figure 8.19. Section overhead contains framing bytes used to mark the beginning of each STS-1 frame; an STS-1 identification byte; a bit-interleaved parity (BIP) byte; a byte used for section orderwire; a byte reserved for the user; and 3 bytes for a data communications channel. Line overhead includes pointer bytes used to locate the path overhead; a bit-interleaved parity (BIP) byte; automatic protection switching

(APS) bytes; 9 bytes for a data communications channel; 2 bytes for future growth; and a line orderwire byte. Path overhead includes a trace byte used to connect a transmitter to its receiver; a bit-interleaved parity byte; a byte identifying the type of signal carried in the SPE; a path status byte carrying maintenance information; a user byte; a multiframe alignment byte; and bytes reserved for future growth. The line and section overhead is read, interpreted, and modified at each line and section termination, but the path overhead is part of the SPE payload and remains with it from PTE to PTE.

8.7.3   SONET Signal Synchronization

SONET equipment must contend with the same synchronization problems faced by TDM equipment, namely that tributary and STS signals may not be timed by the same clock. Thus some form of frequency justification is needed, just as the case with asynchronous multiplexers examined in Chapter 4. The existing hierarchy of pulse stuffing multiplexers, as we saw in Chapter 4, requires that each level of multiplexing and pulse stuffing must be removed by demultiplexing and destuffing, to gain access to the lowest tributary. The SYNTRAN DS-3 multiplexer solved this problem by using bit or byte synchronous multiplexing, but its disadvantage is that the tributaries must then be synchronous. SONET accommodates asynchronous signals by using byte stuffing combined with a variable mapping of the SPE into the STS, which allows lower level signals to be directly extracted from higher level signals without the need to demultiplex and destuff various intermediate layers and without the need to buffer the signals for frame synchronization.

To accommodate frequency differences, the SPE is allowed to float within the STS frame. Hence an SPE can straddle the boundary between two STS frames, as shown in Figures 8.20, 8.21, and 8.22. To keep track of the SPE's location, the payload pointer contained in the line overhead (H1 and H2 bytes) designates the location of the first byte of the SPE, which can be anywhere within the STS. The pointer gives a count of the number of bytes between H3 and J1, the first byte of the SPE. The transport overhead bytes (first three bytes in each row) are not included in the count, so for example, if the pointer number is 0, the first SPE byte immediately follows H3, whereas a pointer number of 87 indicates that the first SPE byte immediately follows K2. In either case, the first column of the SPE is path overhead.

As the STS-1 signal is passed from section to section, and line to line, timing at each location will vary slightly, resulting in some movement of the SPE relative to section and line timing. At each terminal, then, H1 and H2 must be recalculated to account for changes in local timing. An increment in the pointer is accompanied by a positive stuff byte, and a decrement in the pointer is accompanied by a negative stuff byte. For example, if the STS payload data rate is high relative to the STS frame rate, the pointer is decremented by one and the H3 overhead byte is used to carry the extra byte for

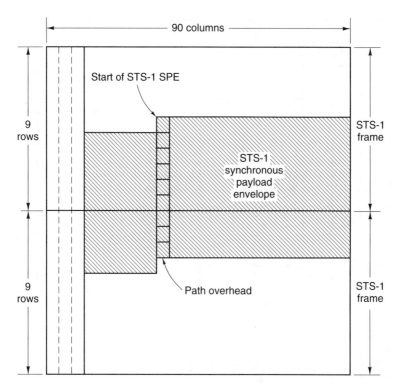

FIGURE 8.20   SONET SPE Contained Within Two STS-1 Frames (Adapted from [1])

that frame. The effect of decrementing the pointer is to create one SPE that has 782 bytes rather than 783, as shown in Figure 8.21. If the STS payload data rate is lower than the STS frame rate, the pointer is incremented by one and an extra dummy byte is added, creating a 784-byte SPE. This extra byte is placed in the SPE byte immediately following H3, as shown in Figure 8.22, and must be deleted during SPE recovery.

As described earlier, multiple STS-1 signals can be multiplexed by byte interleaving to form STS-N signals. The individual STS-1s must first be frame aligned before interleaving. Frame alignment is accomplished by using the local clock to regenerate new line and section overhead and by recalculating the pointer values for the individual SPEs. The SPEs themselves do not have to be aligned, since the pointers will enable each SPE to be uniquely located. It is also possible to concatenate several STS-1 signals to carry a higher rate signal such as the European 139.264 Mb/s rate. These higher rates (above a STS-1) are transported as a *concatenated STS-Nc* signal within a STS-M signal ($M \geq N$). The STS-Nc signal is treated as a single entity, with the constituent STS-1 signals transported together throughout

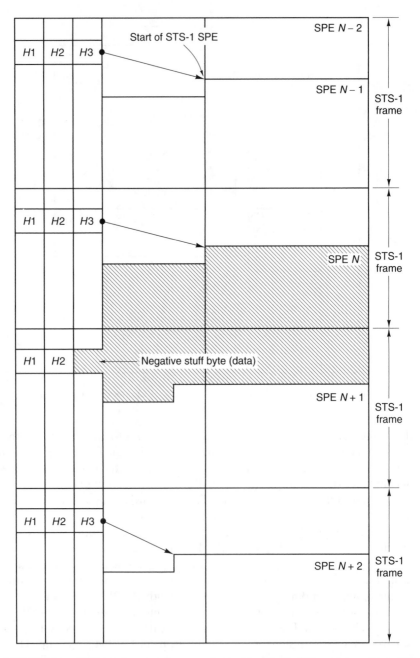

FIGURE 8.21   Pointer Decrement Accompanied by Negative Stuff Byte (Adapted from [1])

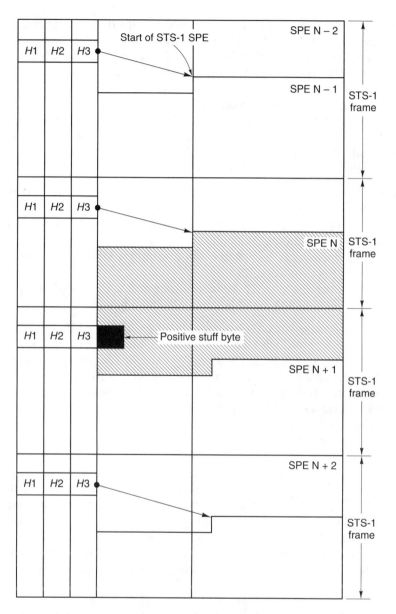

FIGURE 8.22   Pointer Increment Accompanied by Positive Stuff Byte (Adapted from [1])

the network. A concatenation indication contained in a particular STS-1 payload pointer means that the STS-1 is part of a STS-Nc. The pointer of the first STS-1 of a STS-Nc applies to all STS-1s in the group, while all subsequent STS-1 pointers contain the concatenation indication.

8.7.4   Virtual Tributary Structure

Standard hierarchical rates below an STS-1, such as a DS-1, DS-1C, DS-2, or E-1 signal, are carried within an STS-1 via a *virtual tributary* (*VT*). VTs are available in four different sizes, corresponding to the rate being carried, and occupy a different number of columns within the 9-row SPE:

- A VT1.5 operates at 1.728 Mb/s, can transport a DS-1, and occupies 3 columns (27 bytes) per SPE
- A VT2 operates at 2.304 Mb/s, can transport an E-1, and occupies 4 columns (36 bytes) per SPE
- A VT3 operates at 3.456 Mb/s, can transport a DS-1C, and occupies 6 columns (54 bytes) per SPE
- A VT6 operates at 6.912 Mb/s, can transport a DS-2, and occupies 12 columns (108 bytes) per SPE

To combine these virtual tributaries, A *VT group* is used that occupies 12 columns of 9 rows each (108 bytes) and is able to transport 4 VT1.5s, 3 VT2s, 2 VT3s, or 1 VT6, as shown in Figure 8.23. The SPE may contain up to 7 VT groups, in which case the 7 groups occupy 84 columns, 2 unused columns are filled with stuff bytes, and the remaining column contains the path overload. Different types of VT groups may be carried within the same SPE, but a VT group may contain only one type of VT.

Two different modes, *floating* and *locked,* are used to transport payloads within a VT. The floating mode allows the VT SPE to move within the VT payload structure, which is analogous to the operation of the floating SPE within a STS-1. In this floating mode a 500-μsec structure is created known as the VT superframe, which consists of four consecutive 125-μsec frames of the STS-1 SPE. There are four different versions of the VT superframe corresponding to the four different types of VTs. The VT superframe contains the VT pointer and the VT synchronous payload envelope (SPE), as shown in Figure 8.24. The VT pointer consists of four bytes, V1, V2, V3, and V4, one for each of the four frames in the VT superframe. The VT pointer value indicates the offset from V2 to the first byte of the VT SPE. Frequency justification between the VT SPE and local clock is also accomplished by the VT pointer, similar to use of the STS-1 pointer for STS-1 SPE frequency justification, by using positive byte stuffing immediately following the V3 byte or negative byte stuffing using V3 as a data byte. The VT SPE contains one byte of path overhead (V5), with all remaining bytes allocated to the VT payload capacity. To provide payloads greater than a VT6 within a STS-1,

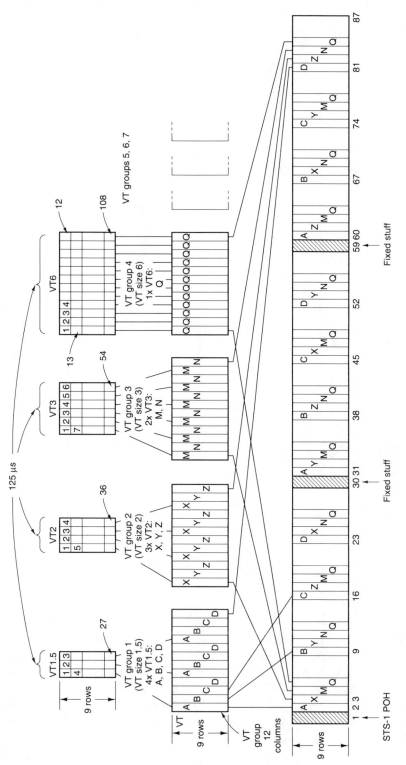

FIGURE 8.23 Example of STS-1 SPE Carrying Virtual Tributary (VT) Groups (Adapted from [1])

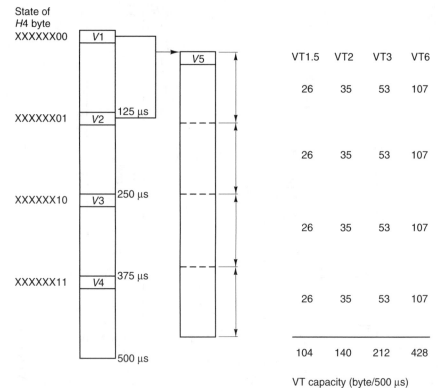

V1 VC pointer 1
V2 VC pointer 2
V3 VC pointer 3 (action)
V4 Reserved

FIGURE 8.24   SONET Virtual Tributary (VT) Superframe (Adapted from [1])

VT6s may be concatenated to form a VT6-Nc that accommodates N VT6s. A concatenation indication is then used in the VT pointers, similar to the convention used for STS-Nc.

The locked mode provides a fixed mapping of the VT payload into the STS-1 SPE, thus eliminating the need for VT pointers. Hence all 108 bytes of a VT group are available for the payload. Because no VT overhead is available, mappings are applicable to synchronous signals, and performance monitoring is limited to the STS-1. A mix of floating and locked modes is not permissible within a STS-1, but there is a conversion available for each of the four VT groups [1]. Each mode also allows different types of payload mappings for each hierarchical rate, which are summarized in Table 8.5 and described in the following subsections.

TABLE 8.5   SONET Mappings of Sub-STS-1 Signals

Mappings	Virtual Tributary (VT) Modes	
	Floating VT	Locked VT
Asynchronous	DS-1, DS-1C, DS-2	
	E-1	
Byte-Synchronous	DS-1	DS-1
	E-1	E-1
		SYNTRAN
Bit-Synchronous	DS-1	DS-1
	E-1	E-1

Asynchronous Mapping

An asynchronous mapping handles an asynchronous signal, derived from any clock source, as a bulk transport. Only the floating mode of VT operation is used, with positive pulse stuffing. Applicable rates are DS-1, E-1, DS-1C, and DS-2. As an example, a VT superframe based on an asynchronous DS-1 signal is shown in Figure 8.25. The two sets $(C1, C2)$ of stuff control bits provide 3-bit codes for the two stuffing opportunities $(S1, S2)$. The advantages of asynchronous mapping are an unrestricted payload, timing transparency, and minimum mapping delay, while its main disadvantage is the inability to access DS-0s within the DS-1.

Bit-Synchronous Mapping

Bit-synchronous mappings are used with clear (unframed) signals that are synchronized with the SONET transport signal. Both the floating and locked modes are applicable, but rates are limited to DS-1 and E-1. The floating mode applied to a DS-1 signal results in a mapping similar to Figure 8.25, except that the stuff control bits (C and S bits) are unnecessary and are replaced by overhead and fixed stuff bytes. The locked mode for DS-1 signals has a 26-byte frame, with two overhead and fixed stuff bytes followed by 24 bytes of data. Bit-synchronous mapping does not require DS-1 or E-1 framing but does require synchronous signals, while allowing either channelized or nonchannelized DS-1 and E-1 signals.

Byte-Synchronous Mapping

Byte-synchronous mappings provide DS-0 visibility and are therefore appropriate where DS-0 signals are to be added, dropped, or cross-connected directly from an STS signal. Both floating and locked modes are used, for rates of DS-1 and E-1. As Figure 8.26 indicates, the DS-0 signals and their associated signaling within a DS-1 are individually accessible

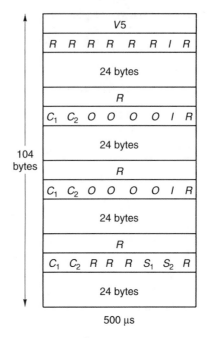

FIGURE 8.25   SONET Mapping of Asynchronous DS-1 (Adapted from [1])

within either the floating or locked modes. In the case of the floating mode, the S bits contain signaling for the 24 DS-0 channels, the F bit contains the DS-1 frame bit, and the P bits provide VT superframe synchronization. In the locked mode, a multiframe indicator byte (H4) provides synchronization of the signaling bits. For both the floating and locked VT modes, superframe and extended superframe formats are available along with 2-, 4-, or 16-state signaling. Robbed-bit signaling on any incoming DS-1 signal is replaced with dedicated signaling bits. Byte-synchronous mapping requires both DS-0 channelization and framed signals to allow access to the DS-0 channels and signaling bits.

8.7.5   Higher Level SONET Mapping
Several mappings have been defined for signals that occupy an entire STS-1 and for signals that occupy an entire STS-3c. Both asynchronous DS-3 and

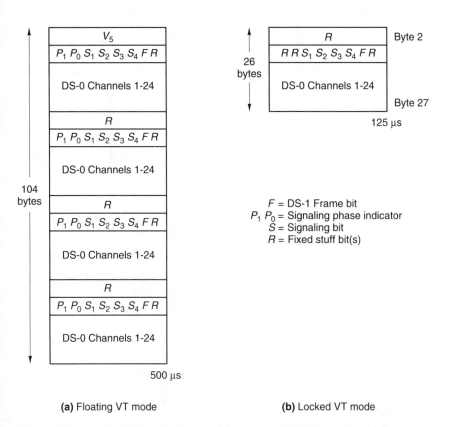

(a) Floating VT mode   **(b)** Locked VT mode

FIGURE 8.26   SONET Byte-Synchronous Mapping for DS-1 (Adapted from [1])

SYNTRAN formats can be mapped into a STS-1. The DS-3 mapping into an STS-1 is shown in Figure 8.27. A set of nine subframes is specified within a 125-μs frame, with each subframe consisting of one row (87 bytes) of the STS-1 SPE. One stuffing opportunity (S bit) and five associated stuff control (C) bits are included in each subframe, together with fixed stuff (R) and overhead (O) bits.

Figure 8.28 is the mapping of a 139.264 Mb/s signal (known as the E-4 or DS-4NA) into a STS-3c. Each row of the STS-3c SPE consists of a column for the path overhead followed by 260 columns of payload. To accommodate the 139.264-Mb/s signal, the payload capacity has each row divided into 20 blocks of 13 bytes each. The first byte of each block consists of either information (I) bits, stuff (S) bits, stuff control (C) bits, fixed stuff (R) bits, or overhead (O) bits. The last 12 bytes of a block consists of information bits. Each row has a single stuff opportunity and 5 stuff control bits associated with it.

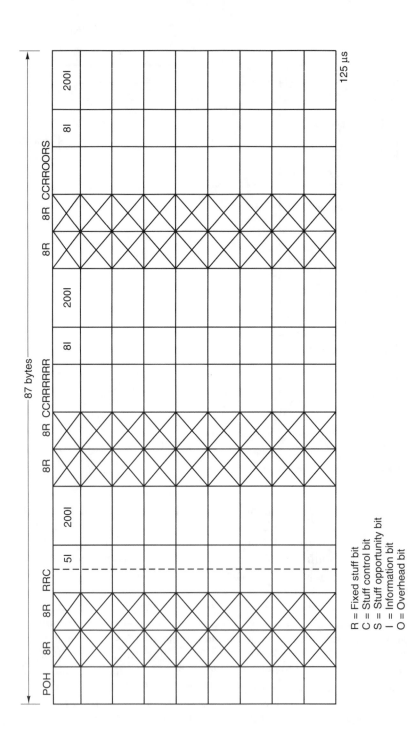

R = Fixed stuff bit
C = Stuff control bit
S = Stuff opportunity bit
I = Information bit
O = Overhead bit

FIGURE 8.27   SONET Mapping of Asynchronous DS-3 (Adapted from [1])

530

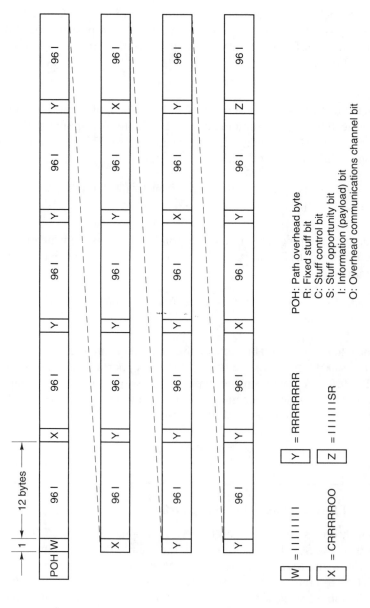

FIGURE 8.28 SONET Mapping of Asynchronous 139.264-Mb/s Signal (Adapted from [1])

POH: Path overhead byte
R: Fixed stuff bit
C: Stuff control bit
S: Stuff opportunity bit
I: Information (payload) bit
O: Overhead communications channel bit

W = I I I I I I I I
X = CRRRROO
Y = RRRRRRR
Z = I I I I I I SR

Example 8.6 _____

For the asynchronous DS-3 mapping shown in Figure 8.27, find (a) the minimum possible DS-3 data rate; (b) the maximum possible DS-3 data rate; (c) the stuff opportunity rate; and (d) the stuffing ratio for a 44,736,000-b/s data rate.

Solution
(a) Each row carries a total of 696 bits: 621 data bits, one stuff bit, and the rest overhead bits. The minimum possible DS-3 rate corresponds to the case where all stuff bits are used, so that there are exactly 621 data bits per row, yielding a minimum rate of

(621 data bits per row) × (9 rows per frame) × (8000 frames per sec)
= 44.712 Mb/s

(b) The maximum possible rate corresponds to the case in which none of the stuff bits are used, that is, they all carry data bits, which yields a maximum rate of

(622 data bits per row) × (9 rows per frame) × (8000 frames per sec)
= 44.784 Mb/s

(c) Since there is one stuff bit per row, the stuff opportunity rate is (9 rows per frame) × (8000 frames per second) = 72,000 stuff opportunities per second.
(d) The difference between the maximum possible rate and the nominal rate is 48,000 b/s, which is the number of bit locations which must be stuffed. Using (4.27), we have that the stuffing ratio is

$$\rho = \frac{48,000 \text{ b/s}}{72,000 \text{ b/s}} = 0.667$$

8.8   CCITT Synchronous Digital Hierarchy

The Synchronous Digital Hierarchy (SDH) adopted by the CCITT in 1988 is based on the SONET structure, with the main difference being that the first level of the SDH is at the 155.52-Mb/s (STS-3) rate rather than the SONET STS-1 rate of 51.84 Mb/s. CCITT Recs. G.707, G.708, and G.709 prescribe the bit rates, signal formats, multiplexing structure, and payload mappings for the Network Node Interface (NNI) of the SDH [2]. The NNI is the interface between the transmission facility, such as a fiber optic or digital radio link, and the network node which performs termination, switching, cross-connection, and multiplexing. The main feature of the SDH is the integration of various user rates and digital hierarchies (North American, European, and Japanese). As with SONET, the SDH provides synchronous multiplexing, digital cross-connects, advanced network management and performance monitoring, and flexibility to accommodate existing and future services. The following description of SDH is based on CCITT recommendations, which use different terminology than used here for SONET. However, the two standards agree on frame structure and byte assignments at those levels that are common to both, allowing SONET and SDH to be compatible with one another.

The basic building block of the SDH is the Synchronous Transport Module (STM-N) with $N = 1, 4, 16$, and so on. The STM-1 has a 9-row-by-270-column structure based on the SONET STS-3 and operates at the STS-3 rate of 155.52 Mb/s. The STM-N structure is formed by single-byte interleaving N STM-1 signals. The basic frame structure of the STM-1 is shown in Figure 8.29. The section overhead and administration unit (AU) pointers occupy the first 9 columns of the 9-row frame. The section overhead provides framing, error checking, a data communications channel, protection switch control, and maintenance. The AU pointer occupies the fourth row of columns 1–9; additional AU pointers may be located in rows 1–3 of columns 11–14. The first column of the payload is dedicated to path overhead (POH), which performs error checking and maintenance. The byte assignments for the section and path overhead are identical to the SONET section, line, and path overhead shown in Figure 8.19.

The multiplexing structure of SDH shown in Figure 8.30 is based on several elements defined as follows:

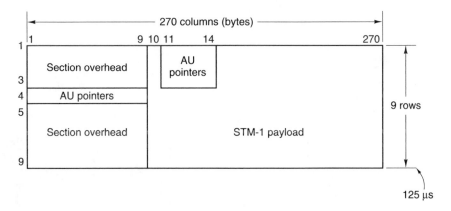

FIGURE 8.29   Synchronous Digital Hierarchy STM-1 Frame Structure (Courtesy CCITT [2])

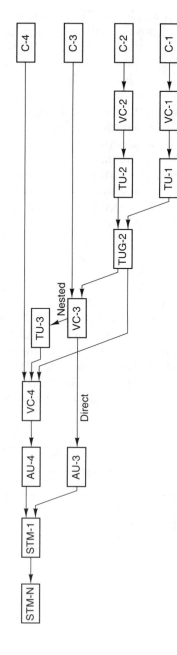

FIGURE 8.30 Multiplexing Structure of Synchronous Digital Hierarchy (Courtesy CCITT [2])

- A *container C-n* ($n = 1$ to 4) is equivalent to standard rates from the North American, European, or Japanese hierarchies, for hierarchical levels (n) one through four.
- A *virtual container VC-n* ($n = 1$ to 4) consists of a single container plus its path overhead, or an assembly of tributary unit groups together with the path overhead.
- A *tributary unit TU-n* ($n = 1$ to 3) consists of a virtual container plus a tributary unit pointer. The pointer specifies the starting point of a VC-n within a VC-n + 1.
- An *administrative unit AU-n* ($n = 3,4$) consists of a VC-n ($n = 3,4$) plus an administrative unit (AU) pointer. The AU pointer, which has a fixed location within the STM-1, specifies the starting point of the VC-n within the STM-1.
- A *tributary unit group TUG-n* ($n = 2,3$) is a group of tributary units formed by multiplexing.

The STM-1 payload can support 1 AU-4 (user rate of 139.264 Mb/s), 3 North American level AU-3s (equivalent to 3 44.736 Mb/s signals), or 4 European level AU-3s (equivalent to 4 34.368 Mb/s signals). Each VC-n associated with a AU-n does not have a fixed-phase relationship with the STM-1 frame but instead is allowed to float within its AU. The AU pointer provides the offset in bytes between the pointer position and the first byte of the VC. Frequency differences between the frame rate of the section overhead and that of the VC are compensated with stuff bytes. Further nesting is possible of tributary units within a virtual container. For example, a VC-4 within a AU-4 may carry 3 TU-3s each associated with the 44.736-Mb/s rate or 4 TU-3s each associated with the 34.368-Mb/s rate. The TU-3 pointer is located at a fixed position within the VC-4, and this pointer identifies the location of the first byte of the VC-3.

To illustrate the equivalence of SONET and SDH at comparable levels, we will consider the two SONET examples shown in Figures 8.27 and 8.28 for DS-3 and DS-4NA mappings. For the mapping of an asynchronous DS-3, the SDH uses a VC-3 whose frame structure is identical to Figure 8.27 except that two columns of fixed stuff bytes are missing, resulting in a 85-column rather than 87-column payload. When the VC-3 is mapped into its AU-3, however, two columns of fixed stuff bytes are inserted into the VC-3 payload, making it equal to the AU-3 payload capacity and also equal to the 87-column payload of the STS-1 SPE. The mapping of an asynchronous DS-4NA into a VC-4 is identical to Figure 8.28. In addition to these two examples, the other mappings defined for SONET have corresponding mappings in SDH. The virtual tributary modes and mappings shown in Table 8.5 all have identical, corresponding modes (floating and fixed) and mappings (asynchronous, bit-synchronous, and byte-synchronous) within SDH. Because the SDH is an international standard, there are additional features not found in SONET, making SONET a subset of the SDH.

Optical interfaces and line systems for SDH are also the subject of CCITT recommendations [27]. Performance requirements are given in Rec. G.958, including jitter, error, and availability, along with other specifications on optical line systems. Optical interfaces are prescribed in Rec. G.957 for three general applications, intraoffice (2 km or less), short haul up to 15 km, and long haul. Single-mode fiber is specified for each of these three applications, using Recs. G.652, G.653, and G.654 [3]. Rec. G.957 also prescribes power, pulse shape, attenuation, spectral width, dispersion, and receiver sensitivity characteristics. Particular choices for optical sources and receivers are avoided to allow freedom in implementation and innovation in technology.

8.9 Summary

Fiber optic systems offer greater bandwidths, lower attenuation, and no crosstalk or electrical interference compared to metallic cable, and these advantages have led to dramatic growth in fiber optic systems worldwide. Low-data-rate, short-haul fiber optic systems tend toward multimode cable, LED transmitters, and PIN diode receivers; high-data-rate, long-haul systems tend toward single-mode cable, laser diode transmitters, and avalanche photodiode receivers. Tables 8.6 and 8.7 provide a summary of salient characteristics for optical sources and detectors.

Repeater spacing in fiber optic systems is determined by dispersion and attenuation characteristics. A power budget (see Table 8.1) accounts for attenuation effects by summing the loss or gain of each system component, including transmitter, connectors, fiber, splices, and receiver. The net difference between gains and losses is equivalent to a system margin, which protects against degradation not included in the power budget. Like attenuation, dispersion increases with distance; thus dispersion reduces the available

TABLE 8.6   Characteristics of Optical Sources

Characteristic	Light Emitting Diode (LED)	Laser Diode (LD)
Output power, mW	1 to 10	5 to 100
Coupling loss, dB	−10 to −30	−2 to −10
Spectral width, nm	35 to 100	2 to 5
Rise time, ns	2 to 50	<1
Life, hrs	10^5 to 10^6	10^4 to 10^5
Cost	Lower	Higher
Application	Short distances, low bit rates	Long distances, high bit rates

TABLE 8.7   Characteristics of Optical Detectors

Characteristic	PIN Photodiode	Avalanche Photodiode
Receiver sensitivity, dBm	−30 to −45	−40 to −60
Responsivity, A/W	0.4 to 0.7	0.6 to 0.7
Quantum efficiency, %	50 to 80	80
Avalanche gain	1.0	80 to 150
Rise time, ns	1 to 10	<1
Operating temperature, °C	−55 to 125	−40 to 70
Cost	Lower	Higher
Application	Short distances, low bit rates	Long distances, high bit rates

bandwidth and allowed repeater spacing. The calculation of repeater spacing as a function of dispersion is based on the required bandwidth (see Table 8.2) and system rise time (see Table 8.3). Figure 8.31 illustrates the relationship of bit rate (bandwidth) and repeater spacing as a function of attenuation, dispersion, and source linewidth, for a system consisting of a 1540-nm transmitter, a dispersion-shifted fiber with zero dispersion at 1550 nm, and a direct detection receiver. Attenuation tends to be the limiting factor for low-data-rate applications, while dispersion sets the limit for higher speed applications.

Latest generation fiber optic systems have introduced innovations that have significantly improved the bandwidth and repeater spacing possible.

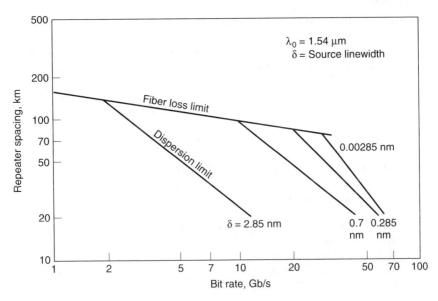

FIGURE 8.31   Repeater Spacing versus Data Rate (Courtesy Alcatel)

Coherent detection via either homodyne or heterodyne techniques allows much greater bandwidths to be realized. Several wavelengths can be transmitted simultaneously in wavelength-division multiplexing, analogous to frequency-division multiplexing used in telephony. Optical amplifiers are now available that eliminate electronics and instead use specially doped fiber or semiconductor laser devices. The use of optical amplifiers will allow a fiber optic system to be upgraded in bit rate without changeout of the repeaters. Optical amplifiers have also been used to achieve ultralong distances via soliton transmission, which is the transmission of an idealized pulse without loss of pulse shape.

Standards for fiber optic transmission have been developed, initially in North America under the name Synchronous Optical Network (SONET) and later by the CCITT using the name Synchronous Digital Hierarchy (SDH). These standards offer several important advantages. First, existing digital hierarchies (North American, European, and Japanese) are incorporated as standard interfaces, thus providing an integration of these diverse digital bit rates. Flexibility in the SONET/SDH allows a synchronous network to be created from the existing asynchronous environment. New services may be easily introduced, including broadband high-speed data rates from 1.544 Mb/s to 150 Mb/s. Finally, greater overhead capacity provides improved maintenance, performance monitoring, control, and network management.

References

1. ANSI T1.105-1990, *Digital Hierarchy—Optical Interface Rates and Formats Specification*, 1990.
2. CCITT Blue Book, vol. III.4, *General Aspects of Digital Transmission Systems; Terminal Equipment* (Geneva: ITU, 1989).
3. CCITT Blue Book, vol. III.3, *Transmission Media Characteristics* (Geneva: ITU, 1989).
4. MIL-HDBK-415, *Design Handbook for Fiber Optic Communications Systems*, U.S. Department of Defense, February 1, 1985.
5. Charles Kittel, *Introduction to Solid State Physics* (New York: John Wiley, 1968).
6. J. Salz, "Modulation and Detection for Coherent Lightwave Communications," *IEEE Comm. Mag.*, June 1986, pp. 38–49.
7. R. A. Linke, "Optical Heterodyne Communications Systems," *IEEE Comm. Mag.*, October 1989, pp. 36–41.
8. K. Nosu, "Advanced Coherent Lightwave Technologies," *IEEE Comm. Mag.*, February 1988, pp. 15–21.
9. R. S. Vodhanel and others, "10 Gbit/s Modulation Performance of Distributed Feedback Lasers for Optical Fiber Communication Systems," *1990 Global Telecommunications Conference*, pp. 803.6.1–803.6.7.

10. J. H. Winters, "Equalization in Coherent Lightwave Systems Using a Fractionally-Spaced Equalizer," *1990 Global Telecommunications Conference,* pp. 503.5.1–503.5.5.

11. G. Winzer, "Wavelength Multiplexing Components—A Review of Single-Mode Devices and Their Applications," *Journal of Lightwave Technology,* vol. LT-2, no. 8, August 1984, pp. 369–378.

12. H. Ishio and others, "Review and Status of Wavelength-Division Multiplexing Technology and Its Applications," *Journal of Lightwave Technology,* vol. LT-2, no. 8, August 1984, pp. 448–463.

13. S. S. Wagner and H. Kobrinski, "WDM Applications in Broadband Telecommunication Networks," *IEEE Comm. Mag.,* March 1989, pp. 22–30.

14. H. Kobrinski and K. W. Cheung, "Wavelength-Tunable Optical Filters: Applications and Technologies," *IEEE Comm. Mag.,* October 1989, pp. 53–63.

15. K. W. Cheung and others, "Wavelength-Selective Circuit and Packet Switching Using Acoustic-Optic Tunable Filters," *1990 Global Telecommunications Conference,* pp. 803.7.1–803.7.7.

16. W. Warzanskyj, F. Heisman, and R. C. Alferness, "Polarization-Independent Electro-Optically Tunable Narrowband Wavelength Filter," *Applied Phys. Lett* 53 (1988):13–15.

17. A. Shibukawa and M. Kobayashi, "Optical TE-TM Mode Conversion in Double Epitaxial Garnet Waveguides," *Applied Optics* (1981):24–44.

18. P. R. Trischitta and D. Chen, "Repeaterless Undersea Lightwave Systems," *IEEE Comm. Mag.,* March 1989, pp. 16–21.

19. M. J. O'Mahony, "Semiconductor Laser Optical Amplifiers for Use in Future Fiber Systems," *Journal of Lightwave Technology,* vol. LT-6, no. 4, April 1988, pp. 531–544.

20. M. J. O'Mahony, "Developments in Optical Amplifier Technology and Systems," *1990 Global Telecommunications Conference,* pp. 706A.4.1–706A.4.5.

21. S. Ryu and others, "Long-Haul Coherent Optical Fiber Communication Systems Using Optical Amplifiers," *Journal of Lightwave Technology,* vol. LT-9, no. 2, February 1991, pp. 251–260.

22. S. Saito, T. Imai, and T. Ito, "An Over 2200-km Coherent Transmission Experiment at 2.5 Gb/s Using Erbium-Doped Fiber In-Line Amplifiers," *Journal of Lightwave Technology,* vol. LT-9, no. 2, February 1991, pp. 161–169.

23. L. F. Mollenauer, J. P. Gordon, and M. N. Islam, "Soliton Propagation in Long Fibers with Periodically Compensated Loss, *IEEE Journal on Quantum Electronics,* vol. QE-22, no. 1, January 1986, pp. 157–173.

24. L. F. Mollenauer, S. G. Evangelides, and H. A. Haus, "Long-Distance Soliton Propagation Using Lumped Amplifiers and Dispersion Shifted Fiber," *Journal of Lightwave Technology,* vol. LT-9, no. 2, February 1991, pp. 194–197.

25. L. F. Mollenauer, S. G. Evangelides, and J. P. Gordon, "Wavelength Division Multiplexing with Solitons in Ultra-Long Distance Transmission Using Lumped Amplifiers," *Journal of Lightwave Technology,* vol. LT-9, no. 3, March 1991, pp. 362–367.

26. ANSI T1.106-1988, *Telecommunications—Digital Hierarchy Optical Interface Specifications: Single Mode*, 1988.

27. R. Balcer and others, "An Overview of Emerging CCITT Recommendations for the Synchronous Digital Hierarchy: Multiplexers, Line Systems, Management, and Network Aspects," *IEEE Comm. Mag.*, August 1990, pp. 21–25.

Problems

8.1 Using Snell's Law, derive: (a) Equation (8.2); and (b) Equation (8.3).

8.2 A step-index fiber has a core and cladding with indices of refraction 1.43 and 1.39, respectively. Find: (a) The numerical aperture; (b) the acceptance angle; and (c) the critical angle.

8.3 This problem considers the calculation of multimode dispersion using the difference of minimum versus maximum propagation delay.

(a) Derive an expression for the mode propagating with angle zero (minimum delay mode) as a function of the length of the cable D and speed of light in the fiber c/n.

(b) Derive an expression for the mode propagating at the critical angle (maximum delay mode) as a function of D, c/n, and θ_c.

(c) Using results from parts (a) and (b), find the multimode dispersion for 1000 meters of the fiber given in Problem 8.2.

8.4 A 1-ns pulse is transmitted over a single-mode, 50-km fiber and has a received pulse width of 1.02 ns. If the fiber is used with a laser having linewidth 3 nm, find the intramodal dispersion coefficient of the fiber.

8.5 A typical graded-index fiber has a parabolic profile whose index of refraction varies with r, the distance from the fiber center, according to:

$$n(r) = n_1 \left\{ 1 - \left[(NA/n_1)^2 (r/a)^2 \right] \right\}^{1/2}$$

where a is the core radius. For $n_1 = 1.45$ and $n_2 = 1.4$, find the index of refraction at (a) $r = 0$; (b) $r = a$; and (c) $r = a/2$.

8.6 Consider a Lambertian optical source operating with a fiber having $NA = 0.21$.

(a) Derive an expression that gives the coupling loss in decibels due to the mismatch between the source and fiber core diameters. Assume that the optical source diameter is greater than the fiber core diameter.

(b) Using the expression found in part (a), calculate the coupling loss for a source with diameter 75 μm and a fiber core with diameter 50 μm.

(c) Calculate the coupling loss due to NA mismatch, and find the total coupling loss due to diameter and NA mismatches.

8.7 When optical fibers of different characteristics are joined together, losses will occur due to mismatches in the core diameters and in the NAs.

(a) Derive an expression for the loss in decibels due to core diameter mismatch.

(b) Derive an expression for the loss in decibels due to NA mismatch.

(c) Using results from parts (a) and (b), find the total loss in joining two fibers, one with core of 50 µm and NA = 0.22 and the other fiber with core of 62.5 µm and NA = 0.275.

8.8 Thermal noise is known to have noise density given by

$$N(v) = \frac{hv}{e^{hv/kT} - 1}$$

where k is Boltzman's constant (1.38×10^{-23} joules per degree Kelvin) and T is the temperature in degrees Kelvin. Assuming room temperature with $T = 290°$:

(a) Find the noise density at the optical frequency corresponding to a wavelength of 1.3 µm.

(b) Simplify this expression for frequencies in the microwave band by using the approximation $e^x \simeq 1 + x$ for small x. Calculate the noise density for a frequency of 2 GHz.

(c) Compare hv with kT in parts (a) and (b). What conclusion can you draw concerning the effects of thermal noise at optical frequencies versus microwave frequencies?

8.9 The incident optical power to a photodiode is 2×10^{-9} watts. The wavelength is 1550 nm and the quantum efficiency is 80 percent. Find the maximum current output of the photodiode.

8.10 A 10-Gb/s signal is to be transmitted at a wavelength of 1550 nm. Find the received power necessary to receive 1000 photons per bit.

8.11 A single-mode fiber with intramodal dispersion of 0.03 ns/nm-km is to be used with a laser which has a linewidth of 0.5 nm.

(a) Ignoring the contribution to rise time from other components, what bandwidth is supported by the fiber if NRZ modulation is assumed?

(b) What bandwidth-distance product is required of the fiber for a 100-km link if the length dependence factor γ is 0.9?

8.12 An LED and fiber combination has a power efficiency of 10 percent and a coupling loss of 20 dB. If the fiber has an attenuation of 0.5 dB/km, what will be the received power if the input power to the LED is 10 mw and the link distance is 30 km?

8.13 Calculate the system rise time and bandwidth for a fiber optic system that uses components with the following specifications:

LED rise time = 5 ns
PIN photodiode rise time = 3 ns
Coefficient of multimode dispersion = 1 ns/km
Coefficient of material dispersion = 50 ps/nm-km
Distance = 20 km
Length dependence factor = 0.85
Line spectrum = 40 nm
Modulation = NRZ

8.14 For a choice of return-to-zero as the optical modulation technique:
  (a) Find an expression for the maximum allowed rise time as a function of the bit rate.
  (b) Calculate the rise time for a bit rate of 500 Mb/s.

8.15 A fiber optic link is designed to operate at a rate of 140 Mb/s over a distance of 50 km, using NRZ modulation, for the following component values:

Connector at transmitter and receiver = 1.5-dB loss each
Loss per unit length of fiber = 0.6 dB/km
Number of splices = 50
Average loss per splice = 0.2 dB
Minimum input to receiver to achieve BER objective = –58 dBm
Allowance for degradation with temperature variation and aging = 2 dB
  each for transmitter, fiber, and receiver
Length dependence factor = γ = 1.0
Transmitter power = –2 dBm

  (a) Calculate the resulting system margin (in dB).
  (b) Calculate the bandwidth-distance product required to meet the end-to-end bandwidth of the system.
  (c) Find the distance possible if the fiber loss is reduced to 0.4 dB/km and the receiver threshold is improved by 3 dB. Assume the same system margin found in part (a) and that the number of splices remains unchanged.

8.16 Consider a fiber optic system with a wavelength = 0.85 microns, a core index of refraction = 1.5, and a cladding index of refraction = 1.485.

(a) What is the maximum angle at which a ray may enter the fiber and be totally reflected?

(b) If the transmit level is 0 dBm, the receiver threshold is –40 dBm, the coupling losses at the transmitter and receiver are 2 dB each, and the fiber loss is 0.25 dB/km, what distance is possible between repeaters?

(c) For the receiver threshold = –40 dBm, find the number of received photons per bit if the bit rate is 150 Mbps.

8.17 A fiber optic link is designed to operate at a rate of 600 Mb/s over a distance of 70 km using the following components:

> Modulation = NRZ
> Connector at transmitter and receiver = 0.5-dB loss each
> Average loss per unit length of fiber = 0.2 dB/km
> Number of splices = 15
> Average loss per splice = 0.1 dB
> Minimum input to receiver to achieve BER objective = –48 dBm
> Allowance for degradation with temperature variation and aging = 1 dB each for transmitter, fiber, and receiver
> Length dependence factor = γ = 1.0
> System margin = 3 dB

(a) Calculate the required optical transmitter power (in dBm).

(b) If the receiver power of –48 dBm provides a BER of 10^{-6}, calculate the BER for a receiver power of –51 dBm.

(c) Calculate the bandwidth-distance product required to meet the end-to-end bandwidth of this system.

8.18 Consider a wavelength-division multiplexing system with the following parameters:

> Transmitter power = 0 dBm
> Receiver threshold = –32 dBm
> Filter loss = 5 dB
> Fiber transmission loss = 5 dB
> System margin = 5 dB

If a star coupler is used to connect each input to every fiber, determine how many inputs can be supported for this system.

8.19 In the SONET STS-1 frame, positive and negative stuff bytes are used to compensate for frequency differences between the frame rates of the transport overhead and SPE. The use of these stuff bytes is signaled by a 5-bit control word transmitted in the pointer.

(a) What is the range of data rates allowed by this strategy?

(b) Calculate the time between stuff control errors for a transmission bit error rate of 1×10^{-4}, assuming the use of majority vote on each 5-bit code.

8.20 Consider the SONET asynchronous mapping for 1.544 Mb/s shown in Figure 8.25. (a) Find the maximum and minimum data rates that can be supported. (b) Find the stuff opportunity rate. (c) Find the stuffing ratio for the 1,544,000 b/s rate.

8.21 Consider the SONET asynchronous mapping for 139.264 Mb/s shown in Figure 8.28. (a) Find the maximum and minimum data rates that can be supported. (b) Find the stuff opportunity rate. (c) Find the stuffing ratio for the 139,264,000-b/s rate.

9

Digital Radio Systems

OBJECTIVES

- Considers free-space propagation and discusses deviations from it due to effects of terrain, atmosphere, and precipitation
- Describes multipath fading due to ground reflection or atmospheric refraction and characterizes it with a statistical model
- Explains how the effects of multipath fading on digital radio performance can be mitigated by such techniques as diversity and adaptive equalization
- Discusses the effects of frequency allocations on digital radio capacity and system planning
- Points out techniques for minimizing intrasystem and intersystem RF interference
- Describes the components of a digital radio—including transmitter, receiver, antennas, diversity combiner, adaptive equalizer, power amplifiers, and filters
- Lists the procedures for calculating link performance or determining design parameters

9.1 Introduction

The basic components required for operating a radio over a line-of-sight (LOS) link are the transmitter, towers, antennas, and receiver (Figure 9.1). Transmitter functions typically include multiplexing, encoding, modulation, up-conversion from baseband or intermediate frequency (IF) to radio frequency (RF), power amplification, and filtering for spectrum control. Antennas are placed on a tower or other tall structure at sufficient height to

545

provide a direct, unobstructed path between the transmitter and receiver sites. Receiver functions include RF filtering, down-conversion from RF to IF, amplification at IF, equalization, demodulation, decoding, and demultiplexing. Some of these transmitter and receiver functions have been described in earlier chapters, including multiplexing (Chapter 4), coding (Chapter 5), and modulation (Chapter 6). The other functions that are unique to digital radio are described in detail in this chapter. Further, various phenomena associated with line-of-sight propagation are described here as they affect digital radio performance.

In describing radio and link design, the focus is on digital line-of-sight microwave systems (from about 1 to 30 GHz). The design of millimeter-wave (30 to 300 GHz) radio links is also considered, however. Further, much of the material on line-of-sight propagation, including multipath and interference effects, and link design methodology also applies to the design of mobile and analog radio systems. Many of the design and performance characteristics considered here apply to satellite communications as well.

9.2   Line-of-Sight Propagation

The modes of propagation between two radio antennas may include a direct, line-of-sight (LOS) path but also a ground or surface wave that parallels the earth's surface, a sky wave from signal components reflected off the troposphere or ionosphere, a ground reflected path, and a path diffracted from an obstacle in the terrain. The presence and utility of these modes depend on the link geometry, both distance and terrain between the two antennas, and the operating frequency. For frequencies in the microwave band, the LOS propagation mode is the predominant mode available for use; the other modes may cause interference with the stronger LOS path. Line-of-sight links are limited in distance by the curvature of the

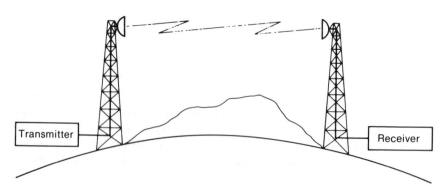

FIGURE 9.1   Line-of-Sight Radio Configuration

earth, obstacles along the path, and free-space loss. Average distances for conservatively designed LOS links are 25 to 30 mi, although distances up to 100 mi have been used. The performance of the LOS path is affected by several phenomena addressed in this section, including free-space loss, terrain, atmosphere, and precipitation. The problem of fading due to multiple paths is addressed in the following section.

9.2.1 Free-Space Loss

Consider a radio path consisting of isotropic antennas at the transmitter and receiver. An isotropic transmitting antenna radiates its power P_{ta} equally in all directions. In the absence of terrain or atmospheric effects (that is, free space), the radiated power density is equal at points equidistant from the transmitter. One can imagine a sphere of radius D where the power density on the surface of the sphere is given by

$$P_D = \frac{P_{ta}}{4\pi D^2} \tag{9.1}$$

If a receiving antenna has an effective area A_r and is located a distance D from the transmitting antenna, the received power P_{ra} is equal to

$$P_{ra} = \frac{P_{ta}A_r}{4\pi D^2} \tag{9.2}$$

The effective area of an isotropic antenna is known to be [1]

$$A_r = A_{isotropic} = \frac{\lambda^2}{4\pi} \tag{9.3}$$

where λ is the wavelength of the radio signal. Combining (9.2) and (9.3), we can express the free-space path loss between isotropic antennas as

$$\frac{P_{ta}}{P_{ra}} = \left(\frac{4\pi D}{\lambda}\right)^2 \tag{9.4}$$

When expressed in decibels, path loss L_p is

$$L_p = 96.6 + 20 \log f + 20 \log D \tag{9.5}$$

where f is radio frequency in gigahertz and D is path length in miles. Plots of L_p for representative frequencies and distances are shown in Figure 9.2. Note that the doubling of either frequency or distance causes a 6-dB increase in path loss.

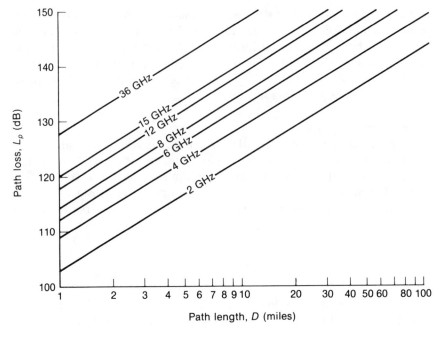

FIGURE 9.2   Free-Space Loss Between Isotropic Antennas

9.2.2   Terrain Effects

Obstacles along a line-of-sight radio path can cause the propagated signal to be reflected or diffracted, resulting in path losses that deviate from the free-space value. This effect stems from electromagnetic wave theory, which postulates that a wave front diverges as it advances through space. A radio beam that just grazes the obstacle is diffracted, with a resulting obstruction loss whose magnitude depends on the type of surface over which the diffraction occurs. A smooth surface, such as water or flat terrain, produces the maximum obstruction loss at grazing. A sharp projection, such as a mountain peak or even trees, produces a knife-edge effect with minimum obstruction loss at grazing. Most obstacles in the radio path produce an obstruction loss somewhere between the limits of smooth earth and knife edge.

Reflections

When the obstacle is below the optical line-of-sight path, the radio beam can be reflected to create a second signal at the receiving antenna. Reflected signals can be particularly strong when the reflection surface is smooth terrain or water. Since the reflected signal travels a longer path than the direct signal, the reflected signal may arrive out of phase with the direct signal. The degree of interference at the receiving antenna from the reflected signal depends on the relative signal levels and phases of the direct and reflected signals.

At the point of reflection, the indirect signal undergoes attenuation and phase shift, which is described by the reflection coefficient R, where

$$R = \rho e^{j\phi} \qquad (9.6)$$

The magnitude ρ represents the change in amplitude and ϕ is the phase shift on reflection. The values of ρ and ϕ depend on the wave polarization (horizontal or vertical), angle of incidence, dielectric constant of the reflection surface, and the wavelength λ of the radio signal. The mathematical relationship has been developed elsewhere [2] and will not be covered here. For microwave frequencies, however, two general cases should be mentioned:

1. For horizontally polarized waves with small angle of incidence, $R = -1$ for all terrain, such that the reflected signal suffers no change in amplitude but has a phase change of 180°.
2. If the polarization is vertical with grazing incidence, $R = -1$ for all terrain. With increasing angle of incidence, the reflection coefficient magnitude decreases, reaching zero in the vicinity of $\phi = 10°$.

To examine the problem of interference from reflection, we first simplify the analysis by neglecting the effects of the curvature of the earth's surface. Then, when the reflection surface is flat earth, the geometry is as illustrated in Figure 9.3a, with transmitter (Tx) at height h_1 and receiver (Rx) at height h_2 separated by a distance D and the angle of reflection equal to the angle of incidence ψ. Using plane geometry and algebra, the path difference δ between the reflected and direct signals can be given by

$$\delta = (r_1 + r_2) - r$$
$$\approx \frac{2h_1h_2}{D} \qquad (9.7)$$

The overall phase change experienced by the reflected signal relative to the direct signal is the sum of the phase difference due to the path length difference δ and the phase ϕ due to the reflection. The total phase shift is therefore

$$\gamma = \frac{2\pi\delta}{\lambda} + \phi \qquad (9.8)$$

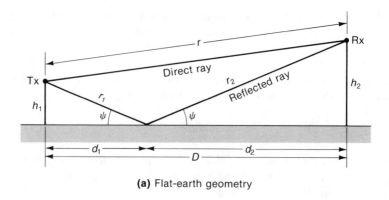

(a) Flat-earth geometry

(b) Two-ray geometry

FIGURE 9.3   Geometry of Two-Path Propagation

At the receiver, the direct and reflected signals combine to form a composite signal with field strength E_C. By the simple geometry of Figure 9.3b and using the law of cosines,

$$E_C^2 = E_D^2 + E_R^2 + 2E_D E_R \cos \gamma \qquad (9.9)$$

or

$$E_C = E_D\sqrt{1 + \rho^2 + 2\rho \cos \gamma} \qquad (9.10)$$

where E_D = field strength of direct signal
E_R = field strength of reflected signal
ρ = magnitude of reflection coefficient = E_R/E_D
γ = phase difference between direct and reflected signal as given by (9.8)

The composite signal is at a minimum, from (9.10), when $\gamma = (2n + 1)\pi$, where n is an integer. Similarly, the composite signal is at a maximum when $\gamma = 2n\pi$. As noted earlier, the phase shift ϕ due to reflection is usually around 180° for microwave paths since the angle of incidence upon the reflection surface is typically quite small. For this case, the received signal minima, or nulls, occur when the path difference is an even multiple of a half-wavelength, or

$$\delta = 2n\left(\frac{\lambda}{2}\right) \quad \text{for minima} \tag{9.11}$$

The maxima, or peaks, for this case occur when the path difference is an odd multiple of a half-wavelength, or

$$\delta = \frac{(2n + 1)}{2}\lambda \quad \text{for maxima} \tag{9.12}$$

When the spherical surface is substituted for the plane reflecting surface of Figure 9.3a, the reflected signal diverges at a greater rate such that the field strength is further reduced. To allow for the effects of round earth, it is necessary to introduce a *divergence factor,* defined as the ratio of the field strength obtained after reflection from a spherical surface to that obtained after reflection from a plane surface. The analysis required in calculating the divergence factor is not simple but has been done for smooth earth [2]. This calculated value is then multiplied by the reflection coefficient in (9.6) to describe the reflected signal characteristics.

Fresnel Zones
The effects of reflection and diffraction on radio waves can be more easily seen by using the model developed by A. Fresnel for optics. Fresnel accounted for the diffraction of light by postulating that the cross section of an optical wave front is divided into zones of concentric circles separated by half-wavelengths. These zones alternate between constructive and destructive interference, resulting in a sequence of dark and light bands when diffracted light is viewed on a screen. When viewed in three dimensions, as necessary for determining path clearances in line-of-sight radio systems, the Fresnel zones become concentric ellipsoids. The first Fresnel zone is that locus of points for which the sum of the distances between the transmitter and receiver and a point on the ellipsoid is exactly one half-wavelength longer than the direct path between the transmitter and receiver. The nth Fresnel zone consists of that set of points for which the difference is n half-wavelengths.

The radius of the nth Fresnel zone at a given distance along the path is given by

$$F_n = 17.3 \sqrt{\frac{nd_1d_2}{fD}} \quad \text{meters} \qquad (9.13)$$

where d_1 = distance from transmitter to a given point along the path (km)

d_2 = distance from receiver to the same point along the path (km)

f = frequency (GHz)

D = path length (km)$(D = d_1 + d_2)$

As an example, Figure 9.4a shows the first three Fresnel zones for an LOS path of length (D) 40 km and frequency (f) 8 GHz. The distance h represents the clearance between the LOS path and the highest obstacle along the terrain.

Using Fresnel diffraction theory, the effects of path clearance on transmission loss can be calculated as shown in Figure 9.4b. The three cases shown correspond to different reflection coefficient values as determined by differences in terrain roughness. The curve marked $R = 0$ represents the case of knife-edge diffraction, where the loss at a grazing angle (zero clearance) is equal to 6 dB. The curve marked $R = -1.0$ illustrates diffraction from a smooth surface, which produces a maximum loss equal to 20 dB at grazing. In practice, most microwave paths have been found to have a reflection coefficient magnitude of 0.2 to 0.4; thus the curve marked $R = -0.3$ represents the ordinary path [3]. For most paths, the signal attenuation becomes small with a clearance of 0.6 times the first Fresnel zone radius. Thus microwave paths are typically sited with a clearance of at least $0.6F_1$.

The fluctuation in signal attenuation observed in Figure 9.4b is due to alternating constructive and destructive interference with increasing clearance. Clearance at odd-numbered Fresnel zones produces constructive interference since the delayed signal is in phase with the direct signal; with a reflection coefficient of -1.0, the direct and delayed signals sum to a value 6 dB higher than free-space loss. Clearance at even-numbered Fresnel zones produces destructive interference since the delayed signal is out of phase with the direct signal by a multiple of $\lambda/2$; for a reflection coefficient of -1.0, the two signals cancel each other. As indicated in Figure 9.4b, the separation between adjacent peaks or nulls decreases with increasing clearance, but the difference in signal strength decrease with increasing Fresnel zone numbers.

9.2.3 Atmospheric Effects

Radio waves travel in straight lines in free space, but they are bent, or *refracted,* when traveling through the atmosphere. Bending of radio waves is caused by changes with altitude in the index of refraction, defined as the ratio

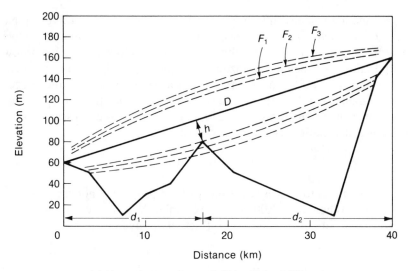

(a) Fresnel zones for an 8-GHz, 40-km LOS path

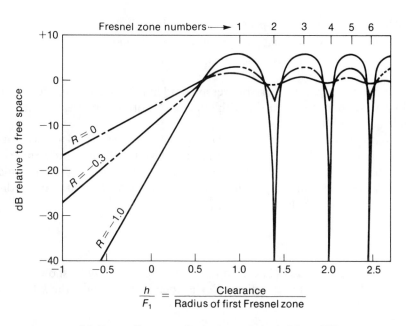

$$\frac{h}{F_1} = \frac{\text{Clearance}}{\text{Radius of first Fresnel zone}}$$

(b) Attenuation vs. path clearance (adapted from [8])

FIGURE 9.4   Fresnel Zones

of propagation velocity in free space to that in the medium of interest. Normally the refractive index decreases with altitude, meaning that the velocity of propagation increases with altitude, causing radio waves to bend downward. In this case, the radio horizon is extended beyond the optical horizon.

The index of refraction n varies from a value of 1.0 for free space to approximately 1.0003 at the surface of the earth. Since this refractive index varies over such a small range, it is more convenient to use a scaled unit, N, which is called **radio refractivity** and defined as

$$N = (n - 1)10^6 \qquad (9.14)$$

Thus N indicates the excess over unity of the refractive index, expressed in millionths. When $n = 1.0003$, for example, N has a value of 300. At microwave frequencies, the radio refractivity is given by

$$N = \frac{77.6p}{T} + \frac{3.73 \times 10^5 w}{T^2} \qquad (9.15)$$

where
p = atmospheric pressure (millibars)
w = partial pressure of water vapor (millibars)
T = absolute temperature (°K)

Owing to the rapid decrease of pressure and humidity with altitude and the slow decrease of temperature with altitude, N normally decreases with altitude and tends to zero.

To account for atmospheric refraction in path clearance calculations, it is convenient to replace the true earth radius a by an **effective earth radius** a_e and to replace the actual atmosphere with a uniform atmosphere in which radio waves travel in straight lines. The ratio of effective to true earth radius is known as the k **factor:**

$$k = \frac{a_e}{a} \qquad (9.16)$$

By application of Snell's law in spherical geometry, it may be shown that as long as the change in refractive index is linear with altitude, the k factor is given by

$$k = \frac{1}{1 + a(dn/dh)} \qquad (9.17)$$

where dn/dh is the rate of change of refractive index with height. It is usually more convenient to consider the gradient of N instead of the gradient of *n.* Making the substitution of dN/dh for dn/dh and also entering the value of 6370 km for a into (9.17) yields the following:

$$k = \frac{157}{157 + (dN/dh)} \tag{9.18}$$

where dN/dh is the N gradient per kilometer. Under most atmospheric conditions, the gradient of N is negative and constant and has a value of approximately

$$\frac{dN}{dh} = -40 \text{ units/km} \tag{9.19}$$

Substituting (9.19) into (9.18) yields a value of $k = 4/3$, which is commonly used in propagation analysis. An index of refraction that decreases uniformly with altitude resulting in $k = 4/3$ is referred to as **standard refraction.**

Anomalous Propagation
Weather conditions may lead to a refractive index variation with height that differs significantly from the average value. In fact, atmospheric refraction and corresponding k factors may be negative, zero, or positive. The various forms of refraction are illustrated in Figure 9.5 by presenting radio paths over both true earth and effective earth. Note that radio waves become straight lines when drawn over the effective earth radius. Standard refraction is the average condition observed and results from a well-mixed atmosphere. The other refractive conditions illustrated in Figure 9.5—including subrefraction, superrefraction, and ducting—are observed a small percentage of the time and are referred to as **anomalous propagation.**

  Subrefraction ($k < 1$) leads to the phenomenon known as inverse bending or earth bulge, illustrated in Figure 9.5b. This condition arises because of an increase in refractive index with altitude and results in an upward bending of radio waves. Substandard atmospheric refraction may occur with the formation of fog, as cold air passes over a warm earth, or with atmospheric stratification, as occurs at night. The effect produced is likened to the bulging of the earth into the microwave path that reduces the path clearance or obstructs the line-of-sight path. To characterize the phenomenon known as **obstruction fading,** the geometry and refractivity conditions of the path must be known. Consider the path in Figure 9.6 shown for two different k factors, where the trajectory of the refracted ray may be described by the parameter

$$H_d = \frac{d_1 d_2}{12.6} (1 - k^{-1}) \tag{9.20}$$

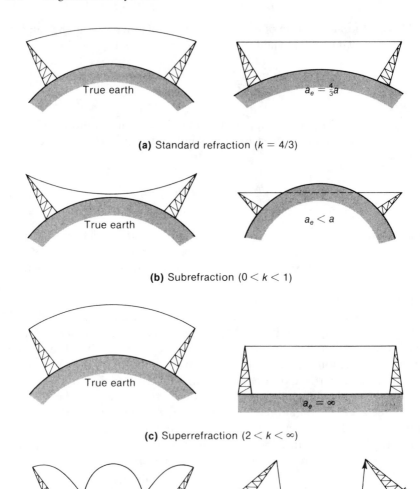

(a) Standard refraction ($k = 4/3$)

(b) Subrefraction ($0 < k < 1$)

(c) Superrefraction ($2 < k < \infty$)

(d) Ducting ($k < 0$)

FIGURE 9.5   Various Forms of Atmospheric Refraction

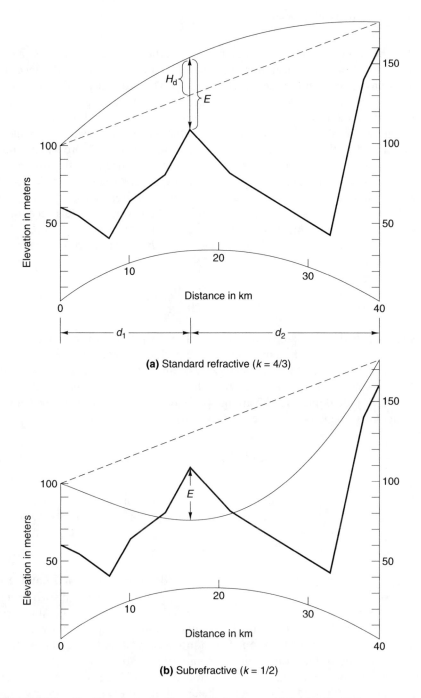

FIGURE 9.6   Ray Trajectories for Standard Refraction and Subrefractive Paths

H_d is the vertical separation in meters between the refracted ray and a straight line drawn from the transmitter to the receiver, and d_1 and d_2 are distances in km from a given point to the ends of the path. The proximity of the ray to the terrain is usually described by a normalized vertical distance, given by the ratio of H_d to F_1, the radius of the first Fresnel zone as given by (9.13) with $n = 1$. Fading due to obstruction is controlled by the terrain point with the smallest normalized vertical distance to the trajectory of the ray. In Figure 9.6 the controlling terrain obstruction is shown to have a clearance E to the trajectory of the ray. In well-designed microwave links, antenna heights are selected to provide a normalized clearance E/F_1 of unity or greater with $k = 4/3$. On paths that exhibit subrefractive conditions, however, the clearance may become negative leading to obstruction fading, as shown in Figure 9.6b.

For deep obstruction fading, the clearance E is related to the observed fade level by [4]

$$E = \frac{F_1(M + 10)}{20} \tag{9.21}$$

where M, with $M < -20$, is the fade level in dB relative to free-space loss. The clearance E is generally negative, signifying blockage. Knowing E and the antenna heights, the ray trajectory H_d can be found by plane geometry. By then solving (9.20) for k, we know the critical value of k that produces a particular fade depth M. The annual probability that the k factor exceeds a particular value k' is composed of four seasonal parts

$$P(k > k') = (1/4) \sum_{i=1}^{4} P_i(k > k') \tag{9.22}$$

For each season, the distribution is a combination of two distributions corresponding to mixed and stratified atmospheric conditions, according to

$$P_i(k > k') = 0.8 P_{m,i} + 0.2 P_{s,i} \tag{9.23}$$

where the subscript m refers to the mixed (daytime) atmosphere and the subscript s refers to a stratified (nightime) atmosphere. Both distributions are gaussian and therefore are characterized by their mean and standard deviation. The mean is the same for both distributions, although it varies by season and geographic location. The standard deviation for P_m is the same for all seasons and locations, but the standard deviation for P_s exhibits considerable variation by season and location [5]. A database containing values of mean and standard deviation for the distributions of refractivity gradient dN/dh, which is related to k factor by (9.18), is found in [6].

Superrefraction ($2 < k < \infty$) causes radio waves to refract downward with a curvature greater than normal. The result is an increased flattening of the effective earth. For the case illustrated in Figure 9.5c the effective earth radius is infinity ($k = \infty$)—that is, the earth reduces to a plane. From Equation (9.18) it can be seen that an N gradient of -157 units per kilometer yields a k equal to infinity. Under these conditions radio waves are propagated at a fixed height above the earth's surface, creating unusually long propagation distances and the potential for overreach interference with other signals occupying the same frequency allocation. Superrefractive conditions arise when the index of refraction decreases more rapidly than normal with increasing altitude, which is produced by a rise in temperature with altitude, a decrease in humidity, or both. An increase in temperature with altitude, called a temperature inversion, occurs when the temperature of the earth's surface is significantly less than that of the air, which is most commonly caused by cooling of the earth's surface through radiation on clear nights or by movement of warm dry air over a cooler body of water.

A more rapid decrease in refractive index gives rise to more pronounced bending of radio waves, in which the radius of curvature of the radio wave is smaller than the earth's radius. As indicated in Figure 9.5d the rays are bent to the earth's surface and then reflected upward from it. With multiple reflections, the radio waves can cover large ranges far beyond the normal horizon. In order for the radio wave's bending radius to be smaller than the earth's radius, the N gradient must be less than -157 units per kilometer. Then, according to (9.18), the k factor and effective earth radius both become negative quantities. As illustrated in Figure 9.5d the effective earth is approximated by a concave surface. This form of anomalous propagation is called **ducting** because the radio signal appears to be propagated through a waveguide, or duct. A duct may be located along the earth's surface or it may be elevated above the earth's surface. The meteorological conditions responsible for either surface or elevated ducts are similar to conditions causing superrefractivity. With ducting, however, a transition region between two differing air masses creates a trapping layer. In ducting conditions, refractivity N decreases with increasing height in an approximately linear fashion above and below the transition region, where the gradient departs from the average. In this transition region, the gradient of N becomes steep (Figure 9.7).

Atmospheric Absorption

For frequencies above 10 GHz, attenuation due to atmospheric absorption becomes an important factor in radio link design. The two major atmospheric gases contributing to attenuation are water vapor and oxy-

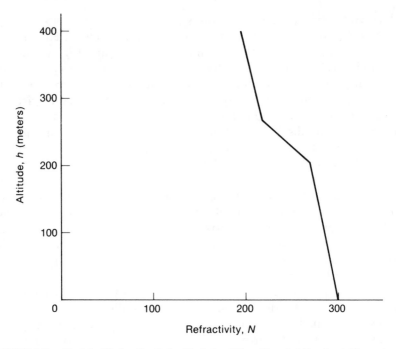

FIGURE 9.7   Model of Refractive Index Variation When Elevated Ducting is Present

gen. Studies have shown that absorption peaks occur in the vicinity of 22.3 and 187 GHz due to water vapor and in the vicinity of 60 and 120 GHz for oxygen [7]. These peaks with their attenuation are illustrated in Figure 9.8. The total attenuation due to atmospheric absorption may be written as

$$A_a = (\Gamma_o + \Gamma_w)D \qquad (9.24)$$

where Γ_o and Γ_w are the specific attenuations in dB/km for oxygen and water vapor, respectively, and D is the path length in km. The calculation of specific attenuation produced by either oxygen or water vapor is complex, requiring computer evaluation for each value of temperature, pressure, and humidity. Formulas that approximate Γ_o and Γ_w may be found in CCIR Rep. 721-2 [10]. The plot of Figure 9.8 has been drawn for a temperature of 15°C, pressure of 1013 mb, and absolute humidity of 7.5 g/m³.

At millimeter wavelengths (30 to 300 GHz), atmospheric absorption becomes a significant problem. To obtain maximum propagation range, frequencies around the absorption peaks are to be avoided. As can be seen in

Figure 9.8, certain frequency bands have relatively low attenuation. In the millimeter wave range, the first two such bands, or *windows,* are centered at approximately 36 and 85 GHz.

Rain Attenuation
Attenuation due to rain and suspended water droplets (fog) can be a major cause of signal loss, particularly for frequencies above 10 GHz. Rain and fog cause a scattering of radio waves that results in attenuation. Moreover,

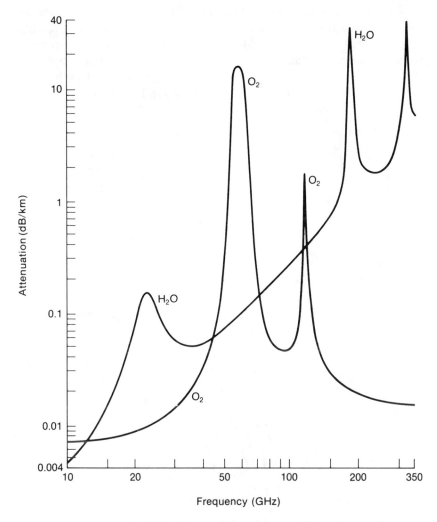

FIGURE 9.8   Absorption by Water Vapor and Oxygen

for the case of millimeter wavelengths where the raindrop size is comparable to the wavelength, absorption occurs and increases attenuation. The degree of attenuation on an LOS link is a function of (1) the point rainfall rate distribution; (2) the specific attenuation, which relates rainfall rates to point attenuations; and (3) the effective path length, which is multiplied by the specific attenuation to account for the length of the path. Figure 9.9 indicates specific attenuation calculated as a function of frequency for several rainfall rates [8]. Note that heavy rain, as found in thunderstorms, produces significant attenuation, particularly for frequencies above 10 GHz. The point rainfall rate distribution gives the percentage of a year that the rainfall rate exceeds a specified value. Rainfall rate distributions depend on climatological conditions and vary from one location to another. To relate rainfall rates to a particular path, measurements must be made by use of rain gauges placed along the propagation path. As an example, Figure 9.10 shows the path average rainfall rate in Washington,

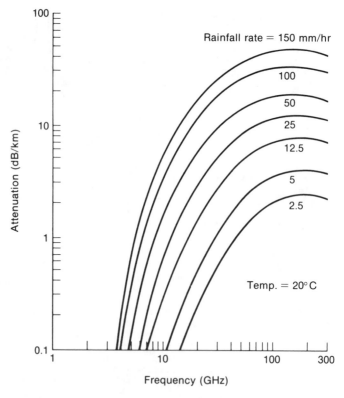

FIGURE 9.9   Calculated Values of Rain Attenuation in dB/km [8]

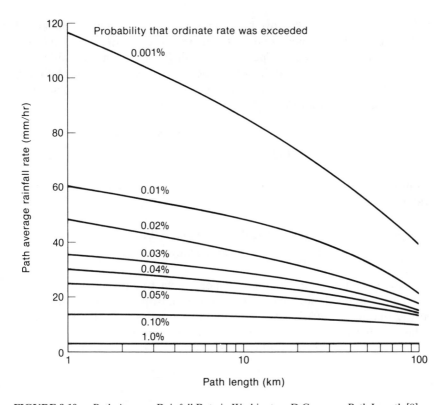

FIGURE 9.10   Path Average Rainfall Rate in Washington, D.C., versus Path Length [9]

D.C., as a function of path length for various probability levels [9]. In the absence of such rainfall data along a specific path, it becomes necessary to use maps of rain climate regions, such as those provided by CCIR Rep. 563.3 [10].

A comparison of Figures 9.9 and 9.10 with Figure 9.8 indicates that rainfall attenuation often exceeds the combined oxygen and water vapor attenuation. To counter the effects of rain attenuation, it should first be noted that neither space nor frequency diversity is effective as protection against rainfall effects. Measures that are effective, however, include increasing the fade margin, shortening the path length, and using a lower frequency band.

Example 9.1 _____

Consider a millimeter-wave radio link to operate at 36 GHz over a path of length 5 km (3.1 mi) in the Washington, D.C., area. Calculate the total propagation losses allowed to meet a path availability of 99.99 percent.

Solution   For millimeter-wave radio links, the factors that affect path availability are free-space loss and atmospheric absorption (which are here assumed to be constant in time) and rainfall (which is time-variable). For short paths at 36 GHz, varying atmospheric conditions (multipath fading) have negligible effect on path loss variation compared to rainfall attenuation. From Equation (9.5) or Figure 9.2, the free-space loss is

$$L_p = 96.6 + 20 \log(36) + 20 \log(3.1)$$
$$= 137.5 \text{ dB}$$

From Figure 9.8, atmospheric absorption due to the presence of oxygen and water vapor is

$$O_2: \quad (0.02 \text{ dB/km})(5 \text{ km}) = 0.1 \text{ dB}$$
$$H_2O: \quad (0.05 \text{ dB/km})(5 \text{ km}) = 0.25 \text{ dB}$$

To provide a path availability of 99.99 percent we find from Figure 9.10 that the rainfall rate will not exceed 50 mm/hr for 0.01 percent of the time. The corresponding attenuation due to 50 mm/hr rainfall is found from Figure 9.9 to be 12.6 dB/km, or 63 dB for a 5-km path. Summing up propagation losses, we find the total propagation loss to be 201 dB.

A mathematical model for rain attenuation is found in CCIR Rep. 721-3 [10] and summarized here. The specific attenuation is the loss per unit distance that would be observed at a given rain rate, or

$$\gamma_R = kR^\beta \tag{9.25}$$

where γ_R is the specific attenuation in dB/km and R is the point rainfall rate in mm/hr. The values of k and β depend on the frequency and polarization, and may be determined from tabulated values in CCIR Rep. 721-2. Because of the nonuniformity of rainfall rates within the cell of a storm, the attenuation on a path is not proportional to the path length; instead it is determined from an effective path length given in [11] as

$$L_{eff} = \frac{L}{1 + (R - 6.2)L/2636} \tag{9.26}$$

Now, to determine the attenuation for a particular probability of outage, the CCIR method calculates the attenuation $A_{0.01}$ that occurs 0.01 percent of a year and uses a scaling law to calculate the attenuation A_p at other probabilities,

$$A_{0.01} = \gamma_R L_{eff} \qquad (9.27)$$

where γ_R and L_{eff} are determined for the 0.01 percent rainfall rate. The value of the 0.01 percent rainfall rate is obtained from world contour maps of 0.01 percent rainfall rates found in CCIR Rep. 563-4. The attenuation for other percentages of time in the range 0.001 percent to 1 percent is given by the CCIR as

$$A_p = 0.12 A_{0.01} p^{-(0.546 + 0.043 \log p)} \qquad (9.28)$$

where p is the outage percentage. Alternatively, the percentage of time between 0.001 percent and 1 percent that a specified attenuation is exceeded may be calculated by inverting Equation (9.28) to yield

$$p = 10 \exp\left[6.34884\left(\sqrt{1 + 0.57696 \log(0.12\, A_{0.01}/A_p)} - 1\right)\right] \qquad (9.29)$$

9.2.4 Path Profiles

In order to determine tower heights for suitable path clearance, a profile of the path must be plotted. The path profile is obtained from topographical maps that should have a scale of 1:50,000 or less. For line-of-sight links under 70 km in length, a straight line may be drawn connecting the two end-points. For longer links, the great circle path must be calculated and plotted on the map. The elevation contours are then read from the map and plotted on suitable graph paper, taking special note of any obstacles along the path. The path profiling process may be fully automated by use of CD-ROM technology and an appropriate computer program. CD-ROM data storage discs are available that contain a global terrain elevation database from which profile points can be automatically retrieved, given the latitudes and longitudes of the link endpoints [12].

The path profile may be plotted on special graph paper that depicts the earth as curved and the transmitted ray as a straight line or on rectilinear graph paper that depicts the earth as flat and the transmitted ray as a curved line. The use of linear paper is preferred because it eliminates the need for special graph paper, permits the plotting of rays for different effective earth radius, and simplifies the plotting of the profile. Figure 9.11 is an example of a profile plotted on linear paper for a 40-km, 8-GHz radio link; Figure 9.12 is the same example plotted on $k = \frac{2}{3}$ earth radius paper.

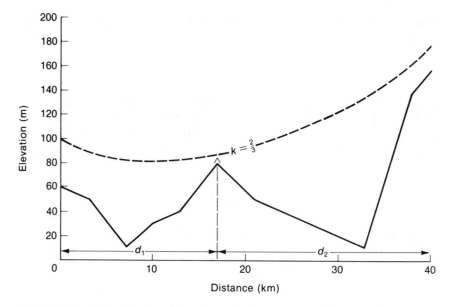

FIGURE 9.11   Example of an LOS Path Profile Plotted on Linear Paper

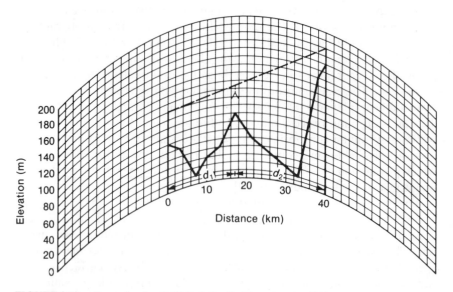

FIGURE 9.12   Example of an LOS Path Profile Plotted on $k = \frac{2}{3}$ Paper

The use of rectilinear paper, as suggested, requires the calculation of the earth bulge at a number of points along the path, especially at obstacles. This calculation then accounts for the added elevation to obstacles due to curvature of the earth. Earth bulge in meters may be calculated as

$$h = \frac{d_1 d_2}{12.76} \qquad (9.30)$$

where d_1 = distance from one end of path to point being calculated (km)
$\quad\quad d_2$ = distance from same point to other end of the path (km)

As indicated earlier, atmospheric refraction causes ray bending, which can be expressed as an effective change in earth radius by using the k factor. The effect of refraction on earth bulge can be handled by adding the k factor to (9.30):

$$h = \frac{d_1 d_2}{12.76k} \qquad (9.31)$$

or, for d_1 and d_2 in miles and h in feet,

$$h = \frac{d_1 d_2}{1.5k} \qquad (9.32)$$

To facilitate path profiling, Equation (9.31) or (9.32) may be used to plot a curved ray template for a particular value of k and for use with a flat earth profile. Alternatively, the earth bulge can be calculated and plotted at selected points that represent the clearance required below a straight line drawn between antennas; when connected together, these points form a smooth parabola whose curvature is determined by the choice of k.

In path profiling, the choice of k factor is influenced by its minimum value expected over the path and the path availability requirement. With lower values of k, earth bulging becomes pronounced and antenna height must be increased to provide clearance. To determine the clearance requirements, the distribution of k values is required; it can be found by meteorological measurements [10]. This distribution of k values can be related to path availability by selecting a k whose value is exceeded for a percentage of the time equal to the availability requirement.

Apart from the k factor, the Fresnel zone clearance must be added. Desired clearance of any obstacle is expressed as a fraction, typically 0.3 or 0.6, of the first Fresnel zone radius. This additional clearance is then plotted on the path profile, shown as a small tick mark on Figures 9.11 and 9.12, for each point being profiled. Finally, clearance should be provided for trees (nominally 15 m) and additional tree growth (nominally 3 m) or, in the absence of trees, for smaller vegetation (nominally 3 m).

The clearance criteria can thus be expressed by specific choices of k and fraction of first Fresnel zone. Here is one set of clearance criteria that is commonly used for highly reliable paths [13]:

1. Full first Fresnel zone clearance for $k = \frac{1}{3}$
2. 0.3 first Fresnel zone clearance for $k = \frac{2}{3}$

whichever is greater. Over the majority of paths, the clearance requirements of criterion 2 will be controlling. Even so, the clearance should be evaluated by using both criteria along the entire path.

Adequate tower heights may now be determined by plotting the radio ray for the proper value of k and superimposing this ray on the terrain profile such that proper first Fresnel zone and any vegetation clearance is achieved. Figure 9.13 shows the relationship between tower height and path length for $k = \frac{2}{3}$ and 0.6 first Fresnel zone clearance. Note that the tower

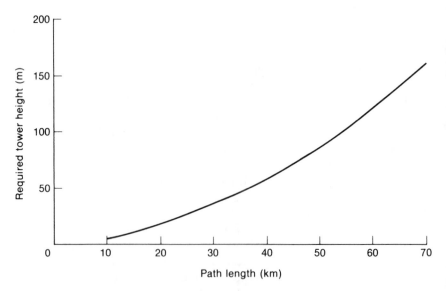

FIGURE 9.13   Tower Height Required for Smooth Earth Clearance for $k = \frac{2}{3}$, 0.6 First Fresnel Zone Clearance, and Equal Antenna Heights

height required for clearance over a smooth earth increases as the square of the path length.

9.2.5   Reflective Paths

Paths with strong reflection areas will give rise to interference, which can severely degrade digital radio performance. Statistical models that would allow calculation of this degradation do not yet exist, but potential reflective paths and methods to combat their effect can be identified. The means of identifying and mitigating reflective paths uses information contained in the path profile. An accurate and up-to-date path profile is critical, since any errors in terrain data can lead to problems in path design.

Unless a particular path has been tested to identify and characterize the presence of reflections, it becomes necessary to use the path profile with Fresnel zones superimposed. From Figure 9.4b note that minima in the received signal level occur with even-numbered Fresnel zones. For smooth earth with $R = -1.0$, deep nulls are observed, indicating that reflections are a particular problem for paths over flat surfaces. The existence of reflections can be checked by plotting a set of even-numbered Fresnel zones on the path profile for several k factors, ranging from $k = \frac{2}{3}$ to infinity. The criteria for identifying reflections are based on the extent of areas along the profile that have tangency with even-numbered Fresnel zones and are visible (not blocked) at both ends of the path. One set of tests used by AT&T to identify reflective paths is [14]:

1. Tangency with an extent of 5 km or more, or
2. Tangency with an extent of 0.8 to 5 km and at least one extenuating terrain or climate factor. Terrain that is highly reflective, such as water or desert, and extreme k-factor variability in the area of the path are examples of these extenuating factors.

An example of a path meeting the first criterion is shown in Figure 9.14, which shows tangency of the profile with Fresnel zone number 14.

Since it is not always possible to avoid the use of a path with known reflections, techniques have been developed to reduce the effects of reflections. By suitable choice of antenna heights, it is sometimes possible to find an obstruction near the low end of the path that blocks the reflected signal. If blocking of the reflected path is impossible, then the antenna heights should be selected to shift reflection areas away from highly reflective surfaces toward rougher terrain or to reduce the range of

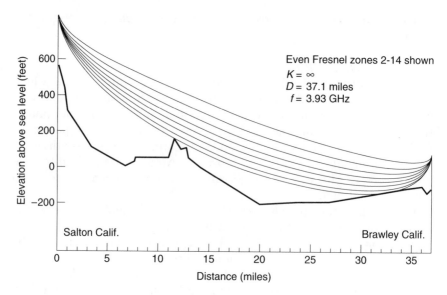

FIGURE 9.14   Example of Reflective Path (Adapted from [14]) (© 1990 IEEE)

path clearance significantly below the second Fresnel zone boundary. In selecting antenna heights, however, adequate clearance must be maintained at low k factors to avoid obstruction fading. The choice of antenna size and alignment can also be used to mitigate effects of reflections by selecting a configuration for which the reflected signal is at the side of the main beam or in the side lobes.

Where antenna configuration and placement alone are not sufficient to reduce the effects of reflections, diversity techniques (described in Section 9.5.3) can also be effective. A choice of appropriate spacing between space diversity antennas can ensure that the two received signals will be out of phase. To find this spacing, consider the geometry of Figure 9.3a in which a reflection point is located a distance d_1 from the transmitter. Now add a second receive antenna, located a distance Δh_2 above the first antenna height of h_2. To provide an out-of-phase condition between the two antennas, the increment Δh_2 is

$$\Delta h_2 = 75D/f(h_1 - h) \tag{9.33}$$

where h is the earth bulge given by (9.31) at the reflection point; Δh_2, h_1, and h are in meters; D is the path length in kilometers; and f is the frequency in GHz. Since h and therefore Δh_2 are a function of k, (9.33) yields the optimum antenna spacing for one value of k factor only. For protection over a large range of k, a more suitable choice for antenna separation would be

$$S = 75D/fh_1 \qquad (9.34)$$

which corresponds to Δh_2 for $k = \infty$. On some reflective paths, angle diversity has been found to outperform space diversity [15]. Paths for which angle diversity may be superior are those that are highly dispersive or where there is a significant angular separation between the main and reflected rays at the receiving antenna.

9.3   Multipath Fading

Fading is defined as variation of received signal level with time due to changes in atmospheric conditions. The propagation mechanisms that cause fading include refraction, reflection, and diffraction associated with both the atmosphere and terrain along the path. The two general types of fading are referred to as **multipath** and **power** fading and are illustrated by the recordings of RF received signal levels shown in Figure 9.15.

Power fading, sometimes called attenuation fading, results mainly from anomalous propagation conditions, such as (see Figure 9.5) *subrefraction* $(k < 1)$, which causes blockage of the path due to the effective increase in earth bulge, *superrefraction* $(k > 2)$, which causes pronounced ray bending and decoupling of the signal from the receiving antenna, and *ducting* $(k < 0)$, in which the radio beam is trapped by atmospheric layering and directed

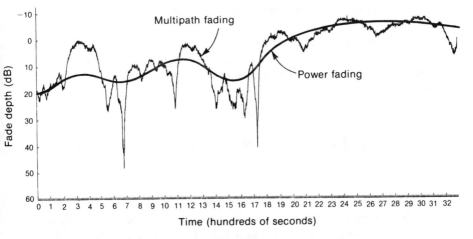

FIGURE 9.15   Example of Multipath and Power Fading for LOS Link

away from the receiving antenna. Rainfall also contributes to power fading, particularly for frequencies above 10 GHz. Power fading is characterized as slowly varying in time, usually independent of frequency, and causing long periods of outages. Remedies include greater antenna heights for subrefractive conditions, antenna realignment for superrefractive conditions, and added link margin for rainfall attenuation.

Multipath fading arises from destructive interference between the direct ray and one or more reflected or refracted rays. These multiple paths are of different lengths and have varied phase angles on arrival at the receiving antenna. These various components sum to produce a rapidly varying, frequency-selective form of fading. Deep fades occur when the primary and secondary rays are equal in amplitude but opposite in phase, resulting in signal cancellation and a deep amplitude null. Between deep fades, small amplitude fluctuations are observed that are known as **scintillation;** these fluctuations are due to weak secondary rays interfering with a strong direct ray.

Multipath fading is observed during periods of atmospheric stratification, where layers exist with different refractive gradients. The most common meteorological cause of layering is a temperature inversion, which commonly occurs in hot, humid, still, windless conditions, especially in late evening, at night, and in early morning. Since these conditions arise during the summer, multipath fading is worst during the summer season. Multipath fading can also be caused by reflections from flat terrain or a body of water. Hence multipath fading conditions are most likely to occur during periods of stable atmosphere and for highly reflective paths. Multipath fading is thus a function of path length, frequency, climate, and terrain. Techniques used to deal with multipath fading include the use of diversity, increased fade margin, and adaptive equalization.

9.3.1   Statistical Properties of Fading

The random nature of multipath fading suggests a statistical approach to its characterization. The statistical parameters commonly used in describing fading are:

- Probability (or percentage of time) that the line-of-sight link is experiencing a fade below threshold
- Average fade duration and probability of fade duration greater than a given time
- Expected number of fades per unit time

The terms to be used are defined graphically in Figure 9.16. The threshold L is the signal level corresponding to the minimum acceptable signal-to-noise ratio or, for digital transmission, the maximum acceptable probability of error. The difference between the normal received signal level and

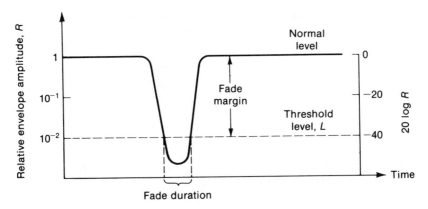

FIGURE 9.16   Definition of Fading Terms

threshold is the **fade margin.** A fade is defined as the downward crossing of the received signal through the threshold. The time spent below threshold for a given fade is then the fade duration.

For line-of-sight links, the probability distribution of fading signals is known to be related to and limited by the Rayleigh distribution, which is well known and is found by integrating the curve shown in Figure 9.17. The Rayleigh probability density function is given by

$$p(r) = \frac{r}{\sigma^2} e^{-(r^2/2\sigma^2)} \tag{9.35}$$

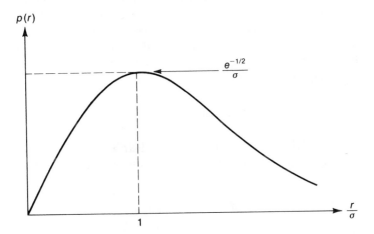

FIGURE 9.17   Rayleigh Probability Density Function

for envelope amplitude r ($r \geq 0$) and mean square amplitude σ^2. The Rayleigh distribution function has the form

$$P(r \leq r_0) = \int_0^{r_0} p(r)dr = \int_0^{r_0} \frac{r}{\sigma^2} e^{-(r^2/2\sigma^2)}\, dr \tag{9.36}$$

$$= 1 - e^{-r_0^2/2\sigma^2} \qquad r_0 \geq 0$$

For relative envelope amplitude $R = r/\sigma\sqrt{2}$ and relative threshold amplitude $L = r_0/\sigma\sqrt{2}$, the distribution function becomes

$$P(R < L) = 1 - e^{-L^2} \tag{9.37}$$

An approximation to (9.37) valid for small values of L (representing deep fades) is

$$P(R < L) \approx L^2 \qquad \text{for } L < 0.1 \tag{9.38}$$

Fading probabilities are more conveniently expressed in terms of the fade margin F in decibels by letting $F = -20 \log L$. Then

$$P(R < L) = 10^{-F/10} \tag{9.39}$$

Actual observations of multipath fading indicate that in the region of deep fades, amplitude distributions have the same slope as the Rayleigh distribution but displaced. This characteristic corresponds to the special case of the Nakagami–Rice distribution where a direct (or *specular*) component exists that is equal to or less than the Rayleigh fading component [16]. Thus, for deep fading, the distribution function becomes

$$P(R < L) = d(1 - e^{-L^2}) \tag{9.40}$$

The parameter d that modifies the Rayleigh distribution has been termed a multipath occurrence factor. Experimental results of Barnett [17] show that

$$d = \frac{abD^3 f}{4} \times 10^{-5} \tag{9.41}$$

where $\qquad$ $D =$ path length in miles

$f =$ frequency in GHz

$a =$ terrain factor

$$= \begin{cases} 4 \text{ for overwater or flat terrain} \\ 1 \text{ for average terrain} \\ \frac{1}{4} \text{ for mountainous terrain} \end{cases}$$

$b =$ climate factor

$$= \begin{cases} \frac{1}{2} \text{ for hot, humid climate} \\ \frac{1}{4} \text{ for average, temperate climate} \\ \frac{1}{8} \text{ for cool, dry climate} \end{cases}$$

Combining this factor with the basic Rayleigh probability of (9.39) results in the following overall expression for probability of outage due to fading deeper than the fade margin:

$$P(o) = d10^{-F/10} = \left(\frac{abD^3 f}{4} \times 10^{-5} \right) (10^{-F/10}) \qquad (9.42)$$

As an example, values of $P(o)$ as a function of fade margin are plotted in Figure 9.18 for a 30-mi path with average terrain and climate. The Rayleigh distribution given by (9.39) is also shown to indicate the limiting value for multipath fading. Note that the distributions all have a slope of 10 dB per decade of probability.

Vigants [18, 19] observed that fade durations have an average value $\bar{t}$, proportional to L and independent of frequency, given by

$$\bar{t} = \frac{L}{c} \qquad \text{for } L < 0.1$$
$$= \frac{10^{-F/20}}{c} \qquad (9.43)$$

where c is an experimental constant equal to approximately 2.22×10^{-3} s^{-1}. The probability distribution function for fade durations is also based on empirical results from Vigants and is given by

$$P(t) = e^{-1.15(t/\bar{t})^{2/3}} \qquad (9.44)$$

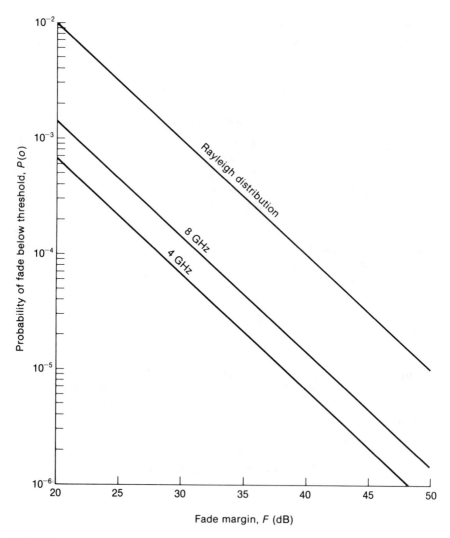

FIGURE 9.18   Probability of Fading Below the Fade Margin on Typical LOS Link (30-mi Path with Average Terrain and Climate)

An expression for expected number of fades per unit time, N, can now be written directly in terms of $P(o)$ and $\bar{t}$:

$$N = \frac{P(o)}{\bar{t}} \qquad (9.45)$$

Substituting (9.42) and (9.43) into (9.45), N becomes

$$N = (2.5 \times 10^{-6})(abcD^3 f)10^{-F/20} \qquad (9.46)$$

For digital transmission applications, it is often more useful to express the fading statistics in terms of error rate—for example as:

- Probability of exceeding a specified error rate
- Average duration of error rate exceeding a specified level
- Frequency that a specified error rate is reached

These error statistics are derivable by a change of variables, from signal level to error rate, in the Rayleigh or Nakagami–Rice distribution functions given in (9.37) and (9.40). The modulation technique must be specified, however, to state the error rate as a function of signal level or signal-to-noise ratio first. D. J. Kennedy has provided these error rate statistics for various modulation techniques [20]. For the Nakagami–Rice channel, the probability distribution function was found to be

$$P(o) = d\left[1 - (2p)^{1/\alpha\gamma_0}\right] \qquad (9.47)$$

where d = multipath occurrence factor

p = bit error rate

γ_0 = signal-to-noise (E_b/N_0) averaged over fading channel

$$\alpha = \begin{cases} 0.25 & \text{for binary noncoherent ASK} \\ 0.5 & \text{for binary noncoherent FSK} \\ 1.0 & \text{for binary differential PSK} \end{cases}$$

The expression for $P(o)$ is plotted in Figure 9.19 as a function of γ_0 for $\alpha = 0.5$ and $d = 0.25$. For other values of α, the factor of $-10 \log(2\alpha)$ must be added to the abscissa, while for other values of d the ordinate must be multiplied by the factor $d/0.25$. A similar set of curves can be derived to show the rate of fading or average duration of fading to a specified error rate. Caution is advised in using these error statistics in predicting performance of wideband digital LOS links, however, since intersymbol interference due to multipath dispersion may be the dominant source of degradation unless adaptive equalization is employed (see Section 9.3.3).

9.3.2 Diversity Improvement
Diversity is used in line-of-sight radio links to protect against either equipment failure or multipath fading. Here we consider the improve-

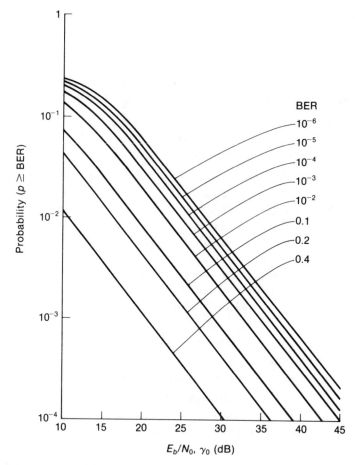

FIGURE 9.19   Probability of Specified BER or Worse for Typical LOS Link [13] © 1973 IEEE.

ment in multipath fading afforded by the two most commonly used diversity techniques:

- **Space diversity,** which provides two signal paths by use of vertically separated receiving antennas
- **Frequency diversity,** which provides two signal frequencies by use of separate transmitter-receiver pairs

The degree of improvement provided by diversity depends on the degree of correlation between the two fading signals. In practice, because of limitations in allowable antenna separation or frequency spacing, the fading correlation tends to be high. Fortunately, improvement in link availability remains quite significant even for high correlation. To derive the diversity improvement, we begin with the joint Rayleigh probability distribution function to describe the fading correlation between diversity signals, given by

$$P(R_1 < L, R_2 < L) \approx \frac{L^4}{1 - k^2} \quad \text{(for small } L\text{)} \tag{9.48}$$

where R_1 and R_2 are signal levels for diversity channels 1 and 2 and k^2 is the correlation coefficient. By experimental results, empirical expressions for k^2 have been established that are a function of antenna separation or frequency spacing, wavelength, and path length.

Space Diversity Improvement Factor
Vigants [21] has developed the following expression for k^2 in space diversity links:

$$k^2 = 1 - \frac{S^2}{2.75D\lambda} \tag{9.49}$$

where S = antenna separation, D = path length, λ = wavelength, and S, D, and λ are in the same units. A more convenient expression is the space diversity improvement factor, given by

$$I_{sd} = \frac{P(R_1 < L)}{P(R_1 < L, R_2 < L)} \approx \frac{L^2}{L^4/(1 - k^2)}$$
$$= \frac{1 - k^2}{L^2} \tag{9.50}$$

Using Vigants' expression for k^2 (Equation 9.49), we obtain

$$I_{sd} = \frac{S^2}{2.75D\lambda L^2} \tag{9.51}$$

When D is given in miles, S in feet, and λ in terms of carrier frequency f in gigahertz, I_{sd} can be expressed as

$$I_{sd} = \frac{(7.0 \times 10^{-5})fS^2}{D} 10^{F/10} \tag{9.52}$$

where F is the fade margin associated with the second antenna.

Frequency Diversity Improvement Factor
Using experimental data and a mathematical model, Vigants and Pursley
[22] have developed an improvement factor for frequency diversity,
given by

$$I_{fd} = \left(\frac{50}{fD}\right)\left(\frac{\Delta f}{f}\right)10^{F/10} \tag{9.53}$$

where
$$f = \text{frequency (GHz)}$$
$$\Delta f = \text{frequency separation (GHz)}$$
$$F = \text{fade margin (dB)}$$
$$D = \text{path length (mi)}$$

Effect of Diversity on Fading Statistics
The effect of diversity improvement, I_d, on probability of outage due to fading can be expressed as

$$P_d(o) = \frac{P(o)}{I_d} \tag{9.54}$$

where $P_d(o)$ is the probability of simultaneous fading in the two diversity
signals. For space diversity, substituting Equations (9.42) and (9.52) into
(9.54) we obtain

$$P_{sd}(o) = \frac{abD^4}{28S^2}10^{-F/5} \tag{9.55}$$

where the fade margins on the two antennas are assumed equal. Likewise,
for frequency diversity we obtain

$$P_{fd}(o) = (5 \times 10^{-8})\left(\frac{abD^4 f^3}{\Delta f}\right)10^{-F/5} \tag{9.56}$$

Vigants [18] and Lin [23] observed in their experimental results that
diversity reduces the average fade duration by a factor of 2, so that Equation (9.43) is modified to read

$$\bar{t}_d = \frac{L}{2c} = \frac{10^{-F/20}}{2c} \tag{9.57}$$

The probability distribution function for fade durations has the same form as the expression for nondiversity (Equation 9.44), but with t_d replacing t so that

$$P(t_d) = e^{-1.15(t_d/\bar{t_d})^{2/3}}$$ (9.58)

The expected number of fades per unit time is also reduced with diversity, according to

$$N_d = \frac{P_d(o)}{\bar{t_d}}$$ (9.59)

For the case of space diversity, N_{sd} is found by substituting (9.55) and (9.57) into (9.59) to yield

$$N_{sd} = \frac{abD^4}{(6.3 \times 10^3)S^2} 10^{-3F/20}$$ (9.60)

where the fade margin F is assumed to be equal for both antennas. Similarly, for frequency diversity we find

$$N_{fd} = 2.22 \times 10^{-10} \left(\frac{abD^4 f^3}{\Delta f} \right) 10^{-3F/20}$$ (9.61)

Example 9.2 _____

Consider the design of an 8-GHz line-of-sight microwave link for a path with length 30 mi, average terrain, and moderate climate. You are to characterize the effects of fade margin and diversity on the probability of fade outage, fade duration, and expected number of fades per unit time. Assume the use of space diversity with equal fade margins on the two receiving antennas.

Solution To obtain general insight into the effect of fade margin and antenna separation, the fading statistics can be plotted by using the equations given earlier. For the link conditions given, and using Equation (9.55), Figure 9.20 shows the probability of fading below threshold as a function of the fade margin. Plots are provided for antenna separations of 20, 30, and 40 ft. Using Equation (9.60), Figure 9.21 shows the expected number of fades per second, again as a function of fade margin and antenna separation.

Using Equations (9.57) and (9.58), the average fade duration is given in Figure 9.22; the probability distribution of fade durations is given in Figure 9.23 for fade margins of 20, 30, and 40 dB. Note that the fading duration statistics are applicable to both space and frequency diversity and are independent of the path parameters.

Example 9.3 _____

Using the same link characteristics given in Example 9.2, compare the fade outage probabilities and link availabilities for (a) nondiversity, (b) space diversity with antenna separation of 30 ft, and (c) frequency diversity with 5 percent spacing. Assume a fade margin of 40 dB.

Solution

(a) Using Equation (9.42) or Figure 9.18, we obtain

$$P_{nd}(o) = 1.35 \times 10^{-5}$$

or, in terms of link availability (A),

$$A = 1 - P(o) = 0.9999865$$

(b) Using Equation (9.55) or Figure 9.20, we obtain

$$P_{sd}(o) = 8.0 \times 10^{-8}$$

or, in terms of link availability,

$$A = 0.99999992$$

(c) Using Equation (9.56) we obtain

$$P_{fd}(o) = 1.3 \times 10^{-7}$$

or, in terms of link availability,

$$A = 0.99999987$$

For this example, we see that the improvement factor of space diversity is greater than that of frequency diversity.

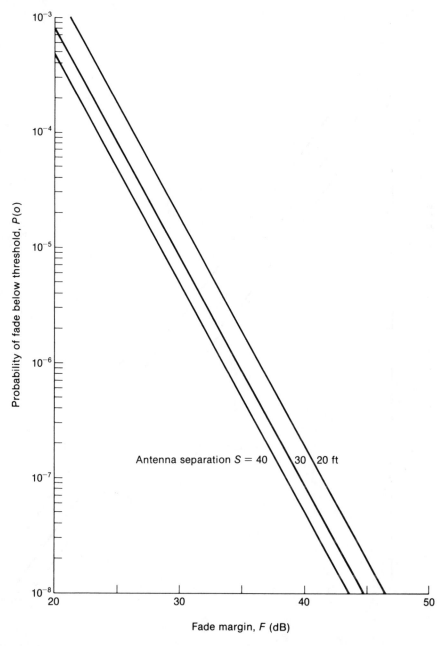

FIGURE 9.20   Probability of Fading for LOS Space Diversity Link versus Fade Margin (30-mi Path with Average Terrain and Climate)

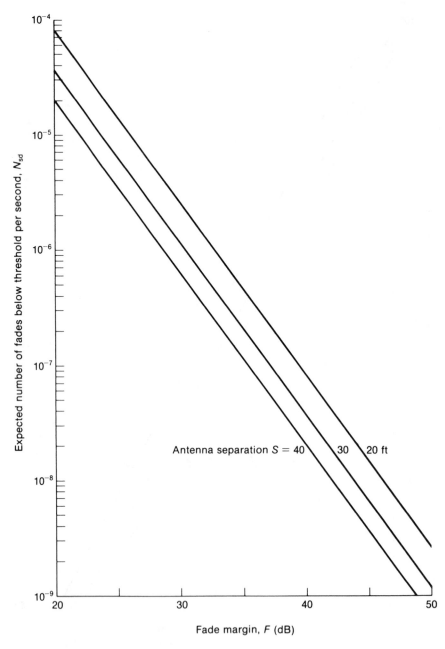

FIGURE 9.21   Expected Number of Fades Below Threshold Per Second versus Fade Margin for LOS Space Diversity Link (30-mi Path with Average Terrain and Climate)

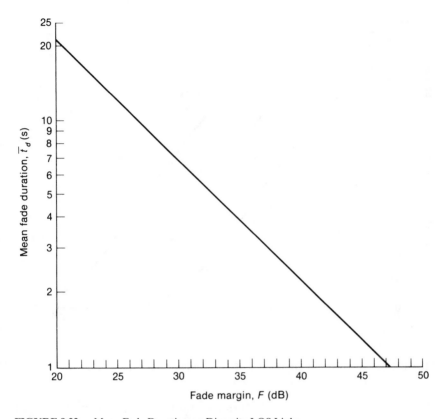

FIGURE 9.22   Mean Fade Duration on Diversity LOS Links

9.3.3   Frequency-Selective Fading

The first experiences with wideband digital radios revealed that measured error performance fell far short of the performance predicted by the **flat fading** model assumed in our discussions so far. This result is due to the presence of **frequency-selective fading** during which the amplitude and group delay characteristics become distorted. For digital signals, this distortion leads to intersymbol interference that in turn degrades the system error rate. This degradation is directly proportional to system bit rate, since higher bit rates mean smaller pulse widths and greater susceptibility to intersymbol interference. Previously, in analog radio transmission, frequency-selective fading caused intermodulation distortion, but this effect was always secondary when compared to the received signal power. For digital radio systems, however, the traditional fade depth is found to be a poor indicator of error rate.

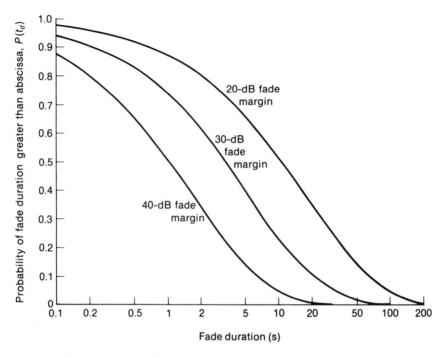

FIGURE 9.23   Probability Distribution of Fade Durations for LOS Diversity Links

Multipath geometry and its frequency-selective nature is illustrated in Figure 9.24. When the amplitude of the received signal is plotted versus frequency, deep amplitude notches appear when the direct ray is out of phase with the indirect rays. These notches are separated in frequency by $1/\tau$, where τ is the time delay between the direct and indirect rays. The notch depth is determined by the relative amplitude of the direct and indirect rays. When an amplitude notch or slope appears in the band of a radio channel, degradation in the error rate can be expected. This variation of amplitude with frequency is known as **amplitude dispersion** and is often the main source of degradation in digital radio systems.

Amplitude dispersion can be measured by recording the amplitudes across the RF or IF band of the received radio signal. Usually dispersion is calculated by taking the difference of spectral amplitudes at the two band edges and dividing that difference by the corresponding RF or IF bandwidth. Figure 9.25 is such a recording of amplitude dispersion together with received signal level for an 8-GHz, 56-mi path. These recordings indicate the expected correlation of dispersion and signal level.

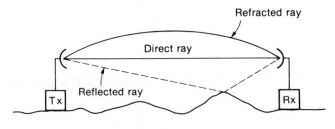

(a) Refraction or reflection geometry resulting
in multipath propagation

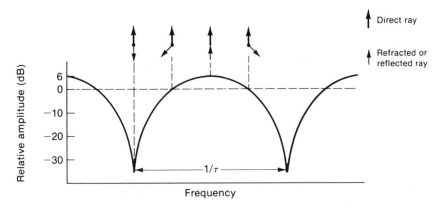

(b) Amplitude characteristics in multipath

FIGURE 9.24   Multipath Fading Effect

Note that during the three deepest fades, the slope of the dispersion changes signs, which indicates that an amplitude notch has passed through the radio channel.

Channel Models
Both low-order power series [24] and multipath transfer functions [25] have been used to model the effects of frequency-selective fading. Several multipath transfer function models have been developed, usually based on the presence of two [26] or three [25] rays. In general, the multipath channel transfer function can be written as

$$H(\omega) = 1 + \sum_{i=1}^{n} \beta_i \, e^{j\omega \tau_i} \qquad (9.62)$$

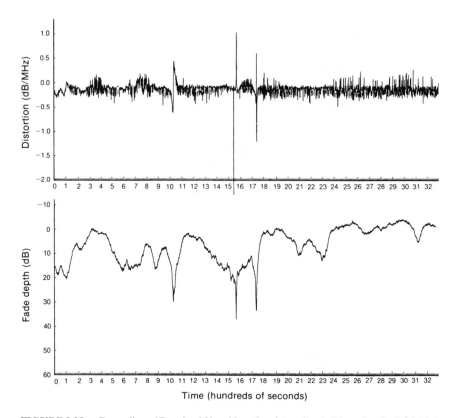

FIGURE 9.25   Recording of Received Signal Level and Amplitude Distortion for LOS Link

where the direct ray has been normalized to unity and the β_i and τ_i are amplitude and delay of the interfering rays relative to the direct ray. The two-ray model can thus be characterized by two parameters, β and τ. In this case, the amplitude of the resultant signal is

$$R = (1 + \beta^2 + 2\beta \cos \omega \tau)^{1/2} \tag{9.63}$$

and the phase of the resultant is

$$\phi = \arctan\left(\frac{\beta \sin \omega \tau}{1 + \beta \cos \omega \tau}\right) \tag{9.64}$$

The group delay is then

$$T(\omega) = \frac{d\phi}{d\omega} = \beta\tau\left(\frac{\beta + \cos\omega\tau}{1 + 2\beta\cos\omega\tau + \beta^2}\right) \tag{9.65}$$

The deepest fade occurs with

$$\omega_d\tau = \pi(2n - 1) \qquad (n = 1, 2, 3, \ldots) \tag{9.66}$$

where both R and T are at a minimum, with

$$R_{\min} = 1 - \beta \tag{9.67a}$$

$$T_{\min} = \frac{\beta\tau}{1 - \beta} \tag{9.67b}$$

The frequency defined by ω_d is known as the *notch frequency* and is related to the carrier frequency ω_c by

$$\omega_d = \omega_c + \omega_o \tag{9.68}$$

where ω_o is referred to as the *offset frequency*.

Typical amplitude and group delay curves are shown in Figure 9.26 for 5-dB and 20-dB fades—that is for two different ratios of direct to interfering rays. Note that the delay peaks and amplitude nulls repeat for a frequency separation of $1/\tau$. When the amplitude of the direct ray is stronger than the interfering ray, the sign of the group delay is the same as the amplitude response. This case is illustrated in Figure 9.27a and is known as *minimum phase*. If the interfering ray is stronger than the direct ray, the sign of the group delay is opposite from the amplitude response and is known as the *nonminimum phase* case (Figure 9.27b). The significance of the group delay's sign is that amplitude equalizers actually increase the envelope delay distortion for nonminimum phase. Measurements of amplitude and group delay by Martin [27] indicate the presence of nonminimum fades for approximately 50 percent of fades greater than 20 dB but less than 16 percent of fades smaller than 20 dB.

Although the two-ray model described here is easy to understand and apply, most multipath propagation research points toward the presence of three (or more) rays during fading conditions. Out of this research, Rummler's three-ray model [25] is the most widely accepted.

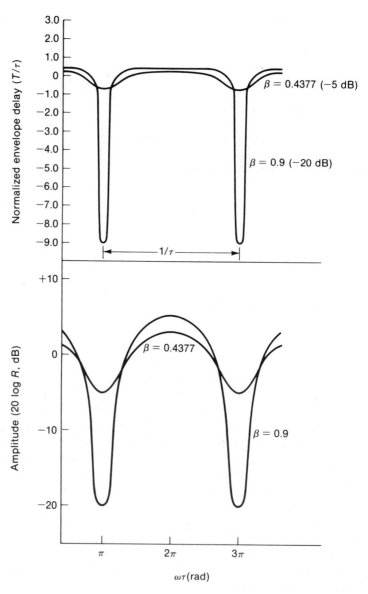

FIGURE 9.26   Two-Ray Envelope Delay and Amplitude Distortion for Fades of 5 and 20 dB

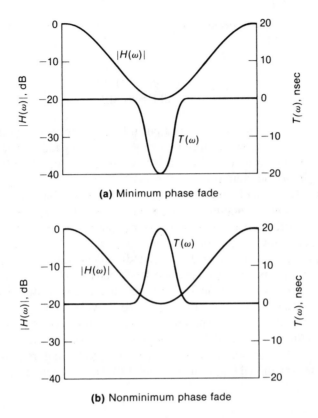

(a) Minimum phase fade

(b) Nonminimum phase fade

FIGURE 9.27   Fading Event Showing Minimum versus Nonminimum Phase

Dispersive Fade Margin

The effects of dispersion due to frequency selective fading are conveniently characterized by the dispersive fade margin (DFM), defined as in Figure 9.16 but where the source of degradation is distortion rather than additive noise. The DFM depends on both the radio performance for given amounts of dispersive fading and the relative occurrence of those dispersive effects for a particular radio link. In equation form [28]

$$DFM = DFMR - 10 \log (DR/0.36) \qquad (9.69)$$

where the Reference Dispersive Fade Margin (DFMR) is the DFM that would be measured for a particular radio under reference fading conditions, and the Dispersion Ratio (DR) describes the relative degree of dispersion observed on different paths. The Dispersion Ratio is defined as (from CCIR Rep. 784-3)

$$DR = \frac{P\ [\text{IBPD} > 10\ \text{dB}]}{P\ [\text{SFF} > 30\ \text{dB}]} \tag{9.70}$$

that is, the ratio of the probability that in-band power difference (IBPD) exceeds 10 dB to the probability that single frequency fades (SFF) exceed 30 dB [10, 29]. Geographical databases have been developed for the dispersion ratio [30] to facilitate the calculation of DFM.

The DFMR is best obtained from field measurements, but can also be estimated from **M-curve signatures.** This signature method of evaluation has been developed to determine radio sensitivity to multipath dispersion using analysis [31], computer simulation, or laboratory simulation [32]. The signature approach is based on an assumed fading model (for example, two ray or three ray). The parameters of the fading model are varied over their expected ranges, and the radio performance is analyzed or recorded for each setting. To develop the signature, the parameters are adjusted to provide a threshold error rate (say, 1×10^{-3}) and a plot of the parameter settings is made to delineate the outage area. In the two-ray model, for example, only two parameters are needed: notch depth $(1 - \beta)$ and offset frequency (f_o). A typical signature developed by using the two-ray model is shown in Figure 9.28. The area under the signature

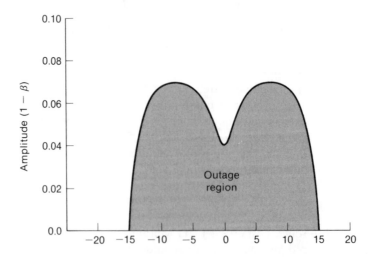

curve corresponds to conditions for which the bit error rate exceeds the threshold error rate; the area outside the signature corresponds to a BER that is less than the threshold value. The *M* shape of the curve (hence the term *M curve*) indicates that the radio is less susceptible to fades in the center of the band than fades off center. As the notch moves toward the band edges, greater notch depth is required to produce the threshold error rate. If the notch lies well outside the radio band, no amount of fading will cause an outage.

Ideally, calculation of the DFMR should be based on an average of four *M* curves to take into account hysteresis effects of adaptive receiver subsystems and minimum versus nonminimum phase fades, since the shape and size of the *M* curve may vary under these conditions. State-of-the-art digital radio receivers are equipped with several adaptive circuits, such as adaptive equalizers, adaptive timing recovery, and adaptive carrier recovery. The hysteresis of these receiver circuits will cause variation in the *M* curve, depending on whether the selective fade is increasing from a small value while the adaptive circuits are properly working or decreasing from a large value after the receiver has lost synchronization and is attempting to regain synchronization. Moreover, the selective fade itself may be minimum or nonminimum, which will affect the behavior of these adaptive circuits. For each of the four cases, the integral of the joint probability density of β and f_0 over the area under each *M* curve corresponds to the probability of outage for that case. The overall reference dispersive fade margin in dB may then be calculated as [33]

$$\text{DFMR} = 18.75 - 10 \log\left(\sqrt{S_{Ma}S_{Mb}/158.4}\right) \tag{9.71}$$

where

$$S_{Ma} = \int_{-39.6}^{39.6} \left\{ \exp\left[-\frac{B_{na}}{5x}\right] + \exp\left[-\frac{B_{ma}}{5x}\right] \right\} df \tag{9.72}$$

and

$$S_{Mb} = \int_{-39.6}^{39.6} \left\{ \exp\left[-\frac{B_{nb}}{5x}\right] + \exp\left[-\frac{B_{mb}}{5x}\right] \right\} df \tag{9.73}$$

The expressions in (9.73) and (9.72) represent the area under the *M* curve corresponding to respectively before and after loss of synchronization due to receiver hysteresis. The integration is done over the four signatures:

B_{ma} = Minimum phase and after loss of synchronization
B_{na} = Non-minimum phase and after loss of synchronization

B_{mb} = Minimum phase and before loss of synchronization
B_{nb} = Nonminimum phase and before loss of synchronization

The factor x in (9.72) and (9.73) is experimentally derived, varies with time and location, and has a value of approximately 1 [34, 35].

Probability of Outage
A method for predicting digital link performance in the presence of frequency-selective fading has not yet emerged that is as simple and well established as that for flat fading channels. Most of the outage prediction techniques assume a two-ray or three-ray multipath model and then measure or infer statistics of each parameter in the model. The probability of outage can then be calculated by integrating the probability density function (pdf) of the channel parameters over the radio outage region as established by the M curve. Emshwiller [31], for example, assumed a two-ray model with a joint pdf of $p(\beta, \tau)$ and a probability of outage given by

$$P(o) = \int_{\tau} \int_{\beta} p(\beta, \tau)\, d\beta\, d\tau \qquad (9.74)$$

This integral was evaluated by assuming $p(\beta, \tau) = p(\beta)p(\tau)$, assuming a form for $p(\beta)$ and $p(\tau)$, and replacing the M curve by a rectangle with the same area. Rummler and Lundgren [32] assumed a three-ray model, developed statistics of the joint pdf based on experimental data, and evaluated the probability of outage using an integral similar in form to Equation (9.74).

Another approach to predicting digital radio system performance is based on the assumption that amplitude dispersion is the principal cause of outages. The probability of outage can then be expressed as the product of a dispersion occurrence factor and the multipath occurrence factor. The dispersion occurrence factor is conveniently expressed as a probability distribution function of in-band amplitude dispersion. This approach has the advantage of simplicity and independence from the ray model assumed. Amplitude dispersion is defined here as the difference in amplitude, ΔA, for two in-band frequencies, usually located at the two band edges; thus, in terms of decibels,

$$\Delta A = |20 \log \alpha| \qquad (9.75)$$

where $\alpha = R_1/R_2$ and R_1 and R_2 are the amplitudes of frequencies located at the band edges. Note that when ΔA is expressed in dB, it is equivalent to the in-band power difference, IBPD. Assuming that R_1 and R_2 are correlated

Rayleigh-distributed signals, Vigants [36] has shown the probability that α exceeds some value α_0 to be

$$P(\alpha > \alpha_0) = \frac{100\Delta f \alpha_0^2}{f^2 D(\alpha_0^2 - 1)^2} \qquad (\alpha_0 \gg 1) \qquad (9.76)$$

where Δf = frequency separation of two signals and also the transmission bandwidth (GHz)
f = radio frequency (GHz)
D = path length (mi)

The probability of outage can now be written

$$P(o) = P(\alpha > \alpha_0) \cdot d \qquad (9.77)$$

where α_0 now represents the threshold of in-band amplitude dispersion for a specified BER and where the multipath occurrence factor d is defined by (9.41). Substituting (9.41) and (9.76) into (9.77) we obtain the final form as

$$P(o) = (2.5 \times 10^{-4})\left(\frac{abD^2\Delta f}{f}\right)\left[\frac{\alpha_0^2}{(\alpha_0^2 - 1)^2}\right] \qquad (\text{for } \alpha_0 \gg 1) \qquad (9.78)$$

The theoretical predictions of (9.78) show good agreement with measured results of outages for 8-PSK and 16-QAM radio systems as well as measured data of amplitude dispersion distributions [37].

Example 9.4 _____

Consider an 8-PSK radio operating at a radio frequency of 6 GHz over a path of average terrain, and length 26.4 mi. The threshold of in-band amplitude dispersion for a BER of 10^{-3} is known to be 8 dB over a measured bandwidth of 23.1 MHz. Find the probability of outage (BER $\geq 10^{-3}$), for a heavy fading month (climate factor $b = 1$).

Solution   From (9.78) we obtain

$$P(o) = \frac{(2.5 \times 10^{-4})(0.0231)(26.4)^2(2.5)^2}{6(5.25)^2}$$
$$= 1.5 \times 10^{-4}$$

This result compares favorably with the measured result of 4.1×10^{-4} found by Barnett [38].

A third and simpler approach [39] to outage prediction for selective fading is based on the assumption that outages due to flat and dispersive fading are independent and can therefore be added, so that

$$P(o) = P_f(o) + P_s(o) \tag{9.79}$$

where $P_f(o)$ is the probability of outage for flat fading and $P_s(o)$ is the probability of outage due to selective fading. Using the same arguments as Section 9.3.1, we can immediately write

$$P(o) = d10^{-FFM/10} + d10^{-DFM/10} \tag{9.80}$$

where FFM is the flat fade margin and DFM is given by (9.69).

Improvements Due to Diversity and Equalization
Both diversity and adaptive equalization can be used, separately or together, to improve digital radio performance in the presence of frequency-selective fading. Diversity reduces the probability of in-band dispersion. Adaptive equalization reduces the in-band difference between the minimum and maximum amplitude values and, depending on the type of equalizer, reduces the in-band difference between group delay values also. In many instances, both diversity and equalization have been necessary to meet performance objectives.

Tests of digital radio links indicate that the diversity improvement factors are larger than those predicted for flat fading, which has led to the development of new models for diversity improvement [40]. Using the model assumed by (9.80), we may expand the probability of outage given in (9.54) to also include the effect of diversity on selective fading,

$$P_d(o) = \frac{P_f(o)}{I_d^f} + \frac{P_s(o)}{I_d^s} \tag{9.81}$$

where I_d^f and I_d^s are the diversity-improvement factors for flat fading and selective fading, respectively. Exact expressions for space and frequency diversity-improvement factors for flat fading were given by (9.52) and (9.53), but improvement factors for selective fading are known only in a general form for space diversity [30]:

$$I_{sd}^s = (k/D) \, 10^{DFM/10} \tag{9.82}$$

and for frequency diversity

$$I_{fd}^s = (w/D) \, 10^{DFM/10} \qquad (9.83)$$

where k and w are experimentally derived constants.

Interestingly, the combined improvement obtained by simultaneous use of diversity and equalization has been found to be larger than the product of the individual improvements. This synergistic effect has been reported in several experiments [41, 42], where the added improvement has resulted from the diversity combiner's ability to replace in-band notches with slopes that are easier to equalize. Giger and Barnett [43] have derived a formula for this improvement, given by

$$I_t = I_d \cdot I_e^2 \qquad (9.84)$$

where
$\quad I_t =$ total improvement factor
$\quad I_d =$ diversity improvement factor
$\quad I_e =$ equalization improvement factor

These improvement factors are measured or calculated as the ratio of the nondiversity, unequalized system outage probability to the total, diversity, or equalized system outage probability.

Field measurements of BER distributions for a 6-GHz, 90-Mb/s, 8-PSK radio on a 37.3-mi link are shown in Figure 9.29 [44]. This system was tested in four configurations: unprotected; with adaptive equalization; with space diversity; and with equalization plus diversity. These results are tabulated in Table 9.1 and compared to other results; all sets of data indicate a synergistic effect, and large improvement is observed with the combination of equalization and diversity.

9.4   Frequency Allocations and Interference Effects

The design of a radio system must include a frequency allocation plan, which is subject to approval by the local frequency regulatory authority. In the United States, radio channel assignments are controlled by the Federal Communications Commission (FCC) for commercial carriers, by the National Telecommunications and Information Administration (NTIA) for government systems, and by the Military Communications and Electronics Board (MCEB) for military systems. The FCC's regulations for use of microwave spectrum establish eligibility rules, permissible use rules, and technical specifications. There are four principal users who either share or

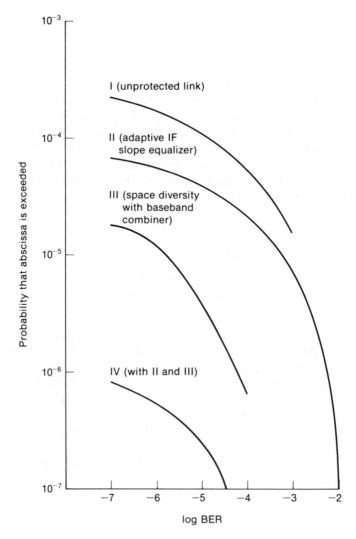

FIGURE 9.29   BER Distributions for 6-GHz, 90-Mb/s, 8-PSK, 37.3-mi Link Showing Effects of Equalization and Diversity [44]

exclusively use a particular spectrum allocation: common carriers (regulated by Part 21 of the FCC Rules), broadcasters (Part 74), cable TV operators (Part 78), and private companies (Part 94). Technical specifications are intended to protect against interference and to promote spectral efficiency. Equipment type acceptance regulations include transmitter power limits, frequency stability, out-of-channel emmission limits, and antenna

TABLE 9.1   Summary of Space Diversity and Adaptive Equalizer Performance for Digital Radio Links

	Improvement Factors[a]		
Radio System	Space Diversity	Adaptive Equalizer	Combined
1[b]	38	2	770
2[c]	6	3–5	175
3[d]	14.2	2.6	320

[a]Improvement factors are based on fraction of time that BER ≥ threshold, with threshold = 10^{-4} for radio system 1, 10^{-3} for radio system 2, and 10^{-5} for radio system 3.
[b]8-GHz, 91-Mb/s, QPR, 32-mi link with IF combining and linear adaptive equalizer [41].
[c]6-GHz, 78-Mb/s, 8-PSK, 26.4-mi link with IF combining and IF amplitude slope equalizer [38,42].
[d]6-GHz, 90-Mb/s, 8-PSK, 37.3-mi link with baseband combining and IF amplitude slope and notch equalizer (Figure 9.29).

directivity. For digital microwave, specifications are added for digital modulation, spectral efficiency in bps/Hz, and voice channel capacity [45]. A standard has also been developed for the U.S. military, MIL-STD-188-322, for the application of digital microwave to military frequency bands [46]. Table 9.2 lists the microwave frequency bands that are authorized for digital microwave in the United States together with the allowable bandwidths and required spectral efficiencies.

The International Radio Consultative Committee (CCIR) issues recommendations on radio channel assignments for use by national frequency allocation agencies. Although the CCIR itself has no regulatory power, it is important to realize that with the exception of the United States, CCIR recommendations are usually adopted on a worldwide basis. With regard to digital microwave systems, the CCIR has issued recommendations beginning with the 1982 Plenary Assembly. Table 9.3 lists reports and recommendations of CCIR Study Group 9 pertinent to the planning of frequency allocations for digital microwave systems [47, 29].

The usual practice in frequency channel assignments for a particular frequency band is to separate transmit ("go") and receive ("return") frequencies by placing all go channels in one half of the band and all return channels in the other half. With this approach, all transmitters on a given station are in either the upper or lower half of the band, with receivers in the remaining half. Within each half-band, adjacent channels must be spaced far enough apart to avoid energy spillover between channels. A common scheme used to increase adjacent channel discrimination is to alternate between vertical and horizontal polarization. The isolation provided by cross-polarizing adjacent channels is on the order of 20 dB or more. At the edges of the band, a guard spacing is necessary to protect against interference into and from adjacent bands.

TABLE 9.2   Authorized Frequency Bands in the U.S. for Digital Microwave Systems [45, 46]

Use	Frequency Band (GHz)	Allowable Bandwidth (MHz)	Spectral Efficiency Required (bps/Hz)
Common carrier	2.110–2.130	3.5	1.8
	2.160–2.180	3.5	1.8
	3.700–4.200	20	3.7
Military	4.400–5.000	3.5	0.9
		7.0	0.9
		10.5	0.9
		14	0.9
Common carrier	5.925–6.425	30	2.5
Military	7.125–8.400	3.5	0.9
		7.0	0.9
		10.5	0.9
		14	0.9
		20	0.92
Common carrier, private company	10.55–10.68	2.5, 5	1.0
Common carrier	10.700–11.700	40	1.8
Military	14.400–15.250	20	0.92
		28	0.9
		40	0.92
Common carrier, broadcasters, cable TV, private company	17.7–19.7	2, 5, 6, 10, 20, 40, 80	1.0

As an illustration of the frequency planning procedures described above, Figure 9.30 shows a suggested RF channel arrangement for a 16-QAM, 140-Mb/s digital radio system operating in the 6-GHz band. The band is first separated into halves to accommodate transmitting and receiving frequencies. Recommended spacings are shown for the band edge (ZS), adjacent channels (XS), and adjacent transmitter and receiver (YS). Note that adjacent channels also have opposite polarization, alternating between vertical (V) and horizontal (H).

Another important consideration in radio link design is RF interference, which may occur from sources internal or external to the radio system. The system designer should be aware of these interference sources in the area of each radio link, including their frequency, power, and directivity. Certain steps can be taken to minimize the effects of interference: good site selection, use of properly designed antennas and radios to reject interfering signals, and use of a properly designed frequency plan.

Figure 9.31 illustrates internal sources of RF interference, which are classified as overreach, adjacent station, and spur interference. (The solid lines

TABLE 9.3   CCIR Frequency Allocations Applicable to Digital Radio

Frequency Band (GHz)	Frequency Range (GHz)	Channel Spacing (MHz)	Modulation Technique	Spectral Efficiency (bps/Hz)	Bit Rate (Mb/s)	CCIR Rec. or Report
2	1.7–2.7	29	8-PSK	2.1	70	Report 934-2
		29	16-QAM	2.7	90	and Rec. 283-5
		14	64-QAM	5	70	
4	3.7–4.2	29	8-PSK	2.1	70	Report 934-2
		29	16-QAM	2.7	90	and Rec. 382-5
		40	16-QAM	3.16	140	Report 934-2
		10 m (m = 1,2,3)			90, 140, 200	Rec. 635-1
6	5.925–6.425	20	64-QAM	4.32	90	Report 934-2
		29.65	8-PSK	2.24	70	Report 934-2
		29.65	16-QAM	2.88	90	Report 934-2
		29.65	64-QAM	4.48, 4.98	140, 155	Rec. 383-4 and Report 934-2
		40	16-QAM	3.3	140, 155	Rec. 384-5 and Report 934-2
8	6.430–7.110	29.65	8-PSK	1.78	70	Report 934-2
	7.725–8.275	29.65	16-QAM	2.3	90	Report 934-2
11	10.7–11.7	67	4-PSK		140	Report 782-3
		60	8-PSK		140	Report 782-3
		48	16-QAM		140	Report 782-3
		40			140, 155	Rec. 387-5
13	12.75–13.25	28			34	Rec. 497-3
		3.5m, 7m (m = 1,2,3 … 8)				Rec. 497-3
		35			70	Rec. 497-3
15	14.4–15.35	28			70-140	Rec. 636-1
		14			70-140	Rec. 636-1
		7				Rec. 636-1
18	17.7–19.7	220			280	Rec. 595-2
		110			140	Rec. 595-2
		27.5			34	Rec. 595-2
23	21.2–23.6	2.5n, 3.5n				Rec. 637 and
		50				Report 936-2

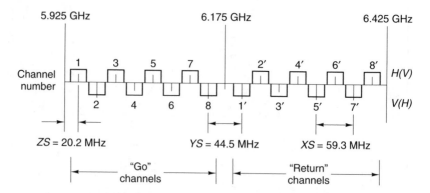

FIGURE 9.30   RF Channel Arrangement for a 140-Mb/s Digital Radio-Relay System Operating in the 6-GHz Band [47]

in the figure signify the desired signal; the dashed lines represent the interfering signal.) Overreach may occur when radio links in tandem are positioned along a straight line. In Figure 9.31a, overreach interference becomes significant when energy transmitted from site *A* arrives at site *D*—which may occur during superrefractive conditions—while there is fading present on the signal received from *C*. This problem can be reduced by staggering links to avoid a straight-line sequence of paths or by using earth blockage on the overreach path. Adjacent station and spur interference is more complex to analyze but is a function of antenna performance parameters, such as front-to-back ratio and side lobes.

Other sources of interference that may arise outside the radio system include radar, troposcatter, satellite, and other LOS systems. Radar systems often propagate energy at high levels and in a 360° arc and are therefore a significant source of interference for microwave receivers. Earth blocking or filtering of the interfering signals is recommended to minimize interference from radar. Certain frequency bands are shared among line-of-sight, troposcatter, and satellite systems. Because of the large power used by troposcatter and satellite transmitters, LOS receivers are susceptible to interference from these transmitters even at distances well beyond the horizon. High-power transmitters for satellite and troposcatter transmission are usually located in isolated areas, however, which reduces this interference problem. Existing LOS systems, either analog or digital, may parallel or intersect the new link. Information needed of existing systems includes transmit power, frequency, distance, and antenna discrimination.

The effect of RF interference on a radio system depends on the level of the interfering signal and whether the interference is in an *adjacent* channel

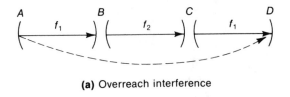

(a) Overreach interference

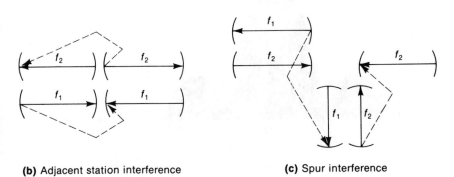

(b) Adjacent station interference

(c) Spur interference

FIGURE 9.31   Intrasystem Sources of RF Interference [13]

or is *cochannel.* A cochannel interferer has the same nominal radio frequency as that of the desired channel. Cochannel interference arises from multiple use of the same frequency without proper isolation between links. Adjacent channel interference results from the overlapping components of the transmitted spectrum in adjacent channels. Protection against this type of interference requires control of the transmitted spectrum, proper filtering within the receiver, and orthogonal polarization of adjacent channels. Both types of interference are illustrated in Figure 9.32.

The performance criteria for digital radio systems in the presence of interference are usually expressed in one of two ways: allowed degradation of the S/N threshold or allowed BER. Both criteria are stated for a given signal-to-interference ratio (S/I). For example, AT&T requirements for 6-GHz digital radio are a maximum allowed BER of 10^{-3} with an S/I of 24 dB for cochannel interference and the same BER with an S/I of 10 dB for adjacent channel interference [48]. In another example, Bellcore requirements for digital radio systems are that a carrier-to-interference ratio (C/I) of $X + 3$ dB shall produce a DS-3 bit error rate of 10^{-3} or less, where X is the S/N under thermal noise conditions producing a DS-3 BER of 10^{-3}. The interfering signal is specified to be cochannel, either a single tone or a signal with modulation identical to the desired system [50]. As a further illustration, the effect of various cochannel interference levels, with S/I of

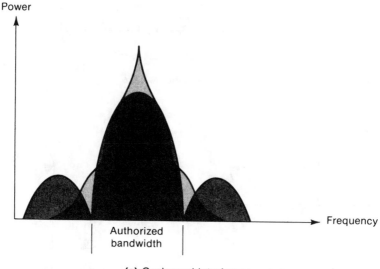

(a) Cochannel interference

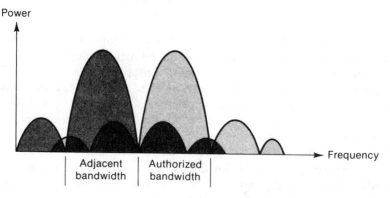

(b) Adjacent channel interference

FIGURE 9.32   Types of RF Interference

20, 25, and 30 dB, on the BER and S/N of coherent phase-shift-keyed (PSK) systems is shown in Figure 9.33 and 9.34 [49]. In Figure 9.33, the S/N ratio has been adjusted to give a BER of 1×10^{-8} with $S/I = \infty$.

9.5   Digital Radio Design

A block diagram of a digital radio transmitter and receiver is shown in Figure 9.35. The traffic data streams at the input to the transmitter are usually

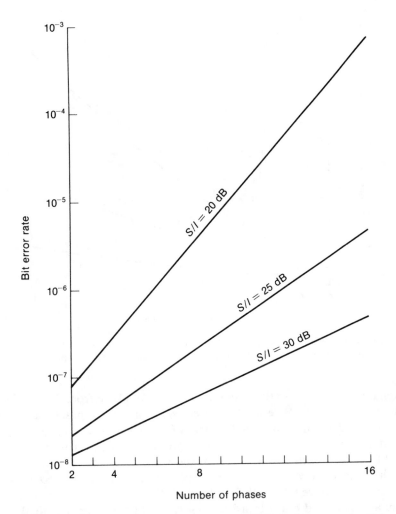

FIGURE 9.33   Effects of Cochannel Interference on BER Performance of a Coherent Detection PSK System as a Function of Number of Phases (Adapted from [49])

in coded form, for example bipolar, and therefore require conversion to an NRZ signal with an associated timing signal. The multiplexer combines the traffic NRZ streams and any auxiliary channels used for orderwires into an aggregate data stream. This step is accomplished either by using pulse stuffing, which allows the radio clock rate to be independent of the traffic data, or by using a synchronous interface, which requires the radio and traffic data to be controlled by the same clock. The aggregate signal is scrambled to obtain a smooth radio spectrum and ensure recovery of the timing signal at the receiver. For phase modulation, some form of differential encoding is

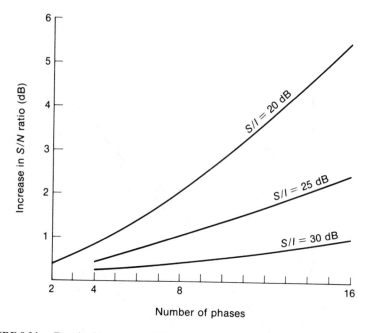

FIGURE 9.34   Required Increase in S/N Ratio to Maintain a 1×10^{-8} BER Performance of a Coherent Detection PSK System as a Function of Number of Phases (Adapted from [49])

often employed to map the data into a change of phase from one signaling interval to the next. The modulator converts the digital baseband signals into a modulated intermediate frequency (IF), which is typically at 70 MHz when the final frequency is in the microwave band. The RF carrier is generated by a local oscillator, which is mixed with the IF modulated signal to produce the microwave signal. The RF power amplification is accomplished by a traveling-wave tube (TWT) or by a solid-state amplifier such as the gallium arsenide field-effect transistor (GaAs FET) amplifier. The final component of the transmitter is the RF filter, which shapes the transmitted spectrum and helps control the signal bandwidth.

At the receiver, the RF signal is filtered and then mixed with the local oscillator to produce an IF signal. The IF signal is filtered and amplified to provide a constant output level to the demodulator. Automatic gain control (AGC) in the IF amplifier provides variable gain to compensate for signal fading. Because the AGC voltage is a convenient indicator of received signal level, it is often used for performance monitoring or diversity combining. Fixed equalization is required to compensate for static amplitude or delay distortion from radio components, such as a TWT or filter, or to build out differential delay between RF channels in diversity operation. Adaptive

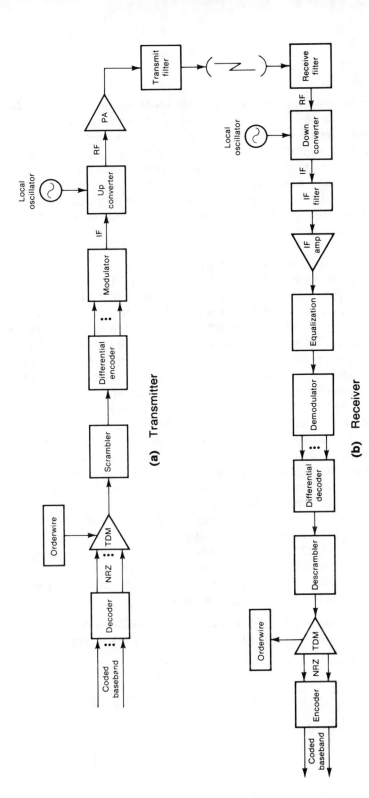

FIGURE 9.35   Block Diagram of a Digital Radio

equalization may also be required to deal with frequency-selective fading on the transmission path. Using the amplified and equalized IF signal, the demodulator recovers data and corresponding timing signals. Some type of performance monitoring is also commonly found in the demodulator, often based on eye pattern opening or pseudoerror techniques. The recovered baseband signal is next decoded and descrambled to reconstruct the aggregate data stream. The demultiplexer recovers the traffic data streams and auxiliary channels. Finally, in the baseband encoder the standard data interface is generated.

9.5.1   Effects of Nonlinear Amplifiers

In order to maximize efficiency (DC to RF) and RF power output, the microwave power amplifier (PA) shown in Figure 9.35a is often operated near saturation. An amplifier output is referred to as saturated when the output power is no longer increased by an increase in input level. For small input level, the output of a linear amplifier increases linearly with increasing input power. Near saturation, however, the input/output relationship becomes nonlinear. Although FM analog radios operate satisfactorily with saturated amplifiers, with digital modulation schemes, particularly those having an AM component, amplifiers operating in a nonlinear region are frequently a source of performance degradation and spectral spreading.

In order to examine these effects, we will consider the amplitude and phase characteristics of a typical TWT [51]. In Figure 9.36a, the power output increases linearly with increasing power input until saturation begins to occur where the output becomes nonlinear. This amplitude characteristic also produces intermodulation (IM) products that can be harmful if they fall within the radio channel. The most harmful IM product is third-order, since it falls in-band and contains more power than higher-ordered products. For Figure 9.36b, zero phase shift is observed with increasing power input until saturation begins to occur where the phase shift increases in a nonlinear fashion. These amplitude and phase characteristics of Figure 9.36 give rise to two forms of nonlinear distortion:

• Amplitude modulation to amplitude modulation (AM/AM) conversion
• Amplitude modulation to phase modulation (AM/PM) conversion

To model these nonlinear effects, let the TWT input signal be of the form

$$x(t) = A(t)\cos\left[\omega_0 t + \theta(t)\right] \tag{9.85}$$

Then the TWT output signal may be expressed by

$$y(t) = G\left[A(t)\right]\cos\left\{\omega_0 t + \theta(t) + F\left[A(t)\right]\right\} \tag{9.86}$$

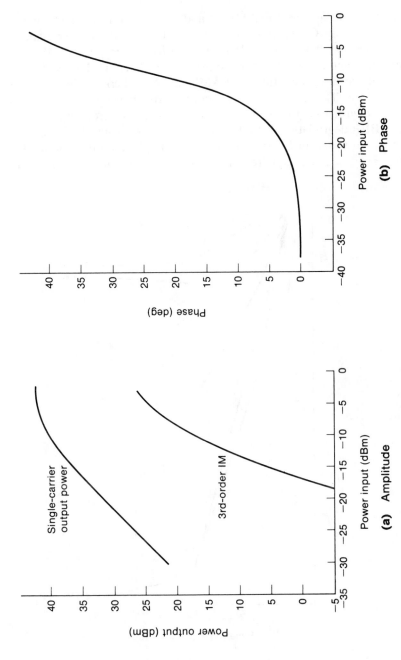

FIGURE 9.36 Amplitude and Phase Characteristics of Traveling-Wave Tube (English Electric Valve Model N10022) [51]

609

where $G(\cdot)$ and $F(\cdot)$ are the AM/AM and AM/PM conversion characteristics, respectively. Considerable efforts have been made to derive analytic expressions that characterize TWT nonlinearity [52]. Experimental investigations [53] and computer simulations have also been used to examine the effects of traveling-wave tube AM/AM and AM/PM conversion on system performance. The effects of TWT nonlinearity have been found to be twofold: degradation of the bit error rate and spectral spreading.

The BER degradation of a QPR modulated signal, caused by a nonlinear TWT, is illustrated in Figure 9.37. In this example, the effect of amplifier nonlinearity becomes apparent for any power output above 2.5 watts,

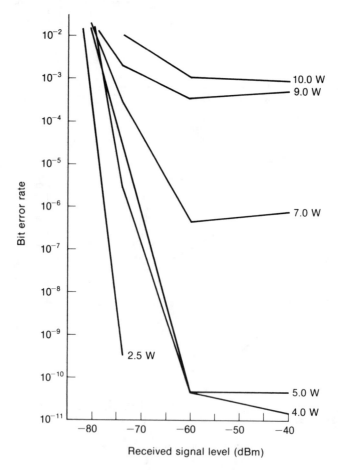

FIGURE 9.37   Effect of TWT Nonlinearity on BER for QPR Modulation and Different Output Power [54]

where an irreducible error rate is observed. The effect of TWT nonlinearity on the transmitted spectrum of a QPR signal is illustrated in Figure 9.38. Digital modulation techniques such as QPR have spectra that decay slowly compared with FM spectra. Filtering to remove higher-order side lobes and reduce the width of the main lobe is usually done at baseband or IF, but nonlinearity in the RF amplifier can cause the restoration of out-of-band side lobes. This effect may cause interference with adjacent radio channels and violation of bandwidth allocations.

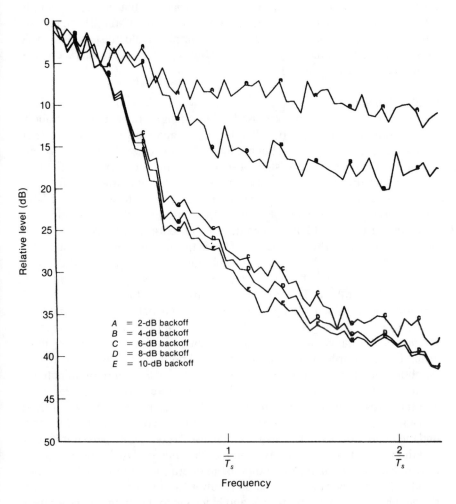

FIGURE 9.38   Spectrum of QPR Modulation for Different TWT Backoffs Using TWT Characteristics of Figure 9.36

Figures 9.37 and 9.38 also illustrate one remedy for the nonlinear effects of a TWT—that is, to reduce, or back off, the output power so that the amplifier operates in a more linear region. To maintain desired output power, an alternative solution is to provide any spectrum control filtering at RF, after the nonlinear amplifier. This approach permits nearly saturated operation of the amplifier but at the expense of the insertion loss characteristic of an RF filter. Further, this approach means more difficult design and higher cost for RF filtering compared to IF or baseband filtering. As a third alternative, the amplifier characteristics can be compensated by IF predistortion so that the overall characteristic approaches that of a linear amplifier. With this approach the amplitude and phase of the IF signal are adjusted to cancel the nonlinearity of the amplifier. Postdistortion can be used in the same manner as predistortion but is performed at the receiver IF rather than the transmitter IF.

9.5.2   Antennas

Two basic types of antennas are used in line-of-sight transmission: parabolic reflectors and horn reflectors. Parabolic antennas are the simplest, least expensive, and most popular, but are limited to low-capacity systems. Where multiple RF channels are to share a single antenna, the horn antenna offers improved performance because of a more favorable antenna pattern.

Typical parabolic antennas have an aperture diameter of two to fifteen feet and use reflectors made of a conductive material such as aluminum. The reflecting surface must have a high degree of accuracy with respect to a true parabola to achieve the desired gain. Hence the antenna must be designed to maintain the correct shape during installation and under expected environmental conditions. The feed point for these antennas is located at the focal length of the parabola. Two types of feeds are used, either a "button-hook," which brings the feed though the reflector from behind, or a front feed, which is installed along the axis of the tower. To improve the noise and interference performance, a metallic shield is often added, forming a cylinder around the periphery of the reflector. Another improvement to the parabolic antenna uses a second reflector at the focal point to illuminate the parabola. There are two types of dual reflector antennas, the Cassegrainian and the Gregorian, as shown in Figure 9.39. The difference between these two antennas is the shape of the second reflector used between the feedhorn and the main reflector. The Cassegrainian uses a convex-shaped subreflector and the Gregorian has a concave-shaped subreflector. Dual reflector antennas are used where high gain or dual-band operation is required. In addition to added cost, dual reflector antennas suffer from spillover of energy from the feed horn past the subreflector and blockage to the main reflector due to the presence of the subreflector.

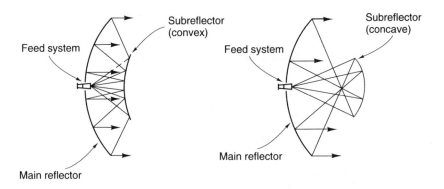

(a) Cassegrainian antenna system   **(b)** Gregorian antenna system

FIGURE 9.39   Dual Reflector Antennas

In the parabolic antennas described so far, the antenna feed system is positioned at the focal point of the reflector and therefore within the aperture of the antenna. As a result, performance is degraded due to blockage to the reflector as well as scattering from the feed support structure. The horn reflector antenna, shown in Figure 9.40, avoids this problem by using a sector of an offset paraboloid and by placing the feed system outside the antenna's aperture. The horn reflector antenna uses a pyramidal or conical shaped horn positioned at the focal point of the offset paraboloid. The horn reflector antenna has better suppression of side and back lobes of the antenna pattern and greater bandwidth than a parabolic antenna with comparable gain, but has the disadvantages of greater cost and larger antenna wind loads.

Wind load and precipitation accumulation, which will affect the performance of the antenna, can be minimized by enclosing the reflector aperture with a radome. Radomes may be planar or geometrically shaped, for example, parabolic or conical. The radome is constructed of a material such as woven fiberglass that does not affect the antenna's performance. Ice and snow accumulation may be minimized through use of special chemical coatings that cause water, ice, and snow to be readily shed, or by use of resistance heaters distributed through the radome surface. The loss introduced by a radome, under dry conditions, varies from less than a dB to several dB, depending on the type of radome. Grid antennas consisting of tubular elements that form a grid can also be used to minimize wind load effects. Grid antennas, however, can be used only in single polarized applications operating below 2.7 GHz [55].

Microwave antennas used in line-of-sight transmission (terrestrial or satellite) provide high gain because of the focusing of radiated energy

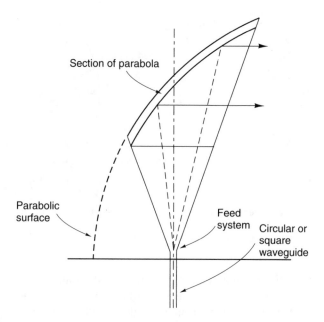

FIGURE 9.40   Horn Reflector Antenna

within a small angular region. The power gain of a parabolic antenna is directly related to both the aperture area of the reflector and the frequency of operation. Relative to an isotropic antenna, the gain can be expressed as

$$G = 4\pi A\eta/\lambda^2 \tag{9.87}$$

where λ is the wavelength of the frequency, A is the actual area of the antenna in the same units as λ, and η is the efficiency of the antenna. The antenna efficiency is a number between 0 and 1 that reduces the theoretical gain because of physical limitations such as nonoptimum illumination by the feed system, reflector spillover, feed system blockage, and imperfections in the reflector surface. Single reflector antennas have efficiencies in the range of 50 to 65 percent, while dual reflector antennas can achieve efficiencies as high as 80 percent. Converting (9.87) to more convenient units, the gain may be expressed in dB as

$$G_{dB} = 20 \log f + 20 \log d + 10 \log \eta - 49.92 \tag{9.88}$$

where f is the frequency in MHz and d is the antenna diameter in feet. Antenna gains for typical antenna sizes and frequency bands are shown in

Table 9.4. Radiation patterns are commonly plotted in the form shown in Figure 9.41. Here the center of the graph represents the location of the antenna, and field strength is plotted along radial lines, outward from the center. The line at 0° shows the direction of maximum radiation, while in this example the radiated power is down 3 dB at ± 30°. In this case the antenna has a 3-dB beamwidth $\theta_{1/2}$ of 60° (for sake of illustration this example exaggerates the 3-dB beamwidth, which is typically only a few degrees). The null-to-null beamwidth θ_B describes the beamwidth of the main lobe, which is larger than the 3-dB beamwidth. In equation form,

$$\theta_{1/2} = \left(142/\sqrt{G}\ \right)^{\circ} \tag{9.89}$$

and

$$\theta_B \approx (2.4\ \theta_{1/2})^{\circ} \tag{9.90}$$

where $\sqrt{G} = \log^{-1}(G_{dB}/20)$. From (9.89) we note that the antenna beamwidth decreases with increasing frequency and antenna diameter. As a result, tower rigidity requirements go up with increasing antenna size and frequency because of greater weight and smaller beamwidth of the antenna. Another important measure of antenna performance is the ratio of power radiated from the front lobe to that radiated from the back lobe. Figure 9.41 shows a small lobe extending from the rear of the antenna, which may cause interference with another nearby antenna. Parabolic antennas attain front-to-back ratios of 50 to 60 dB, while horn antennas can achieve 60 to 70 dB ratios.

Waveguide or coaxial cable is required to connect the top of the radio rack to the antenna. Coaxial cable is limited to frequencies up to the 2-GHz

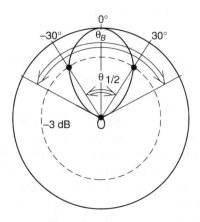

FIGURE 9.41   Typical Antenna Radiation Pattern

TABLE 9.4   Gain of Parabolic Antennas (decibels)

Parabolic Antenna Diameter (ft)	2 GHz		4 GHz		6 GHz		8 GHz	
	G_A	$2G_A$	G_A	$2G_A$	G_A	$2G_A$	G_A	$2G_A$
4	25.3	50.6	31.3	62.6	34.8	69.6	37.3	74.6
6	28.8	57.6	34.8	69.6	38.3	76.6	40.8	81.6
8	31.3	62.6	37.3	74.6	40.8	81.6	43.3	86.6
10	33.3	66.6	39.3	78.6	42.8	85.6	45.3	90.6
12	34.8	69.6	40.8	81.6	44.3	88.6	46.8	93.6
15	36.8	73.6	42.8	85.6	46.3	92.6	48.8	97.6

Note: $G_A = 20 \log f + 20 \log d - 52.75$, where f = frequency in megahertz and d = antenna diameter in feet.

band. Three types of waveguide are used—rectangular, elliptical, and circular—which vary in cost, ease of installation, attenuation, and polarization. Rectangular waveguide has a rigid construction, comes in standard lengths, and is typically used with short runs requiring only a single polarization. Standard bends, twists, and flexible sections are available, but multiple joints can cause reflections. Elliptical waveguide is semirigid, comes in any length, and is therefore easy to install. Elliptical waveguide only accommodates a single polarization, but has lower attenuation than rectangular waveguide. Circular waveguide has the lowest attenuation, about one half that of rectangular, and also provides dual polarization and multiple band operation in a single waveguide. Its disadvantages are higher cost and more difficult installation. The kind of antenna used will also influence the choice of waveguide type. Typically, with parabolic antennas, elliptical waveguide is used for the vertical run, while rectangular waveguide is often used between the radio and the base of the tower. Circular waveguide is commonly used with horn antennas because of its ability to carry orthogonal polarizations and operate in dual bands. A dual-polarization single band radio system may also be implemented with two individual rectangular or elliptical waveguide runs. Table 9.5 provides attenuation factors as a function of frequency for commonly used coaxial cable and waveguide types.

9.5.3   Diversity Design

Diversity in LOS links is used to increase link availability by reducing the effects of multipath fading, improving the combined output S/N ratio, and protecting against equipment failure. The most common forms of diversity use two parallel paths, separated in frequency or space, to provide one-for-one (1:1) protection on each link. The improvement afforded each link depends on the degree of correlation in fading between the two paths and the ability of a combiner to recognize and mitigate the effects of fading or equipment failure.

TABLE 9.5   Transmission Line Loss Factors (decibels/meter) [13, 68]

Transmission Line Type	Frequency Band (GHz)			
	2	4	6	8
Rectangular waveguide (WR)	—	0.027 (WR 229)[a]	0.068 (WR 137)[a]	0.087 (WR 112)[a]
Elliptical waveguide (EW)	—	0.028 (EW 37)[a]	0.039 (EW 52)[a]	0.058 (EW 77)[a]
Circular waveguide (WC)	—	0.013 (WC 269)[a]	0.030 (WC 166)[a]	0.022 (WC 166)[a]
Coaxial (⅞″, air dielectric)	0.062 (HJ 5)[b]	—	—	—

[a]Designates type of waveguide.
[b]Designates type of coaxial cable.

A typical arrangement for space diversity is shown in Figure 9.42. The two transmitters operate on the same frequency and can be switched for output to a common antenna. One transmitter can operate in a hot standby mode while the other is on-line, as shown in Figure 9.42; but as an alternative configuration, they can be combined to provide a 6-dB increase over the power available from a single transmitter. In this latter case, the failure of one transmitter power amplifier causes a 6-dB drop in output power but the link remains operational. The two receivers are connected to different antennas that are physically separated to provide the desired space diversity effect. The receiver outputs are fed to the combiner, which combines the two received signals. Space diversity, unlike frequency diversity, does not require an additional frequency assignment and is therefore more efficient in the use of spectrum. Its disadvantage is that additional antennas and waveguide are required, making it more expensive than frequency diversity arrangements.

Figure 9.43 illustrates a typical arrangement for frequency diversity. The two transmitters operate continuously on different frequencies but carry identical traffic. The receivers are connected to the same antenna but are tuned to separate frequencies. The combiner function is identical to that of the space diversity configuration. The use of frequency diversity doubles the spectrum amount required—a significant disadvantage in congested frequency bands. Unlike space diversity, however, frequency diversity provides two complete, independent paths, allowing testing of one path without interrupting service, while requiring only a single antenna per link end.

Angle diversity is a newer form of diversity that has been shown to have cost and performance advantages over the more conventional diversity techniques already discussed. The basic idea behind angle diversity is that most deep fades, which are caused by two rays interfering with one another, can be mitigated by a small change in the amplitude or phase of the individ-

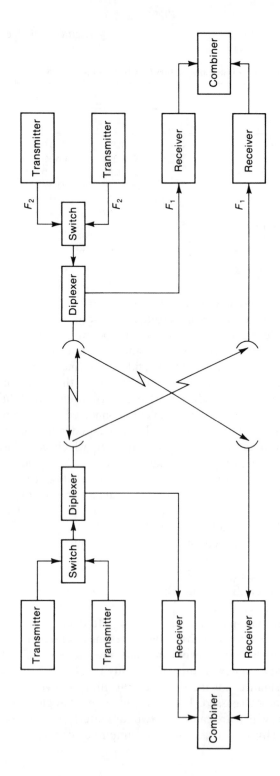

FIGURE 9.42　Space Diversity System with Hot Standby Transmitter

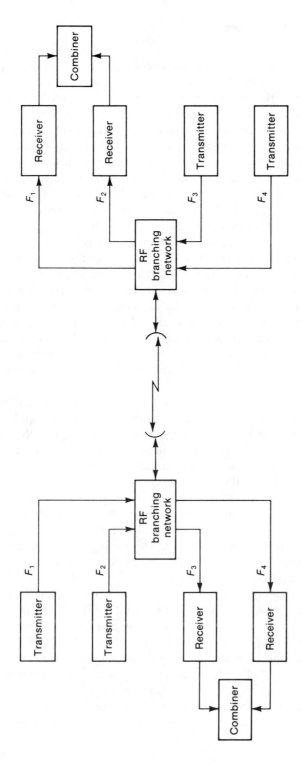

FIGURE 9.43 Frequency Diversity System (Single Antenna)

ual rays. Therefore, if a deep fade is observed on one beam of an angle diversity receiving antenna, switching to (or combining with) the second beam will change the amplitude or phase relationship and reduce the fade depth. Two techniques have been used to achieve angle diversity: a single antenna with two feeds that have a small angular displacement between them, or two separate antennas having a small angular difference in vertical elevation. The dual beam, single antenna technique, shown in Figure 9.44, is the more popular and economical of the two techniques. In one approach, shown in Figure 9.44a, the individual beams are directly fed to a combiner or switch. With this approach, the lower beam is usually set at boresight for $k = \frac{4}{3}$, and the upper beam is aimed above boresight; the angular separation between beams is selected such that the patterns intersect at the 3-dB points. Alternatively, as shown in Figure 9.44b, each feedhorn output can be fed to a sum and difference hybrid with the hybrid outputs then connected to the combiner or switch. Similar results have been reported using either single antenna approach [39], but the use of a hybrid provides greater discrimination between the two patterns and more predictable results [56]. Performance comparisons with space diversity found in the literature are numerous but not conclusive, although angle diversity clearly outperforms space diversity for highly dispersive channels but requires a larger flat fade margin [57, 58].

Variations of these protection arrangements include hot standby, hybrid diversity, and *M:N* protection. Hot standby arrangements apply to those cases where a second RF path is not deemed necessary, as with short paths where fading is not a significant problem. Here both pairs of transmitters and receivers are operated in a hot standby configuration to provide protection against equipment failure. The transmitter configuration is identical to that of space diversity. The receiving antenna feeds both receivers, tuned to the same frequency, through a power splitter. Hybrid diversity is provided by using frequency diversity but with the receivers connected to separate antennas that are spaced apart. This arrangement, which combines space and frequency diversity, improves the link availability beyond that realized with only one of these schemes.

For more efficient use of equipment and spectrum, diversity techniques are sometimes applied to a section of one or more links. In its simplest form, frequency diversity is used per section, with one protection channel used for *N* operational channels. This method can be extended to provide *M:N* protection, where *M* protection channels are shared by *N* operational channels. Further protection can be provided by using space diversity on a per hop basis and frequency diversity on a section basis.

A diversity combiner performs the combining or selection of diversity signals. This function can be performed at RF [59], IF [60], or baseband

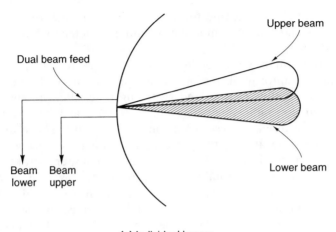

(a) Individual beams

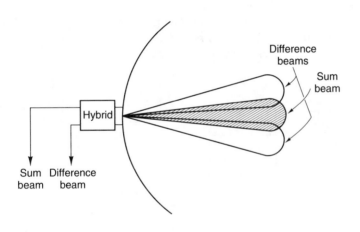

(b) Sum and difference beams

FIGURE 9.44   Dual Beam Antenna Patterns

[51]. Combiner techniques used in analog radio transmission [16] are generally applicable to digital radio. Phase alignment of the diversity signals becomes more important in digital radio systems, however, because of the potential occurrence of error bursts or loss of timing synchronization when combining misaligned diversity signals. Since delay equalization is simpler at baseband than at IF or RF, baseband "hitless" switching is a popular choice in digital radio combiners. This selection combiner uses

some form of in-service performance monitor in each receiver to select the output signal after demodulation and data detection, as illustrated in Figure 9.45.

The performance of diversity combiners is greatly affected by the performance monitor used to sense signal quality and by delayed combiner action due to slow sensing and switching times. Performance monitors used in analog radio diversity combiners, such as receiver AGC, pilot tone, or out-of-band noise detection, are not sufficient in digital radio applications. Because of the presence of frequency-selective fading, the performance monitor must be responsive to the dispersion of the received signal and not just the total power received. Moreover, the performance monitor must be able to respond to other forms of signal degradation, including flat fading, interference, and additive noise, before the onset of an outage. To meet these requirements the BER of each receiver must be accurately estimated by the monitoring scheme; such estimation is possible by monitoring overhead (frame) bits or using the pseudoerror techniques described in Chapter 11.

The time required to assess each diversity channel and effect diversity selection combining can also adversely affect diversity performance. Multipath fading is a relatively slow process, however, so that a switching action

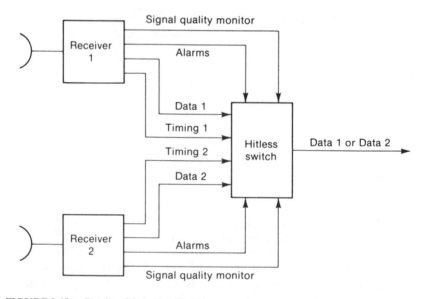

FIGURE 9.45   Baseband Selection Combiner

can be initiated and completed before the channel has faded to threshold. Nevertheless, the response time of the performance monitor must match the speed of fading on LOS links.

Ideally, the selection combiner will always instantaneously select the received signal with the highest S/N ratio. Some inaccuracy in the performance monitor is inevitable, however, and leads to the use of *hysteresis* [61], defined as the minimum ratio (h^2) of received signal levels required to initiate a switch from the signal with lower S/N to the signal with higher S/N. The hysteresis (H) in decibels is related to the ratio h^2 by

$$H = 10 \log h^2 \tag{9.91}$$

The additional outage caused by hysteresis is given by the multiplicative factor $0.5(h^2 + h^{-2})$. A plot of the hysteresis outage factor is given in Figure 9.46. This factor modifies the expression for probability of fade outage for space diversity (9.55) and for frequency diversity (9.56) to account for the

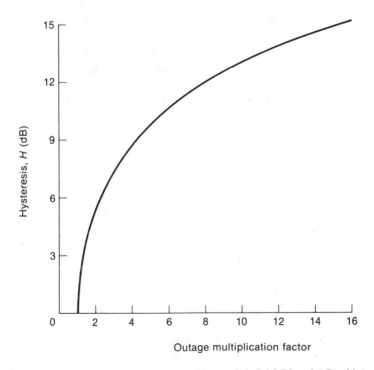

Outage multiplication factor

FIGURE 9.46   Outage Multiplication Due to Hysteresis in LOS Diversity Combiners

effect of hysteresis on total outage. The hysteresis should be small near threshold to minimize outage extension. Typical values are in the range of 3 to 6 dB near threshold.

The control of diversity combining is usually based on a multitude of alarms and monitors ordered according to some priority. Switching criteria are based first on signal continuity and second on signal quality. Alarms resulting from equipment failure or loss of frame synchronization are assigned the highest priority in switchover control. Signal quality monitors, such as parity errors, frame errors, pseudoerrors, or eye pattern closure, are then secondary although essential during periods of fading.

9.5.4   Adaptive Equalizer Design

Initial applications of wideband digital radios revealed that dispersion due to frequency-selective fading was the dominant source of multipath outages. These experiences led to the development and use of adaptive equalization so that outage requirements could be met. Since the introduction of the first adaptive equalizers to digital radio, these devices have undergone a rapid evolution. The degree of sophistication required of the equalizer depends primarily on the bit rate and path length. The types of equalizers in use today range from simple amplitude slope equalizers to complex transversal filter equalizers.

The slope equalizer was designed to detect the presence of amplitude slope in the spectrum and perform slope correction. This objective is achieved by comparing the amplitude at the high end of the channel with the amplitude at the low end. A linear slope correction circuit then provides the desired slope across the passband. This type of equalizer performs well when the notch is outside the passband but is severely limited in the vicinity of amplitude notches. The next improvement was the addition of a notch detector that controls a notch correction circuit. The notch detection circuit operates by comparing the energy at midband with the total energy in the passband. Typical implementations of such an equalizer can correct up to ±12 dB of amplitude slope and 12 dB of amplitude notch [62]. This performance is demonstrated in Figure 9.47, which shows the effects on the radio signature of adaptive equalization for simulated two-ray multipath. Here the delay of the interfering signal is fixed at 6.3 ns, the amplitude is adjusted to yield a BER of 1×10^{-3}, and the offset frequency is adjusted to sweep the notch across the passband [63].

The slope plus notch equalizer corrects linear and parabolic distortion, but it does not correct higher-order distortion. Moreover, it is ineffective for notches as they move away from midband, and it actually degrades group delay for nonminimum phase notches. A means of correcting multiple notches or a single notch as it moves through the passband is to place bump

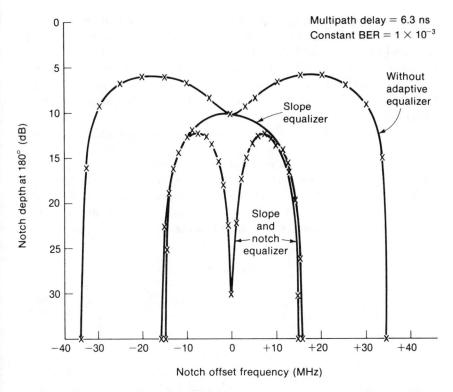

FIGURE 9.47   Radio Signature for Collins MDR-6 at 10^{-3} BER Showing Effects of Adaptive Equalizer [63] (Reprinted by Permission of Collins Transmission Systems Division, Rockwell International)

equalizers at selected frequencies across the passband. Each bump equalizer consists of a filter tuned to a selected frequency and matched to the inverse shape (*bump*) of the notch. The transfer function of a bump equalizer may be expressed as

$$H(\omega) = \frac{1}{1 + \beta e^{\pm j\omega\tau}} \tag{9.92}$$

where a two-ray model has been assumed as the source of multipath fading. The sign of the term $j\omega\tau$ in (9.92) is set to yield either minimum phase (minus sign) or nonminimum phase (plus sign). Because detection of the group delay sign is difficult, the settings are usually preset and left fixed. A minimum phase bump equalization would then completely correct minimum phase notches whose frequency is the same as a bump frequency, but it would increase the group delay due to a nonminimum phase notch. Figure 9.48 is a block diagram of a generic N-bump equalizer for an IF passband with bandwidth B centered at 70 MHz [64, 65]. The IF signal is equalized by a bank of N bump circuits whose frequencies are spaced at B/N intervals across the passband. The gain of each bump is controlled by a feedback loop in which the IF signal is sampled at the bump frequencies and compared with an undistorted reference signal. The error signal from each comparator then controls the gain of the corresponding equalizer section.

To obtain better equalizer performance, particularly for group delay distortion, transversal equalization techniques are now commonly applied to digital radio [66]. Figure 9.49 shows a block diagram of a QAM demodulator equipped with a transversal filter equalizer. The demodulator outputs are equalized by a forward baseband equalizer whose tap weights are con-

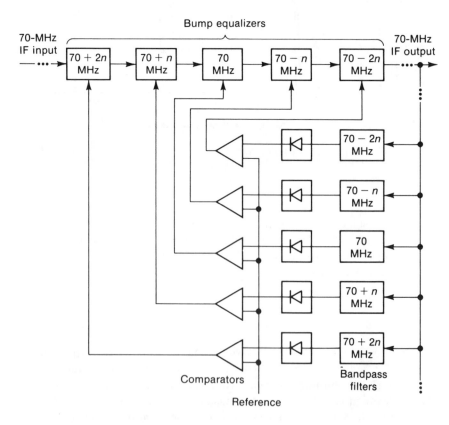

FIGURE 9.48   Block Diagram of IF Bump Equalizer

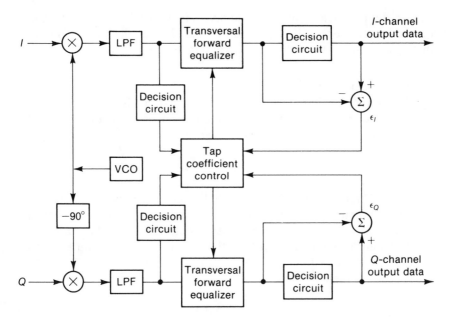

FIGURE 9.49   QAM Demodulator Equipped with Decision-Directed Transversal Filter Equalizer

trolled by decision feedback. The tap weights are continuously adjusted to correct channel distortion and provide the desired pulse response. The number of taps and the tap spacing in the transversal filters are design parameters that are chosen according to the system bit rate, channel fading characteristics, and radio outage requirements. Field tests of a baseband adaptive transversal filter applied to a 90-Mb/s, 16-QAM radio show outage improvement by more than a factor of 3 compared with the same radio equipped with an IF slope equalizer only [67].

9.6   Radio Link Calculations

Procedures for allocating radio link performance based on end-to-end system requirements have already been discussed in Chapter 2. There it was noted that outages occur due to both equipment failure and propagation effects. It was noted, moreover, that the allocations for equipment and propagation outage are usually separated because of their different effect on the user and different remedy by the system designer. The subject of equipment reliability has been treated in Chapter 2. Here we will consider procedures for calculating values of key radio link parameters, such as transmitter power, antenna size, and diversity design, based on propagation outage requirements. These procedures include the calculation of intermediate parameters such as system gain and fade margin.

9.6.1   System Gain

System gain (G_s) is defined as the difference, in decibels, between the transmitter output power (P_t) and the minimum receiver signal level required to meet a given bit error rate objective (RSL_m):

$$G_s = P_t - \text{RSL}_m \tag{9.93}$$

The minimum required RSL, also called receiver threshold, is determined by the receiver noise level and the signal-to-noise ratio required to meet the given BER. Noise power in a receiver is determined by the noise power spectral density (N_0), the amplification of the noise introduced by the receiver itself (noise figure N_f), and the receiver bandwidth (B). The total noise power is then given by

$$P_N = N_0 B N_f \tag{9.94}$$

The source of noise power is thermal noise, which is determined solely by the temperature of the device. The thermal noise density is given by

$$N_0 = k T_0 \tag{9.95}$$

where k = Boltzman's constant $(1.38 \times 10^{-23}$ joule/$^\circ$K)
               T_0 = absolute temperature in degrees Kelvin

The reference for T_0 is normally assumed to be room temperature, 290°K, for which $kT_0 = -174$ dBm/Hz. The minimum required RSL may now be written as

$$\begin{aligned}\text{RSL}_m &= P_N + \frac{S}{N} \\ &= kT_0 B N_f + \frac{S}{N}\end{aligned} \tag{9.96}$$

It is often more convenient to express (9.96) as a function of data rate R and E_b/N_0. Using (6.5) we can rewrite (9.96) as

$$\text{RSL}_m = kT_0 R N_f + \frac{E_b}{N_0} \tag{9.97}$$

The system gain may also be stated in terms of the gains and losses of the radio link:

$$G_s = L_p + F + L_t + L_m + L_b - G_t - G_r \qquad (9.98)$$

where G_s = system gain in decibels

L_p = free space path loss in dB, given by (9.5)

F = fade margin in dB

L_t = transmission line loss from waveguide or coaxials used to connect radio to antenna, in dB (see Table 9.5)

L_m = miscellaneous losses such as minor antenna misalignment, waveguide corrosion, and increase in receiver noise figure due to aging, in dB

L_b = branching loss due to filter and circulator used to combine or split transmitter and receiver signals in a single antenna

G_t = gain of transmitting antenna (see Table 9.4)

G_r = gain of receiving antenna (see Table 9.4)

The system gain is a useful figure of merit in comparing digital radio equipment. High system gain is desirable since it facilitates link design—for example, by easing the size of antennas required. Conversely, low system gain places constraints on link design—for example, by limiting path length.

Example 9.5 _____

Determine the system gain of an 8-GHz, 45-Mb/s, 16-QAM digital radio with a transmitter output power of 30 dBm and a receiver noise figure of 7 dB. Assume that the desired bit error rate is 1×10^{-6}.

Solution   To find the minimum required RSL, the required value of E_b/N_0 for 16-QAM is first found from Figure 6.21. Note that this figure plots probability of *symbol* error $P(e)$ versus E_b/N_0, so that we must convert $P(e)$ to BER according to (6.72):

$$P(e) = (\log_2 4)(\text{BER}) = 2\,\text{BER}$$

The E_b/N_0 for $P(e) = 2 \times 10^{-6}$ is 14.5 dB. The required RSL is found from (9.97):

$$\text{RSL}_m = -174 + 10\log(45 \times 10^6) + 7 + 14.5 = -76\,\text{dBm}$$

The system gain then follows from (9.93):

$$G_s = 30 - (-76) = 106 \text{ dB}$$

9.6.2   Fade Margin

The traditional definition of fade margin is the difference, in decibels, between the nominal RSL and the threshold RSL as illustrated in Figure 9.16. An expression for the fade margin F required to meet allowed outage probability $P(o)$ may be derived from (9.42) for an unprotected link, from (9.55) for a space diversity link, and from (9.56) for a frequency diversity link, with the following results:

$$
\begin{aligned}
F &= 30 \log D + 10 \log(abf) \\
&- 56 - 10 \log P(o) \quad \text{(unprotected link)}
\end{aligned}
\tag{9.99}
$$

$$
\begin{aligned}
F &= 20 \log D - 10 \log S + 5 \log\!\left(\frac{h^2 + h^{-2}}{2}\right) \\
&+ 5 \log(ab) - 7.2 - 5 \log P(o) \quad \text{(space diversity link)}
\end{aligned}
\tag{9.100}
$$

$$
\begin{aligned}
F &= 20 \log D + 15 \log f + 5 \log\!\left(\frac{h^2 + h^{-2}}{2}\right) + 5 \log(ab) - 5 \log\Delta f \\
&- 36.5 - 5 \log P(o) \quad \text{(frequency diversity link)}
\end{aligned}
\tag{9.101}
$$

with combiner hysteresis effects included for space and frequency diversity. This definition of fade margin has traditionally been used to describe the effects of fading at a single frequency for radio systems that are unaffected by frequency-selective fading or to radio links during periods of flat fading. For wideband digital radio without adaptive equalization, however, dispersive fading is a significant contributor to outages so that the flat fade margins do not provide a good estimate of outage and are therefore insufficient for digital link design. Several authors have introduced the concept of effective [41], net [69], or composite [70] fade margin to account for dispersive fading. The effective (or net or composite) fade margin is defined as that fade depth which has the same probability as the observed probability of outage. The difference between the effective fade margin measured on the radio link and the flat fade margin measured with an attenuator is then an indication of the effects of dispersive fading. Since digital radio outage is usually referenced to a threshold error rate, BER_t, the effective fade margin (EFM) can be obtained from the relationship

$$P(A \geq \text{EFM}) = P(\text{BER} \geq \text{BER}_t) \tag{9.102}$$

where A is the fade depth of the carrier. The results of Equations (9.99) to (9.101) can now be interpreted as yielding the effective fade margin for a probability of outage given by the right-hand side of (9.102). Conversely, note that (9.42), (9.55), and (9.56) are now to be interpreted with F equal to the EFM.

As an example, Figure 9.50 gives results for the same 90-Mb/s, 8-PSK system shown in Figure 9.29 [44]. Here observed fade outage probabilities have been plotted along with observed error rate probabilities for an unprotected link and the same link with adaptive equalization. The effective fade margin for a BER threshold of 10^{-6} is shown by the dashed lines indicating an EFM of 33 dB for the unprotected link and 38.5 dB for the equalized link. Figure 9.50 also plots the theoretical performance for flat fading, which indicates a flat fade margin (FFM) of 39 dB for the 10^{-6} BER threshold. The large difference between the unprotected-link EFM and the theoretical FFM indicates that multipath outage is dominated by dispersive effects. Use of equalization virtually eliminates the effects of dispersive fading, however, improving the EFM to within 0.5 dB of the theoretical FFM.

The effective fade margin is derived from the addition of up to three individual fade margins that correspond to the effects of flat fading, dispersion, and interference [70]:

$$\text{EFM} = -10 \log(10^{-\text{FFM}/10} + 10^{-\text{DFM}/10} + 10^{-\text{IFM}/10}) \qquad (9.103)$$

Here the flat fade margin (FFM) is given as the difference in decibels between the unfaded signal-to-noise ratio (S/N_u) and the minimum signal-to-noise ratio (S/N_m) to meet the error rate objective, or

$$\text{FFM} = \left(\frac{S}{N_u}\right) - \left(\frac{S}{N_m}\right) \qquad (9.104)$$

Similarly, we define the dispersive fade margin (DFM) and interference fade margin (IFM) as

$$\text{DFM} = \left(\frac{S}{N_d}\right) - \left(\frac{S}{N_m}\right) \qquad (9.105)$$

and

$$\text{IFM} = \left(\frac{S}{I}\right) - \left(\frac{S}{I_c}\right) \qquad (9.106)$$

where N_d represents the effective noise due to dispersion and S/I_c represents the critical S/I below which the BER is greater than the threshold

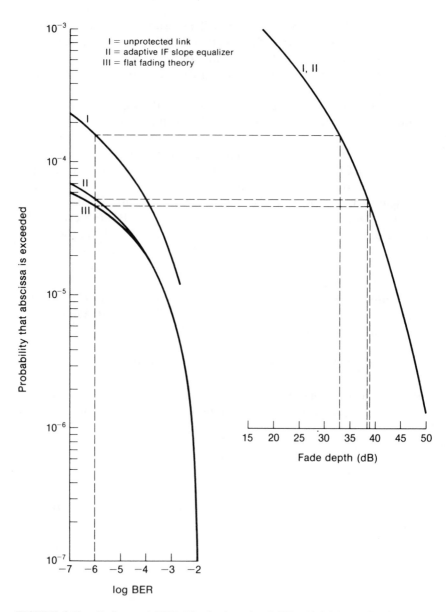

FIGURE 9.50   Fading and BER Distributions for 6-GHz, 90 Mb/s, 8 PSK System on 37.3 mi Path

BER. Each of the individual fade margins can also be calculated as a function of the other fade margins by using (9.103), as for example

$$FFM = -10 \log(10^{-EFM/10} - 10^{-DFM/10} - 10^{-IFM/10}) \qquad (9.107)$$

9.6.3   Link Calculation Procedure

Step-by-step procedures for calculating values for link design parameters are given in Table 9.6. For each parameter listed, a source of information is given for each entry made in the path calculation process. The following assumptions are made in Table 9.6:

1. An adequate path clearance is provided on the path under consideration (see Section 9.2.4).
2. Rain attenuation and atmospheric absorption have been ignored, which is appropriate for frequencies below 10 GHz. These two effects are easily added for use of higher frequencies (see Section 9.2.3).
3. As is often the case, antenna size is the final design parameter. Hence antenna size is calculated from other design parameters whose values are already selected.

The procedures shown can be rearranged to make any design parameter the last value to be determined. Often iteration is required in order to determine the best combination of design values. The following design changes are commonly made to meet performance objectives:

- Increase transmitter power.
- Use lower loss transmission line.
- Use lower noise receiver.
- Add diversity paths to the unprotected path.
- Increase antenna spacing for existing space diversity or frequency spacing for existing frequency diversity.
- Increase antenna size.
- Add an amplitude equalizer or, for greater improvement, add an adaptive transversal equalizer.

Table 9.6 also illustrates the use of these design procedures by giving values for a representative radio system. This example is the 6-GHz, 90-Mb/s, 8-PSK radio system whose performance characteristics were shown earlier in Figures 9.29 and 9.50. This radio system employs space diversity with baseband selection combining and IF adaptive amplitude slope and notch equalization.

Automated design of digital line-of-sight links through use of computer programs is now standard in the industry [12, 40].

TABLE 9.6 Design Procedures for Digital Radio Links

Parameter	Source	Sample Value for 6-GHz, 90-Mb/s, 8-PSK System
1. Frequency, f	System plan	6 GHz
2. Path length, D	Map or site records	37.3 mi
3. Transmitter line losses		
a. Antenna height above ground	Site records or path profile	190 ft
b. Horizontal transmission line length	Site records	55 ft
c. Total transmission line length	Parameters 3a + 3b	245 ft
d. Transmission line type	System design:	Circular waveguide
	rectangular waveguide	
	elliptical waveguide	
	circular waveguide	
	coaxial cable	
e. Transmission line loss factor	Manufacturer's data (see Table 9.5)	0.01 dB/ft
f. Total transmitter transmission line loss	Parameters 3c × 3e	2.45 dB
4. Receiver line losses		
a. Antenna height above ground (upper antenna if space diversity is used)	Site records or path profile	390 ft
b. Horizontal transmission line length	Site records	55 ft
c. Total transmission line length	Parameters 4a + 4b	445 ft
d. Transmission line type	Same as parameter 3d	Circular waveguide
e. Transmission line loss factor	Manufacturer's data (see Table 9.5)	0.01 dB/ft
f. Total receiver transmission line loss	Parameters 4c × 4e	4.45 dB

TABLE 9.6 Design Procedures for Digital Radio Links

Parameter	Source	Sample Value for 6-GHz, 90-Mb/s, 8-PSK System
5. Free space path loss, L_p	$20 \log f + 20 \log D + 96.6$ (f in GHz, D in miles)	143.6 dB
6. Branching loss, L_b	System design	0.5 dB
7. Miscellaneous losses, L_m	System design	3 dB
8. Total losses, L	Parameters $3f + 4f + 5 + 6 + 7$	154 dB
9. Transmitter power, P_t	System design	39 dBm
10. Received signal level threshold, RSL_m	Equipment specification or following procedure	
a. Thermal noise density, N_o	kT_o	−174 dBm/Hz
b. Receiver noise figure, N_f	Equipment specification	8 dB
c. Bit rate, R	System plan	90 Mb/s
d. Required E_b/N_o for BER $= 10^{-6}$	Equipment specification or theory with appropriate implementation factor	22.5 dB
e. Threshold, RSL_m	$kT_o + N_f + 10 \log R + E_b/N_o$	−64 dBm
11. System gain, G_s	$P_t - RSL_m$	103 dB

(Continued)

TABLE 9.6 Design Procedures for Digital Radio Links (*Continued*)

Parameter	Source	Sample Value for 6-GHz, 90-Mb/s, 8-PSK System
12. Effective fade margin		
a. Terrain factor, a	Path profile	1
b. Climate factor, b	Weather map	1/4
c. Antenna spacing, S (space diversity)	System design	40 ft
d. Frequency spacing, Δf (frequency diversity)	System design	N.A.
e. Combiner hysteresis, H	Equipment specification	3 dB
f. Outage allowed	System design	0.00001
g. Effective fade margin, EFM	See equations (9.99), (9.100), or (9.101)	30.7 dB
13. Flat fade margin required		
a. Dispersive fade margin, DFM	Equipment specification or field measurement	40.3 dB
b. Interference fade margin, IFM	Equipment specification or field measurement	59 dB
c. Flat fade margin, FFM	See Equation (9.107)	31.2 dB
14. Total antenna gain required, $2G_A$	$FFM + L - G_s$	82.2 dB
15. Required antenna diameter	Table 9.4	10 ft

9.7   Summary

The design of digital radio systems for line-of-sight links involves engineering of the path to provide proper clearance, evaluation of the effects of propagation on performance, development of a frequency allocation plan, and proper selection of radio and link components. This design process must ensure that outage requirements are met on a per link and system basis.

Propagation losses occur primarily due to free-space loss, which is proportional to the square of both path length and frequency. Terrain and climate also have an effect, causing losses to vary from free-space predictions. Reflection or diffraction from obstacles along the path can cause secondary rays that may interfere with the direct ray. The use of Fresnel zones from optics facilitates the design of links by determining the clearances required over obstacles to minimize interference. Varying index of refraction caused by atmospheric structural changes can cause decoupling of the signal from the receiving antenna or fading due to interference from multiple rays (multipath). At frequencies above 10 GHz, atmospheric absorption and rain attenuation become dominant sources of loss, limiting the path length or availability that can be achieved with millimeter-wave frequencies.

Because of its dynamic nature, multipath fading is characterized statistically—usually by its probability of occurrence, average duration, and rate of fading. Work done primarily by Bell Telephone Laboratories has provided empirical models for multipath fading statistics based on analog radio links. Space or frequency diversity is used to protect radio links against multipath fading by providing a separate (diversity) path whose fading is somewhat uncorrelated from the main path. The introduction and testing of wideband digital radio has led researchers to conclude that outages in digital radio are due principally to dispersion caused by frequency-selective fading. Models for multipath channels have been developed for digital radio application, and radio performance is now characterized by a signature that plots the outage area for a given set of multipath channel model parameters. The use of diversity and adaptive equalization has been found to greatly mitigate the effects of dispersion, allowing performance objectives to be met for digital radio.

The frequency allocation plan is based on four elements: the local frequency regulatory authority requirements, selected radio transmitter and receiver characteristics, antenna characteristics, and potential intrasystem and intersystem RF interference.

Key components of a digital radio include the modulator/demodulator (already described in Chapter 6), power amplifier, antenna/diversity arrangement, and adaptive equalizer. When choosing a traveling-wave tube as the RF power amplifier, the system designer must carefully examine the effects of saturated operation on the error performance and spectral spreading due to nonlinearities. The type of diversity is a function of the path and system availability requirements. The choice of combiner type can significantly affect the performance expected from diversity systems, depending on the performance monitors used in controlling the combining functions. Adaptive equalizers are now common in digital radio design, although simple amplitude slope equalizers have been found to be quite effective. More elaborate designs involving transversal filters have evolved to meet stringent performance objectives.

Finally, a step-by-step procedure has been presented for calculating values of digital radio link parameters. This procedure is easily adapted to calculate any given parametric value as a function of given values for other link parameters.

References

1. J. D. Kraus, *Antennas* (New York: McGraw-Hill, 1950).
2. Henry R. Reed and Carl M. Rusell, *Ultra High Frequency Propagation* (London: Chapman & Hall, 1966).
3. Kenneth Bullington, "Radio Propagation Fundamentals," *Bell System Technical Journal* 36(May 1957):593–626.
4. A. Vigants, "Microwave Radio Obstruction Fading," *Bell System Technical Journal* 60(May 1957): 785–801.
5. J. A. Schiavone, "Prediction of Positive Refractivity Gradients for Line-of-Sight Microwave Radio Paths," *Bell System Technical Journal* 60(May 1957): 803–822.
6. C. A. Samson, "Refractivity Gradients in the Northern Hemisphere," Office of Telecommunications Report 75-59, U.S. Department of Commerce, April 1975.
7. J. H. Van Vleck, *Radiation Lab. Report 664,* MIT, 1945.
8. D. E. Setzer, "Computed Transmission Through Rain at Microwave and Visible Frequencies," *Bell System Technical Journal* 49(October 1970):1873–1892.
9. H. E. Bussey, "Microwave Attenuation Statistics Estimated from Rainfall and Water Vapor Statistics," *Proc. IRE* 38(July 1950):781–785.
10. Reports of the CCIR, 1990, Annex to Vol. V, *Propagation in Non-Ionized Media* (Geneva: ITU, 1990).
11. S. H. Lin, "Nationwide Long-Term Rain Rate Statistics and Empirical Calculation of 11-GHz Microwave Rain Attenuation," *Bell System Technical Journal* 56(November 1977): 1581–1604.
12. D. R. Smith, "A Computer-Aided Design Methodology for Digital Line-of-Sight Links," *1990 International Conference on Communications,* pp. 59–65.
13. *Engineering Considerations for Microwave Communications Systems* (San Carlos, CA: GTE Lenkurt Inc., 1981).
14. G. D. Alley, C. H. Bianchi, and W. A. Robinson, "Angle Diversity and Space Diversity Experiments on the Salton/Brawley Hop," *1990 GLOBECOM,* pp. 504.5.1–504.5.11.

15. D. Vergeres, P. Jordi, and A. Loembe, "Simultaneous Error Performance of Antenna Pattern Diversity and Vertical Space Diversity on a 64QAM Radio Link," *1990 GLOBECOM*, pp. 504.4.1–504.4.5.
16. M. Schwartz, W. R. Bennett, and S. Stein, *Communication Systems and Techniques* (New York: McGraw-Hill, 1966).
17. W. T. Barnett, "Multipath Propagation at 4, 6, and 11 GHz," *Bell System Technical Journal* 51(February 1972):321–361.
18. A. Vigants, "The Number of Fades and Their Durations on Microwave Line-of-Sight Links With and Without Space Diversity," *1969 International Conference on Communications*, pp. 28-25–28-31.
19. A. Vigants, "Number and Duration of Fades at 6 and 4 GHz," *Bell System Technical Journal* 50(March 1971):815–841.
20. D. J. Kennedy, "Digital Error Statistics for a Fading Channel," *1973 International Conference on Communications*, pp. 18-17 to 18-23.
21. A. Vigants, "Space-Diversity Performance as a Function of Antenna Separation," *IEEE Trans. on Comm. Tech.*, vol. COM-16, no. 6, December 1968, pp. 831–836.
22. A. Vigants and M. V. Pursley, "Transmission Unavailability of Frequency Diversity Protected Microwave FM Radio Systems Caused by Multipath Fading," *Bell System Technical Journal* 58(October 1979):1779–1796.
23. S. H. Lin, "Statistical Behavior of a Fading Signal," *Bell System Technical Journal* 50(December 1971):3211–3270.
24. L. C. Greenstein and B. A. Czekaj, "A Polynomial Model for Multipath Fading Channel Responses," *Bell System Technical Journal* 59(September 1980):1197–1205.
25. W. D. Rummler, "A New Selective Fading Model: Application to Propagation Data," *Bell System Technical Journal* 58(May/June 1979):1037–1071.
26. W. C. Jakes, Jr., "An Approximate Method to Estimate an Upper Bound on the Effect of Multipath Delay Distortion on Digital Transmission," *1978 International Conference on Communications*, pp. 47.1.1–47.1.5.
27. L. Martin, "Phase Distortions of Multipath Transfer Functions," *1984 International Conference on Communications*, pp. 1437–1441.
28. W. D. Rummler, "Characterizing the Effects of Multipath Dispersion on Digital Radios," *1988 GLOBECOM*, pp. 52.5.1–52.5.7.
29. Reports of the CCIR, 1990, Annex to Vol. IX, *Fixed Service Using Radio Relay Systems* (Geneva: ITU, 1990).
30. H. Kostal, G. N. Elkhouri, and H. Smullen, "Advances in Line-or-Sight Link Design for DoD Applications," *1990 MILCOM*, pp. 17.4.1–17.4.7.
31. M. Emshwiller, "Characterization of the Performance of PSK Digital Radio Transmission in the Presence of Multipath Fading," *1978 International Conference on Communications*, pp. 47.3.1–47.3.6.
32. C. W. Lundgren and W. D. Rummler, "Digital Radio Outage Due to Selective Fading—Observation vs. Prediction from Laboratory Simulation," *Bell System Technical Journal* 58(May/June 1979):1073–1100.

33. A. Ranade, "Statistics of the Time Dynamics of Dispersive Multipath Fading and its Effect on Digital Microwave Radio," *1985 International Conference on Communications*, pp. 47.7.1–47.7.4.

34. P. Balaban, "Statistical Model for Amplitude and Delay of Selective Fading," *AT&T Technical Journal* 64(1985): 2525–2550.

35. L. J. Greenstein and M. Shafi, "Outage Calculation Methods for Microwave Digital Radio," *IEEE Comm. Mag.*, February 1987, pp. 30–39.

36. A. Vigants, "Distance Variation of Two-Tone Amplitude Dispersion in Line-of-Sight Microwave Propagation," *1981 International Conference on Communications*, pp. 68.3.1–68.3.5.

37. Y. Serizawa and S. Takeshita, "A Simplified Method for Prediction of Multipath Fading Outage of Digital Radio," *IEEE Trans. on Comm.*, vol. COM-31, no. 8, August 1983, pp. 1017–1021.

38. W. T. Barnett, "Multipath Fading Effects on Digital Radio," *IEEE Trans. on Comm.*, vol. COM-27, no. 12, December 1979, pp. 1842–1848.

39. S. H. Lin, T. C. Lee, and M. F. Gardina, "Diversity Protections for Digital Radio—Summary of Ten-Year Experiments and Studies," *IEEE Comm. Mag.*, February 1988, pp. 51–64.

40. T. C. Lee and S. H. Lin, "The DRDIV Computer Program—A New Tool for Engineering Digital Radio Routes," *1986 GLOBECOM*, pp. 51.5.1–51.5.8.

41. C. W. Anderson, S. G. Barber, and R. N. Patel, "The Effect of Selective Fading on Digital Radio," *IEEE Trans. on Comm.*, vol. COM-27, no. 12, December 1979, pp. 1870–1876.

42. T. S. Giuffrida, "Measurements of the Effects of Propagation on Digital Radio Systems Equipped with Space Diversity and Adaptive Equalization," *1979 International Conference on Communications*, pp. 48.1.1–48.1.6.

43. A. J. Giger and W. T. Barnett, "Effects of Multipath Propagation on Digital Radio," *IEEE Trans. on Comm.*, vol. COM-29, no. 9, September 1981, pp. 1345–1352.

44. D. R. Smith and J. J. Cormack, "Improvement in Digital Radio Due to Space Diversity and Adaptive Equalization," *1984 Global Telecommunications Conference*, pp. 45.6.1–45.6.6.

45. Code of Federal Regulations, *Telecommunications*, Chapter I—Federal Communications Commission, Part 21, U.S. Government Printing Office, 1990.

46. MIL-STD-188-322, "Subsystem Design/Engineering and Equipment Technical Design Standards for Long Haul Line-of-Sight (LOS) Digital Microwave Radio Transmission," U.S. Department of Defense, November 1, 1976.

47. Recommendations of the CCIR, 1990, Vol. IX—Part 1, *Fixed Service Using Radio-Relay Systems* (Geneva: ITU, 1990).

48. Bell System Technical Reference, *6 GHz Digital Radio Requirements and Objectives*, Pub. 43501, AT&T, December 1980.

49. V. K. Prabhu, "Error Rate Considerations for Coherent Phase-Shift Keyed Systems with Co-Channel Interference," *Bell System Technical Journal* 48(March 1969):743–768.

50. Bellcore Technical Reference TR-TSY-000752, *Microwave Digital Radio Systems Criteria*, October 1989.

51. C. M. Thomas, J. E. Alexander, and E. W. Rahneberg, "A New Generation of Digital Microwave Radios for U.S. Military Telephone Networks," *IEEE Trans. on Comm.*, vol. COM-27, no. 12, December 1979, pp. 1916–1928.

52. R. G. Lyons, "The Effect of a Bandpass Nonlinearity on Signal Detectability," *IEEE Trans. on Comm.*, vol. COM-21, no. 1, January 1973, pp. 51–60.

53. D. Chakraborty and L. S. Golding, "Wide-Band Digital Transmission Over Analog Radio Relay Links," *IEEE Trans. on Comm.*, vol. COM-23, no. 11, November 1975, pp. 1215–1228.

54. J. E. Hamant, O. P. Connell, and H. S. Walczyk, "Digital Transmission Evaluation Project, DR8A Test Final Report." Report Number CCC-CED-77-DTEP-012, U.S. Army Communications-Electronics Engineering Installation Agency, Ft. Huachuca, Arizona.

55. C. M. Skarpiak, "Selecting Antennas for Terrestrial Microwave Systems," *Microwave Systems News* 19(November 1989): pp. 66–70.

56. E. W. Allen, "Angle Diversity Test Using a Single Aperture Dual Beam Antenna," *1988 International Conference on Communications*, pp. 50.1.1–50.1.7.

57. E. W. Allen, "Angle Diversity at 6 GHz: Methods of Alignment and Test Results," *1989 International Conference on Communications*, pp. 756–761.

58. R. Valentin, H. G. Giloi, and K. Metzger, "More on Angle Diversity for Digital Radio Links," *1990 GLOBECOM*, pp. 504.3.1–504.3.5.

59. I. Horikawa, Y. Okamoto, and K. Morita, "Characteristics of a High Capacity 16 QAM Digital Radio System on a Multipath Fading Channel," *1979 International Conference on Communications*, pp. 48.4.1–48.4.6.

60. G. deWitte, "DRS-8: System Design of a Long Haul 91 Mb/s Digital Radio," *1978 National Telecommunications Conference*, pp. 38.1.1–38.1.6.

61. A. Vigants, "Space-Diversity Engineering," *Bell System Technical Journal* 54(January 1975):103–142.

62. P. R. Hartmann and E. W. Allen, "An Adaptive Equalizer for Correction of Multipath Distortion in a 90 Mb/s 8 PSK System," *1979 International Conference on Communications*, pp. 5.6.1–5.6.4.

63. P. R. Hartmann and E. W. Allen, "Transmission Engineering Considerations," Collins Transmission Engineering Symposium, Dallas, November 1982, pp. 3-1–3-48.

64. T. P. Murphy and others, "Practical Techniques for Improving Signal Robustness," *1981 National Telecommunications Conference*, pp. C3.3.1–C3.3.7.

65. E. R. Johnson, "An Adaptive IF Equalizer for Digital Transmission," *1981 International Conference on Communications*, pp. 13.6.1–13.6.4.

66. C. A. Siller, Jr., "Multipath Propagation," *IEEE Comm. Mag.*, February 1984, pp. 6–15.

67. G. L. Fenderson, S. R. Shepard, and M. A. Skinner, "Adaptive Transversal Equalizer for 90 Mb/s 16-QAM Systems in the Presence of Multipath Propagation," *1983 International Conference on Communications*, pp. C8.7.1–C8.7.6.

68. R. L. Freeman, *Telecommunication Transmission Handbook* (New York: Wiley, 1975).

69. P. Dupuis, M. Joindet, A. Leclert, and M. Rooryck, "Fade Margin of High Capacity Digital Radio System," *1979 International Conference on Communications*, pp. 48.6.1–48.6.5.
70. W. D. Rummler, "A Comparison of Calculated and Observed Performance of Digital Radio in the Presence of Interference," *IEEE Trans. on Comm.*, vol. COM-30, no. 7, July 1982, pp. 1693–1700.

Problems

9.1 Calculate the free space loss to a geostationary satellite at 6105 Mhz. Assume the distance to the satellite is 39,256 km.

9.2 A transmitter operating at 800 Mhz has a power output of 1 watt. If the distance to the receiver is 5 mi, calculate the received signal level in dBm, assuming use of isotropic antennas at both transmitter and receiver.

9.3 Beginning with Equation (9.4), derive Equation (9.5)

9.4 In the United States, satellite systems share the 4- and 6-GHz bands with terrestrial microwave systems used by common carriers. Using the lower end of these two bands, as shown in Table 9.2, calculate the difference in free space loss between the two bands. Suggest a reason why satellite up-links use the 6-GHz band and satellite down-links use 4 GHz.

9.5 Using plane geometry and algebra, derive Equation (9.13). Then show that the area of all Fresnel zones is given by

$$A = \frac{\pi \lambda d_1 d_2}{D}$$

Hint: $\sqrt{1 + x} \cong 1 + x/2$ for $x \ll 1$

9.6 Find the radius of the first Fresnel zone when $f = 4, 6$, and 11 Ghz, $d = D/2$ and $D/10$, and $D = 40$ km. (Do all combinations.)

9.7 Using plane geometry and algebra, and the same hint given in Problem 9.5, derive Equation (9.7). Then show that the phase difference between the direct and reflected paths is

$$\theta = \frac{4 \pi h_1 h_2}{\lambda D}$$

9.8 Consider a line-of-sight microwave link operating over a theoretically flat earth with transmitter height 50 m, receiver height 25 m, distance 30 km, and frequency 6 GHz.

(a) Compute the angle of specular reflection.
(b) Compute the phase difference between the direct and specularly reflected paths.
(c) Assuming a 180° phase shift upon reflection, will the reflected signal interfere constructively or destructively with the direct signal? Why?
(d) Calculate the earth bulge due to earth curvature at the point of reflection for a $k = 4/3$.
(e) Calculate the radius of the first Fresnel zone at the point of reflection.

9.9 Plot k-factor versus dN/dh for $k = -6$ to $+6$ and $dN/dh = -300$ to $+100$ N-units/km on linear paper. Show regions of ducting, subrefraction, superrefraction, and standard refraction, with dN/dh on the abscissa.

9.10 Using plane geometry and algebra, prove that the maximum distance between transmitter and receiver over a smooth earth is given by

$$d_{max} = \sqrt{2a_e h_1} + \sqrt{2a_e h_2}$$

where a_e is the effective earth radius, h_1 is the transmitter height, and h_2 is the receiver height, all distances being in the same units. (Hint: $\cos \theta \simeq 1 - \theta^2/2$ for small θ). With h_1 and h_2 in feet, and d_{max} and a_e in miles, show that

$$d_{max} = \sqrt{12.76kh_1} + \sqrt{12.76kh_2}$$

9.11 Consider a 50-km radio path across a smooth open sea with antenna heights above sea level given by $h_1 = 40$ m and $h_2 = 120$ m. Determine the minimum value of k for which the radio path is unobstructed. Hint: Use the expression for d_{max} given in Problem 9.10.

9.12 Consider a radio system operating over a Rayleigh fading channel.
(a) Show that the Rayleigh probability distribution function can be written as

$$P(A < A_0) = 1 - \exp(-0.693 A_0{}^2)$$

where $A = r/r_m$
$r = $ instantaneous signal amplitude
$r_m = $ median value of signal amplitude

Hint: Note that r_m is that value of r for which $P(r) = 0.5$.
(b) Plot the expression of part (a) versus $20 \log A$ on probability paper.

9.13 This problem concerns the fading statistics of over-the-horizon radio systems, which have a long-term distribution that is log-normal.

(a) If $y = e^x$, and x has a gaussian probability density function, show that y has a log-normal probability density function given by

$$p(y) = \frac{\exp\left(-\{ln\, y - m\}^2/2\sigma^2\right)}{\sigma y \sqrt{2\pi}}$$

where m = mean value of x
         σ = standard deviation of x

(b) Plot $p(y)$ versus $20 \log y$ on linear paper, and comment on the shape of the plot.

(c) Derive an expression for the log-normal distribution function. Express your answer in terms of the complementary error function.

(d) Plot the log-normal distribution function on probability paper for $m = 0$ and $\sigma = 0, 2, 4, 6, 8$, and 10.

9.14 A digital radio system is designed to transmit 135 Mb/s in the presence of additive white gaussian noise with a one-sided power spectral density of 1.67×10^{-20} watts per Hertz. Assume use of Gray coding, a transmitter power of 10 watts, total system losses of 110 dB, and coherent detection.

(a) For a choice of 64-PSK, calculate the bit error rate (BER) of this radio system.

(b) Calculate the probability of outage due to multipath fading, assuming a Rayleigh distribution of fading, for a fade margin of 35 dB. How much would the fade margin have to be increased (in dB) to decrease the probability of outage by a factor of 10?

(c) Now assume a transmitted signal bandwidth of 30 MHz, for a spectral efficiency of 4.5 bps/Hz. For a choice of raised-cosine filtering, what rolloff factor α is required to provide this spectral efficiency?

9.15 Consider a digital radio that uses noncoherent binary FSK and operates over a Rayleigh fading channel. The receiver E_b/N_0 varies randomly and has a probability density function given by

$$f(\gamma) = \frac{1}{\gamma_0} \exp(-\gamma/\gamma_0)$$

where γ is the instantaneous E_b/N_0 and γ_0 is the average E_b/N_0.

(a) Show that the probability of error is given by $P(e) = 1/(2 + \gamma_0)$.

(b) Now assume that selection diversity combining is applied to improve performance. Two receivers are used that have independently varying received signals with identical statistics. The receiver with the instantaneously larger E_b/N_0 is selected in making each binary decision. What is the $P(e)$ now?

9.16 Consider a 6-GHz line-of-sight link, with distance 50 km, average terrain, temperate climate, and space diversity with antenna separation of 40 ft.

(a) Plot probability of outage versus fade margin.

(b) Plot average fade duration versus fade margin.

(c) Plot average fade rate versus fade margin.

(d) Now assume a frequency diversity system for this link, and calculate the frequency separation needed to provide the same probability of outage as part (a) for a fade margin of 40 dB.

9.17 Maximal ratio and **selection** combining are two popular combining techniques applied to microwave radio systems. The maximal ratio combiner sums the input S/N from each of M received signals. A selection combiner picks the best of M received signals and uses that signal alone. When operating over a Rayleigh fading channel, in which each received signal has the same identical S/N, the probability distribution function for maximal ratio combining is given by

$$P_M(A < A_0) = 1 - \exp(-\gamma_0) \sum_{k=0}^{M-1} \gamma_0^k/k!$$

and for selection combining,

$$P_M(A < A_0) = \left[1 - \exp(-\gamma_0)\right]^M$$

where
$$\gamma_0 = 0.693A_0^2$$
$$M = \text{order of diversity}$$

and A is as defined in Problem 9.12.

(a) Derive the expressions for probability distribution function shown above for the maximal ratio and selection combiners.

(b) For the maximal ratio combiner, plot $P_M(A < A_0)$ versus $20 \log A$ on probability paper for $M = 1, 2,$ and 4.

(c) For the selection combiner, plot $P_M(A < A_0)$ versus $20 \log A$ on probability paper for $M = 1, 2,$ and 4.

(d) Compare performance of the two combiners by noting the improvement in dB for $M = 2$ and 4 at the median.

9.18 For radio fading with Rice–Nakagami statistics, we know that the probability of error (p_e) statistics for digital radio applications employing differentially coherent QPSK can be given by

(1) Probability distribution

$$P_M(p_e) = d\left[1 - q \sum_{k=1}^{M} w^{k-1}/(k-1)!\right]$$

where

$q = (2p_e)^{(1/\gamma_0)}$

$w = -\ln q$

d = multipath occurrence factor

$\gamma_0 = E_b/N_0$ averaged over fading channel

M = order of diversity

(2) Average fade rate

$$N_M(p_e) = \frac{M N_1(p_e) P_M(p_e)}{P_1(p_e)}$$

where

$N_1(p_e) = 1.34 f_c q w^{1/2}$

f_c = carrier frequency

(3) Average fade duration

$$t_M = t_1/M$$

where

$\bar{t}_1 = P_1(p_e)/N_1(p_e)$

For a choice of dual diversity, $d = 0.25$, BER $= 10^{-6}$, and $f_c = 6$ GHz:
(a) Plot $P_M(p_e)$ versus γ_0 for $\gamma_0 = 0$ to 30 dB on semi-log paper.
(b) Plot $N_M(p_e)$ versus γ_0 for $\gamma_0 = 0$ to 30 dB on semi-log paper.
(c) Plot $\bar{t}_M$ versus γ_0 for $\gamma_0 = 0$ to 30 dB on semi-log paper.

9.19 The **diversity gain** of Mth order diversity over nondiversity is the ratio of S/N for nondiversity to that for diversity required to achieve a specified probability of outage.

(a) Show that for selection combining with all received signals having equal S/N, operating over a Rayleigh fading channel, the expression for diversity gain is

$$G = \frac{ln\,(1 - p^{1/M})}{ln\,(1 - p)}$$

where p is the probability of outage. Hint: Start with the expression for probability distribution function of a selection combiner given in Problem 9.17.

(b) Plot diversity gain (in dB) versus order of diversity ($M = 1$ to 8) for probabilities of 0.01 percent, 0.1 percent, and 1.0 percent, assuming use of selection combining with all paths having equal S/N.

9.20 The availability (defined as the percentage of time the threshold SNR is exceeded) of a certain nondiversity radio link is 90 percent for a Rayleigh fading channel. This link is to be upgraded using triple diversity with selection combining.

(a) What will be the new availability? Hint: See Problem 9.17 for the expression for the probability distribution function of a selection combiner.

(b) What is the diversity gain in dB? Hint: See Problem 9.19 for an expression for diversity gain of a selection combiner.

(c) What increase (in dB) in system gain would be required in the nondiversity link to achieve the same improvement as triple diversity?

9.21 Statistics for diversity improvement generally assume independence of fading among the received paths. However, a high degree of correlation often exists in diversity channels. The effect of correlation is to reduce the S/N γ to an effective value γ_{eff}, given by

$$\gamma_{eff} = \gamma\sqrt{1 - \rho^2} \qquad \text{for } \rho < 1$$

where ρ is the correlation coefficient. On probability paper, plot curves of the probability distribution of the combiner output for dual-diversity selection combining for correlation values of $\rho = 0, 0.5, 0.9,$ and 0.95, assuming that the SNRs are equal on the two diversity paths. Hint: See Problem 9.17 for the probability distribution of a selection combiner.

9.22 For a selection combiner used in a digital radio system,

(a) Plot the hysteresis (in dB) versus the outage multiplication factor for values of hysteresis up to 12 dB.

(b) For hysteresis equal to 6 dB, what increase in fade margin would be required to compensate for the outage multiplication due to hysteresis in a dual diversity system?

9.23 Assume a two-ray model for LOS frequency-selective fading, with β and τ equal to the amplitude and delay of the indirect path relative to the direct path.

(a) Derive the expressions given in Equations (9.63) and (9.64) for the amplitude and phase of the resultant signal.

(b) Plot normalized group delay, T/τ, and amplitude, 20 log R, versus $\omega\pi$ (between 0 and 4π radians) for $\alpha = -10$ dB and -25 dB, where $\alpha = 20$ log $(1 - \beta)$.

9.24 According to the CCIR, the 4-GHz band has the following characteristics:

Center frequency = 4.0035 GHz
Bandwidth = 400 MHz
Adjacent channel separation = 29 MHz
Spacing to band edge = 21 MHz
Spacing between lower half and upper half of band = 68 MHz
A total of 6 pairs of frequencies

(a) Show a frequency plan of go and return channels for the 4-GHz band.

(b) Indicate expressions for the lower band center frequencies f_n and upper band center frequencies f_n'.

9.25 A 16-QAM digital radio link is subject to two interfering signals, with S/I of 18 and 22 dB, respectively. The signal to gaussian noise ratio (S/N) is 24 dB.

(a) If the effect of the interfering signals is assumed to be the same as gaussian noise, what is the resulting probability of error?

(b) What increase in transmitter power is necessary to offset the effect of the two interfering signals?

9.26 Using the mathematical properties of a parabola, for a parabolic reflector antenna:

(a) Show that the distance from the feedhorn to the antenna surface is equal to $f + x$, where f = focal length and x = distance along the x-axis.

(b) Show that rays originating at the focal point will after reflection be parallel to the axis of the parabola. (Hint: All path lengths measured from the focus along rays to the reflecting surface and then along reflected rays out to the focal plane are equal to $2f$.)

9.27 Consider an antenna system with $d = 10$ ft, $f = 8$ GHz, and $\eta = 0.5$. (a) Calculate the antenna gain. (b) Calculate the 3-dB beamwidth. (c) Calculate the null-to-null beamwidth.

9.28 Consider a transmitter with power output of 2 watts operating at 6 GHz. An antenna with diameter 6 ft and efficiency 70 percent is used with this transmitter.

(a) Calculate the EIRP of the transmitted signal in dBm and in dBW. Note: The effective isotropic radiated power (EIRP) is the product of the transmitter power and antenna gain, or when using decibels, the sum of the transmitter power and antenna gain.

(b) Now double the diameter of the antenna and the transmitter power, and recalculate the EIRP of the transmitted signal in dBm and in dBW.

9.29 A digital radio system employs differentially coherent QPSK and transmits 1.544 Mbps.

(a) Determine the E_b/N_o necessary to achieve a BER of 1×10^{-6}.

(b) Determine the received signal level (in dBm or dBW) at the radio receiver corresponding to the E_b/N_o found in part (a) if the radio has an input noise temperature of 290° K and noise figure of 10 dB. Note that Boltzman's constant is 1.38×10^{-23} joule/degree K.

9.30 A digital line-of-sight has the following characteristics:

Distance = 40 mi	Fade margin = 40 dB
Frequency = 4 GHz	Transmit power = 40 dBm
Receiver threshold = –80 dBm	Multipath occurrence factor $(d) = 0.25$
Transmit antenna gain = 39.3 dB	Receive antenna gain = 39.3 dB
Transmit line loss = 5 dB	Receive line loss = 5 dB
Transmit branching loss = 3 dB	Receive branching loss = 3 dB

(a) Find the free space loss in dB.
(b) Find the antenna diameter in ft.
(c) Find the system gain in dB.
(d) Using above results and the given characteristics, find the allowed miscellaneous losses.
(e) Find the probability of outage due to multipath.

9.31 You are given a digital microwave system with the following parameters:

Frequency = 6 GHz   Transmitter height = 200 ft
Tx and Rx antenna diameter = 8 ft   Receiver height = 100 ft
Average climate (b = 1/4)   Modulation technique =
                                                    8-PSK
Average terrain (a = 1)   Bit rate = 90 Mbps
Probability of outage = 0.0001
Distance = 30 mi

(a) What is the required fade margin to meet the probability of outage?
(b) Find the antenna gain if the antenna efficiency = 0.55.
(c) For a transmitter power of 1 watt, calculate the received signal level in dBm at the receiver, in the absence of any atmospheric or terrain effects. What is the receiver threshold, defined as the received signal level at a fade depth equal to the fade margin?
(d) Now assume a two-ray channel operating over flat earth, in which the reflected ray has an angle of incidence equal to the angle of reflection. Find the total phase difference between the reflected and direct rays.
(e) What is the probability of *bit error* at the receiver threshold in the presence of additive white gaussian noise with one-sided power spectral density of 1×10^{-18} watts/Hz?

9.32 Consider a LOS link with the following conditions:

Frequency = 35 GHz
Transmitter power = 2 watts
Transmit antenna gain = 48 dB with vertical polarization
Receive antenna gain = 48 dB with vertical polarization
k = rain attenuation coefficient = 0.233
β = rain attenuation coefficient = 0.963
Total line losses = 6.5 dB
Threshold RSL = −72 dBm
Required reliability = 99.999 percent
0.01 percent rainfall rate = 10 mm/hr

(a) Considering free space loss, atmospheric attenuation, and rain attenuation, find the maximum distance for this link. Ignore the effects of multipath fading. Hint: A computer program will facilitate the solving of this problem.
(b) Determine the transmitter output power required to increase the link distance by 50 percent.

9.33 Consider the use of a passive repeater in a transhorizon link. The passive repeater consists of a flat, rectangular reflector located within line-of-sight of the two radio sites which are to be connected. The geometry of a path employing a single flat reflector is shown below

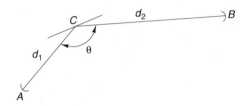

where radios are located at sites A and B; a flat reflector is located at site C at distances d_1 and d_2 to sites A and B, respectively; and θ is the angle between sites A, C, and B.

(a) The flat reflector behaves like a combination of receive and transmit antennas, with the gain in each direction given by

$$G = \frac{4\pi\, A \cos\phi}{\lambda^2}$$

where λ = wavelength of the radio frequency f
         A = area of flat reflector, in the same units as
              the squared wavelength, λ^2
         $\phi = \theta/2$.

Write an expression for the gain in dB as a function of ϕ, frequency f in MHz, and area A in m^2.

(b) To assure stability of the radio beam, the horizontal width of a reflector is twice the vertical height. Using this characteristic, find the dimensions of a flat rectangular reflector designed to yield a gain of 54.44 dB for $f = 11000$ MHz and $\theta = 48°$.

(c) Determine the net loss in dB of the path ACB by taking into account the gains of the flat reflector in both the receive and transmit directions and the free space loss on the two paths, given that $d_1 = 1.62$ km, $d_2 = 33$ km, $f = 11000$ MHz, $\theta = 48°$, and G = gain of reflector from part (b).

(d) Consider a change in the location of the reflector to be at midpath so that $d_1 = d_2 = 17.3$ km. Assuming that the angle remains unchanged, calculate the increase or decrease in reflector size necessary to provide the same net loss calculated in part (c). From your result, what can you conclude about the best choice for placement of the reflector?

9.34 A 7.4-GHz nondiversity LOS link is to be installed in an area with hot, humid climate and flat terrain. A data rate of 45 Mb/s is required and differentially coherent QPSK is to be used, having a 3-dB degradation from theory, to provide a BER = 1×10^{-8}. Assume transmit power = 32 dBm; transmit and receive antennas with 10-ft diameters; receiver noise figure = 7 dB; sum of line losses, branching losses, and miscellaneous losses = 10 dB; and negligible losses from atmospheric absorption, rainfall attenuation, and dispersion. Find the maximum distance possible if 99.99 percent availability is required.

9.35 Consider a space diversity LOS link operating at 6 GHz with the following characteristics:

Average climate
Average terrain
Probability of outage = 0.0001
Antenna separation = 40 ft

(a) Plot fade margin versus distance (up to 100 mi).
(b) Calculate the maximum link distance if the system gain = 105 dB, and the transmit and receive antennas are both 8 ft in diameter. Assume transmission line, branching, and miscellaneous losses total 10 dB, and that atmospheric and rainfall attenuation can be ignored.

10

Network Timing and Synchronization

OBJECTIVES

- Develops the concept of time as the basic unit for timing and synchronization
- Discusses the types of clocks commonly found in timing systems
- Describes pertinent clock parameters with examples
- Considers candidate systems for accurate network timing and synchronization
- Examines time and frequency dissemination systems used by communication networks
- Cites examples of network synchronization schemes

10.1 Introduction

The evolution of digital transmission toward end-to-end digitally switched voice and data channels has led to new timing and synchronization requirements. These timing requirements did not exist in analog transmission networks, and they differ greatly from most present-day digital transmission networks. In the past, store-and-forward data networks have used magnetic tape or punched paper tapes with large storage capabilities to eliminate the need for accurate network timing. Today's emerging data networks, however, rely on some form of network synchronization to provide greater efficiency and performance. Initial implementation of digital voice transmission using PCM employed point-to-point application of PCM channel banks in which all voice channels are converted to analog form for switching, thus avoiding the need for accurate network synchronization. In this configuration, shown in Figure 10.1a, timing is supplied by clock circuits within each PCM channel

653

bank, and received timing is slaved to the distant transmitter by use of clock recovery circuits. In the second phase of PCM applications, higher-level multiplexers are used to combine several lower-level PCM bit streams, as shown in Figure 10.1b. In this case pulse stuffing is conventionally used to convert each incoming PCM bit stream to a rate synchronous with the higher-level multiplexer. With the third phase of PCM applications, analog switches are replaced with digital switches that require each incoming PCM bit stream to be synchronized in frequency and in frame. Buffering is provided to compensate for small differences in clock frequencies throughout the network and to facilitate frame alignment, as indicated in Figure 10.1c.

Digital transmission inherently requires that each signal, no matter where it originates in the network, must be available at a precise time for multiplexing or switching functions. Because of variations in transmission times and timing inaccuracies and instabilities in equipment, there is some variation

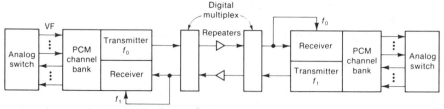

(a) PCM transmission and analog switching

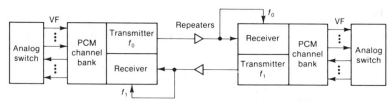

(b) Digital multiplexing and analog switching

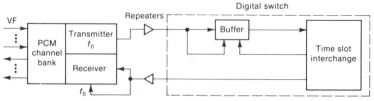

(c) Digital transmission and digital switching

FIGURE 10.1   Evolution of PCM Network Synchronization

in the arrival time of individual bits. The velocity of propagation in coaxial cable varies with temperature. Terrestrial transmission paths of over-the-horizon tropospheric scatter, ionospheric scatter, or line-of-sight radio links vary with meteorological conditions. Path lengths through synchronous satellites vary because of cyclic position changes of the satellite relative to earth stations. Clocks used to time bits within transmission equipment vary in accuracy and stability from equipment to equipment. Therefore the number of bits in transit between two given nodes varies with time.

This variation in transmission paths and clock parameters is accommodated by storage buffers in which incoming bits are held until needed by the time-division multiplexer or switch. Some degree of timing precision is required for clocks in order to assure that buffers of finite size infrequently (or never) empty or overflow, thereby causing a loss of bit count integrity. Such losses of BCI that do not result in loss of alignment are known as **timing slips.** Acceptable slip rates depend on the services provided in the network, but in any case they should be accounted for in the development of BCI performance objectives as described in Chapter 2. The CCITT recommends that the end-to-end mean slip rate of an international digital connection should be not greater than five slips in 24 hrs [1], an objective that has been adopted by AT&T [2] and ANSI [3].

10.2   Time Standards

Time is one of four independent standards or base units of measurement (along with length, mass, and temperature). The real quantity involved in this standard is the time *interval.* Today time is based on the definition of a second, given as "the duration of 9, 192, 631, 770 periods of the radiation corresponding to the transition between the two hyperfine levels of the ground state of the cesium atom 133" [4].

Time scales have been developed to keep track of dates or number of time intervals (seconds). Two basic measurement approaches are astronomical time and atomic time [5]. Methods for reconciling these two times also exist. A nationally or internationally agreed time scale then enables precise time and frequency dissemination to support a number of services including synchronization of communication networks.

10.2.1   Mean Solar Time

For many centuries, mean solar time was based on the rotation of the earth about its axis with respect to the sun. Because of irregularities in the earth's speed of rotation, however, the length of a solar day varies by as much as 16 min through the course of a year. Early astronomers understood the laws of motion and were able to correct the apparent solar time to obtain a more uniform *mean solar day,* computed as the average length of all the apparent days in the year. Mean solar time corrects for two effects:

- *Elliptical orbit:* The earth's orbit around the sun is elliptical. When nearer to the sun, the earth travels faster in orbit than when it is farther from the sun.
- *Tilted axis:* The axis of the earth's rotation is tilted at an angle of about $23\frac{1}{2}°$ with respect to the plane that contains the earth's orbit around the sun.

10.2.2   Universal Time

Universal Time (UT), like mean solar time, is based on the rotation of the earth about its axis. The time scale UT0 designates universal time derived from mean solar time. As better clocks were developed, astronomers began to notice additional irregularities in the earth's rotation, leading to subsequent universal time scales, UT1 and UT2. The UT1 time scale corrects UT0 for wobble in the earth's axis. The amount of wobble is about 15 m; if left uncorrected, it would produce a discrepancy as large as 30 ms from year to year. The UT2 time scale is UT1 with an additional correction for seasonal variations in the rotation of the earth. These variations are apparently due to seasonal shifts in the earth's moment of inertia—shifts that occur, for example, with changes in the polar ice caps as the sun moves from the southern to the northern hemisphere and back again in the course of a year.

10.2.3   Coordinated Universal Time

With the advent of atomic clocks [6, 7], improved timing accuracies led to the development of Coordinated Universal Time (UTC). The problem in using atomic clocks is that even with the refinements and corrections that have been made in UT (earth time), UT and atomic time will get out of step because of the irregular motion of the earth. Prior to 1972, UTC was maintained by making periodic adjustments to the atomic clock frequency that allowed the clock rate to be nearly coincident with UT2. Because a correction factor to UTC was necessary, this method of timekeeping was a problem for navigators who needed solar time. Since 1972, UTC has been generated not by offsetting the frequency of the atomic clock but by adding or subtracting "leap seconds" to bring atomic time into coincidence with UT1. Since the rotation of the earth is not uniform we cannot exactly predict when leap seconds will be added or subtracted, but to date these adjustments, if required, have been made at the last day, last second, of 31 December or 30 June. Positive leap seconds have been made at an average of about one per year since 1972, indicating that the earth is slowing down by approximately 1 s per year. To the user of time or frequency, UTC guarantees accuracy to within 1 s of UT1, sufficient for most users' requirements. Perhaps as important, since the frequency of the atomic clocks used for UTC is no longer periodically adjusted, UTC can also provide users a source of precise frequency limited only by intrinsic characteristics of the atomic clocks.

Numerous national and international organizations are responsible for maintaining standards of time, time interval, and frequency [8]. In the United States there are two organizations primarily responsible for providing time and frequency information: the National Institute for Standards and Technology (NIST) and the U.S. Naval Observatory (USNO) [9]. The NIST develops and maintains the atomic frequency and time interval standards for the United States. The NIST also disseminates time and frequency via radio broadcasts. The USNO makes astronomical observations for determining UT1 and keeps accurate clocks running for use by the U.S. Department of Defense. The USNO also controls distribution of precise time and time interval (PTTI) from navy radio stations, satellites, and radio navigation (LORAN C) systems operated by the U.S. Coast Guard.

Time and frequency information from national organizations is collected by the Bureau International de l'Heure (BIH) in Paris, which is the agency responsible for international coordination. At the BIH this information is evaluated and used to determine corrections for each contributing clock. By international agreement, all UTC time scales must agree with the UTC time scale operated by the BIH to within ±1 ms [10]. Relationships among the NIST, USNO, and BIH are illustrated in Figure 10.2.

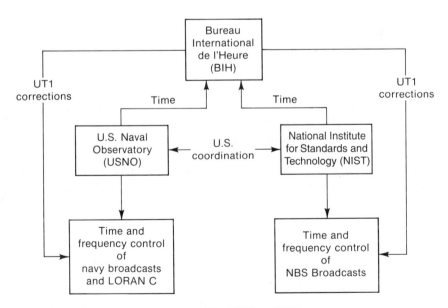

FIGURE 10.2   Relationships Among NIST, USNO, and BIH

10.3   Frequency Sources and Clocks

The term **frequency** can be defined as the number of events per unit of time. One definition of the unit of time is thus a specific number of periods of a well-defined event generator. For example, the second is today defined on the basis of a number of periods emitted by a certain resonant frequency of the element cesium. Thus a **clock** can be defined as a device that counts the number of seconds occurring from an arbitrary starting time. From this definition it appears that a clock needs three basic parts as shown in Figure 10.3: a source of events to be counted, a means of accumulating these events, and a means of displaying this accumulation of time. In a frequency source there are two major sources of error: accuracy and stability. In a clock there are these two major sources of error plus the accuracy of the initial time setting [11].

10.3.1   Definitions

Certain terms are of primary interest in network timing and synchronization.

Accuracy defines how close a frequency agrees with its designated value. A 1-MHz frequency source that has accuracy of 1 part in 10^6 can deviate ± 1 Hz from 1,000,000 Hz. The accuracy of a source is often specified with physical conditions such as temperature. **Fractional frequency difference** is the relative frequency departure of a frequency source, f_0, from its desired (nominal) value, f_D, defined as

$$\frac{\Delta f}{f} = \frac{f_0 - f_D}{f_D} = \frac{f_0}{f_D} - 1 \tag{10.1}$$

and referred to variously as fractional, relative, or normalized frequency difference. The frequency difference Δf has units in hertz, so that $\Delta f/f$ is dimensionless. The abbreviated form $\Delta f/f$ is most commonly used in the literature and has been adopted here.

Settability is the degree to which frequency can be adjusted to correspond with a reference frequency.

Reproducibility is the degree to which independent devices of the same design can be adjusted to produce the same frequency or, alternatively, the

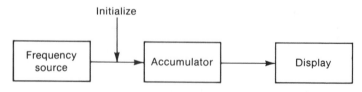

FIGURE 10.3   Basic Clock

degree to which a device produces the same frequency from one occasion to another.

Stability [12] specifies the rate at which a device changes from nominal frequency over a certain period of time. Stability is generally specified over several measurement periods, which are loosely divided into long-term and short-term stability. Usually the stability of a frequency source improves with longer sampling time, but there are exceptions to this rule. Random noise produces a jitter in the output frequency and is the dominant source of instability for short-term measurements. **Systematic drift** is the dominant source of long-term instability. Systematic drift results from slowly changing physical characteristics, which cause changes in frequency that over a long period of time tend to be in one direction. Quartz crystal oscillators, for example, exhibit a linear frequency drift, the slope of which reveals the quartz crystal's aging rate. A typical frequency stability curve is shown in Figure 10.4. Theoretically the units of drift are in hertz per second. In practice, however, **relative drift** is more commonly used, given in units as $\Delta f/f$ per unit time. Since $\Delta f/f$ is dimensionless, the unit of relative drift is time^{-1}.

Time interval error (TIE) is the variation in time delay of a given timing signal with respect to an ideal timing signal over a particular time period. **Maximum time interval error (MTIE)** is the maximum TIE for all possible measurement intervals over the measurement period. Figure 10.5 illustrates

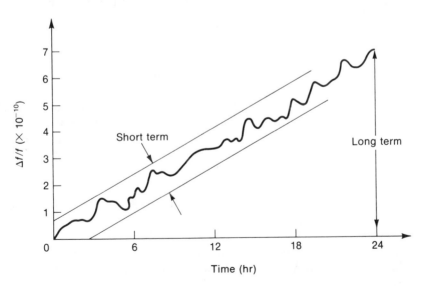

FIGURE 10.4   Typical Frequency Stability Curve

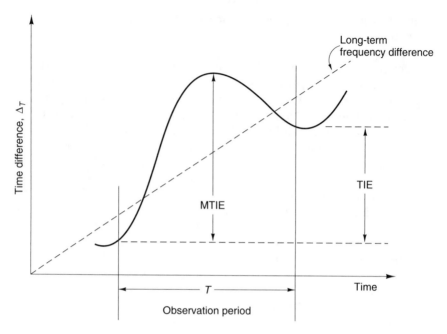

FIGURE 10.5   Time Interval Error (TIE) and Maximum Time Interval

both TIE and MTIE. To calculate the time error, let the frequency at any time t be

$$f_t = f_0 + \alpha t \tag{10.2}$$

as illustrated in Figure 10.6 and where f_0 is the initial frequency at time $t = 0$ and where α is the drift rate in hertz per second. It should be noted that a general representation of f_t would include higher-order terms in time. When one is considering precision oscillators, however, the coefficients for these higher-order terms can be considered zero [13]. The phase accumulation in true time T is then

$$\phi(T) = \int_0^T f_t dt = f_0 T + \tfrac{1}{2}\alpha T^2 + \phi_0 \tag{10.3}$$

where true time is defined as that time determined by the international definition of the second. We assume the phase error at time $t = 0$ to be ϕ_0. The apparent clock time after true time T is

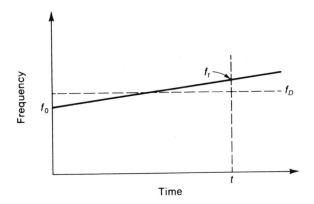

FIGURE 10.6   Oscillator Frequency vs. Time

$$C(T) = \frac{\phi(T)}{f_D} \tag{10.4}$$

where f_D is the desired frequency. Thus the time difference between apparent clock time and true time is

$$\Delta_T = C(T) - T = \frac{\phi(T)}{f_D} - T$$
$$= \left(\frac{f_0}{f_D}T + \frac{1}{2}aT^2 + \varepsilon_0 \right) - T \tag{10.5}$$

where $\quad a = \alpha/f_D = $ relative drift rate
$\quad\quad\quad \varepsilon_0 = \phi_0/f_D = $ initial time error (setting error)

Simplifying (10.5), we get

$$\Delta_T = \varepsilon_0 + \left(\frac{f_0}{f_D} - 1 \right)T + \frac{1}{2}aT^2 \tag{10.6}$$

or from (10.1)

$$\Delta_T = \varepsilon_0 + \frac{\Delta f}{f_D}T + \frac{1}{2}aT^2 \tag{10.7}$$

This relationship of frequency error (due to setting error and drift) and time error is illustrated in Figure 10.7.

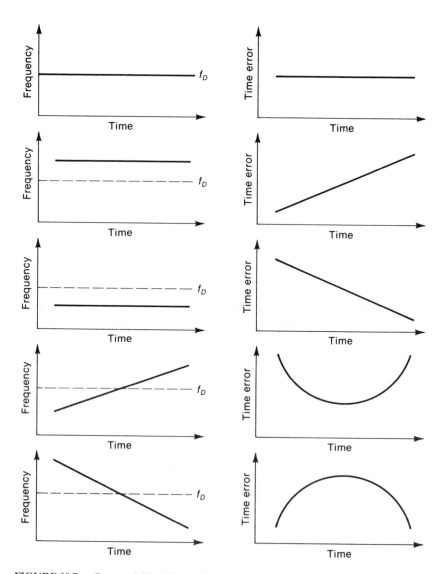

FIGURE 10.7   Causes of Clock Error (Reprinted by permission of Austron, Inc.)

Example 10.1 ————————————————————————————————

A 1-MHz crystal oscillator has a setting accuracy of 1×10^{-8} and a relative drift rate of 1×10^{-8} per day. Assuming the initial time error to be zero, the time base error in one 24-hr period is

$$\Delta_T = (1 \times 10^{-8})(86,400 \text{ s}) + \frac{1}{2} \frac{(1 \times 10^{-8})(86,400)^2}{86,400}$$

$$= (864 + 432) \times 10^{-6} \text{ s}$$

$$= 1296 \, \mu s$$

Clocks can be operated in a number of configurations or modes. In a **free-run mode,** the oscillator of the clock is neither locked to an external synchronization reference nor is capable of holding to a previously established reference. In a **holdover mode** the oscillator of the clock is not locked to an external reference but uses storage techniques to maintain its accuracy with respect to the last known frequency comparison with a synchronization reference. The **hold-in range** is a measure of the maximum frequency deviation of an input reference from the nominal clock rate that a clock can accommodate and still hold itself into synchronization with another clock. The holdover specification indicates the time it would take for a clock to gradually transition to its free-running accuracy once the reference signal is lost. The **pull-in range** is a measure of the maximum input frequency deviation from the nominal clock rate that can be overcome by a clock to pull itself into synchronization with the reference. Based on their performance, clocks used in network synchronization can be classified into **stratum levels.** In the United States four levels are used in public networks, with stratum 1 being the highest and stratum 4 being the lowest level of performance. Table 10.1 gives specifications for these four strata.

10.3.2 Frequency Sources

The many kinds of frequency-determining devices can be grouped into three classes: mechanical, electronic, and atomic. Mechanical resonators like the pendulum and the tuning fork, and electronic resonators such as LC tank circuits and microwave cavities, are of little importance in today's high-performance frequency sources and will not be discussed here. Atomic frequency sources used in communication network synchronization and quartz crystal oscillators used as backup in network synchronization schemes or as an internal time base in communication equipment will be discussed extensively. Characteristics of the clocks described here are summarized in Table 10.2.

TABLE 10.1 Stratum Clock Requirements

	Stratum 1	Stratum 2	Stratum 3	Stratum 4
Accuracy	$\pm 1 \times 10^{-11}$	$\pm 1.6 \times 10^{-8}$	$\pm 4.6 \times 10^{-6}$	$\pm 3.2 \times 10^{-5}$
Holdover (Stability)	Not applicable	1×10^{-10} per day	≤255 DS1 frame slips in 1st 24 hours	Not applicable
Availability	No requirement	Duplicated Clock Hardware	Duplicated Clock Hardware	Single Clock Hardware
MTIE During Rearrangement	Not applicable	MTIE ≤ 1 μsec; Phase Change Slope: ≤81 nsec in any 1.326 msec	MTIE ≤ 1 μsec; Phase Change Slope: ≤81 nsec in any 1.326 msec	No requirement
Pull-in Range	Not applicable	3.2×10^{-8}	9.2×10^{-6}	6.4×10^{-5}
External Timing Interface	Must be traceable to UTC if used as Primary Reference Source	Yes	Yes	Optional

TABLE 10.2 Comparison of Frequency Sources (Adapted from [5])

Characteristic	Frequency Source				
	Quartz	Quartz (Temp. Compensated)	Quartz (Single Oven)	Cesium	Rubidium
Basic resonator frequency		10 kHz to 100 MHz		9,192,631,770 Hz	6,834,682,613 Hz
Output frequencies provided		10 kHz to 100 MHz		1, 5, 10 MHz typical	1, 5, 10 MHz typical
Relative frequency drift, short term, 1 s	1×10^{-9} typical	1×10^{-9} typical	1×10^{-9} to 1×10^{-10}	5×10^{-11} to 5×10^{-13}	2×10^{-11} to 5×10^{-12}
Relative frequency drift, long term, 1 day	1×10^{-7} typical	1×10^{-8} typical	1×10^{-7} to 1×10^{-9}	1×10^{-13} to 1×10^{-14}	5×10^{-12} to 5×10^{-13}
Relative frequency drift, longer term	5×10^{-6} per year	1×10^{-8} to 5×10^{-7} per year	1×10^{-9} to 5×10^{-11} per year	$<5 \times 10^{-13}$ per year	1×10^{-11} per month
Principal environmental effects	Motion, temperature, crystal drive level			Magnetic field, accelerations, temperature change	Magnetic field, temperature change, atmospheric pressure
Principal causes of long-term instability	Aging of crystal, aging of electronic components, environmental effects			Component aging	Light source aging, filter and gas cell aging, environmental effects

Quartz

Crystalline quartz has great mechanical and chemical stability, a most useful characteristic in a frequency source [14, 15]. The quartz crystal clock is actually a mechanical clock, since such crystals vibrate when excited by an electric potential. Conversely, if the crystal is made to vibrate, an electric potential is induced in nearby conductors. This property in which mechanical and electrical effects are linked in a crystal is known as the **piezoelectric effect** [16].

In building a quartz oscillator, the quartz crystal is first cut and ground to create the desired resonant frequency. To produce the piezoelectric effect, metal electrodes are attached to the crystal surfaces. Using the crystal, an oscillator is produced by adding an amplifier with feedback as shown in Figure 10.8. Its frequency in turn is determined by the physical dimensions of the crystal together with the properties of the crystalline quartz used.

Drift, or aging, and dependence on temperature are common traits of all crystal oscillators. When switched on, the crystal oscillator exhibits a

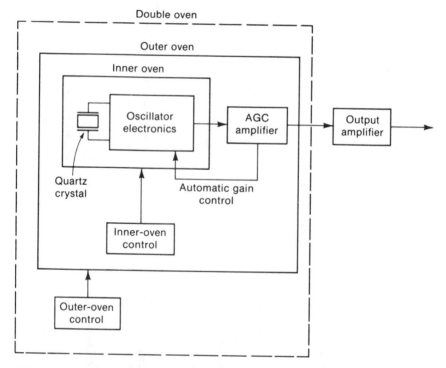

FIGURE 10.8   Typical Quartz Crystal Oscillator (Reprinted by permission of the Efratom Division of the Ball Corp. [17])

"warm-up time" due to temperature stabilization. After this initial aging period of a few days to a month, the aging rate can be considered constant and usually remains in one direction. Thus periodic frequency checks and corrections are needed to maintain a quartz crystal frequency standard. The temperature dependence is caused by a slight change in the elastic properties of the crystal. If large temperature fluctuations are to be tolerated or if high-frequency precision is required, the crystals can be enclosed in electronically controlled ovens that maintain constant temperatures (Figure 10.8). An alternative solution to the temperature problem is the so-called temperature-compensated crystal oscillator (TCXO). A temperature sensor and varactor are added to cancel any changes in the resonant frequency due to temperature. The overall performance of the quartz oscillator is also affected by the line voltage, gravity, shock, vibration, and electromagnetic interference.

Performance of a crystal oscillator is determined primarily by the aging rate and temperature dependence. The most stable crystal oscillators utilize ovens and exhibit an aging characteristic as good as 10^{-11} per day. Crystal oscillators manufactured for use at room temperature exhibit aging rates of typically 5×10^{-7} per month. The good short-term stability characteristic and frequency control capability make crystal oscillators ideal for use as slave clocks with, for example, atomic clocks. Table 10.2 compares the various crystal oscillators and their typical specifications.

Cesium (Cs)
Use of atomic clocks such as the cesium standard is based on the quantum-mechanical phenomenon that atoms release energy in the form of radiation at a specific resonant frequency. The atom is a natural resonator whose natural frequency is immune to the temperature and frictional effects that beset mechanical clocks. In an atomic clock the oscillator is always a quartz oscillator, and the high – Q resonator is based on some natural frequency of an atomic source. The frequency-determining element is the atomic resonator, which consists of a cavity containing the atomic material and some means of detecting changes in the state of the atoms. The output of a voltage-controlled crystal oscillator (VCXO) is used to coherently generate a microwave frequency that is very close to resonance of the atomic resonator. A control signal, related to the number of atoms that changed their state due to the action of the microwave signal, is fed back to the crystal oscillator. Atomic resonances are at gigahertz frequencies whereas crystal oscillator frequencies are at megahertz frequencies, so that multiplication is required by a factor of about 1000. When the atomic standard is in lock, the VCXO output is exactly at its resonant frequency (typically 5 MHz) to within the accuracy tolerance provided by the atomic resonator [18].

For a cesium oscillator, atomic resonance is at 9,192,631,770 Hz. The so-called cesium-beam frequency standard is shown in Figure 10.9. The cesium metal is housed in an oven. When heated to about 100°C, cesium gas forms an atomic beam that is channeled into the vacuum chamber. The magnetic field at the two ends of the tube separates the atoms according to their energy state. When passing through the microwave signal, which is set very near to the cesium resonant frequency, cesium atoms change energy state and are detected at the far end of the tube. The detector produces a control voltage that is related to the number of atoms reaching it, and this signal is fed back to control the microwave frequency through the crystal oscillator in order to maximize the number of atoms reaching the detector—which means that the radio signal is at the cesium atom's natural frequency.

Cesium frequency standards are used extensively where high reproducibility and long-term stability are needed. Cesium standards exhibit no systematic long-term drift. Frequency stability of parts in 10^{14} is possible at sampling times of less than 1 hr to days. The very short term stability of cesium standards is governed by the stability of the quartz oscillators within

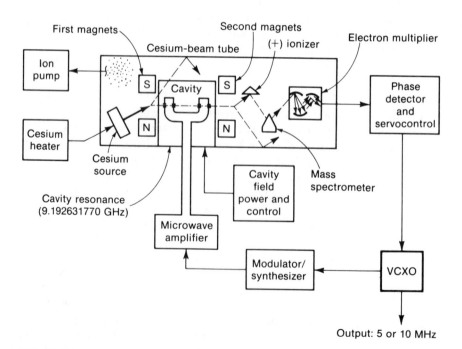

FIGURE 10.9   Cesium-Beam Frequency Standard (Reprinted by permission of the Efratom Division of the Ball Corp. [17])

them. The time at which the longer-term stability properties of the atomic standard become dominant depends on the time constant of the quartz oscillator control loop, which is typically in the range 1 to 60 s. Used as a clock, cesium oscillators provide accuracy to a few microseconds per year. Cesium standards are the predominant source in today's precise time and frequency dissemination systems.

Rubidium (Rb)

The atomic resonator for the rubidium standard is a gas cell that houses rubidium (Rb^{87}) gas at low pressure [19]. As shown in Figure 10.10, the rubidium gas interacts with both a light source that is generated by a rubidium lamp and a microwave signal that is generated by a VCXO. One energy state is excited by the beam of light, while another state is excited by the RF signal. The microwave signal, when at the resonant frequency 6,834,682,613 Hz of rubidium, converts atoms into an energy state that will absorb energy from the light source. A photodetector monitors the amount of light absorbed as a function of the applied microwave frequency. The microwave

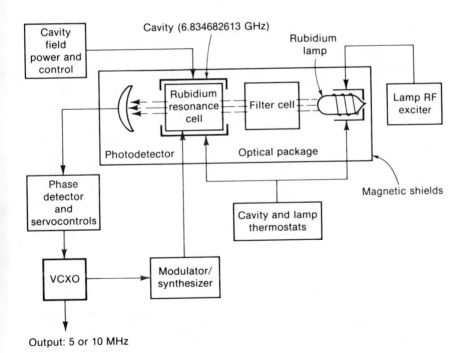

FIGURE 10.10   Rubidium Frequency Standard (Reprinted by permission of the Efratom Division of the Ball Corp. [17])

signal is derived by multiplication of the quartz oscillator frequency. A servoloop controls the oscillator frequency so that the oscillator is locked to the atomic resonance.

Rubidium oscillators vary in their resonance frequency by as much as 10^{-9} because of differences in gas composition, temperature, and pressure and in the intensity of the light. Therefore rubidium oscillators require initial calibration and also recalibration because they exhibit a frequency drift or aging like crystal oscillators. This aging is due to such factors as drift in the light source and absorption of rubidium in the walls of the gas chamber. Nevertheless, stability performance of rubidium oscillators is quite good. As with cesium standards, the very short term stability of rubidium standards is governed by the stability of the slaved quartz oscillator. At 1-s sampling times, they display a stability of better than 10^{-11} and nearly 10^{-13} for sampling times of up to a day. For longer averaging times, the frequency stability is affected by long-term drift, which is typically 1 part in 10^{11} per month. Rubidium oscillators are used wherever excellent medium-term stability—minutes to a day—is required and where reduced cost and size, as compared with cesium oscillators, are important.

10.4   Network Synchronization Techniques

Candidate systems for providing digital transmission network synchronization fall into two major categories: nondisciplined and disciplined techniques. All nondisciplined clocks are asynchronous since all clocks run independent of control from a reference. The slip rate is a function of accuracy between any two clocks and buffer size. For disciplined techniques, each nodal clock is synchronized to a reference signal. One of the principal differences between the disciplined techniques is the choice of reference. This difference has a direct impact on the ability of the technique to accommodate transmission delay variations and to adapt to a loss of reference signal. Other differences in disciplined techniques are the capabilities of providing precise time and clock error correction. Conceptual illustrations of both nondisciplined and disciplined network synchronization techniques are shown in Figures 10.11 and 10.12.

10.4.1   Pulse Stuffing

Pulse stuffing can be used as a form of network synchronization, allowing asynchronous interface at each node. Analog switching networks that use digital transmission generate TDM signals that are not normally synchronous with each other. The individual TDM signals (for example, 1.544 Mb/s for North American systems and 2.048 Mb/s for European systems) are synchronized to a common clock in a higher-level multiplexer via pulse

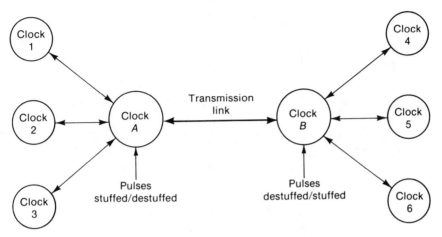

(a) Pulse stuffing synchronization concept

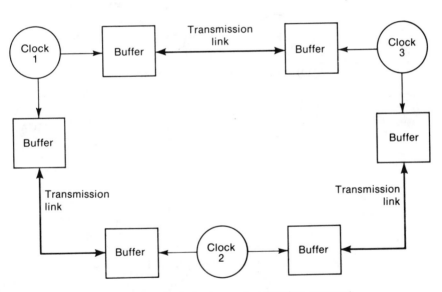

(b) Independent clock synchronization concept

FIGURE 10.11   Nondisciplined Network Synchronization Concepts

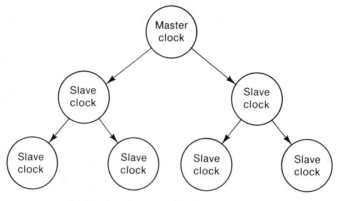

(a) Master-slave synchronization concept

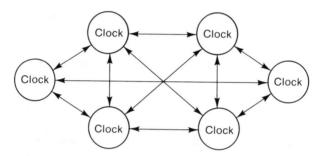

(b) Mutual synchronization concept

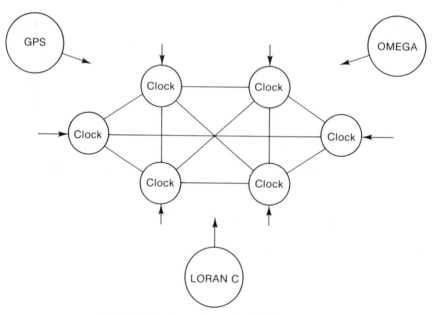

(c) External reference synchronization concept

FIGURE 10.12   Disciplined Network Synchronization Concepts

stuffing. This technique has the advantage of not requiring stringent tolerance on the frequency of each multiplexer clock.

Pulse stuffing techniques were originally designed for operation over cable
(such as T carrier) and later line-of-sight radio. When operated over links
such as tropospheric scatter that are subject to bursts of errors or interruptions due to deep fades, however, errors may occur in the stuff control bits
resulting in a loss of BCI on the individually derived bit streams. Moreover,
timing jitter introduced by the pulse destuffing circuits can result in degraded
performance in equipment that must extract timing from the jittered data or
clock signals. Pulse stuffing also involves a considerable amount of hardware
and somewhat reduces the efficiency of a multiplexer. For these reasons, and
with the evolution toward all-digital switched systems, pulse stuffing has
served chiefly as an interim means of network synchronization.

10.4.2 Independent Clock

The independent clock approach is based on a **plesiochronous** (nearly synchronous) concept of network synchronization in which data buffers are
used to compensate for the differential phase in the system clocks. Highly
stable references such as atomic clocks are used at each node of the network
with a buffer large enough to maintain BCI over a predetermined period.
Buffer length depends on differential clock drift, path length variation, link
data rate, and buffer reset period. Since the length of the buffer is fixed, the
buffer maintains synchronization (BCI) over some time period T. The rate
of buffer reset $(1/T)$ can be made acceptably low with proper choice of clock
stability and buffer length.

Conceptually, independent clocks constitute the simplest of all techniques for network synchronization. The chief advantage lies in system sur-
vivability, where link degradation or failure does not affect nodal timing
and clock failure of one node does not affect other nodal clocks. There are
two significant disadvantages, however, for the independent clock technique vis-à-vis any of the disciplined clock approaches. First, without some
form of discipline, greater stability must be demanded from individual
clocks, implying a higher cost penalty. Second, periodic slips due to buffer
overflow or underflow must be tolerated along with attendant resynchronization periods.

Buffer Length
The buffer length (B) required for a single link under independent clock
operation is given by

$$B = 2R(T_{CE} + T_{PD}) \quad \text{bits} \tag{10.8}$$

where R = data rate in bits per second
T_{CE} = clock error ($\pm$) in seconds accumulated during
            buffer reset period
T_{PD} = path delay variation ($\pm$) in seconds accumulated during
            buffer reset period

The factor of 2 in (10.8) results from a requirement that the buffer be set initially at center position and be able to shift in either the positive or negative direction.

Calculation of Clock Error (T_{CE}).   Time (phase) difference between two clocks can accumulate due to differences in initial frequency (accuracy) and stability. If $f_i(0)$ is the frequency of a clock at the ith node at time zero, then the frequency after time t is

$$f_i(t) = f_i(0) + \alpha_i t \tag{10.9}$$

where α_i is the drift rate in hertz per second of the ith nodal clock and where higher-order terms of time are assumed to have negligible contribution. The factor α_i is actually a function of time, but here α_i will be assumed constant and equal to the worst resulting offset due to instability. The difference in frequency between any two clocks is then

$$f_i(t) - f_j(t) = f_i(0) - f_j(0) + (\alpha_i - \alpha_j)t \tag{10.10}$$

Letting $f_i(t) - f_j(t) = \Delta f(t)$ and dividing by the desired (nominal) frequency f, we get

$$\frac{\Delta f(t)}{f} = \frac{\Delta f(0)}{f} + at \tag{10.11}$$

where $\Delta f(0)/f$ and $\Delta f(t)/f$ are the relative frequency differences between two clocks at times zero and t, respectively, and where

$$a = \frac{\alpha_i - \alpha_j}{f} \tag{10.12}$$

is the relative (difference) drift rate between the two clocks. To determine the accumulated clock error T_{CE}, the frequency difference given in (10.11) is integrated over some time T:

$$T_{\mathrm{CE}} = \int_0^T \left[\frac{\Delta f(0)}{f} + at \right] dt$$

$$= \frac{\Delta f(0)}{f} T + \frac{1}{2} a T^2$$

(10.13)

Figure 10.13 is a family of plots of accumulated phase error in time versus operating time T for various initial frequency accuracies between two clocks. Figure 10.14 shows a similar family of plots for selected long-term (per day) drift rates; these drift rates are based on an assumed temperature range of 0 to 50°C. Total clock error, given by Equation (10.13), is determined by summing contributions from initial frequency inaccuracy (Figure 10.13) and from frequency drift (Figure 10.14). Figure 10.15 shows buffer fill in bits versus accumulated time error for standard data rates. To determine buffer lengths required to compensate for accumulated clock error, first select the buffer reset period T. For a given T, Figures 10.13, 10.14, and 10.15 can be used to yield the required buffer length. The factor of 2 in (10.8) doubles the buffer length calculated here to allow opposite direction (sign) of drift and initial inaccuracy. The underlying, worst-case assumption here is that both frequency inaccuracy and drift are biased in opposite directions between the two clocks. If two clocks are settable within ± 1 part in 10^{11}, for example, the maximum initial frequency offset is 2 parts in 10^{11}. In practice, accuracy and drift differences between two clocks are not always biased in opposite directions; rather, they differ from one pair of clocks to another.

Example 10.2 _____

Assume a buffer reset period (T) of 1 week and the use of independent cesium frequency standards. Determine the required buffer length for a data rate of 1.544 Mb/s. From Figures 10.13 and 10.14 we have

$$\text{Phase error:} \quad \frac{\Delta f}{f} T \approx 6 \times 10^{-6} \text{ s}$$

$$\text{Phase error:} \quad \frac{1}{2} a T^2 \approx 2 \times 10^{-6} \text{ s}$$

$$\overline{\text{Total phase error} \qquad \approx 8 \times 10^{-6} \text{ s}}$$

Using Figure 10.15, we find the required buffer length to be approximately ± 15 (30) bits.

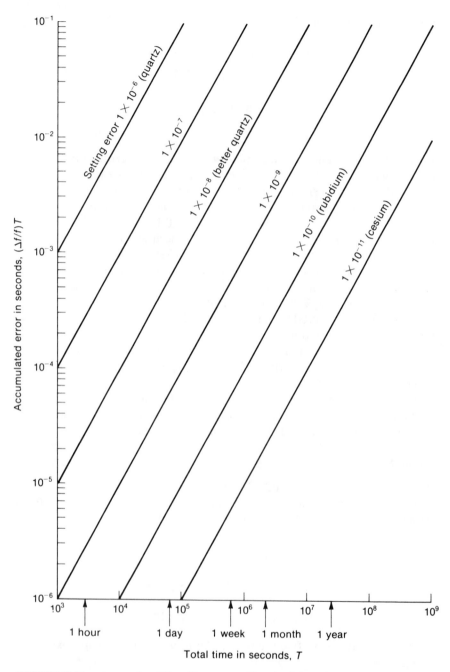

FIGURE 10.13   Accumulated Time Error versus Time for Selected Frequency Offsets

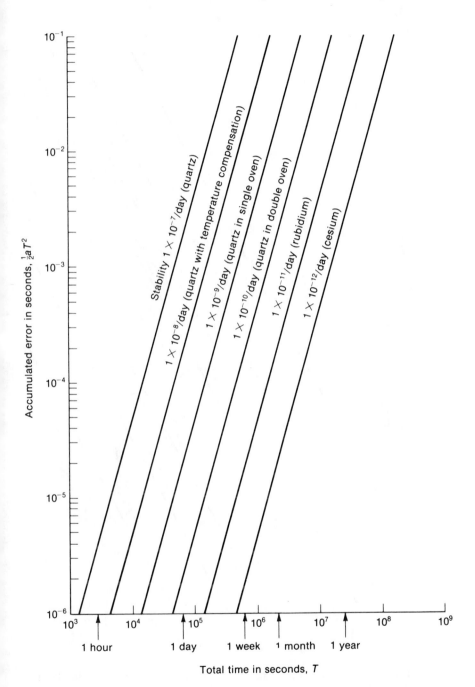

FIGURE 10.14   Accumulated Time Error versus Time for Selected Drift Rates

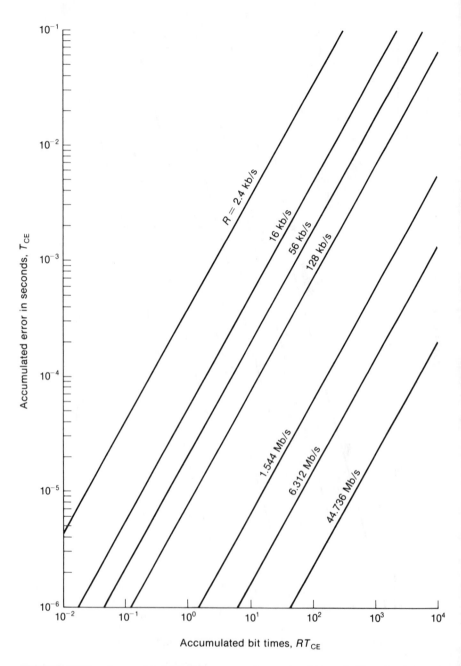

FIGURE 10.15   Accumulated Bit Times versus Accumulated Error in Time for Selected Bit Rates

Determination of Path Delay Variation (T_{PD}).   Fixed delay over a transmission path does not affect clock synchronization between nodes. Variations in transmission delay, however, can cause **doppler shift,** or frequency wander in timing derived from the received signal. The effect of path delay variations on the independent clock technique is to shift the "pointer" of the affected buffer by the amount of the phase change.

Delay variations on *line-of-sight microwave links* are due to meteorological parameters such as atmospheric pressure, temperature, relative humidity, wind velocity, and solar radiation. Measured data indicate that the long-term component of delay variation for LOS microwave transmission is typically 0.1 to 0.2 ns/km over a 24-hr period [20, 21].

Tropospheric scatter links are subject to daily variations that are several orders of magnitude larger than variations observed in line-of-sight. This increase in delay variation is due to a continuously varying index of refraction in the troposphere with respect to both space and time. The delay variations depend on the path length and the intersection angles of the scattering volume. For typical troposcatter links, the worst peak-to-peak delay variation that may be expected is on the order of 1.0 μs [21, 22].

Geostationary satellite paths experience path delay variation, primarily due to satellite orbit inclination, satellite orbit eccentricity, and atmospheric and ionospheric variations [23]. Unless corrected, the orbital plane changes inclination by approximately 0.86° per year due to lunar and solar gravitational effects. This orbital inclination causes the point directly below the satellite to move in a figure eight pattern, completing the pattern every 24 hr. The dimensions of the figure eight increase with the inclination. For an inclination angle I in radians, relative to the equatorial plane, the diurnal path delay variation causes a time change of

$$\Delta T \approx \frac{4Ir_e}{c} \qquad \text{for } I \ll 1 \qquad (10.14)$$

where r_e is the earth's radius and c is the speed of light. Thus for $I = 0.015$ rad ($\approx 0.86°$), there is $\Delta T \approx 1.3$ ms peak-to-peak variation. The eccentricity e due to an elliptical orbit causes the path distance to an observer on the equator to vary by

$$\pm e (r_e + H)$$

for an altitude H. For a typical maximum of $e = 0.01$ and for $H = 22,300$ mi, the variation is ±263 mi or ±1.4 ms. Communication satellite eccentricities are typically on the order of 10^{-4}, however, and produce a small path delay change compared to the inclination. There are also the usual atmospheric and ionospheric mechanisms that cause change in the index of refraction. Resulting changes in delay are independent of orbital parameters and are smaller in magnitude; ±1 ms is an upper bound.

Buffer Location
The choice of buffer placement in a demultiplexer hierarchy depends on a number of considerations, such as performance, network topology, cost, and reliability.

Performance.   When buffering is carried out at one of the lower levels, all higher-level equipment must be clocked from the recovered clock, which exhibits degradation due to jitter and other media effects (such as diurnal variations for satellites). To avoid passing on this clock degradation, buffers can be placed at higher levels in the hierarchy. Another influence on performance is the resetting of buffers due to underflow or overflow; data are lost (or repeated), resulting in loss of BCI and the need to resynchronize downstream equipment. Here the choice of buffer location determines the number of users affected by a single buffer reset.

Network Topology.   At some nodes, switching and multiplexing of data from several remote sources may be required. This is usually done with synchronously clocked data, implying that the data must be buffered prior to switching or multiplexing. The number of users (circuits) requiring synchronous operation also affects the choice of buffer location. If only a few low-speed users require synchronous operation, it may be less costly and more flexible to place buffers only at the low-data-rate location. Placement of the buffer at the highest level in the data hierarchy provides synchronous data to all users, independent of their needs, and facilitates transition to an all-synchronous switched system. However, one topological disadvantage of high-level buffer placement is that for many networks high-level multiplexers are used in greater proportion than lower-level multiplexers—implying the need for more frequent buffering at the higher level and hence a larger buffer reset period per link in order to meet end-to-end slip rate requirements.

Cost.   Present technology suggests that the cost of buffers is somewhat insensitive to the size (length) of the buffer. A large buffer with high operating speed is probably cheaper than a number of smaller ones with similar total capacity, each operating at a slower speed, unless reliability requirements dictate a more expensive approach. Multiplex and switching equipment inherently requires buffers to perform their function, however, suggesting that such built-in buffers could be made sufficiently longer to provide a network synchronization function also. With this assumption, separate buffers can be avoided by using buffers built directly into transmission and switching equipment.

Reliability. A disadvantage of the high-level buffer placement is the addition of another series component that is critical to system reliability. If the high-level buffer fails, all downstream lower-level outputs are affected. The magnitude of such a failure forces the use of a high-reliability design or a standby buffer with automatic switchover. When the lower-level buffer location is used, the impact of a failed buffer is less and therefore a lower-reliability design could be used.

Example 10.3 _____

Table 10.3 presents buffer length requirements for both terrestrial and satellite applications and both low-level and high-level buffer location. Buffer length is specified as plus and minus so that the buffer can initially be set at mid position and allowed to move in either direction to its specified maximum. The timing source is assumed to be a cesium clock. The low-level buffer placement corresponds to the lowest level in the multiplex hierarchy, or a user level, perhaps a single voice or data channel as indicated by the bit rates shown in Table 10.3. The higher-level placement corresponds to the highest multiplex level, equal to the total transmission rate. Buffer reset periods, here set at 10 and 50 days for lower and higher buffer placement respectively, are determined by the relative density of multiplex breakouts at low and high levels in the hierarchy. The conclusions that can be drawn from the examples in Table 10.3 are that phase error due to clock differences is much greater than phase error due to propagation delay variations in terrestrial systems but much less than phase error due to satellite doppler.

_____

10.4.3 Master-Slave System

The simplest and most utilized [24] form of disciplined network synchronization is the **master-slave system,** which distributes a reference or master clock to all nodes via a tree-type structure. As with most disciplined systems, timing signals are distributed along with data over all transmission links. Timing is then recovered from the incoming data, either from bit transitions or from frame information. The most important feature is the maintenance of timing distribution in the event of link failure. According to this feature, several types of master-slave systems can be envisioned.

Fixed Hierarchy

With the fixed-hierarchy approach, when the normal clock distribution path has failed, restoration at each slave node is accomplished via a predetermined algorithm [25]. All nodal clocks are prioritized, and each node is capable of locking itself to the highest-rank clock in

TABLE 10.3 Typical Buffer Length Requirements for Independent Clock Synchronization

Buffer Location	Clock Accuracy	Long-Term Clock Stability	Propagation Delay Variations (T_{PD})	Buffer Reset Period	Clock Error (T_{CE})	Buffer Length Requirements	
						Bit Rate (kb/s)	Buffer Length (bits)
Lower level (terrestrial link)	1×10^{-11}	1×10^{-13}	$\leq 1\ \mu s$	10 days	$9\ \mu s$	512	±6
						256	±3
						128	±2
						64	±1
Higher level (terrestrial link)	1×10^{-11}	1×10^{-13}	$\leq 1\ \mu s$	50 days	$54\ \mu s$	44,736	±2460
						6,312	±347
						1,544	±85
Lower level (satellite link)	1×10^{-11}	1×10^{-13}	10 ms	10 days	$9\ \mu s$	128	±1282
						64	±641
						32	±321
						16	±161

operation. If a clock distribution path fails, the node changes to the next priority source. In this system, there is no need to transfer control signals between nodes.

Self-Organizing
To implement the self-organizing synchronization scheme, control signals must be transferred between nodes to allow automatic rearrangement of the clock distribution network. This control information describes the performance of the clock signal being used at a given transmitting node, and this information is transmitted to all other connected nodes. Typically three kinds of control signals are transmitted [26, 27]:

- Designation of the node used as the master reference for the local clock
- The number and quality of links that have been encountered in the path from the reference clock to the local clock
- The rank of the local clock

The third type of control signal refers to a preassigned ranking to each node that depends, for example, on the quality of its clock.

Loosely Coupled
In the loosely coupled approach [28], each node maintains high-frequency accuracy even in the event of a fault. Thus no control is needed to rearrange the clock distribution network. This approach is made possible by use of a phase-locked oscillator that stores the previous input frequency in memory [29], by use of high-quality backup oscillators that allow plesiochronous operation [2], or by use of external reference, such as LORAN C, as a backup source of timing [30]. This version of the master-slave system may be viewed as the simplest from the viewpoint of network control and maintenance.

10.4.4   Mutual Synchronization
The technique known as **mutual synchronization,** or frequency averaging, is one in which each nodal clock is adjusted in frequency in order to reduce the timing error between itself and some weighted average of the rest of the network. The basic approach is illustrated in Figure 10.16, where a weighted sum of phase errors for all incoming clocks is used as a control signal to adjust the local clock. Figure 10.16 depicts the circuitry at one node where $f_1, f_2, \ldots, f_i$ represent clock signals regenerated through the use of a phase-locked loop from each incoming

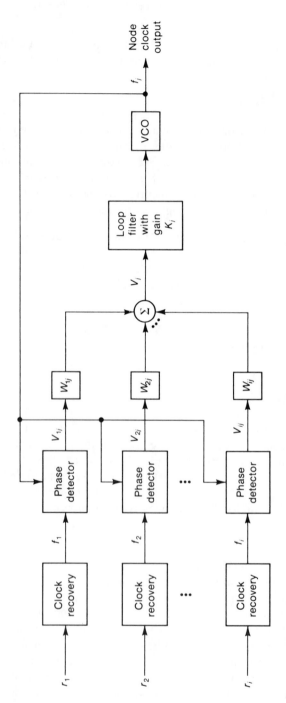

FIGURE 10.16 Implementation of Mutual Synchronization Technique

nodal bit stream, $r_1, r_2, \ldots, r_i$. The phase detector is modeled as a flip-flop with output

$$V_{1j} = \phi_j - \phi_1$$
$$V_{2j} = \phi_j - \phi_2$$
$$\vdots$$
$$V_{ij} = \phi_j - \phi_i$$

(10.15)

where ϕ_j and $\phi_1, \phi_2, \ldots, \phi_i$ represent the nodal clock and incoming clock phases. The output of the summing network is

$$V_j = \sum_i W_{ij} V_{ij}$$

(10.16)

where $\sum_i$ denotes summation over all nodes connected to node j. If the free-running clock of the jth node has frequency denoted as f_{0j}, then since the local frequency is controlled by voltage V_j through the filter with gain K_j, the nodal clock output is given as

$$f_j = f_{0j} - K_j V_j$$

(10.17)

where K_j is a constant with units of radians per volt.

The version of mutual synchronization described above is called *single-ended* because the error control signals at a given node are derived only from timing signals available at the node. In a *double-ended* mutual synchronization system, timing information from all neighboring nodes is transmitted to the given node and used in determining the local frequency. The advantage of this technique over the single-ended control technique is that the system frequency of the entire interconnected network is not a function of transmission delay but rather depends only on the parameters of the clock at each node.

Mutual synchronization is a widely studied technique [31 to 37], but it has been used operationally in few applications and then only outside the United States. One advantage is provided by the multinode source of timing, which assures that no single clock or transmission path is essential. Another basic advantage is its ability to adapt to system changes and clock drift. As a result, a less accurate reference clock can be used than with other approaches. Some disadvantages of the mutual synchronization technique are:

- The eventual system frequency is difficult to predict, since it is a function of VCO frequencies, network topology, link delays, and weighting coefficients.

- Changes in link delay or nodal dropout can cause significant perturbations in nodal frequencies and a permanent change in system frequency.
- Lack of a fixed reference results in offset with respect to any external network or other external source of time or frequency, making mutual synchronization incompatible with a UTC reference.

10.4.5   External Reference

The **external reference** technique obtains a timing or frequency reference for all nodal clocks from a source external to the communication network. There exist several such sources based on worldwide systems providing precise time and frequency to navigational and communication systems. Table 10.4 summarizes the characteristics of the major time and frequency dissemination systems that are described in more detail in the next section.

10.5   Time and Frequency Dissemination Systems

Numerous systems have been developed for the dissemination of precise time and frequency. These systems work on the basis of continuous or periodic radio broadcasts of time and frequency information, using terrestrial or satellite transmission techniques. Here we will examine those time and frequency dissemination systems that are used by communication networks.

10.5.1   LF and VLF Radio Transmission

Radio waves at low frequency (LF) and very low frequency (VLF) are used for precise time and frequency dissemination because of their excellent stability and potential for good earth coverage. Accuracies of 1 part in 10^{11} in frequency and 500 μs or better for time can be achieved using LF and VLF broadcasts. Some 20 individual LF–VLF stations exist worldwide that provide dissemination of time and frequency information. Here, however, two systems will be described that utilize several stations collectively to provide good earth coverage; one system operates in the VLF band (OMEGA) and the other in the LF band (LORAN C).

OMEGA Navigation System
The OMEGA navigation system is operated by the United States (U.S. Coast Guard) and several other countries and is composed of eight VLF radio stations operating in the 10 to 14 kHz range [38, 39]. These stations radiate a nominal 10 kW of power sufficient for moderate accuracy navigation on a worldwide basis. All stations transmit according to a predetermined time sequence shown in Figure 10.17, where the asterisks signify the

TABLE 10.4   Characteristics of the Major Time and Frequency Dissemination Systems (Adapted from [5])

Dissemination Techniques	Accuracy for Frequency Comparison or Calibration	Accuracy for Time Transfer	Ambiguity	Coverage	Reference
VLF radio					
OMEGA	1×10^{-11}	Envelope: 1–10 ms	1 cycle	Nearly global	UTC
LF radio					
LORAN C	5×10^{-12}	1 μs (ground) 50 μs (sky)	30–50 ms	Most of northern hemisphere	UTC
HF/MF radio					
WWV and WWVH	1×10^{-6}	1000 μs	Code:year Voice:1 day Tick:1 s	Hemisphere	NIST master clock
TV (VHF/UHF radio)					
TV Line-10	1×10^{-11}	1 μs	N.A.	Network coverage	None
Color subcarrier	1×10^{-11}	N.A.	N.A.		
Satellite (SHF radio)					
TRANSIT	3×10^{-10}	30 μs	15 min	Nearly global	UTC
Global Positioning System	5×10^{-12}	100 ns	N.A.	Global	UTC

unique frequency for the station. During each transmission interval only three stations are radiating, each at a different frequency. The duration of each transmission varies from 0.9 to 1.2s, depending on the station's assigned location within the signal pattern. With a silent interval of 0.2s between each transmission, the entire cycle of the signal pattern repeats every 10 s. This signal format allows each station to be identified by its transmission of a particular frequency at a prescribed time.

Each OMEGA transmitting station employs four cesium-beam frequency standards for reliable timing and interstation synchronization. All eight stations are synchronized to each other to about 2 or 3 μs by means of interstation measurements made twice daily. OMEGA clocks are referenced to UTC to the extent that OMEGA system time "tracks" UTC but is not adjusted for yearly leap seconds. Time transfer via OMEGA can provide accuracies of 1 to 10 ms by using envelope detection. Frequency comparison can be made to 1×10^{-11} per day. Daily phase values referenced to UTC standards for OMEGA station transmissions are published weekly by the USNO.

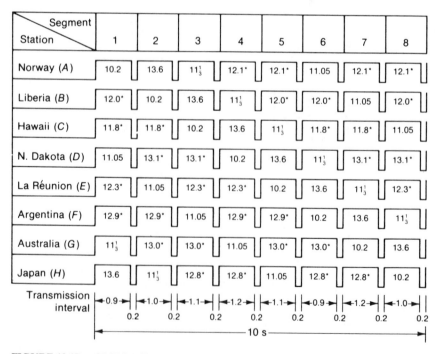

Station \ Segment	1	2	3	4	5	6	7	8
Norway (A)	10.2	13.6	$11\frac{1}{3}$	12.1*	12.1*	11.05	12.1*	12.1*
Liberia (B)	12.0*	10.2	13.6	$11\frac{1}{3}$	12.0*	12.0*	11.05	12.0*
Hawaii (C)	11.8*	11.8*	10.2	13.6	$11\frac{1}{3}$	11.8*	11.8*	11.05
N. Dakota (D)	11.05	13.1*	13.1*	10.2	13.6	$11\frac{1}{3}$	13.1*	13.1*
La Réunion (E)	12.3*	11.05	12.3*	12.3*	10.2	13.6	$11\frac{1}{3}$	12.3*
Argentina (F)	12.9*	12.9*	11.05	12.9*	12.9*	10.2	13.6	$11\frac{1}{3}$
Australia (G)	$11\frac{1}{3}$	13.0*	13.0*	11.05	13.0*	13.0*	10.2	13.6
Japan (H)	13.6	$11\frac{1}{3}$	12.8*	12.8*	11.05	12.8*	12.8*	10.2
Transmission interval	←0.9→	←1.0→	←1.1→	←1.2→	←1.1→	←0.9→	←1.2→	←1.0→
	0.2	0.2	0.2	0.2	0.2	0.2	0.2	0.2

|←——————————— 10 s ———————————→|

FIGURE 10.17   OMEGA Signal Format (frequencies in kHz)

LORAN C

LORAN C (long-range navigation) is a low-frequency radio navigational system operated by the U.S. Coast Guard [40, 41, 42]. Although primarily used for navigation, LORAN C transmission may also be used for time dissemination, frequency reference, and communication networks. LORAN C transmitting stations are grouped into chains of at least three stations. One transmitting station is designated master while the others are called secondaries. Each transmitting station derives its frequency from three cesium standards. Chain coverage is determined by the stations' transmitter power, their orientation and receiver sensitivity, and the distance between stations.

Low-frequency (LF) transmission produces both a groundwave propagation mode, which is extremely stable, and a skywave mode that results from ionospheric reflection and exhibits strong diurnal variation. Because of propagational delays in receiving the skywave signal, interference with the groundwave signal could occur without proper choice of LORAN C signal format. Using groundwave propagation, LORAN C can provide about 1 µs time transfer accuracy and about $\pm 5 \times 10^{-12}$ frequency stability control over a 24-hr period. Groundwave coverage with reliable performance extends to about 1500 mi for presently used transmitter power. Because of ionospheric perturbations, skywave propagation yields poorer time transfer than the groundwave mode and typically provides ±50 µs accuracy. At distances greater than 1500 mi and up to 5000 mi, however, the skywave signal is usually stronger than the groundwave. Figure 10.18 shows groundwave and skywave coverage for the present LORAN C chains.

Signal Characteristics. The LORAN C carrier frequency is 100 kHz with a 20-kHz bandwidth. Phase and coding modulation is used to minimize the effects of skywave interference and to allow identification of a particular station being received. The 100-kHz LORAN C pulse, shown in Figure 10.19, has a fast rise time that achieves high power prior to the arrival of skywaves. The shape of the pulse also allows the receiver to identify one particular cycle of the 100 kHz carrier. In addition to transmitting pulses for separation of groundwave and skywave signals, the pulses are transmitted in groups for station identification. The secondary stations in a LORAN C chain transmit eight pulses spaced 1 ms apart; the master station adds a ninth pulse, 2 ms after the eighth, for identification. To reduce skywave interference, binary phase shifting of the RF carrier of individual pulses in a group is used, according to an alternating code for the master and secondary stations. Figure 10.19 illustrates this signal format for a chain consisting of a master and two secondary stations labeled *X* and *Y.*

Since all LORAN C stations broadcast at 100 kHz, it is possible to receive many different LORAN chains at the same time. Therefore the pulses in a chain are transmitted at slightly different rates, called the *group*

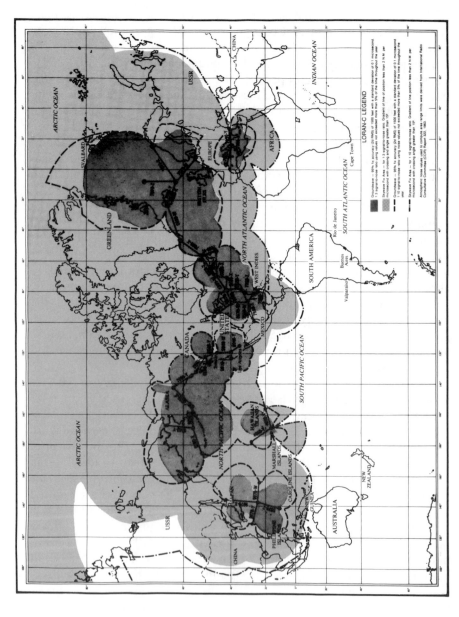

FIGURE 10.18 LORAN C Coverage Diagram (Courtesy U.S. Coast Guard)

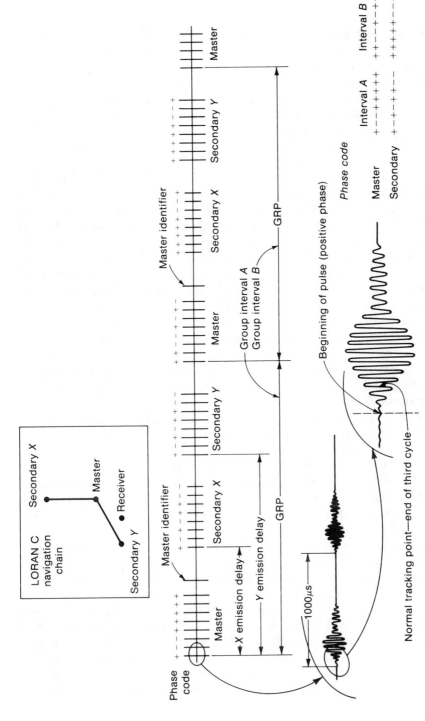

FIGURE 10.19 LORAN C Signal Format and Waveforms (Reprinted by permission of Austron, Inc.)

691

repetition interval (GRI) or *group repetition period* (GRP), so the receiver can distinguish between them. For each chain, a GRI is specified of sufficient length so that it contains time for transmission of the pulse group from each station plus a time delay between each pulse group. Thus the signals from two or more stations cannot overlap in time anywhere in the coverage area. The minimum GRI is then a function of the number of stations and the distance between them.

Time Determination.   The LORAN C system does not broadcast a time code signal. With knowledge of the time of coincidence (TOC) of a particular GRP relative to a second, however, a timing signal can be extracted from a LORAN C signal. This procedure is illustrated in Figure 10.20 for the Gulf of Alaska LORAN C chain, which has a GRP of 79,600 μs. Since this GRP along with most others is not a submultiple of 1 s, there is only periodic coincidence between the group interval and a second. At the TOC interval shown in Figure 10.20, which is 3 min, 19 s, this chain's master station transmits a pulse that is coincident with the UTC second. A LORAN C receiver can then synchronize its internal time signal to UTC when the TOC occurs. Using LORAN C emission and propagation delay figures and the receiver's internal delay, the local clock time can be corrected to be on time. It should be noted that external coarse clock synchronization to within one-half of a repetition period is required to resolve the ambiguity inherent in a pulsed system. For LORAN C the half-period interval varies from 30 to 50 ms. Resolution to this degree is obtainable from standard HF or VLF time transmissions.

With the cooperation of the U.S. Coast Guard, the U.S. Naval Observatory maintains synchronization of LORAN C transmissions to the observatory's master clock. Since the master clock is on UTC, time derived from LORAN C is also on this scale. Reference to the master clock and corrections to LORAN transmitting stations are made by a combination of portable clock and satellite techniques. Tables showing the discrepancy between each LORAN chain and UTC are published periodically by the Naval Observatory.

Frequency Determination.   If the LORAN C receiver is continuously phase-locked to the received signals, it can provide a phase-locked 100 kHz or, more commonly, a 1-MHz output frequency. Since the frequency of LORAN C chains is controlled by a cesium-beam oscillator, the receiver output frequency exhibits the identical long-term stability of the cesium oscillator and can therefore be used as an excellent frequency reference. The short-term stability is determined by the choice of local oscillator (usually good-quality quartz crystal) and by receiver measurement fluctuations due to noise and interference. Since the frequency of each LORAN chain is traceable to UTC, LORAN C transmissions become a reliable frequency

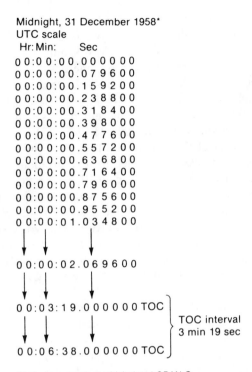

Midnight, 31 December 1958*
UTC scale
Hr: Min: Sec
0 0:0 0:0 0.0 0 0 0 0 0
0 0:0 0:0 0.0 7 9 6 0 0
0 0:0 0:0 0.1 5 9 2 0 0
0 0:0 0:0 0.2 3 8 8 0 0
0 0:0 0:0 0.3 1 8 4 0 0
0 0:0 0:0 0.3 9 8 0 0 0
0 0:0 0:0 0.4 7 7 6 0 0
0 0:0 0:0 0.5 5 7 2 0 0
0 0:0 0:0 0.6 3 6 8 0 0
0 0:0 0:0 0.7 1 6 4 0 0
0 0:0 0:0 0.7 9 6 0 0 0
0 0:0 0:0 0.8 7 5 6 0 0
0 0:0 0:0 0.9 5 5 2 0 0
0 0:0 0:0 1.0 3 4 8 0 0

0 0:0 0:0 2.0 6 9 6 0 0

0 0:0 3:1 9.0 0 0 0 0 0 TOC ⎫
                                  ⎬ TOC interval
                                  ⎭ 3 min 19 sec
0 0:0 6:3 8.0 0 0 0 0 0 TOC

*Reference epoch at which time LORAN C
master stations are assumed to have
transmitted their first pulse.

FIGURE 10.20   Time of Coincidence for Gulf of Alaska LORAN C Chain with GRP =
79,600 μs (Reprinted by permission of Austron, Inc.)

reference. The frequency of the radiated carrier at LORAN C stations is
kept within $\pm 2 \times 10^{-12}$ with respect to UTC.

10.5.2   HF Radio Broadcasts
The long-range capability of high-frequency (HF) propagation makes HF
one of the most commonly used sources of time and frequency dissemina-
tion [43, 44]. Western hemisphere HF stations such as WWV from Ft.
Collins, Colorado, WWVH from Kauai, Hawaii, and CHU from Ottawa,
Canada, provide essentially worldwide coverage.

WWV and WWVH stations broadcast on carrier frequencies of 2.5, 5, 10,
and 15 MHz. WWV also broadcasts on 20 MHz. These two stations derive
carrier frequencies, audio tones, and time of day from three cesium-beam
frequency standards. The station references are controlled to within 1 part

in 10^{12} of the NIST frequency standard in Boulder, Colorado. Time at these stations is controlled to within a few microseconds of NIST UTC.

Time and frequency are encoded on WWV and WWVH transmissions by means of UTC voice announcements, time ticks, and gated tones. The beginning of each hour is identified by an 800-ms tone burst at 1500 Hz. The beginning of each minute is identified by an 800-ms tone burst at 1000 Hz for WWV and 1200 Hz for WWVH. The remaining seconds in a minute are encoded by 5-ms audio ticks, at 1000 Hz for WWV and 1200 Hz for WWVH, which occur at the beginning of each second. The twenty-ninth and fifty-ninth second ticks are omitted to serve as identifiers for the half- and full-minute ticks. Each of the second ticks is preceded by 10 ms of silence and followed by 25 ms of silence to make the ticks for seconds more discernible. The tones or ticks also preempt voice announcements, but this shortcoming causes only small audio distortion in the voice announcements. Voiced time-of-day announcements are broadcast once a minute.

For time calibration the voice announcements can be used if accuracy requirements are low (within 1 s). For higher accuracy, a user must measure the seconds' ticks or decode the WWV/WWVH time code. The time code indicates UTC by day of the year, hour, minute, and second, and it is broadcast once per second in binary coded decimal (BCD) format on a 100-Hz subcarrier. Use of this time code or seconds' ticks with appropriate electronics permits time accuracy to 1 ms. Frequency comparison with WWV or WWVH can usually be accomplished to about 1 part in 10^6 by a variety of methods, including time comparisons to compute the frequency offset according to the relationship*

$$\frac{\Delta f}{f} = -\frac{\Delta t}{T}$$

$$(10.18)$$

10.5.3 Television Transmission

The major television networks in the United States use atomic oscillators (cesium or rubidium) to generate their reference signals [45, 46]. A communication station can calibrate a local precise oscillator by measuring the difference between the local oscillator and the received television signal. Accuracies obtainable by TV signal comparison range from 1 part in 10^9 to a few parts in 10^{11}. The two basic methods of calibrating a frequency source using TV transmission are color burst comparisons and TV Line 10 comparison. The TV Line 10 approach is also used for time transfer and time comparison.

---

*Easily derived by elementary calculus from the definition of frequency as $f = 1/t$.

TV Color Subcarrier Comparison

The major network color subcarrier frequency is approximately 3.58 MHz, derived from network atomic frequency standards. The most straightforward method for use of the color subcarrier frequency in calibration of a local oscillator is to compare, via a phase detector, a locally generated 3.58 MHz signal with a received sample of 3.58 MHz from a color TV set. The dc output of the phase detector, which represents differential frequency for large offsets and differential phase for small offsets, is amplified, plotted, and calibrated in phase change per unit time. Since relative frequency is related to the relative phase (or time) by Equation (10.18), we can derive the frequency error from the phase error. In most cases of TV reception, resolution is limited to about 10 ns in 15 min, which corresponds in frequency resolution to

$$\frac{\Delta f}{f} = \frac{\Delta t}{T} = \frac{10 \text{ ns}}{15 \text{ min}} = 1.1 \times 10^{-11} \qquad (10.19)$$

Some knowledge of network scheduling is required in order to make use of the 3.58-MHz subcarrier for frequency calibration. During station breaks, the color signal often originates from the local station, which uses a lower accuracy source. Tape delay broadcasts also suffer from degraded subcarrier frequency and are therefore invalid as a precision reference.

TV Line 10

In the United States a television picture frame is made up of 525 lines that are scanned or traced to produce an image on the screen. First the odd lines are traced; then the even lines are traced in between the odd. Line 10 is one of the odd lines that make up each picture; it was chosen because it is easy to pick out from the rest of the lines with simple circuitry. To calibrate a local frequency source, a clock pulse from the source is compared to the trailing edge of the Line 10 synchronizing pulse at a certain time each day. The Line 10 technique can be used to determine the frequency stability of a local oscillator compared to another driven by Line 10 measurements. Likewise, frequency or time source calibration between two stations can be accomplished by simultaneously recording the times of arrival of the Line 10 pulse output and comparing them to local clock time. Frequency accuracy to 1 part in 10^{11} over a period of several years and time accuracy to several microseconds have been reported for Line 10.

10.5.4 Satellite Transmission

Time and frequency dissemination via satellite facilitates global coverage while avoiding the ionospheric effects that beset terrestrial systems. Three different modes of satellite time and frequency transfer have been developed and come into practice [47]. One method uses geostationary satellites to broadcast time signals, referenced to the NIST in the United States [48]. A second method uses communications satellites to transfer time [55]. The third method, which is based on navigation satellites carrying time standards, is commonly used for synchronization of communication networks. The following sections describe two navigation satellite systems, TRANSIT and Global Positioning System, as they are used for network timing and synchronization.

TRANSIT

TRANSIT is a satellite-based navigational system operated by the U.S. Navy and consisting of five satellites and associated ground stations [49, 50]. The satellites are in nearly circular, polar orbits at an altitude of about 1100 km; thus any user on the earth will see a TRANSIT satellite about every 90 min. TRANSIT provides worldwide coverage and is, in fact, the first time and frequency dissemination system to work anywhere in the world.

The satellites broadcast ephemeris data continuously on approximately 150 MHz and 400 MHz carriers. Navigational information is broadcast in 2 min intervals, including a time mark every even minute. Each satellite derives its clock time and carrier frequencies from a temperature-stabilized, crystal-controlled 5-MHz oscillator. Satellite clock time is monitored from earth-control stations, and clock corrections are periodically sent to the satellites.

TRANSIT allows a user to recover time signals that can be used as a clock or to steer an oscillator's frequency. Time to better than 30 μs and frequency accuracy to a few parts in 10^{10} per day can be achieved. The instrumentation needed by a user in general consists of a TRANSIT receiver, a frequency source, and a clock. Moreover, it is necessary to obtain the time of day from another source to within 15 min in order to resolve the time ambiguity of TRANSIT.

Global Positioning System

The Global Positioning System (GPS) is a satellite navigation system under development for the U.S. military that will provide highly accurate position and velocity information in three dimensions and precise time and time interval on a global basis continuously [51 to 54]. When fully deployed, the space segment will consist of 21 satellites in six subsynchronous planes, providing continuous global coverage with at least four satellites in view at a given location. The GPS concept is predicated on accurate and continuous knowledge of the spatial position of each satellite in the system. Each satellite will continuously broadcast data on its position. This information will be periodically updated by the master control station based upon data obtained from monitoring earth stations. The USNO will also monitor GPS and provide PTTI data to the GPS control station.

Data are transmitted from each satellite on two carrier frequencies designated $L1$ (1575.42 MHz) and $L2$ (1227.6 MHz), using an onboard atomic clock as a frequency reference. To allow arrival times to be measured with great precision, spread spectrum is used, such that the 50-b/s navigation message is transmitted in about 20-MHz bandwidth. The $L1$ carrier contains a coarse acquisition (C/A) code in addition to a precise (P) code, and the $L2$ carrier broadcasts only the P code. The C/A code provides lower accuracy but is easier to access, while the P code is intended for high-accuracy navigation and time transfer.

For time and frequency users, the receiver configuration is as shown in Figure 10.21. The time transfer unit (TTU) accepts stable reference time (1 pps) and frequency (5 MHz) signals from a local source and the $L1$ and $L2$ signals from any GPS satellite. Ephemeris data from the satellite are detected and processed to determine satellite position and to estimate time of arrival of the satellite signal epoch. The actual time of arrival is recorded

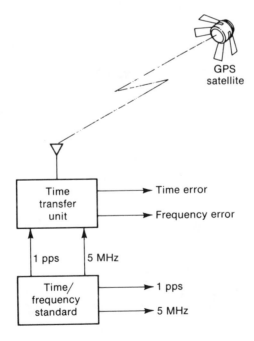

FIGURE 10.21   GPS Time Transfer Unit

and compared with the expected value after correction for atmospheric and relativistic errors. The difference between the calculated and actual time of arrival can be displayed as the local time error or used to correct the local time or frequency standard [52].

The GPS time-measuring concept is shown in Figure 10.22. The difference between GPS time and onboard satellite time is designated time A. The master control system knows A, and this is transmitted to the satellite via uplink telemetry for satellite broadcast. Time A is decoded by the TTU and stored. Time B, the difference between GPS time and the user station time, is an unknown parameter. Time C, the difference between the user station time and the time received from GPS, is measured by the TTU. The satellite-to-user transmit time (T_T) can be estimated from satellite ephemeris and atmospheric propagation correction data. The time error between station clock and the GPS clock (B) is then equal to

$$B = T_T - (A + C) \tag{10.20}$$

Accuracy of time transfer to better than 100 ns and accuracy of frequency determination better than 5 parts in 10^{12} have been reported [53].

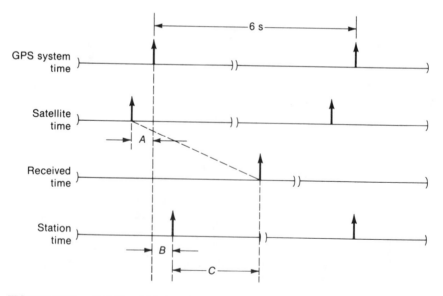

FIGURE 10.22   GPS Time Relationships

10.6   Examples of Network Synchronization Schemes

National networks employ various synchronization systems that are based on master-slave, mutual, or external reference. Plesiochronous synchronization is used to interconnect national systems that make up an international connection, or to interconnect dissimilar national systems within a national connection.

10.6.1   Synchronization of National Networks

CCITT Rec. G.811 prescribes timing requirements for the interconnection of national networks [56]. While the CCITT recognizes that national networks may be synchronized by a variety of internal arrangements, international connections are to be operated plesiochronously. Table 10.5 lists the salient performance characteristics of CCITT Rec. G.811 for plesiochronous operation of international digital links in terms of allowed slip rate, clock inaccuracy, and maximum time interval error (MTIE). These characteristics are based on ideal undisturbed (that is, theoretical) conditions, although in practice performance may be degraded. For example, CCITT Rec. G.822 permits five slips per 24 hr on a 64-kb/s international connection versus one slip in 70 days specified by Rec. G.811; this reduction in the slip rate objective is due to the presence of several plesiochronous interfaces in a typical international connection and to various design, environmental, and operational conditions in international and national sections. The restrictions on time interval error, shown in Table 10.5 and Figure 10.23, are imposed to facilitate the evolution from plesiochronous to synchronous operation within the international network.

TABLE 10.5   CCITT Rec. G.811: Requirements of Primary Reference Clocks Used for Plesiochronous Operation of Digital Links [56]

Occurrence of Slips
    Not greater than one controlled slip in every 70 days per digital international link for any 64-kb/s channel
Frequency Inaccuracy
    Long-term frequency inaccuracy not greater than 1 part in 10^{11}
Maximum Time Interval Error (MTIE)*

Measurement Interval S (s)	Allowed MTIE (ns)
$0.05 < S \le 5$	$(100S)$ ns
$5 < S \le 500$	$(5S + 500)$ ns
$S > 500$	$(10^{-2}S + 3000)$ ns

*See Figure 10.23 for plots of MTIE

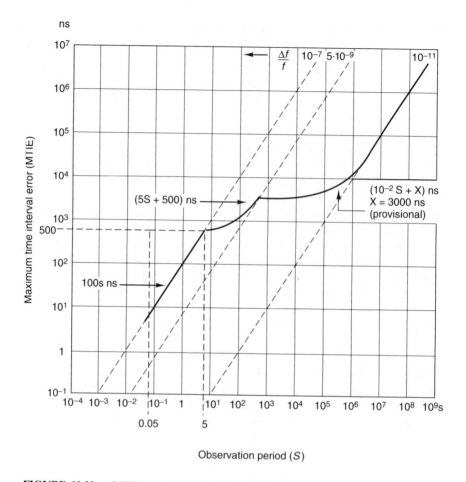

FIGURE 10.23   CCITT Rec. G.811 for Permissible Maximum Time Interval Error vs. Observation Time for a Primary Reference Clock [56] (Courtesy CCITT)

The CCITT also prescribes maximum allowed output jitter and wander for network interfaces. These network limits are found in CCITT recommendations G.823 [57] and G.824 [58] for the European and North American digital hierarchies, respectively; values for other rates are under study. (Note that other CCITT recommendations in the G.700 series apply to jitter and wander at equipment interfaces.) Specification and control of maximum output jitter and wander are necessary for the proper interconnection of network elements to form an end-to-end connection. These limits apply regardless of the number or type of transmission components that make up the connection. The CCITT recommendations for output jitter take the form of Figure 2.20, with a mask of maximum allowed jitter amplitude versus frequency. Since wander is usually characterized by the time error it causes, the CCITT uses time interval error as the wander specification. The MTIE over a period of S seconds is specified to not exceed $(10^{-2}S + 10000)$ ns for values of S greater than 10^4. For values of $S < 10^4$, the MTIE is under study by the CCITT.

10.6.2   Network Synchronization
in the United States

A master-slave hierarchical synchronization method, based on four stratum levels of clocks, is used throughout the public networks in the United States. All clocks must meet industry standards, as given in Table 10.1, and be traceable to a Primary Reference Source [3]. The highest-level clock, stratum 1, acts as the Primary Reference Source for the rest of the network. Stratum 2 clocks, located at toll-switching centers, transmit synchronization to other stratum 2 nodes or to stratum 3 nodes. Stratum 3 clocks are located in a digital cross-connect system or digital end office. Similarly, stratum 3 clocks provide synchronization to other stratum 3 locations or to stratum 4 nodes. Stratum 4 clocks are used in most T1 multiplexers, digital PBXs, and all digital channel banks. Any particular node must receive its synchronization reference signal from another node at the same or higher stratum level. As shown in Figure 10.24, the flow of the synchronization reference is mostly downward, with some horizontal branches but no upward flow being permitted. The actual frequency being transmitted downward is prescribed to be either 2.048 or 1.544 MHz. Existing transmission facilities are used for transmission of these frequency references, without the need for overhead or separate facilities. To ensure high reliability both a primary and secondary transmission facility are often provided between stratum clocks. Transmission facilities are selected to ensure the highest possible availability, to provide diverse routing between primary and secondary facilities, and to minimize the number of nodes in series from the stratum 1 source.

Performance Objectives

The key performance parameters specified for the United States network synchronization scheme are slip rate, maximum time interval error (MTIE), and jitter and wander. The slip rate objective is one slip or less in 5 hr over an end-to-end connection. This allocation is divided in half between transmission facilities and clocks, with both allowed one slip in 10 hr. In trouble-free operation, however, this hierarchical scheme allows zero-slip operation down to the lowest level of the network. In the most extreme trouble condition, that is, when all frequency references have been lost, the slip rate is maintained at or below 255 slips during the first 24 hr.

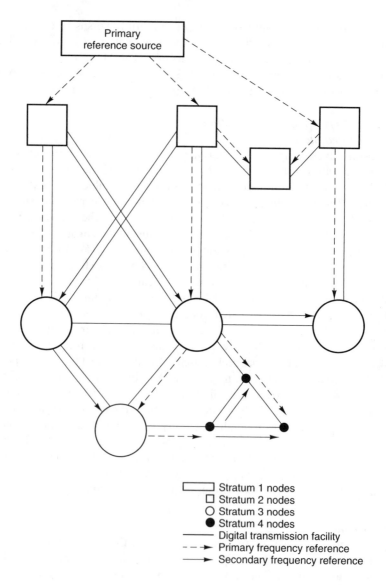

FIGURE 10.24   Hierarchical Network Synchronization as Used in the United States

The use of MTIE allows a practical method of determining slip intervals that will occur due to clock inaccuracies. The permissible MTIE is specified for the Primary Frequency Source and for 2.048- and 1.544-MHz references. For example, in the case of a Primary Frequency Source, the MTIE over an observation period S seconds is prescribed to be $(10^{-2}S + 3000)$ ns for $S > 500$ seconds. For periods less than 500 seconds, MTIE is still under study by the ANSI T1 Committee [3]. Other jitter and wander specifications are likewise under study, but there have been proposed standards for private networks that are similar to the CCITT specifications shown in Figure 2.20 [59].

Stratum Clocks
At the highest level in the hierarchy is the Primary Reference Source (PFS), which is specified to meet stratum 1 specifications and be traceable to UTC. AT&T uses GPS [2], MCI uses LORAN C [3], and Sprint uses a combination of GPS and LORAN C. All three carriers reference their PFSs to UTC and use several PRSs in their networks. For clocks below stratum 1, performance is specified by Table 10.1 and operation is characterized by a strategy of either holdover or switchover to another reference. Each clock is configured so that it will not switch from its reference because of reference frequency deviation unless the pull-in range is exceeded. The hold-in range of each stratum clock is at least as great as its pull-in range, ensuring that switchover occurs prior to loss of synchronization. The particular strategy of holdover versus switchover is different for each stratum. For a stratum 2 clock, loss of the primary reference causes the clock to enter its holdover mode, with manual switching to a secondary reference as a backup. Conversely, stratum 3 clocks are always provided with primary and secondary references and are configured to automatically switch to the backup upon loss of the on-line reference. Stratum 4 clocks have no holdover requirement but may have multiple references, in which case automatic switchover is preferred.

10.7 Summary

The development and widespread use of digital transmission, in combination with digital switching, has led to all-digital networks that require timing and synchronization. Sources of precise time and frequency begin with time standards, both atomic and astronomical. Today, Coordinated Universal Time (UTC) is a universally accepted time scale that is used by communication networks as a source of precise time.

Frequency sources used in communication applications include quartz crystal oscillators for medium-quality requirements and atomic oscillators, such as cesium and rubidium, for high-quality requirements. These frequency sources are characterized by their accuracy, or degree to which the frequency agrees with its specified value, and by their stability, or the rate at which the frequency changes in time. Quartz oscillators are inexpensive and have good short-term stability but exhibit long-term instability. Cesium and rubidium standards both offer the short-term stability of the quartz oscillator but provide greater long-term stability that is important to communication applications.

Network synchronization schemes are divided into two categories: non-disciplined, wherein all nodal clocks are independent of a reference, and disciplined, wherein each nodal clock is synchronized to a reference. Both pulse stuffing (see Chapter 4) and plesiochronous operation are non-disciplined forms. Both schemes suffer from occasional losses of bit count integrity and a requirement for buffering of all signals. Of the disciplined forms of network synchronization, the master-slave system is the simplest and most popular. For this system a master frequency is distributed to all slave nodes via transmission paths. Secondary frequency sources and alternative transmission paths provide backup to the normally available master. Another disciplined scheme, mutual synchronization, uses a weighted average of clock signals derived from incoming transmission paths to obtain each nodal clock frequency. Although widely studied, this scheme has been implemented in very few applications. Another form of disciplined synchronization is available through external reference to one of several time and frequency dissemination systems.

Time and frequency dissemination systems useful in network synchronization are based on radio broadcasting via both terrestrial and satellite media. OMEGA uses eight VLF radio stations to broadcast a sequence of frequencies derived from cesium clocks and referenced to UTC. LORAN C is based on worldwide chains of stations that broadcast codes on a 100-kHz carrier, also derived from cesium clocks and referenced to UTC. Although both OMEGA and LORAN C are navigational systems (operated by the U.S. Coast Guard), both provide nearly global dissemination of time and frequency information. High-frequency radio broadcasts of time and frequency, encoded by means of tones and time ticks, also provide nearly global coverage. WWV (Colorado) and WWVH (Hawaii) HF stations provide full U.S. coverage. The major TV networks in the United States use atomic standards in broadcasting such that certain TV signals (color subcarrier and Line 10) can be used as a source of frequency calibration. Several satellite systems are currently used for dissemination of time and frequency, including the TRANSIT navigational system and the Global Positioning System.

Network synchronization schemes for national systems are based principally on the master-slave concept. For international digital connections, the CCITT has prescribed the use of plesiochronous operation, thus allowing national systems to interoperate using separate network synchronization schemes.

References

1. CCITT Rec. G.822, "Controlled Slip Rate Objectives on an International Digital Connection," Vol. III.5, *Digital Networks, Digital Sections and Digital Line Systems* (Geneva: ITU, 1989).
2. J. E. Abate and others, "AT&T's New Approach to the Synchronization of Telecommunication Networks," *IEEE Comm. Mag.*, April 1989, pp 35–45.
3. ANSI-T1.101-1987, "Synchronization Interface Standards for Digital Networks," December 1986.
4. 13th Annual Conference on Weights and Measures, Paris, 1968.
5. G. Kamas and S. L. Howe, eds., "Time and Frequency Users' Manual," NBS Special Publication 559 (Washington, D.C.: U.S. Government Printing Office, 1979).
6. D. W. Allan, J. E. Gray, and H. E. Machlan, "The National Bureau of Standards Atomic Time Scales: Generation, Dissemination, Stability, and Accuracy," *IEEE Trans. Instr. and Meas.*, vol. IM-21, no. 4, November 1972, pp. 388–391.
7. B. Guinot and M. Granveaud, "Atomic Time Scales," *IEEE Trans. Instr. and Meas.*, vol. IM-21, no. 4, November 1972, pp. 396–400.
8. J. T. Henderson, "The Foundation of Time and Frequency in Various Countries," *Proc. IEEE* 60(May 1972):487–493.
9. J. A. Barnes and G. M. R. Winkler, "The Standards of Time and Frequency in the USA," *Proc. 26th Ann. Symp. on Frequency Control* (Ft. Monmouth, NJ: Electronics Industries Assn., 1972).
10. H. M. Smith, "International Time and Frequency Coordination," *Proc. IEEE* 60(May 1972):479–487.
11. H. Hellwig, "Frequency Standards and Clocks: A Tutorial Introduction," NBS Technical Note 616 (Washington, D.C.: U.S. Government Printing Office, 1972).
12. J. A. Barnes and others, "Characterization of Frequency Stability," *IEEE Trans. Instr. and Meas.*, vol. IM-20, no. 2, May 1971, pp. 105–120.
13. J. A. Barnes, "Atomic Timekeeping and the Statistics of Precision Signal Generators," *Proc. IEEE* 54(February 1966):207–220.
14. J. R. Vig, "Introduction to Quartz Frequency Standards," *Proc. 23rd Annual PTTI Applications and Planning Meeting*, December 1991.
15. "Fundamentals of Quartz Oscillators," Hewlett-Packard Application Note 200-2.
16. W. G. Cady, *Piezoelectricity: An Introduction to the Theory and Applications of Electromechanical Phenomena in Crystals* (New York: Dover, 1964).
17. H. Fruehauf, "Atomic Oscillators Stand Test of Time," *Microwaves and RF* 23 (June 1984):95–124.
18. L. S. Cutler, "The Status of Cesium Beam Frequency Standards," *Proc. 22nd Annual PTTI Applications and Planning Meeting*, December 1990, pp. 19–27.
19. W. J. Riley, "The Physics of the Environmental Sensitivity of Rubidium Gas Cell Atomic Frequency Standards," *Proc. 22nd Annual PTTI Applications and Planning Meeting*, December 1990, pp. 441–452.
20. M. C. Thompson, Jr., and H. B. Jones, "Radio Path Length Stability of Ground-to-Ground Microwave Links," NBS Technical Note 219 (Washington, D.C.: U.S. Government Printing Office, 1964).

40. L. D. Shapiro, "Time Synchronization from LORAN-C," *IEEE Spectrum* 5(August 1968):46–55.

41. C. E. Botts and B. Wiedner, "Precise Time and Frequency Dissemination via the LORAN-C System," *Proc. IEEE* 60(May 1972):530–539.

42. G. R. Westling, M. D. Sakahara, and C. J. Justice, "Synchronizing LORAN-C Master Stations to Coordinated Universal Time," *Proc. 21st Annual PTTI Applications and Planning Meeting,* November 1989, pp. 491–499.

43. N. Hironaka and C. Trembath, "The Use of National Bureau of Standards High Frequency Broadcasts for Time and Frequency Calibrations," NBS Technical Note 668 (Washington, D.C.: U.S. Government Printing Office, 1975).

44. P. P. Viezbicke, "NBS Frequency-Time Broadcast Station WWV, Fort Collins, Colorado," NBS Technical Note 611 (Washington, D.C.: U.S. Government Printing Office, 1971).

45. J. Tolman, V. Ptacek, A. Soucek, and R. Stecher, "Microsecond Clock Comparison by Means of TV Synchronizing Pulses," *IEEE Trans. on Instr. and Meas.,* vol. IM-16, September 1967, pp. 247–254.

46. D. D. Davis, J. L. Jespersen, and G. Kamas, "The Use of Television Signals for Time and Frequency Dissemination," *Proc. IEEE* 58(June 1970):931–933.

47. R. L. Easton and others, "Dissemination of Time and Frequency by Satellite," *Proc. IEEE* 64(October 1976):1482–1493.

48. D. W. Hanson and others, "Time from NBS by Satellite (GOES)," *Proc. 8th Annual PTTI Applications and Planning Meeting,* November 1976.

49. R. E. Bechler and others, "Time Recovery Measurements Using Operational GOES and TRANSIT Satellites," *Proc. 11th Annual PTTI Applications and Planning Meeting,* November 1979.

50. T. D. Finsod, "Transit Satellite System Timing Capabilities," *Proc. 10th Annual PTTI Applications and Planning Meeting,* November 1978.

51. K. D. McDonald, "GPS Status and Issues," *Proc. 23rd Annual PTTI Applications and Planning Meeting,* December 1991.

52. B. Guinot, W. Lewandowski and C. Thomas, "A Review of Recent Advances in GPS Time Comparisons," *Proc. 4th European Frequency and Time Forum,* 1990, pp. 307–312.

53. C. Fox, G. A. Gifford, and S. R. Stein, "GPS Time Determination and Dissemination," *Proc. 23rd Annual PTTI Applications and Planning Meeting,* December 1991.

54. W. J. Klepczynski and L. G. Charron, "The Civil GPS Service," *Proc. 20th Annual PTTI Applications and Planning Meeting,* December 1988, pp. 51–64.

55. L. B. Veenstra, "International Two-Way Satellite Time Transfers Using INTEL-SAT Space Segment and Small Earth Stations," *Proc. 22nd Annual PTTI Applications and Planning Meeting,* December 1990, pp. 383–399.

56. CCITT Recommendation G.811, "Timing Requirements at the Outputs of Primary Reference Clocks Suitable for Plesiochronous Operation of International Digital Links," Vol. III.5, *Digital Networks, Digital Sections and Digital Line Systems* (Geneva: ITU, 1989).

57. CCITT Recommendation G.823, "The Control of Jitter and Wander Within Digital Networks Which Are Based on the 2048 kbit/s Hierarchy," Vol. III.5,

21. Peter Alexander and J. W. Graham, "Time Transfer Experiments for DCS Digital Network Timing," *Proc. 9th Annual PTTI Applications and Planning Meeting*, March 1978, pp. 503–522.

22. P. A. Bello, "A Troposcatter Channel Model," *IEEE Trans. on Comm. Tech.*, vol. COM-17, no. 2, April 1969, pp. 130–137.

23. J. J. Spilker, Jr., *Digital Communications by Satellite* (Englewood Cliffs, NJ: Prentice Hall, 1977).

24. M. Kihara, "Performance Aspects of Reference Clock Distribution for Evolving Digital Networks," *IEEE Comm. Mag.*, April 1989, pp. 24–34.

25. N. Inoue, H. Fukinuki, T. Egawa, and N. Kuroyanagi, "Synchronization of the NTT Digital Network," *ICC 1976 Conference Record*, June 1976, pp. 25.10–25.15.

26. G. P. Darwin and R. C. Prim, "Synchronization in a System of Interconnected Units," U.S. Patent 2,986,732, May 1961.

27. J. G. Baart, S. Harting, and P. K. Verma, "Network Synchronization and Alarm Remoting in the Dataroute," *IEEE Trans. on Comm.*, vol. COM-22, no. 11, November 1974, pp. 1873–1877.

28. B. R. Saltzberg and H. M. Zydney, "Digital Data Network Synchronization," *Bell System Technical Journal* 54(May/June 1975):879–892.

29. F. T. Chen, H. Goto, and O. G. Gabbard, "Timing Synchronization of the DATRAN Digital Data Network," *National Telecommunications Conference Record*, December 1975, pp. 15.12–15.16.

30. R. G. DeWitt, "Network Synchronization Plan for the Western Union All-Digital Network," *Telecommunications* 8(July 1973):25–28.

31. M. B. Brilliant, "The Determination of Frequency in Systems of Mutually Synchronized Oscillators," *Bell System Technical Journal* 45(December 1966):1737–1748.

32. A. Gersho and B. J. Karafin, "Mutual Synchronization of Geographically Separated Oscillators," *Bell System Technical Journal* 45(December 1966):1689–1704.

33. M. B. Brilliant, "Dynamic Response of Systems of Mutually Synchronized Oscillators," *Bell System Technical Journal* 46(February 1967):319–356.

34. M. W. Williard, "Analysis of a System of Mutually Synchronized Oscillators," *IEEE Trans. on Comm.*, vol. COM-18, no. 5, October 1970, pp. 467–483.

35. M. W. Williard and H. R. Dean, "Dynamic Behavior of a System of Mutually Synchronized Oscillators," *IEEE Trans. on Comm.*, vol. COM-19, no. 4, August 1971, pp. 373–395.

36. J. Yamato, S. Nakajima, and K. Saito, "Dynamic Behavior of a Synchronization Control System for an Integrated Telephone Network," *IEEE Trans. on Comm.*, vol. COM-22, no. 6, June 1974, pp. 839–845.

37. J. Yamato, "Stability of a Synchronization Control System for an Integrated Telephone Network," *IEEE Trans. on Comm.*, vol. COM-22, no. 11, November 1974, pp. 1848–1853.

38. U.S. Coast Guard, U.S. Department of Transportation, "OMEGA Global Radionavigation—A Guide for Users," Washington, D.C., November 1983.

39. E. R. Swanson, "Omega," *Proc. IEEE* 71(10)(October 1983):1140–1155.

Digital Networks, Digital Sections and Digital Line Systems (Geneva: ITU, 1989).

58. CCITT Recommendation G.824, "The Control of Jitter and Wander Within Digital Networks Which Are Based on the 1544 kbit/s Hierarchy," Vol. III.5, *Digital Networks, Digital Sections and Digital Line Systems* (Geneva: ITU, 1989).

59. Draft ANSI PN-2198, "Private Digital Network Synchronization," July 1990.

Problems

10.1 Suppose a stratum 2 clock loses its reference.

(a) What would be the initial maximum slip rate?

(b) How long would it take for this clock to reach its free-running accuracy?

10.2 Suppose that two stratum 1 clocks are being used to time the two ends of a digital connection operating at 1.544 Mb/s.

(a) What timing difference would accumulate in a single day if the two clocks had equal accuracies of 1×10^{-11} but with opposite direction?

(b) How many days would elapse before a DS1 frame slip would occur?

10.3 Consider a stratum 2 clock used to time a 1.544-Mb/s digital transmission system. Calculate the maximum slip rate possible after 48 hr.

10.4 During a 24-hr test of Time Interval Error (TIE), a net change of 15 ns is observed when a nominal 1.544-MHz clock is compared to an ideal reference. Calculate the fractional frequency difference in the nominal 1.544-MHz clock that would yield this TIE.

10.5 A certain satellite link has a path delay variation of ± 1 ms, and operates at a bit rate of 1.544 Mb/s. The data signal is timed with a clock having accuracy of 1×10^{-11} and long-term stability of 5×10^{-12}. If the buffer reset interval is to be 10 days, what size buffer is required?

11

Transmission System Testing, Monitoring, and Control

OBJECTIVES

- Discusses techniques for testing the performance of digital transmission systems in terms of bit error rate, jitter, bit count integrity, and equipment reliability
- Describes techniques for continuous monitoring of system and equipment performance
- Covers the techniques for isolating faults in system performance
- Considers various schemes for reporting alarms, monitors, and restoral actions to other stations in the system
- Describes digital cross-connect systems used to consolidate and automate certain functions such as provisioning, reconfiguration, monitoring, and testing

11.1 Introduction

During the process of installation, operation, and maintenance of a digital transmission system, the system operator must be able to test, monitor, and control the system on an end-to-end basis. This capability allows verification of design objectives as part of system installation. After system commission, components must be monitored and controlled to ensure that operation and maintenance (O&M) performance standards are met.

This chapter discusses techniques for testing digital transmission systems for performance-related parameters such as bit error rate, jitter, bit count integrity, and equipment reliability. These methods are suitable for factory or acceptance testing prior to system commission for operation. During system operation, maintenance personnel must be provided alarm and status indications that allow identification of failed equipment, its repair or replacement, and restoral of service. Fault isolation within a network requires performance monitors and alarm indications associated with each piece of equipment and each station. Performance monitors are described here that allow system and equipment margins to be continuously monitored, usually by estimation of the bit error rate. Fault isolation techniques identify the need for prompt maintenance when service has been interrupted or deferred maintenance when automatic switchover to standby units has been used to restore service or when standby units fail leaving no backup to the operational unit. Since maintenance actions taken at one station may affect other stations, it becomes necessary to report alarms, monitors, and restoral actions to other stations, often at a central location. In discussing various schemes used for reporting, the emphasis is on the unattended remote station where it may be necessary to remotely control redundant equipment.

It should be noted that the O&M performance standards discussed here differ from the design objectives discussed in Chapter 2. Various sources of degradation not always included in the design methodology—such as aging, acts of nature, and human error—result in performance limits during the lifetime of the operating system. Further, real circuits differ from the hypothetical reference circuit used in design methodology because of different length, media, equipment, and so forth. These differences must be taken into account when specifying and measuring O&M performance standards.

11.2   Testing Techniques

Performance specifications or standards are verified by testing, whether in the factory or in an operational environment. To facilitate the testing process, these specifications and standards should include descriptions of the tests to be used: test equipment, configurations, procedures, and the like. This section describes commonly used techniques for testing the key performance parameters of digital transmission systems including bit error rate, jitter, and bit count integrity. The testing of individual transmission equipment and links for other performance specifications, such as reliability and radio fade margin, is also discussed.

11.2.1   Bit Error Rate Measurement
The most commonly used performance indicator in the testing of digital transmission systems is bit error rate (BER). As discussed in Chapter 2, error performance can be expressed in many forms, such as errored sec-

onds (see the box on page 713), errored blocks, and average BER. Usually the error parameter used in measuring system performance is selected to match the error parameter used in the system design process to allocate performance. Whatever the error parameter, there are two general approaches to measurement: out-of-service and in-service. In the case of out-of-service measurement, operational traffic is replaced by a known test pattern. Here a pseudorandom binary sequence (PRBS) is used to simulate traffic. The received test pattern is compared bit by bit with a locally generated pattern to detect errors. The repetition period of the test pattern, given by $2^n - 1$ for a shift register of n bits, is selected to provide a sufficiently smooth spectrum for the system data rate. The most common patterns for standard data rates are shown in Table 11.1. Since out-of-service measurement eliminates traffic carrying capability, it is therefore best suited to production testing, installation testing, or experimental systems.

Figure 11.1 is a block diagram of a typical configuration for BER measurement. At the transmitter and receiver, the test pattern of the PRBS generator and detector is selected from available options. At the receiver, a form of error measurement is selected, which might include options for error pulses, average BER, errored seconds, severely errored seconds, degraded minutes, or errored blocks, depending on the design of the bit error rate tester (BERT). The measured error performance is given by visual display and may also be available in printed form. Usually the BER test set includes some diagnostics, such as the ability to force errors at the transmitter and observe them at the receiver and an indication of loss of synchronization at the receiver.

TABLE 11.1   Pseudorandom Binary Sequences Recommended by the CCITT for the Measurement of Error Rate

Applicable Bit Rates	Pattern Length	CCITT Recommendation [1, 2]
50 b/s-19.2 kb/s	2^9-1	O.153, V.52
64 kb/s	$2^{11}-1$	O.152
20 kb/s-168 kb/s	$2^{20}-1$	O.153, V.57
1.544 Mb/s	$2^{15}-1, 2^{20}-1$	O.151
2.048 Mb/s	$2^{15}-1$	O.151
6.312 Mb/s	$2^{15}-1, 2^{20}-1$	O.151
8.448 Mb/s	$2^{15}-1$	O.151
32.064 Mb/s	$2^{15}-1, 2^{20}-1$	O.151
34.368 Mb/s	$2^{23}-1$	O.151
44.736 Mb/s	$2^{15}-1, 2^{20}-1$	O.151
139.264 Mb/s	$2^{23}-1$	O.151

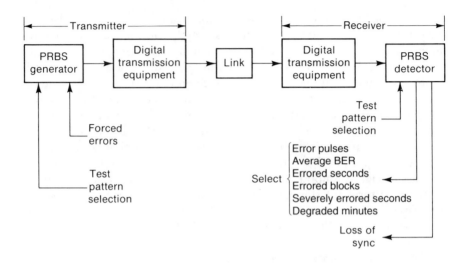

FIGURE 11.1   Block Diagram of BER Measurement

In-service error measurement is possible when the traffic has an inherent repetitive pattern, the line format has inherent error detection, or the received signal is monitored for certain threshold crossings. Thus in-service techniques only estimate the error rate and do not yield a true measurement. These techniques are useful as performance monitors during live system operation. Further, if the error rate estimate can be made quickly enough, in-service schemes can be used to control redundancy or diversity protection switching. Categories of in-service error rate measurement will be described later in this chapter when we consider performance monitoring techniques.

Measured bit error rates have little significance unless a confidence level is stated. For bit error rate measurements, the *confidence level* is defined as the probability that the measured error rate is within an accuracy factor α of the true average BER or, in terms of number of errors,

$$\text{Confidence level} = P(np \leq \alpha k_1) \qquad (11.1)$$

where $k_1 = $ number of measured errors
$p = $ probability of bit error
$n = $ number of trial bits
$np = $ unexpected number of errors

A mathematical model for the confidence level is well known for the case where errors are independently distributed. Here the Bernoulli trials model

---

Error-Free Second Measurements

Error-free seconds, a form of error measurement, is expressed as the percentage of error-free seconds using measurement periods of 1 s. There are two ways of measuring errored seconds: *synchronous* and *asynchronous*. In the synchronous mode an errored second is defined as a 1-s interval following the occurrence of the first error. The advantage of the synchronous mode is that measurements made with different instruments yield the same reading on the same link. Its disadvantage is that the measure of synchronous errored seconds does not directly yield error-free seconds but rather error-free time. In the asynchronous mode each discrete 1-s interval is checked for errors. The advantage of the asynchronous mode is that it directly yields error-free seconds. The disadvantage is that different measurements may be obtained with different equipment. CCITT Rec. G.821 implies the use of asynchronous errored seconds; the emerging industry standard in North America, however, is based on synchronous errored seconds.

---

can be used to yield the probability that αk_1 or fewer errors occur in n independent trials:

$$P(\leq \alpha k_1 \text{ errors in } n \text{ bits}) = \sum_{k=0}^{\alpha k_1} \binom{n}{k} p^k q^{n-k} \qquad (11.2)$$

where $q = 1 - p$. Expression (11.2) is difficult to evaluate for large n, but approximations exist for two common cases. If n is so large that the expected number of errors is also large ($np \gg 1$), the expression may be approximated by

$$P(\leq \alpha k_1 \text{ errors in } n \text{ bits}) = 1 - \text{erfc}\left[(\alpha - 1)\sqrt{k_1}\right] \qquad (np \gg 1) \quad (11.3)$$

where the complementary error function, $\text{erfc}(x)$, was defined earlier in (5.29). In the case where p is small and n is large, so that the expected number of errors is small ($np \approx 1$), this same probability is more nearly Poisson as given by

$$P(np \leq \alpha k_1) = 1 - e^{-\alpha k_1} \sum_{k=0}^{k_1} \frac{(\alpha k_1)^k}{k!} \qquad (np \approx 1) \quad (11.4)$$

Expressions (11.3) and (11.4) can be solved for the probability (confidence level) as a function of the errors measured (k_1) and the actual bit error rate (np), as shown in Figure 11.2 for Expression (11.3). Alternatively, one can fix the probability and solve for the actual BER as a function of measured errors, as shown in Figure 11.3 for 99 percent confidence limits. In both Figures 11.2 and 11.3 the plotted curves indicate probability that the actual BER is less than α times the measured BER, where $\alpha = np/k_1$. With 10 errors recorded, for example, the actual BER is within a factor of two times the measured BER with 99 percent confidence.

Example 11.1 _____

Suppose that for acceptance of a 10-Mb/s system, the actual BER must be less than 1×10^{-9} with a 90 percent confidence. Assuming that the errors are independently distributed, what duration of test would be required and how many errors would be allowable?

Solution   From Figure 11.2 or Expression (11.3) we see that for seven measured errors there exists a 90 percent confidence that the actual BER is less than 1.5 times the measured BER. Therefore if we measure over $(1.5)(7)(10^9) = 1.05 \times 10^{10}$ bits, we are 90 percent confident that the actual BER is less than 10^{-9} if seven or fewer errors are recorded. The measurement period required is then

$$\frac{(1.5)(7)(10^9) \text{ bits}}{10 \text{ Mb/s}} = 1050 \text{ s}$$

11.2.2   Jitter Measurement
Recall from Chapter 2 that jitter performance is characterized in three ways:

- Amount of jitter tolerated at input (tolerable input jitter)
- Output jitter in the absence of input jitter (intrinsic jitter)
- Ratio of output to input jitter (jitter transfer function)

The instrumentation required to measure jitter for these three characteristics is shown in Figure 11.4. For digital transmission equipment, these measurements are all out-of-service and are conducted typically with factory or system commissioning tests. In a network environment, however, the live traffic signal may be used. Network jitter will depend on the particular equipment used in the network and on the bit pattern of the transmitted signal. Because of the random nature of network jitter, a long measurement period is recommended.

The tolerance of digital transmission equipment to input jitter can be measured by the test setup of Figure 11.4a. The source of input jitter con-

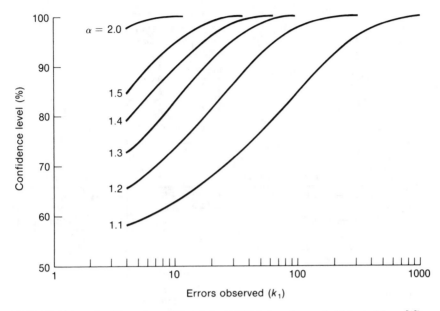

FIGURE 11.2 Confidence Level That Actual BER Is Less Than αk_1 (Adapted from [3])

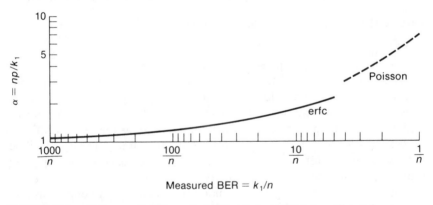

FIGURE 11.3 99 Percent Confidence Level That Actual BER Is Less Than αk_1

sists of a frequency synthesizer, jitter generator, and a pseudorandom test pattern generator. The clock signal produced by the frequency synthesizer is modulated by the jitter generator and then used to clock the pattern generator. The output of the pattern generator is passed through the unit under test from input port to output port. The pattern detector then accepts the output port signal and provides a detected error output to a counter. The

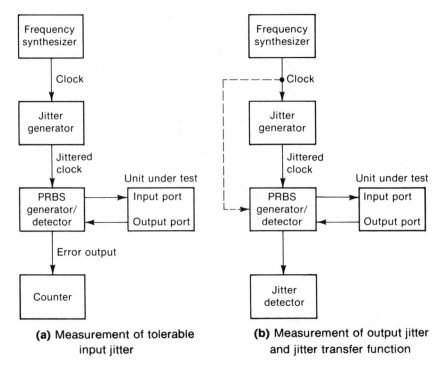

(a) Measurement of tolerable input jitter

(b) Measurement of output jitter and jitter transfer function

FIGURE 11.4   Block Diagram for Jitter Measurement

amplitude of the induced jitter is incrementally increased until bit errors are detected. This test is repeated for a range of jitter frequencies, thus permitting a plot of maximum tolerable input jitter versus jitter frequency to be constructed for the equipment under test. Figure 2.20 is an example of such a plot of maximum tolerable input jitter for several digital transmission hierarchical levels. The type of jitter modulation must also be specified; here sinusoidal jitter is often specified for convenience of testing, even though it is not representative of the type of jitter found in a network. For test purposes, a sufficiently long pseudorandom binary sequence is used as the test pattern, as specified for example in Table 11.1.

A test configuration for measuring intrinsic jitter and the jitter transfer function is shown in Figure 11.4b. This set of test equipment is similar to Figure 11.4a except that a jitter detector replaces the error detector. To measure output jitter in the absence of input jitter, the jitter generator is bypassed so that a jitter-free clock is used to generate the PRBS pattern; the received pattern is applied to the jitter detector, which measures jitter amplitude. The jitter transfer function is characterized by using a PRBS signal modulated by the jitter generator and by measuring the gain or attenuation of jitter in the detected PRBS signal. For both tests, the jitter amplitude, rms or peak-to-peak, is measured for selected bandwidths of interest. To accomplish this, the jitter measuring set may require bandpass, low-pass, or high-pass filters to limit the jitter bandwidth for the measurement of specified jitter spectra. The jitter measurement range is selected to be compatible with the equipment specifications of allowed output jitter or jitter transfer function. Test results are plotted in a manner similar to Figure 2.20. CCITT Rec. 0.171 provides additional information on test configurations for jitter measurement [1].

11.2.3   Bit Count Integrity Testing

Bit count integrity (BCI) performance is determined by various parameters, such as signal-to-noise ratio in a clock recovery circuit, error rate in a multiplexer frame synchronization circuit, or clock accuracy and stability in a plesiochronous network. A test of BCI performance requires stressing the equipment or system to observe conditions under which BCI is lost and regained. Since BCI testing is done out-of-service, it is generally included as part of factory tests or system installation. As an example, the BCI performance of a digital multiplexer is usually specified for a certain error rate. To test this specification, the multiplexer aggregate data stream must be subjected to the specified error rate while measurement is made of the times between losses of BCI and the times to regain BCI. These measured times are compared to specified times in order to determine acceptance or rejection of the unit being tested.

11.2.4   Reliability Testing

Equipment reliability can be verified with a combination of factory testing, analysis, and field results. Nonredundant equipment can be tested directly to verify the mean time between failure (MTBF) specification. Procedures for such tests are well established by industry and government standards [4]. Using enough units under test at one time, it is possible to demonstrate reliability in relatively short periods of time through the use of sequential test plans such as that illustrated in Figure 11.5. Here the cumulative number of failures is plotted versus cumulative test time until an accept or reject decision is reached or until the maximum allowable test time (truncation) is reached.

For redundant equipment, direct testing is not a practical way of verifying mean time between outage (MTBO) because of the inordinate length of time that would be required. A recommended means of verifying the MTBO is to use analysis and MTBF testing. The analysis would produce a list of failure modes and effects, verify that performance sensors will detect

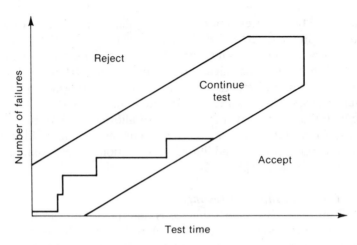

FIGURE 11.5   Reliability Test Plan Showing Accept/Reject/Continue Regions

failures, and verify that switching circuits will cause switchover from failed units to standby units. The MTBO can then be calculated from a demonstrated MTBF and predicted redundancy switching.

Reliability of equipment can also be monitored and recorded in actual field operation. This practice allows high-failure items to be identified and improved. For example, the CCIR has published field reliability results for a number of national radio systems that show equipment failure to be the dominant source of unavailability compared to outages due to human error, propagation, and primary power [5].

11.2.5 Link Testing

Before commission of a cable or radio link, testing is necessary to ensure that link installation has followed the design criteria. A first step is the verification of basic equipment interface compatibility and operation in a nonstressed environment. This type of test is facilitated by operating the equipment in a looped-back configuration rather than over the link. This test period is an appropriate time to identify cable mismatches, faulty grounds, and other defects in equipment installation.

The second phase of link acceptance testing requires operation of the transmission equipment over the link. Although it is impractical to measure and verify link performance objectives directly in a short test, certain design parameters related to link performance can be quickly measured. The signal-to-noise margin at a radio or cable receiver can be directly measured. Error rates can be measured for digital data; voice-channel measurements can be made for 4-kHz analog signals. During quiet periods, with no fading or failures, these design parameters under test should meet their specifications.

The link should also be artificially stressed to verify the design margins. In the case of radio links, for example, the fade margin is determined by the following procedure. First the nominal (nonfaded) received signal level (RSL) is determined by monitoring RSL during stable propagation time (normally around midday). A BER test is then conducted with a calibrated attenuator inserted in the received signal path and set to yield the threshold BER. The attenuator setting is then equal to the link fade margin. The fade margin should be determined separately for each received path of a diversity combiner output as well. Hysteresis of a selection combiner can be tested by fixing the attenuation of one diversity branch at a constant value and then varying the attenuation of the second diversity branch above and below the hysteresis design value. If the measured link fade margin and hysteresis are approximately equal to their design values, the link can be expected to meet performance objectives. Measured fade margins that fall short of design values indicate the possibility of inadequate path clearance or persistent anomalous propagation conditions. If the measured nominal RSL is significantly less than the calculated nominal RSL, antenna misalignment or waveguide losses should be suspected.

A direct measurement of link performance requires considerable test instrumentation and time. In the case of a radio link, the testing of anomalous propagation conditions requires sophisticated techniques to collect meteorological data and indices of refractivity along the full length of the propagation path. Significant test time may be required to collect sufficient data, since anomalous conditions are by definition rare. Testing of the effects of propagation in digital radio links requires instrumentation of both the received signal and BER. An example of a radio test configuration, used to collect data presented in Chapter 9 (see Figures 9.29 and 9.50), is shown in Figure 11.6. This test configuration was designed for one-way transmission, with the transmitter at site A and the receiver at site B. The receiver site has two space diversity radio configurations—one equipped with adaptive equalizers and the other without equalizers. Both BER and RSL are measured simultaneously for both radios; a spectrum analyzer is shared between the two radios. The radio under test used selection combining, so that diversity switch position was also recorded with BER, RSL, and spectrum shape. Radio link testing as described here and shown in Figure 11.6 requires an extended test period to obtain statistically significant results. Both the CCITT and CCIR suggest a month of testing for measurement of bit error rate or unavailability.

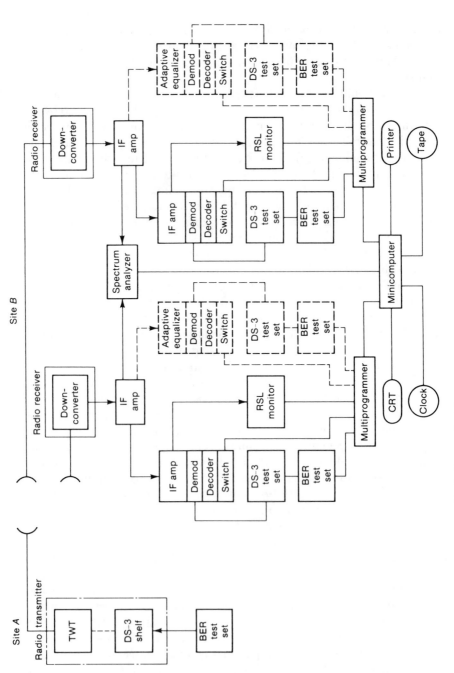

FIGURE 11.6   Configuration for Radio Propagation Tests

11.3 Performance Monitoring Techniques

Digital transmission systems are characterized by rugged performance when operating above their design thresholds and by abrupt degradation below threshold. Both of these characteristics of digital transmission are illustrated in Figure 11.7, which compares the performance of a PCM-derived VF channel versus that of an FDM-derived VF channel for a typical radio link.* In comparing the performance of analog versus digital transmission, note that the FDM-derived VF channel suffers a gradual degradation in performance versus decreasing RSL while the PCM-derived VF channel maintains a high, constant level of performance versus decreasing RSL. Note also that below threshold the digital system performance rapidly degrades over a very small margin of RSL.

The inherent ruggedness and threshold effect of digital transmission create new problems in performance assessment when compared to analog transmission. The gradual degradation of an analog signal is translated directly into degraded user performance and can be easily monitored; however, gradual degradation of a digital signal has no discernable effect on system performance until errors are introduced. Once the digital signal has deteriorated to the point where errors result, only a small margin exists between initial errors and unacceptable performance. An effective performance monitor must be capable of constantly measuring the margin between the state of the transmission channel and the threshold of that channel. A performance monitor indicating such a margin enables the system operator to implement diagnostic and maintenance procedures before user performance degrades.

In the design and use of performance monitors, several attributes should be considered. The following list of attributes applies equally to performance monitors used by system operators in monitoring system health and to those used within transmission equipment to control diversity or redundancy switching automatically:

- *In-service* techniques should be used to allow normal traffic and not affect the end user.
- *Simplicity* should be emphasized to maintain reliability and minimize cost.
- There should be *rapid* and *observable* response to any cause of system degradation.
- The *dynamic range* should extend from threshold to received signal levels that are significantly higher than threshold.

---

*Voice-channel noise in Figure 11.7 is given in units of dBrnC0, which is defined as C-message weighted noise power in decibels referenced to 1 picowatt (−90 dBm) and measured at a zero transmission level point.

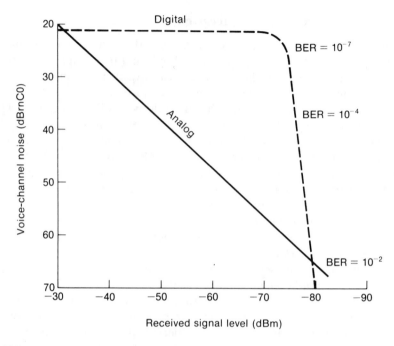

FIGURE 11.7   Performance of Analog vs. Digital Transmission

- Sufficient *resolution* should be provided to allow a continuous quantitative measure of the margin.
- Responses should be *stable* and *repeatable* to minimize or eliminate the need for calibration.
- A functional or empirical *relationship with BER* should be determinable.

The performance monitoring techniques described in the following paragraphs provide many of the attributes listed here. The suitability of each technique depends on the requirements and constraints of the system in question.

11.3.1   Test Sequence Transmission

Often spare channels exist in multiplex equipment or special test channels are provided in auxiliary service channels that allow transmission of a pseudorandom binary sequence at a low bit rate. This test sequence is interleaved with the traffic digital signal at a fraction $1/N$ of the system bit rate. At the receiver the BER is estimated by taking the ratio of errors detected to the number of transmitted test bits. Apart from the added overhead required of this technique, the time required to recognize a specified error rate is N times that required by counting errors of all the transmitted bits. The relatively long times required to estimate low error rates make this technique unsuitable for application to control of protection switching.

11.3.2   Frame Bit Monitoring

Digital multiplexers contain a repetitive frame pattern used for synchronization. Since this pattern is fixed and known beforehand, an estimate of BER can be obtained by measuring frame bit errors. For performance monitoring purposes, multiplex equipment often provides a digital pulse output for each frame bit error, so that only a counter is required to yield the frame bit error rate. In some cases the required counting and averaging are directly incorporated in the multiplexer, resulting in a front panel display of error rate, usually in decade steps—for example, 1×10^{-3}, 1×10^{-4}, 1×10^{-5}, and so forth. As an alternative, the framing signal can be monitored by an instrument separate from the multiplexer [1].

11.3.3   Signal Format Violations

Certain multilevel coding techniques have inherent constraints on transitions from one level to another. A received signal can be monitored for violations of these level constraints to provide error detection. As mentioned earlier in Chapter 5, both bipolar and partial responses codes have intrinsic redundancy and are therefore candidates for this form of error monitoring. The level constraints vary for different codes. Some examples of level violations for standard codes are:

- Bipolar: two consecutive marks of the same polarity
- HDB3: two consecutive bipolar violations of the same polarity
- B6ZS: two consecutive marks of the same polarity excluding violations caused by the zero substitution code

There are several drawbacks to the use of format violations. For example, in the case of bipolar violations, certain combinations of errors such as an even number of errors may not cause a bipolar violation. Also, repeaters and other digital transmission equipment will eliminate bipolar violations as part of signal regeneration, making end-to-end monitoring impossible using bipolar violations. Further, bipolar violations only account for errors in the aggregate signal of a multiplexer and not for any of the errors that may have occurred in the channels making up the aggregate signal. This last limitation occurs since each port of a multiplexer must convert the bipolar format of its input channel to an NRZ format for multiplexing.

11.3.4 Parity Bit Monitoring

Check bits or parity bits are added to blocks of data or to digital multiplexer frames in order to provide error detection. At the transmitter a parity bit is added to each block of data to make the sum of all bits always odd or always even. At the receiver the sum of bits in each block or frame is compared to the value of the received parity bit. Disagreement indicates that an error has occurred. When used with digital radio, two types of parity bits are commonly added to each transmitted frame: a hop parity bit P_H and a section parity bit P_S. The P_H is examined at each radio regenerative repeater to indicate error occurrences on each hop. A new P_H that provides correct parity with the received data is then inserted into the frame for transmission over the next hop. In contrast, the value of P_S is not changed along a section, so that at any station along the radio section P_S can be checked to determine whether transmission errors have occurred up to that point [6].

C-bit parity used in the C-bit DS-3 multiplex also allows end-to-end monitoring, while the P bits simultaneously provide section by section monitoring. Thus C-bit plus P-bit parity overcomes the limitation of bipolar violation monitoring on the B3ZS signal or conventional parity check in the old M13 format. Table 4.9 shows the frame and multiframe structure containing the C- and P-bit parity bits. The two P bits in frames 3 and 4 provide conventional parity check on the payload bits of each M frame, in which these two bits are set to 11 for even parity and 00 for odd. The C bits carried in frame 3 are set to the same value as the two P-bits. The three far-end block error (FEBE) bits in frame 4 indicate a C-bit parity error or loss of frame event from the far end. The near end sets its FEBE bits to 000 to indicate a parity error or loss of frame and to 111 if no such event has occurred. Since both the P bits and FEBE bits are monitored, the performance of both directions can be determined from one end.

Like many other in-service monitoring techniques, parity checking has limitations in its error detection capability. Only an odd number of errors in the block or frame will produce a parity error. Thus parity checking yields an accurate estimate of system error rate only for low error rates (below 10^{-N}, where N is the length of the block or frame).

11.3.5 Monitoring with Cyclic Redundancy Checks

A **cyclic redundancy check** (CRC) provides error detection by appending to each transmitted message a coded sequence that is then used to check for errors at the receiver. Given a k-bit message, the transmitter adds an n-bit sequence, known as a **frame check sequence** (FCS), to create a $(k + n)$-bit frame that is exactly divisible (modulo 2) by some predetermined binary number, P. The receiver then divides each incoming frame by the same binary number P and, if no remainder is found, assumes that no transmission errors occurred. The FCS is generated at the transmitter by shifting the message sequence n bits to the left, dividing the shifted message by P, and using the remainder as the FCS. If no errors occur in transmission, the receiver will have a remainder of 0 after dividing the frame by P. To generate a FCS that is n bits long, the divisor P must be $n + 1$ bits long since in binary division the remainder is always one bit less than the divisor. The sequence P is thus selected to be one bit longer than the desired FCS, with the choice of exact pattern of P depending on the types of errors expected. Standard choices of P provide error detection for all single-bit errors, all double-bit errors, any odd number of errors, any error burst in which the burst length is less than the FCS length, and most larger error bursts.

CRC hardware implementation consists of a shift register and EXCLU-SIVE-OR gates, which act as the divider circuit $P(x)$. At the transmitter the FCS sequence is generated by entering the k-bit message into the shift register (most significant bit first), then entering n additional 0 bits, and using the remaining shift register contents as the n-bit FCS. At the receiver, the ($k + n$)-bit frame is entered into the same divider logic which will contain the remainder after the entire frame has been entered. If no transmission errors have occurred, the shift register will contain all 0's.

A common way to express CRC codes is based on polynomials with a dummy variable x and binary coefficients b_i:

$$b_n x^n + b_{n-1} x^{n-1} + \cdots + b_1 x^1 + b_0 x^0$$

Thus the polynomial $P(x) = x^{12} + x^{11} + x^3 + x^2 + x + 1$ represents the 13-bit sequence 1100000001111. CCITT Rec. G.704 specifies CRC-6, CRC-4, and CRC-5 error detection for 1.544-, 2.048-, and 6.312-Mb/s multiplexers, respectively, using the following versions of $P(x)$:

$$\text{CRC-6} = x^6 + x + 1 \tag{11.5}$$

$$\text{CRC-4} = x^4 + x + 1 \tag{11.6}$$

$$\text{CRC-5} = x^5 + x^4 + x^2 + 1 \tag{11.7}$$

As an example, we shall consider the CRC-6 used in the extended superframe format of DS-1 signals. The CRC-6 bits are calculated over the 4632 bits contained in the 24 frames of each superframe. The resulting CRC is inserted in the next superframe, with the 6 check bits (CB) located in frames 2, 6, 10, 14, 18, and 22, as shown in Table 4.4. At the far end, the CRC-6 is recalculated and compared with the CRC code sent from the transmitter.

Thus the presence of errors that occurred anywhere along the path from transmitter to the point of observation will be detected with high probability. Although CRC-6 does not indicate the number of errors that occur in a superframe and therefore will not reflect accurate bit error rates in a bursty error channel, it does provide a reliable means of estimating errored seconds and other related parameters. Also, the CRC code is not recalculated at intermediate points such as DS-3 multiplexers, fiber optic terminals, or radio terminals, so that end-to-end performance monitoring is possible. Both network and customer equipment can perform this performance monitoring function afforded by the use of CRC-6. Other applications of the CRC-6 contained in ESF include performance trending, protection switching, automatic restoration, and frame resynchronization [7].

Example 11.2

In-service performance monitoring of DS-1 signals depends on the use of overhead bits. To monitor error performance, framing bits are used with the superframe version of DS-1 and CRC-6 bits are used with the ESF version. Table 11.2 shows one industry standard [8] for DS-1 error monitoring based on use of these overhead bits. These parameters are consistent with those used by the CCITT, and may be used to verify CCITT performance objectives based on models that have been developed that relate CCITT Rec. G.821 objectives to the parameter definitions given in Table 11.2 [9].

11.3.6   Eye Pattern Monitoring

The monitoring of eye pattern opening can provide a qualitative measure of error performance. With signal degradation, the eye pattern becomes fuzzy and the opening between levels at the symbol sampling time becomes smaller. Noise and amplitude distortion close the vertical opening; jitter or timing changes close the horizontal opening. One approach in monitoring the eye pattern is to measure the eye closure and compare the measured values with those obtained for the system operating in an ideal environment. The eye closure can be expressed as system degradation in decibels by

$$\text{System degradation} = -20 \log(1 - \text{eye closure}) \tag{11.8}$$

Such a measure may be converted to a BER estimate by simple calibration of the eye pattern monitor. A more general technique for eye pattern monitoring is to perform analysis on selected bit patterns. This technique compares received eye patterns to the ideal eye pattern for a selected bit pattern, which may be useful in isolating the fault (noise, distortion, jitter) causing the eye closure.

TABLE 11.2   Parameter Definitions for DS-1 Performance Monitoring

Parameter	Superframe	Extended Superframe
Errored second (ES)	A second with 1 or more frame bit errors	A second with 1 or more CRC-6 errors
Bursty errored second (BES)	Not applicable	A second with more than 1 but less than 320 CRC-6 errors
Severely errored second (SES)	A second with 3 or more frame bit errors	A second with 320 or more CRC-6 errors
Consecutively SES (CSES)	A period of 3 to 9 consecutive SES	A period of 3 to 9 consecutive SES
Unavailable second (UAS)	Every second following 10 consecutive SES until 10 consecutive non-SES are detected	Every second following 10 consecutive SES until 10 consecutive non-SES are detected

11.3.7   Pseudoerror Detector

The pseudoerror detector is a technique that permits extrapolation of the error rate by modifying the decision regions in a secondary decision circuit to degrade the receiver margin intentionally and create a pseudoerror rate. The pseudoerror technique is implemented by offsetting the sampling instant [10], decision threshold [11], or both [12] from their optimum value and then counting the number of detected bits (pseudoerrors) that fall into the offset region. For a given type of degradation, there is a fixed relationship between the actual BER and pseudoerror rate (PER). This relationship depends on the amount of intentional offset selected for the pseudo decision circuit, but in general the PER is always greater than the BER. This multiplication factor is an important aspect of pseudoerror techniques, since error rate can be estimated in a shorter time, especially for low error rates, than by the other methods considered here, including both out-of-service and in-service. And since the pseudoerror technique detects channel impairment before actual errors are made, it can be used to control diversity switching in digital radio systems [13]. However, the multiplication factor may vary depending on the source of degradation (jitter, noise, multipath). Hence the calibration of PER to BER should be related to the potential sources of degradation for the application.

A functional diagram of a pseudoerror detector based on decision threshold offset is shown in Figure 11.8. The pseudoerror zone is established by two slicers, one at h and the other at $-h$ relative to a full bit of height ± 1.0. The pseudoerror decision is made by using the same clock as the actual data decision. Note that the timing of the pseudoerror decision could be offset from the optimum sampling instant to further increase the probability of a pseudoerror. Likewise, the pseudoerror probability can be increased or decreased by adjustment of the pseudoerror zone height.

A model of the performance of the pseudoerror detector circuit in Figure 11.8a is shown in Figure 11.8b and c. The degradation is assumed to be additive gaussian noise with zero mean. For a transmitted 1, the noise density is centered about the received signal at a level 1. The probability of pseudoerror can be written mathematically by simple modification of the expressions for actual error probability. As an example, consider the case of coherent BPSK or QPSK, where the probability of pseudoerror is given by

$$PER = P[\text{pseudoerror}] = \text{erfc}\sqrt{\frac{2E_b}{N_0}(1 - h)} - \text{erfc}\sqrt{\frac{2E_b}{N_0}(1 + h)} \quad (11.9)$$

Figure 11.9 shows a comparison of the BER and PER as a function of the pseudoerror zone height. Note that over the range of zone heights the PER is roughly parallel to the BER. The pseudoerror rate for an offset $h = 0.2$ is approximately two orders of magnitude worse than the ideal error rate. With increasing values of h, the multiplication factor increases even further, meaning that the time required to estimate the true BER is reduced. As an example, for $h = 0.4$ and a true BER of 10^{-7} the PER will register 10^{-3} and allow the true BER to be estimated 10,000 times faster than the time that would be required to measure BER with conventional out-of-service methods.

11.4 Fault Isolation

Digital transmission systems typically include built-in alarm and status indicators that allow operator recognition of faults and isolation to the failed equipment. Fault and status indicators are displayed on the equipment's front panel, usually by light-emitting diodes (LED), or are provided by teletypewriter output and are available for remote display through relay drivers. Ideally each major function (multiplexer, demultiplexer, timing circuits, and so forth) and each module (individual printed circuit board) have an associated alarm or status indicator. Often, built-in test equipment is used to recognize faults either automatically or with operator assistance. With the advent of the microprocessor, many transmission systems now contain embedded microprocessors that continuously monitor system performance by a sequence of tests designed to check signal paths and modules. The information provided by fault indicators is used to facilitate routine maintenance, to aid in troubleshooting and fault isolation, and to effect automatic equipment protection switching.

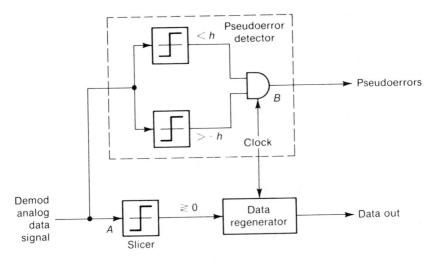

(a) Functional block diagram

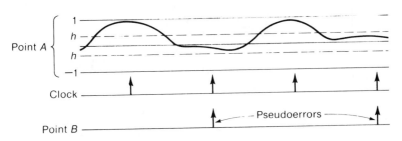

(b) Signal waveforms

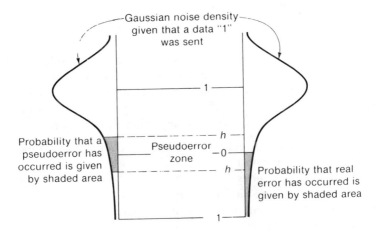

(c) Theory of operation

FIGURE 11.8 Pseudoerror Detection Technique

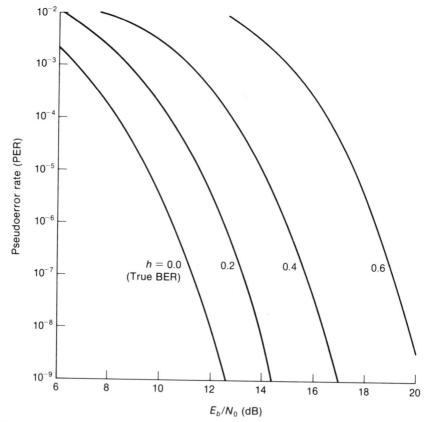

FIGURE 11.9   Pseudoerror Rate for BPSK or QPSK as a Function of Pseudoerror Zone Height h

Alarm and status indicators are displayed and analyzed locally if repair personnel are available. For unmanned sites, however, these signals are transmitted to a manned facility where the decision is made whether to activate remote control or dispatch repair personnel. The following types of alarm and status points are provided either locally or remotely:

- Failure of transmission equipment (isolated to a module or function)
- Failure of primary power
- Failure of alarm circuitry
- Failure of protection switch circuitry
- Status (position) of automatic switching equipment
- Status of off-line equipment
- Status of various performance monitors

The information contained with these indications should be sufficient to isolate a failure to the site, equipment, and even module (circuit board). Further, an alarm should be classified as affecting service (major) or not affecting service (minor). For multiplex equipment, the level of traffic affected by a failure should also be indicated (for example, number of DS-1 signals affected, number of DS-2 signals affected, and so on).

As examples of the level of control, status, and monitoring points used in digital radio and multiplex equipment, consider the front panel displays illustrated in Figures 11.10 and 11.11. The radio [14] used for Figure 11.10 includes a TDM that combines two mission bit streams (MBS) with a service channel bit stream (SCBS). Moreover, both the transmitter and receiver are fully redundant with two independent units, A and B, that can be automatically or manually switched on-line. The alarm LED matrix and meter readings can be used for troubleshooting and preventive maintenance checks. The meter is used to monitor dc voltages from the power supply and significant RF signals. Controls include selection of manual versus automatic protection switching, manual selection of A versus B units, and loopback of the off-line transmitter and receiver.

The digital multiplexer used for Figure 11.11 combines up to eight 1.544-Mb/s channels (or ports) into a single mission bit stream (MBS). The port, multiplexer, and demultiplexer modules are all fully redundant; independent units A and B are controlled by a peripheral module. Alarm indications include loss of frame, individual frame errors, loss of input or output signals at the port or MBS level, and faulty modules. Each fault is isolated to a single module—either one of the port modules or the multiplex, demultiplex, peripheral, or power supply modules. Status displays include switchover control, whether automatic, manual or remote; indication of whether the A or B unit is on-line; and whether the frame search inhibit is enabled or disabled. Controls via the three toggle switches are for switchover from one unit to the other, loopback of the off-line transmitter and receiver, and lamp test.

In addition to the display of local alarms, major alarms should be transmitted to the remote end and to other levels of equipment affected by the alarm condition. Alarm indications may be conveniently transmitted to the remote end by changing prescribed bits of the frame synchronization pattern or other overhead patterns. In typical PCM multiplex equipment, major alarms result in the suppression of analog VF outputs. Similarly, for digital multiplex equipment an alarm indication signal (AIS) is transmitted in the direction affected (downstream). The AIS is typically an all 1's pattern or some other substitute for the normal signal, indicating to downstream equipment that a failure has been detected and that other maintenance alarms triggered by this failure should be ignored or inhibited

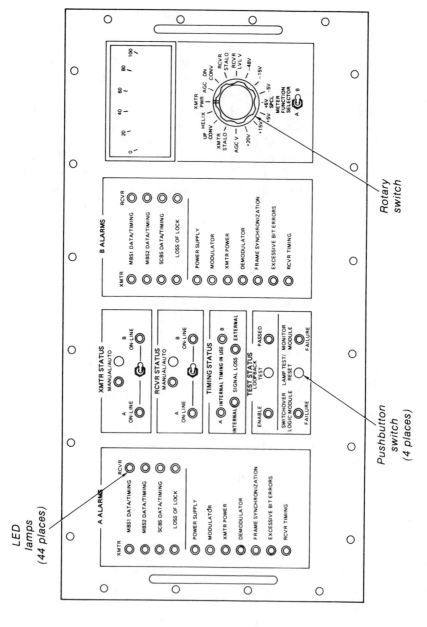

FIGURE 11.10 Display Panel Controls, Monitors, and Status Indicators for a Digital Radio (Reprinted by permission of TRW)

732

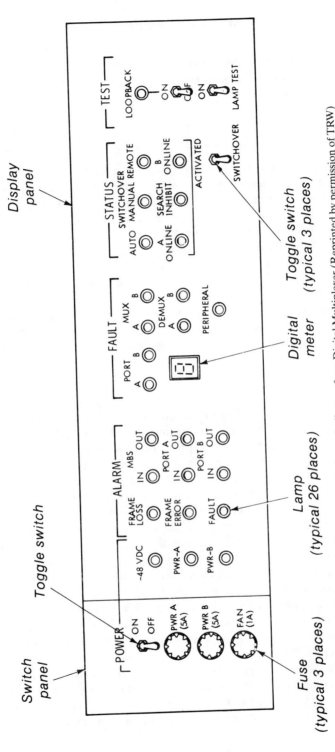

FIGURE 11.11 Display Panel Controls, Monitors, and Status Indicators for a Digital Multiplexer (Reprinted by permission of TRW)

in order to eliminate unnecessary actions. Detection of fault conditions and consequent actions have been recommended by the CCITT for PCM and digital multiplex equipment [15].

Example 11.3 _____

DS-1 signal formats provide three basic carrier failure alarms (CFA) that allow both ends to determine the status of the path. When failure occurs in one direction of transmission, a local *red alarm* is generated by the receiver when it cannot lock to the framing pattern. A *yellow alarm,* also known as a *remote alarm indication,* is transmitted to the distant end when a red alarm is present. This yellow alarm is generated by forcing the second bit to 0 in all DS-0 channels of the superframe format, and using a repetitive 16-bit pattern in the data link of the extended superframe format. The *blue alarm,* also called the *alarm indication signal,* is activated if the equipment is no longer available or if the receiver has lost the incoming signal completely. The blue alarm consists of an unframed all-1's signal that maintains signal continuity and indicates a failure to downstream equipment.

For redundant systems, a failed unit is automatically switched out and replaced with an operational standby unit, thus deferring the need for maintenance action. If the automatic protection switching circuitry itself fails, however, manual switching becomes necessary. Manual switching can be accomplished by local control and usually by remote control as well. Thus manual switching, whether local or remote, serves only as a backup to the preferred method of automatic protection switching.

For nonredundant systems, a failed unit requires manual maintenance action for restoral of service. Depending on the availability of properly trained personnel, test equipment, and spare parts, this restoral action may take considerable time. As an alternative, particularly for high-priority circuits, alternate transmission paths can be planned that are activated upon failure of equipment in the primary path. This approach will probably still require manual action, but limited to simple patching of circuits. In general, the maintenance approach for service restoral depends on the performance requirements of the service.

Knowledge-based *expert systems* have added a degree of automation to the job of fault detection and isolation. Much of the expertise required to perform fault detection can be reduced to a set of rules or patterns suitable for applications of artificial intelligence as an expert system. Two approaches have been used: a conventional rule-based system and a machine learning system. The rule-based approach is empirically derived and requires a set of rules for each new communication system. Conversely, the machine learning approach uses adaptive pattern recognition techniques to train the expert system on each new communication system [16]. There are a number of commercially available expert system tools which have been used as a shell for fault isolation and service restoral in communications systems. A typical expert system then can access test and communication equipment for the purpose of running diagnostic tests and switching spare assets to restore service [17].

11.5 Monitoring and Control Systems

Digital transmission systems provide monitoring and control through both the equipment and the signals that make up the system. Equipment used in digital transmission systems generally has built-in test equipment and displays that facilitate performance monitoring and fault isolation. In addition, separate subsystems specifically designed to provide monitoring and control are often used to supplement the transmission equipment's monitoring and control capabilities. These separate subsystems may be centralized, in which case there will be a master station and multiple remote stations, or distributed, in which case any station may act as a master. Transmission signals contain overhead that also provides an important role in monitoring and control. Examples such as parity and cyclic redundancy codes have already been discussed. These forms of performance monitoring of the transmission line facilitate assessment of performance by both the user and network provider.

11.5.1 User Monitoring and Control

As services and equipment available to the telecommunications customer have expanded, the presence of a monitoring and control capability at the user location has become standard practice. Devices known as data service units (DSU) and channel service units (CSU) allow users to connect to a variety of digital services now available from telephone carriers. The first such service was AT&T's Digital Data System (DDS), which required a DSU to interconnect the customer data terminal equipment (DTE) with a subrate data multiplexer [18]. The DSU provides conversion of the DTE signal format (for instance, EIA-232-D) to the format required by the transmission equipment (for instance, bipolar with proper ones density). The DSU receiver on the network side will recognize idle and troubleshooting codes, and will respond to loopback commands from the network. The DSU also provides equalization, automatic gain control, data detection, and clock recovery. DSUs are available for a variety of rates, from DDS rates up to T1 and E1, and offer a great range of interfaces, operating modes, built-in diagnostics, and other features.

As shown in Figure 11.12, channel service units interconnect the network transmission equipment to the customer premise equipment, usually a multiplexer, channel bank, private branch exchange (PBX), or DSU. For those cases where the CSU interfaces with a DSU, the two devices can be combined into one devise. Furthermore, DSU/CSU functions can also be integrated with other customer premise equipment such as multiplexers. CSUs provide a variety of functions including signal regeneration, performance monitoring, loopback on either the line or payload side, protection switching, and conversion of line formats (for instance, SF to ESF). For DS-1 services, the interface between the carrier and the customer have been standardized [19,20], so that some CSU functions are likewise standardized. When using the ESF format, CSUs utilize the 4-kb/s data link (DL) to relay performance monitoring and control information. Two DL signal formats are used, *bit-oriented signals*, which transport code words, and *message-oriented signals*, which transport messages [19]. The bit-oriented signals are used for alarms, commands, and responses, such as the yellow alarm and loopback commands. The message-oriented signals are used for performance monitoring, including CRC errors, framing bit errors, bipolar violations, and frame slips. Thus performance data are measured at one end and transmitted via the data link to the far end, so that either end can monitor both directions of transmission. The CSU also stores this performance data, which can then be accessed by the customer or the network provider. Two registers are used, one for user access and the other for network access. These features of CSUs allow proactive monitoring, that is, use of performance data to give precursive warning before outages occur.

11.5.2   Network Monitoring and Control

A transmission monitoring and control system block diagram is shown in Figure 11.13 for a simple two-station network. The two types of stations illustrated in Figure 11.13, remote and master, are generic to transmission monitoring and control. The remote station must collect alarm and status signals for transmission to a master station and receive and execute control commands from a master station. The master station displays local alarm and status signals, receives and displays alarm and status signals for each remote site, and generates and transmits control commands to each remote station. In general, a master may interface and control several remote stations. Conversely, any remote station may report to two or more master stations, one of which is designated primary and the others backup. The primary master can request information or initiate control actions at remote stations. Backup masters listen to the primary master's requests and to

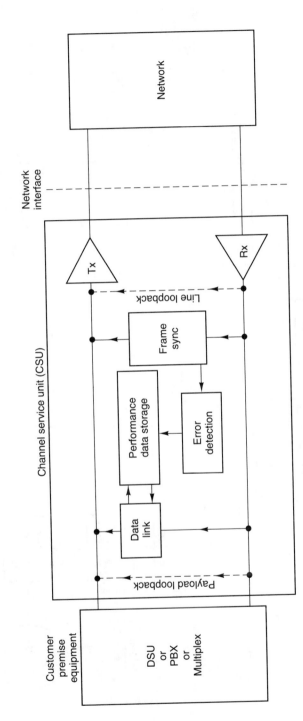

FIGURE 11.12 Typical Channel Service Unit Configuration (Adapted from [20])

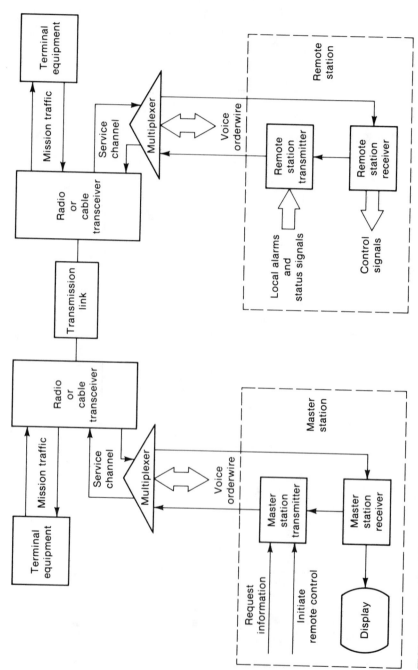

FIGURE 11.13 Block Diagram of Transmission Monitoring and Control System

remote responses and are capable of performing as primary masters on command from the system operator.

Monitoring and control signals are carried by a service channel that is usually separate from the mission traffic. The service channel is composed of voice orderwires and telemetry signals required for supervision and monitoring of the overall network and each individual link. As shown in Figure 11.13, a multiplexer is used to combine voice orderwires and alarm, status, and control signals for transmission via the service channel. The voice orderwire is used by maintenance personnel to coordinate work on the radio or cable link. Equipment alarm and status signals take the form of contact closures, pulse outputs, and analog voltages, which are collected from the remote station's transmission equipment and encoded into a serial, digital form for interface with the service channel multiplexer. Control signals are also distributed via the service channel, emanating from the master station for interface with transmission equipment at the remote station. Other requirements and characteristics of a monitoring and control system are described in the following paragraphs.

Reporting System

Two general types of reporting can be used: continuous and polled. In a continuous reporting system, each remote station transmits its alarm and status signals continuously; a different transmission channel is required for each remote station. The master station receives these alarm and status signals and transmits control data to each remote station via separate transmission channels. As the number of remote stations reporting to a master increases, the number of transmission channels also increases—a disadvantage of the continuous reporting system. Most practical systems use the polling technique, whereby the remote stations all share a common transmission channel on a party-line basis. Each remote station has its own address and responds only when addressed, or *polled*, by the master. All other remote stations remain idle until addressed. When polled, the remote station's reply contains its address and the requested data.

Response Time

In the continuous reporting scheme, alarm and status signals are transmitted without delay so that response time is not of concern. In the polling scheme, however, each remote station must wait its turn to report. Thus information must be stored at each remote station until that station has been polled. The actual waiting time for reporting from a remote station is determined by the number of remote stations, the size of the remote stations, and the capacity of the telemetry channel used for reporting. The required response time is determined by the maintenance philosophy—that is, by what an operator at the master does when an alarm condition is detected. If the operator's response is to dispatch a maintenance team to the remote station, quick response should not be important. Quick response from the reporting system is important only if remote controls are available that will allow removal of the failed unit and restoral with a standby unit. System response time can be improved, if necessary, by increasing the telemetry channel data rate or reducing the number of remote stations reporting to a given master station.

System Integrity

The reporting system must also provide protection against transmission errors, short-term outages, and equipment failure. Some form of error detection is required in order to prevent erroneous reporting of alarms or accidental initiation of remote controls. Moreover, an acknowledge/not acknowledge exchange is required whenever control actions are initiated at a master station for execution at a remote station. Finally, the reporting system must be fail-safe, so that equipment failure does not cause extraneous alarms or control actions.

Transmission Techniques

Techniques for transmission of service channels in digital transmission systems fall into three categories: (1) transmission within the main digital bit stream, (2) transmission in an auxiliary channel that is combined with the main digital bit stream using multiplexing or modulation techniques, and (3) transmission that is separate from the media used for the main bit stream. The first technique simply takes channels of the mission equipment, say the multiplexer, for use as the service channel. This approach has the advantage of not requiring any auxiliary equipment or media, but it does require demodulation and demultiplexing of the mission traffic at each station in order to gain access to the service channels. The third technique requires a separate medium or network, such as the public telephone network, to provide service channel connections among the various stations. This approach provides the desired accessibility and independence from the main traffic signal, but it has the distinct disadvantage of higher cost and potentially less reliability. Only the second technique meets the requirement to drop and insert the service channel at each station easily with only a modest increase in transmission cost and complexity.

Thus the most commonly used technique for service channel transmission is the auxiliary channel that is multiplexed or modulated into the same passband as the mission traffic. This technique can be implemented in two ways: by time-division multiplexing the service channel in digital form with the mission digital bit stream or by modulation of the service channel, in analog or digital form, using a secondary modulation technique (AM, FM, and so on). The TDM approach requires multiplexing circuitry that combines the service channel and mission channels at the transmitter and provides the inverse operation at the receiver. This additional circuitry and transmission overhead is minimal and thus has a negligible impact on system cost and performance. The second implementation of auxiliary channel transmission is generally applicable to radio systems, in which the auxiliary signal is frequency-modulated onto the carrier, amplitude-modulated with the RF signal, or placed on a subcarrier that is inserted above or below the mission spectrum. These service channel transmission techniques for radio must utilize low levels of modulation and small bandwidths in order to avoid degradation of the mission traffic and exceeding of frequency allocation limitations. Because of these restrictions, the TDM approach is more commonly used in radio systems.

11.6   Digital Cross-Connect Systems

The wiring and patching of digital signals once required in a telephone central office are now done in electronic form via a **digital cross-connect system** (DCS). Used at a central location, a DCS permits many functions to be consolidated and automated that previously required separate equipment and several operators. These functions include cross-connect of DS-0 and DS-1 signals within or between higher levels, test access, and performance monitoring. Network management associated with a DCS allows either the customer or carrier to modify and monitor the network at various levels such as DS-0, DS-1, and DS-3.

The first such form of digital cross-connect was the DCS 1/0, which is terminology for a DS-0 cross-connect with DS-1 interfaces. The DCS 1/0 eliminates back-to-back channel banks otherwise required to drop or rearrange DS-0s contained in DS-1s. Variations on the DCS 1/0 have followed, such as the DCS 3/1, which is a DS-1 cross-connect with both DS-1 and DS-3 interfaces. The DCS 3/1 eliminates the need for back-to-back M13 multiplexers to drop or rearrange DS-1s contained in DS-3s. A DCS 3/1/0 combines the functions of the 1/0 and 3/1 so that both DS-0 and DS-1 cross-connects are possible in a single DCS. These digital cross-connect systems have many features and have evolved to include other functions, as will be described in the following sections. Figure 11.14 depicts the various types of digital cross-connects in a hypothetical network, showing typical interconnection of the DCSs to each other and to customer premise equipment.

11.6.1   DCS 1/0

The basic function performed by a DCS 1/0 is the cross-connection of any DS-0 channel in an incoming DS-1 signal to another DS-0 channel on any

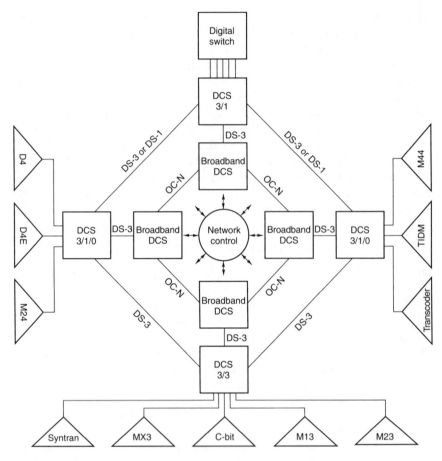

FIGURE 11.14   Typical Interconnection of Digital Cross-Connect Systems in a Hypothetical Network

outgoing DS-1 signal, with zero or negligible blocking probability. Here a zero blocking probability means that a cross-connection can be made regardless of any other cross-connections that already exist. Cross-connects are normally two-way, but the two directions of a DS-0 channel may be independently cross-connected to form one-way or multipoint connections. Digital cross-connection of DS-0 channels allows the DCS to *groom* networks in which all channels with a common destination are placed on a common DS-1. Grooming thus consolidates facilities and increases bandwidth utilization by filling DS-1s. In the same way, the DCS can segregate channels into specific DS-1 signals for termination on local or distant end equipment, such as a channel bank. Hence grooming replaces back-to-back channel banks or multiple channel banks otherwise required to drop and insert DS-0s from multiple DS-1 terminations. Since the cross-connection is accomplished digitally, there are no unnecessary analog-to-digital conversions as would be required with channel banks. From the viewpoint of the network, the DCS provides a hub that allows point-to-point T1s to be replaced with a hubbing configuration that uses the DCS to aggregate DS-0s and more efficiently fill T1s.

Another basic feature in the DCS is its ability to provision, reconfigure, test, and restore circuits. Provisioning of new circuits and reconfiguration of existing circuits are done electronically via a control terminal. With its ability to be controlled remotely, the DCS facilitates rapid reconfiguration at multiple locations to restore service after transmission facility or equipment failure. This automated reconfiguration capability also facilitates the sharing of a circuit between applications, say voice during the day and data transfer at night. Rapid restoration of failed DS-1 facilities can also be done by the DCS using cross-connect commands stored in a backup memory map. Testing of reconfigured or newly provisioned circuits can be done from a local or remote location. Any DS-0 channel may be routed to a DS-1 designated as a test access digroup to allow monitoring and testing of selected circuits to be conducted without interfering with any other DS-0. For monitoring access, the DCS keeps the original cross-connection in service and bridges the two directions of transmission. A split mode breaks the connection and allows independent monitoring and testing of signals associated with each side of the cross-connect. DS-1 test equipment can then be interfaced with the test access digroup(s) to allow monitoring and testing of the selected DS-0s.

Signals that are terminated on a DCS 1/0 can have a variety of formats and line codes that are selectable in the DCS. Selectable frame formats include SF, ESF, and T1DM; line codes include bipolar and *B8ZS*. Signaling modes that are supported include robbed bit (2-state, 4-state, and 16-state), common channel, and transparent. In the transparent mode, the cross-connect of 64-kb/s clear channels is possible. The DCS monitors the 8-kb/s overhead of each incoming DS-1 for framing losses, frame slips, and bit errors. Performance data are accumulated and analyzed to determine if maintenance or out-of-service thresholds have been exceeded. When a carrier failure alarm (CFA) occurs in a DS-1 terminating on the DCS, *trunk conditioning* is applied on a per channel basis so that downstream equipment (for instance, channel bank or digital switch) can drop the connection and remove the trunk from service. The DCS performs trunk conditioning using signaling bits and by sending an idle or out-of-sync code in the 8 bits of the affected channel. CEPT level 1 interfaces are also included in some

DCS 1/0 versions. Frame, line, and signaling formats are based on CCITT Recs. G.703, G.704, and G.732. A gateway function then allows conversion between DS-1 and E-1 signals, including the necessary changes to rates, signaling, and companding laws. Conversions of signaling and companding laws are done on a per channel basis, with the option of inhibiting these conversions for 64 digital data (noncompanded) channels.

Other data rates that may be cross-connected include those used in subrate data multiplexers, channel bank dataports, and transcoders. The subrate cross-connect feature is limited to circuits having the Digital Data System (DDS) format [21]. Functions performed by the DCS subrate feature include cross-connection and test access for 2.4-, 4.8-, 9.6-, and 56-kb/s subrate circuits as well as DS-0A and DS-0B channels. In general, all the functions performed by a T1DM are replaced in the DCS subrate feature. DCS can also cross-connect bundles of multiple DS-0 signals, from 1 to 24. DS-1 signals containing 32-kb/s ADPCM channels can be terminated and cross-connected on a DCS 1/0 in one of two ways. When configured in the bundled format, the ADPCM channels are grouped into 4 384-kb/s bundles each containing 12 ADPCM channels without signaling or 11 ADPCM channels plus a signaling channel. If the bundle contains signaling, the DCS must cross-connect the entire 384-kb/s bundle, but if signaling is not included the DCS may cross-connect DS-0s as pairs of 32-kb/s channels. ADPCM channels that use robbed bit signaling cannot be cross-connected, since a DCS 1/0 would not recognize the signaling bits. Transcoders can be placed either within or outside a DCS 1/0, however, to translate both coding and signaling formats between PCM and ADPCM to then allow cross-connection for all channel types. Other functions such as alarm propagation and test access are available on any individual subrate channel or bundled channels as in the case for any DS-0.

11.6.2   Higher-Level DCS

Higher-level digital cross-connect systems may include interface with synchronous (SYNTRAN) and asynchronous (pulse stuffing) DS-3 signals and with SONET STS-N and OC-N signals. Standard features are similar if not identical to the DCS 1/0, such as grooming, hubbing, provisioning, reconfiguration, restoration, performance monitoring, and test access at various levels, which may include DS-0, DS-1, and DS-3. Compatibility with the M13, M23, and MX3 multiplexers means that back-to-back M13, M23, or MX3 multiplexers can be eliminated. For DS-3 digital cross-connect systems, standard frame formats include *C*-bit parity and M13 framing, and line codes include *B3ZS*. If SYNTRAN interfaces are also included, then synchronous and asynchronous networks can share the same transmission facilities. Cross-connection between a bit-synchronous DS-1 and a byte-synchronous DS-1 in a SYNTRAN DCS is not generally required, but has been provided as an option. Termination of DS-1C and DS-2 signals is an option on some higher-level DCS as well [22].

For cross-connect of DS-1s contained within a DS-3, all information embedded in the DS-1 frame can be cross-connected without modification, so that the DCS is transparent to the contents of the DS-1. Thus, for example, if a DS-1 alarm indication signal appears at the DS-1 interface, the DCS passes the alarm along without making any changes. However, the DCS will monitor DS-1 interfaces for loss-of-frame and excessive bit error rate (based on bipolar violations) and will send an all-1's alarm indication signal in the downstream direction if these conditions arise. As an option, the DS-1 interface can frame on the incoming frame and provide performance monitoring based on SF or ESF overhead. In addition to the monitor and split access test capability, which provides test access to any DS-1 signal, any DS-1 termination can also be looped on itself to facilitate fault isolation. The timing of DS-1 signals within the higher-level DCS depends on the intended cross-connect. If the DS-1 is cross-connected to another DS-1 or to a DS-1 tributary within an asynchronous DS-3 termination, then the DCS does not retime the DS-1 but simply passes the DS-1 timing through. If the DS-1 is cross-connected to a DS-1 tributary within a SYNTRAN DS-3 termination, however, then the DCS must retime the DS-1 so that it meets the SYNTRAN interface requirements.

Cross-connection between DS-3 signals is transparent to DS-3 format and timing. The overhead of the DS-3 signal is not modified, and the DS-3 signal is not retimed by the DCS. For the case of a DCS 3/1, a DS-3 cross-connection does not require the DS-1 switching matrix and instead uses a DS-3 switching matrix. DS-3 testing, performance monitoring, and loop-back are standard features for DS-3 cross-connection. DS-3 frame and overhead bits are monitored and used to identify maintenance and outage conditions that can be defined by software-selectable thresholds.

A digital cross-connect system with either electrical or optical SONET interfaces is referred to as a *wideband* or *broadband* DCS. SONET-based digital cross-connect systems are capable of cross-connecting virtual tributaries (for instance, VT1.5) and synchronous transport signals (for instance, STS-1) as well as DS-1 and DS-3 signals. SONET OC-N and STS-N signals that terminate on a DCS carry individual DS-1 signals as VT1.5s and carry DS-3 signals as STS-1s, so that the DCS must frame on the incoming OC-N or STS-N signal to identify and access the constituent VT1.5 and STS-1 signals or their corresponding DS-1 and DS-3 signals. A SONET DCS must terminate the overhead (OH) and synchronous payload envelope (SPE) of the OC-N or STS-N signal in order to cross-connect the constituent signals. New OH and SPE signals are generated for the outgoing OC-N or STS-N channels. Cross-connection of a DS-1 between a DS-3 or DS-1 and an OC-N termination requires translation of the DS-1 into a VT1.5 in one direction and demultiplexing of the VT1.5 into a DS-1 in the other direction. Likewise, the cross-connection of DS-3 between a DS-3 interface and an OC-N termination requires mapping of the DS-3 into an STS-1 SPE in one direction and demultiplexing the STS-1 SPE into a DS-3 in the other direction. Cross-connection capability for other VT sizes such as VT2, VT3, and VT6 corresponding to E-1, DS-1C, and DS-2, respectively, are available as options in some applications. SONET cross-connect systems monitor the section, line, and path performance parameters on incoming OC-N facilities, monitor appropriate performance parameters of any DS-1 or DS-3 terminations, and report alarms and threshold crossings. Loopback capability for DS-1, VT1.5, and DS-3 signals is standard practice as well as test access to DS-1 and DS-3 signals [23].

11.6.3   Control of DCS

A DCS is controlled with a computer terminal that can be operated locally or remotely via a data link. Several control ports are available on the DCS to allow control from multiple sources. The ability to remotely control a DCS has led to the use of *customer controlled reconfiguration* (*CCR*), in which the customer has control over their portion of the circuits which traverse the DCS. With CCR the user sends reconfiguration instructions from the customer premise to the CCR controller via a data link. The controller checks the received instructions for accuracy and validity and processes the reconfiguration for immediate requests or stores the instruction for reserved requests. Using CCR the customer has the ability to implement alternate network maps that could describe different DS-0 and DS-1 interconnects. CCR also includes a security system that protects the customer network from unauthorized access. Initial applications of CCR were limited to DCS 1/0 but have been expanded to DCS 3/1/0. For example, AT&T's ACCUNET Bandwidth Manager uses DCS 3/1/0 to provide customer controlled reconfiguration over ACCUNET T1.5 (DS-1), ACCUNET T45 (DS-3), and ACCUNET Spectrum of Digital Services (Fractional T1) services [24].

11.7   Summary

During installation and operation of a transmission system, performance must be verified and continuously monitored to ensure that design and operating standards are met. This requirement has led to the development of testing, monitoring, and control techniques to support the introduction of digital transmission systems.

Testing techniques can be classified as in-service, when the mission traffic is unaffected by the test, or out-of-service, when the mission traffic must be removed to conduct the test. During installation and prior to system commission, out-of-service tests allow complete characterization of various digital performance parameters, such as bit error rate, jitter, and bit count integrity. Testing of other performance parameters, such as equipment reliability and multipath fading, requires longer test periods and more sophisticated techniques.

The robustness of digital transmission systems leads to new problems when one is trying to monitor its health during system operation. User performance is unaffected by loss of margin until the "last few decibels," where performance degrades rapidly. Performance monitors must provide a precursive indication of system degradation to allow operator intervention before user service becomes unacceptable. A number of in-service performance monitors are described here—including the pseudoerror monitor, which possesses the attributes necessary for monitoring of digital transmission systems.

Fault isolation is accomplished largely by alarms and status indicators that are built into the transmission equipment. For redundant equipment, most failures are automatically recognized and removed by protection switching. If redundancy is not used or protection switching itself fails, the operator must have adequate alarm and status indications to isolate the fault to the site and equipment. After fault isolation, service is restored by local or remote control of protection switches, by use of spare equipment, or by dispatching a maintenance team for equipment repair.

In addition to test equipment, alarms, and status indicators built into transmission equipment, some ancillary equipment is usually required to monitor and control the transmission line and system as a whole. Data service units and channel service units are used to monitor and control the user interface to the data port or transmission line, respectively. These DSU and CSU devices include the ability to extract overhead bits, calculate performance parameters, and initiate control actions, done either by the user or the carrier.

Monitoring and control of a large number of sites in a network requires a capability to aggregate monitoring and control subsystems from each equipment and report to a central facility. The collection and distribution of monitor, alarm, and status information requires a service channel that consists of telemetry and voice signals. The service channel is usually combined with the mission traffic using TDM or a subcarrier. At each site, the service channel is accessed to allow operator coordination between sites over a voice orderwire or exchange of alarm, monitor, and control signals. The typical monitor and control system employs a master station that periodically polls each of a number of remote stations to collect alarm and status signals. The master station then displays that information to assist the system operator in fault isolation and restoral of service.

Digital cross-connect systems (DCS) interface with standard digital signals, including DS-1, DS-3, and SONET hierarchical rates, and perform a number of functions that facilitate control, monitoring, and testing. The basic feature of a DCS is that of cross-connecting any circuit (for instance, DS-0) to any other circuit that terminates on the DCS. As a popular example, a DCS 1/0 interfaces with multiple DS-1s and allows the interconnection of any DS-0 in any of the DS-1s to any other DS-0 in any of the DS-1s. Other functions that are standard include monitoring of the overhead bits of the aggregate signals and test access to any circuit. DCS applications include hubbing for more efficient circuit routing than point-to-point, consolidation or segregation of circuits to fill and groom transmission facilities (for instance, T1 facilities), elimination of back-to-back channel banks or digital multiplexers, cross-connection of subrate or bundled channels, automated restoration of circuits and facilities, and customer control of circuit reconfiguration.

References

1. CCITT Blue Book, vol. IV.4, *Specifications for Measuring Equipment* (Geneva: ITU, 1989).
2. CCITT Blue Book, vol. VIII.1, *Data Communication Over the Telephone Network* (Geneva: ITU, 1989).
3. W. M. Rollins, "Confidence Level in Bit Error Rate Measurement," *Telecommunications* 11(December 1977):67–68.
4. U. S. MIL-STD-781C, *Reliability Design Qualification and Production Acceptance Tests: Exponential Distribution,* U.S. Department of Defense, October 21, 1977.
5. CCIR XVth Plenary Assembly, Vol. IX, Part 1, *Fixed Service Using Radio-Relay Systems* (Geneva: ITU, 1982).
6. T. L. Osborne and others, "In-Service Performance Monitoring for Digital Radio Systems," *1981 International Conference on Communications,* pp. 35.2.1–35.2.5.
7. K. Stauffer and A. Brajkovic, "DS-1 Extended Superframe Format and Related Performance Issues," *IEEE Comm. Mag.,* April 1989, pp. 19–23.
8. AT&T Technical Reference, TR 62415, *Access Specification for High Capacity (DS1/DS3) Dedicated Digital Services,* June 1989.
9. R. E. Mallon and S. Ravikumar, "Detection of Bursty Error Conditions Through Analysis of Performance Information," 1987 GLOBECOM.
10. D. R. Smith, "A Performance Monitoring Technique for Partial Response Transmission Systems," *1973 International Conference on Communications,* pp. 40.14–40.19.
11. B. J. Leon and others, "A Bit Error Rate Monitor for Digital PSK Links," *IEEE Trans. on Comm.,* vol. COM-23, no. 5, May 1975, pp. 518–525.

12. J. A. Crossett, "Monitor and Control of Digital Transmission Systems," *1981 International Conference on Communications*, pp. 35.6.1–35.6.5.

13. J. L. Osterholz, "Selective Diversity Combiner Design for Digital LOS Radios," *IEEE Trans. on Comm.*, vol. COM-27, no. 1, January 1979, pp. 229–233.

14. C. M. Thomas, J. E. Alexander, and E. W. Rahneberg, "A New Generation of Digital Microwave Radios for U.S. Military Telephone Networks," *IEEE Trans. on Comm.*, vol. COM-27, no. 12, December 1979, pp. 1916–1928.

15. CCITT Blue Book, vol. III.4, *General Aspects of Digital Transmission Systems; Terminal Equipments* (Geneva: ITU, 1989).

16. K. E. Brown and others, "Knowledge-Based Techniques for Fault Detection in Digital Microwave Radio Communication Equipment," *IEEE Journal on Sel. Areas in Comm.* 6 (June 1988): 819–827.

17. H. Heggestad, "Expert Systems in Communications System Control," *1991 Military Communications Conference*, pp. 36.3.1–36.3.7.

18. Bell System Publication 41450, *Digital Data System Data Service Unit Interface Specifications*, March 1973.

19. ANSI T1.403-1989, "Carrier-to-Customer Installation - DS1 Metallic Interface," February 1989.

20. AT&T Technical Reference, TR 54016, "Requirements for Interfacing Digital Terminal Equipment to Services Employing the Extended Superframe Format," September 1989.

21. Bellcore Technical Advisory, TA-TSY-000280, *Digital Cross-Connect System Requirements and Objectives for the Sub-Rate Data Cross-Connect Feature*, May 1986.

22. Bellcore Technical Advisory, TA-TSY-000233, *DS3 Digital Cross-Connect System (DCS 3/X) Requirements and Objectives*, December 1986.

23. Bellcore Technical Reference, TR-TSY-000233, *Wideband and Broadband Digital Cross-Connect Systems Generic Requirements and Objectives*, September 1989.

24. AT&T Technical Reference 62412, *ACCUNET Information Manager, ACCUNET Bandwidth Manager, and Direct Access Services*, July 1990.

Problems

11.1 Consider the use of a 2^4-1 maximal-length pseudorandom binary sequence for the measurement of error rate over a digital transmission channel. The transmitter has an initial state of all 1's in its shift register, and the receiver has an initial state of all 0's in its shift register. The receiver has two modes of operation, one for initial synchronization and the other for error detection. Assume that the synchronization circuit must be activated by an external command if a loss of synchronization occurs after initial synchronization.

(a) Draw a block diagram of the transmitter and receiver, showing how synchronization and error detection are done at the receiver.

(b) By showing the contents of the two shift registers for a specific transmitted pattern, find the time for the receiver to be synchronized to the transmitter.

(c) Show the effect of an error in the received bit stream on the error detector, by using a 10-bit transmitted sequence with one error received at the 7th bit.

11.2 Repeat Problem 11.1 but now assume that the synchronization circuit in the receiver is automatic so that it will resynchronize automatically if a loss of synchronization occurs in the transmission channel.

11.3 The pseudorandom pattern of length 2^{15}-1 recommended in CCITT Rec. 0.151 for error rate measurement at 1.544 Mb/s is created with a 15-stage shift register with feedback at the 14th and 15th stage via an EXCLUSIVE-OR gate to the first stage.

(a) Show the circuit needed to create this pseudorandom pattern.

(b) Now assume that the register is initially loaded with all 1's and then find the contents of the 15 stages of the shift register after 30 bit times.

11.4 The CRC-6 used with extended superframe format in DS-1 signals has a generating polynomial given by $x^6 + x + 1$. After the 4632-bit superframe has been serially fed to the CRC circuit, the register will contain the 6 CRC bits to be transmitted in the next extended superframe.

(a) Show a circuit consisting of a shift register with feedback paths to EX-OR gates required to implement this CRC-6.

(b) Assuming that the 4632 bits fed to the CRC circuit consists of all 1's, generate the 6-bit CRC.

(c) Show the receiver circuit for CRC-6 and demonstrate that the remainder equals 0 for the all-1's message of part (b).

(d) If the received signal contains 2 errors in the last 2 bits, show that the CRC will detect the errors.

11.5 For very low bit error rates, $p(e) \leq 10^{-5}$, the probability of error detection by the CRC-6, p_{CRC}, in the DS-1 extended superframe format is approximately equal to the probability that there is at least one error in the ESF block of 4632 bits. (a) Find an expression that approximates p_{CRC} for very low bit error rates. (b) Plot p_{CRC} as a function of $p(e)$ for $10^{-8} \leq p(e) \leq 10^{-5}$.

11.6 Calculate the system degradation that occurs with eye closures of: (a) one half the peak-to-peak signal; (b) one fourth the peak-to-peak signal; and (c) one eighth the peak-to-peak signal.

11.7 A duobinary signal is to be monitored for its error rate using a pseudoerror detector. An offset in the optimum sampling time is used in a second detector to measure the pseudoerrors. For an offset of $\frac{3}{8}$ of the sampling interval in the pseudoerror detector:

(a) Find the values of the duobinary impulse response $x(t)$ at the sampling time (x_0) and on either side of the sampling time (x_{-1} and x_1).

(b) Find an approximate expression for the pseudoerror rate as a function of signal-to-noise ratio (S/N).

(c) Plot the pseudoerror rate and the true error rate versus S/N.

11.8 Consider a 1.544-Mb/s transmission line that has randomly distributed errors with a bit error rate of 1×10^{-8}.

(a) Using a PRBS signal for an out-of-service BER measurement, what is the required measurement interval to establish a 90 percent confidence that the measured BER is within 10 percent of the true BER?

(b) Now suppose a pseudoerror technique is used that yields a 1×10^{-5} PER when the true BER is 1×10^{-8}. What measurement interval is required to establish a 90 percent confidence that the measured PER is within 10 percent of the true BER?

11.9 In DS-1 transmission systems a yellow alarm is transmitted to the distant end whenever a carrier failure alarm occurs at the near end. In the extended superframe format, the yellow alarm is signaled by a repetitive 16-bit pattern consisting of 1111111100000000 over the ESF data link. At the distant end a yellow alarm is declared if the exact 16-bit pattern is received in 7 out of 10 consecutive 16-bit pattern intervals.

(a) What is the minimum time required for the yellow alarm to be detected?

(b) What is the probability that the yellow alarm will be detected within the first 10 16-bit patterns if the transmission BER is 1×10^{-3}?

11.10 Consider a 1/0 digital cross-connect system (DCS) in which all incoming DS-1 signals are to be synchronized to the clock within the DCS. If the incoming digroups and the DCS are all timed with a clock meeting CCITT Rec. G.811 requirements (1×10^{-11} accuracy), what buffer length is required for each digroup to meet the G.811 slip rate requirements (no more than one slip in 70 days)?

12

Data Transmission Over the Telephone Network

OBJECTIVES

- Describes the techniques used to adapt telephone channels for the transmission of digital signals
- Outlines the levels of the FDM hierarchy from group to jumbogroup and supermastergroup
- Considers the transmission parameters that determine data communication performance over 4-kHz telephone circuits
- Explains how equalization can be used to correct two critical parameters: attentuation distortion and envelope delay distortion
- Discusses the use of voice-band modems and wideband modems for digital transmission
- Points out the advantages of transmultiplexers over the conventional interface between FDM and TDM systems
- Describes the new generation of hybrid transmission systems

12.1 Introduction

The extent to which telephone networks can accommodate data is limited by the presence of an analog interface with the data user. Economics often make it impractical to build digital transmission facilities, separate from the existing telephone network, for transmission of nontelephone digital signals, such as data, facsimile, or visual telephony. This chapter deals with the techniques used to adapt telephone channels for the transmission of digital signals.

Digital signals can be converted to a form suitable for the telephone network by use of the digital modulation techniques described in Chapter 6. The equipment designed to interface data with telephone channels is termed a **modem** (an acronym for modulator/demodulator). The choice of modulation technique depends on the signal bit rate and characteristics of the channel such as bandwidth and signal-to-noise ratio. The channel characteristics found in telephone systems are well defined within the telephone industry for the basic voice channel and for the frequency-division multiplex (FDM) used to combine voice channels.

Today most data transmission takes place over individual voice channels within a bandwidth of 3 kHz or less. The characteristics of voice channels that affect modem performance can have wide-ranging values, depending on the length and routing of a call. For *dial-up, switched* lines, performance characteristics vary from call to call, so the modem must be designed to adapt to these varying conditions. Use of *private, leased* lines permits simpler modem design, since the channel conditions can be controlled by judicious routing and the use of phase and amplitude equalization.

Most modems operate with simultaneous two-way data exchange (**duplex**). When the modulator and demodulator share the same line, however, **half-duplex** transmission is required in which the modem can transmit and receive data but not simultaneously. **Simplex** modems can only transmit or receive data. In general, half-duplex modems interface with two-wire lines and duplex modems interface with four-wire lines. For low-speed transmission, however, modems can be designed to operate duplex over two-wire lines by use of separate frequency bands between transmit and receive carriers. Compatibility between the data terminal equipment (DTE) and modem requires that data, timing, and control lines be specified. Standard interfaces have been developed by the TIA and CCITT (see Table 2.1). Modems operate with either asynchronous or synchronous interface with the DTE. Asynchronous data terminals do not use a clock signal to define data transitions; synchronous data terminals use a clock provided by either the DTE or the modem. In general, low-speed modems (below 1200 b/s) operate asynchronously and high-speed modems synchronously.

Wideband data require a greater transmission bandwidth than the voice channel. Wideband modems provide the necessary bandwidth by using the bandwidths that occur naturally in the FDM hierarchy. If the requirement for data exceeds the total available transmission bandwidth, however, separate digital transmission facilities are required. As a compromise, hybrid transmission systems can also be used for wideband data; in hybrid systems, voice and data signals share the same transmission media. Since today's communication systems are still dominated by the analog telephone interface, the techniques described in this chapter are expected to play a continuing role in future digital transmission systems.

12.2   Frequency-Division Multiplex (FDM)

In frequency-division multiplex (FDM) each signal is allocated a discrete portion of the frequency spectrum. At the transmitter, each baseband signal is shifted in frequency by a modulation process (usually AM). Carrier frequencies are selected such that the modulated signals occupy adjacent, nonoverlapping frequency bands. Filters with sharp cutoff at the band edges are required in order to minimize interference between signals. At the receiver, each baseband signal is recovered by filtering and demodulation.

Here we focus on the use of FDM in telephone channel transmission. Several levels of multiplexing are used today that form an FDM hierarchy. The first level in the standard hierarchy is the **group,** which occupies the frequency band 60 to 108 kHz and contains 12 voice channels, each of nominal 4 kHz bandwidth. Figure 12.1 illustrates the operation of an FDM group. At the transmitter each voice signal is low-pass-filtered and then used to modulate 12 carriers spaced 4 kHz apart. The modulation process produces a double sideband signal. The outputs of the 12 modulators are bandpass-filtered to limit each signal to 4 kHz. By convention the lower sideband signal is selected, as illustrated in Figure 12.1. At the receiver, bandpass filters pick off each of the 12 signals for demodulation. Synchronization of the carrier frequencies between the transmitter and receiver is achieved by use of a pilot tone. The pilot is transmitted as part of the group signal, usually in one of the narrow guard bands used to separate voice channels. At the receiver, the pilot is recovered and used to synchronize the frequencies used in demodulation.

The second level in the FDM hierarchy is the **supergroup,** formed by combining five groups into a 240-kHz signal. Each group modulates a carrier with frequency $372 \pm 48n$ kHz ($n = 1$ to 5). A bandpass filter selects the lower sideband, and the five resulting signals are combined to form the supergroup, which occupies the frequency band 312 to 552 kHz. This modulation and filtering process is illustrated in Figure 12.2 for the supergroup signal.

The third level of the FDM hierarchy consists of the **mastergroup,** made up of five or ten supergroups, depending on the choice of the CCITT or North American standard. Higher levels of FDM also differ in these two

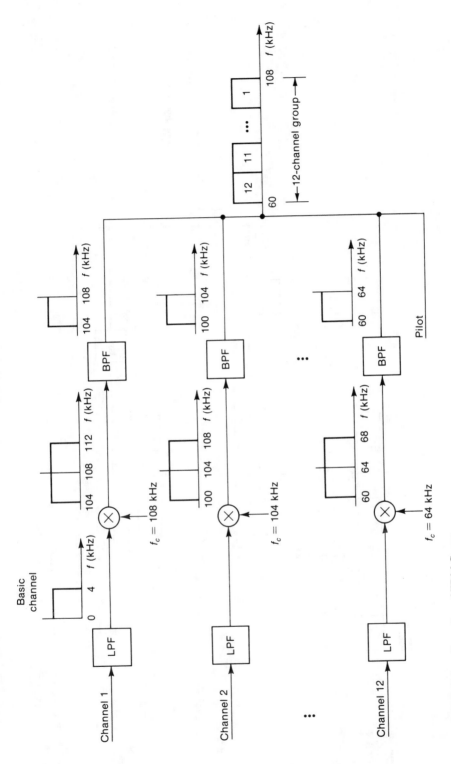

FIGURE 12.1 Operation of FDM Group

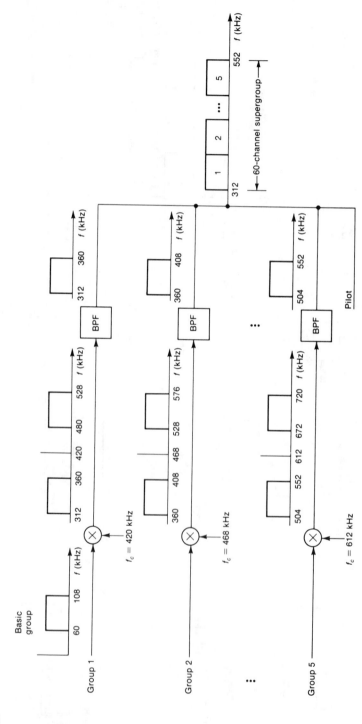

FIGURE 12.2 Operation of FDM Supergroup

TABLE 12.1   CCITT versus North American FDM Hierarchy

Number of Voice Channels	Spectrum (kHz)	CCITT Standard [2]	North American Standard
12	60–108	Group (Rec. G.232)	Group
60	312–552	Supergroup (Rec. G.233)	Supergroup
300	812–2044	Mastergroup (Rec. G.233)	
600	60–2788 (L600) 564–3084 (U600)		Mastergroup
900	8516–12,388	Supermastergroup (Rec. G.233)	
3600	564–17,548		Jumbogroup
10,800	3000–60,000		Jumbogroup multiplex

standard hierarchies. Table 12.1 compares North American and CCITT FDM standards.

12.2.1   North American FDM Hierarchy

Figure 12.3 illustrates the basic building blocks of the North American FDM hierarchy [1]. The basic voice channel has a 200 to 3400 Hz spectrum. Although intended primarily for voice transmission, the voice channel can also be used for data transmission provided that the data spectrum does not exceed the basic voice channel spectrum. Two examples are shown in Figure 12.3: the use of voice-band modems for rates up to 19.2 kb/s and the use of data multiplexers for combining a number of low-speed teletype channels. The 12-channel group is generated by use of single sideband (SSB) modulation by equipment known as an A-type channel bank. The 60-channel supergroup results from combining five groups using a group bank. Again using SSB modulation, the resulting supergroup signal occupies 240 kHz in the frequency range 312 to 552 kHz. The modulation process for the group and supergroup is depicted in Figures 12.1 and 12.2. Any other signal whose spectrum matches that of the group or supergroup may also be multiplexed by using this standard FDM equipment. For example, a 56-kb/s signal can be accommodated by a group bank by use of a group modem, which converts the 56-kb/s signal to a 48-kHz spectrum in the group frequency band. Another example is the transmission of a 256-kb/s signal in a 240-kHz supergroup by means of a supergroup modem.

The supergroup bank combines 10 supergroups into a 600-channel mastergroup. Single sideband modulation is again used to generate the mastergroup. As shown in Figure 12.3, two versions of the mastergroup exist. The

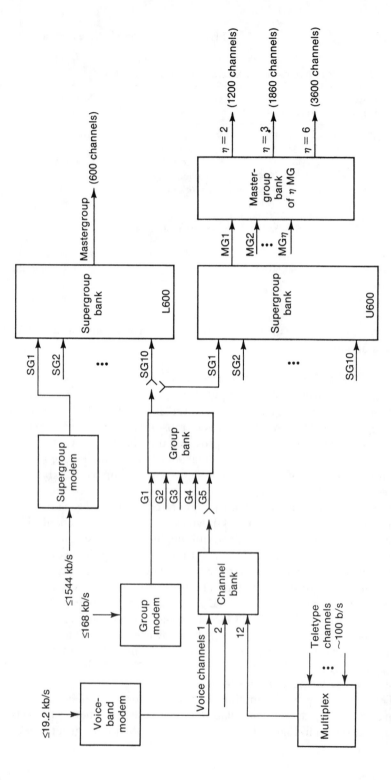

FIGURE 12.3 North American FDM Hierarchy (Adapted from [1])

L600 mastergroup occupies the 60- to 2788-kHz frequency band and is used for wideband transmission over the L1 coaxial cable system and various analog microwave radio systems. The U600 mastergroup occupies the 564- to 3084-kHz frequency range and is used to build larger channel systems. One example of these larger groupings is the L3 carrier consisting of three mastergroups and one supergroup combined to provide 1860 channels. The L4 carrier is formed by combining six U600 mastergroups into a **jumbogroup** comprising 3600 voice channels and occupying the frequency band 564 to 17,548 kHz. The L5 carrier combines three jumbogroups into approximately a 60-MHz bandwidth that contains 10,800 channels.

12.2.2   CCITT's FDM Hierarchy
The standard group and supergroup as defined by the CCITT are the same as the North American standards [2]. However, the CCITT mastergroup contains only five supergroups comprising 300 voice channels. Using SSB modulation, the five supergroups are translated in frequency, resulting in a mastergroup that occupies the spectrum 812 to 2044 kHz. The highest level prescribed by the CCITT's FDM hierarchy is the **supermastergroup,** which contains three mastergroups and occupies the band 8516 to 12,388 kHz.

12.3   Transmission Parameters

In this section we consider the transmission parameters that determine data communications performance over 4-kHz telephone circuits [3, 4]. Depending on the modem design, some of these parameters may have little effect on performance. For most modems, however, the amplitude and phase characteristics are of primary importance due to the intersymbol interference that results from amplitude and phase distortion.

12.3.1   Attenuation Distortion
Ideally, all frequencies of a signal experience the same loss (or gain) in traversing the transmission channel. Typical channels exhibit variation in loss with frequency, however, a form of distortion known as **attenuation** or **amplitude distortion.** Wire-line systems, for instance, are characterized by greater attenuation at high frequencies than at lower ones. Moreover, filters used to bandlimit a signal cause greater attenuation at the band edges than at band center.

Attenuation distortion is specified by a limit placed on the loss at any frequency in the passband relative to the loss at a reference frequency. CCITT Rec. G.132 shown in Figure 12.4 allows up to 9-dB variation from the value expected at an 800-Hz reference frequency. This recommendation is for the case of a four-wire international connection comprising up

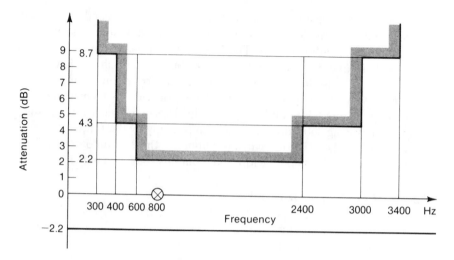

FIGURE 12.4   Attenuation Distortion Allowed by CCITT for a Four-Wire International Telephone Circuit (Courtesy CCITT [2])

to 12 circuits in tandem. AT&T uses a 1004-Hz reference frequency. Attenuation characteristics of the basic AT&T telephone circuit are given in Table 12.2.

12.3.2   Envelope Delay Distortion

The time required for signal transmission through a channel is finite and therefore produces a delay in the arrival time of the signal. For example, a waveform $\sin\omega t$ applied to the input of a transmission channel has an output waveform $\sin(\omega t - \beta)$. The phase shift β can be a function of frequency, so that different phase shifts may occur at various frequencies. The time delay between input and output waveforms is called **phase delay,** defined as

$$T_p = \frac{\beta}{\omega} \frac{\text{rad}}{\text{rad}/\text{s}}$$

(12.1)

If the phase delay varies with frequency, the output signal will be disturbed because of the difference in arrival time of each frequency. The difference between phase delay at two frequencies is termed **delay distortion,** defined by

$$T_d = \frac{\beta_2}{\omega_2} - \frac{\beta_1}{\omega_1}$$

(12.2)

TABLE 12.2   AT&T Conditioning Specifications

Channel Conditioning[a]	Attenuation Distortion (Frequency Response) Relative to 1004 Hz		Envelope Delay Distortion	
	Frequency Range (Hz)[b]	Variation (dB)[c]	Frequency Range (Hz)[b]	Variation (μs)
Basic	500–2500	−2 to +8	800–2600	1750
	300–3000	−3 to +12		
C1	1000–2400[d]	−1 to +3	1000–2400[d]	1000
	300–2700[d]	−2 to +6	800–2600	1750
	300–3000	−3 to +12		
C2	500–2800[d]	−1 to +3	1000–2600[d]	500
	300–3000[d]	−2 to +6	600–2600[d]	1500
			500–2800[d]	3000
C3 (access line)	500–2800[d]	−0.5 to +1.5	1000–2600[d]	110
	300–3000[d]	−0.8 to +3	600–2600[d]	300
			500–2800[d]	650
C3 (trunk)	500–2800[d]	−0.5 to +1	1000–2600[d]	80
	300–3000[d]	−0.8 to +3	600–2600[d]	260
			500–2800[d]	500
C4	500–3000[d]	−2 to +3	1000–2600[d]	300
	300–3200[d]	−2 to +6	800–2800[d]	500
			600–3000[d]	1500
			500–3000[d]	3000
C5	500–2800[d]	−0.5 to +1.5	1000–2600[d]	100
	300–3000[d]	−1 to +3	600–2600[d]	300
			500–2800[d]	600

[a]C conditioning applies only to the attenuation and envelope delay characteristics.
[b]Measurement frequencies will be 4 Hz above those shown. For example, the basic channel will have −2 to +8 dB loss, with respect to the 1004 Hz loss, between 504 and 2504 Hz.
[c](+) means loss with respect to 1004 Hz; (−) means gain with respect to 1004 Hz.
[d]These specifications are FCC-tariffed items. Other specifications are AT&T objectives.
Source: Reprinted from Reference [3] by permission of AT&T.

For modulated waveforms, the envelope of the signal may also suffer distortion due to differences in propagation time between any two specified frequencies. This **envelope delay** or **group delay distortion** is defined as the variation in the slope of the phase shift characteristic:

$$T_e = \frac{d\beta}{d\omega} \tag{12.3}$$

Envelope delay distortion is expressed relative to a reference frequency. For example, CCITT Rec. G.133 uses an 800-Hz reference frequency and

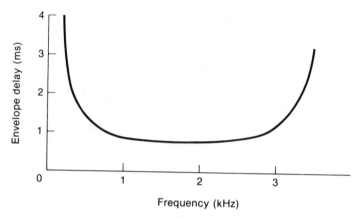

FIGURE 12.5   Envelope Delay Distortion of A5 Channel Bank [3]

AT&T uses 1004 Hz. Envelope delay found in a voice channel of an AT&T A5 channel bank is shown in Figure 12.5. The parabolic shape of envelope delay for the A5 channel bank is typical for FDM equipment, where maximum delays occur at the band edges and minimum delays at the center of the band. Envelope delay distortion is specified by the maximum variation in envelope delay permitted over a band of frequencies. Table 12.2 indicates the limits on envelope delay distortion for the basic AT&T voice channel.

12.3.3   Noise

When amplitude and delay distortion are controlled with equalization, noise often becomes the fundamental limitation to data transmission performance. Noise found in telephone circuits can be classified as gaussian, impulse, or single frequency. Gaussian noise is due to thermal agitation of electrons in a conductor. Its spectrum is uniform over the whole frequency range; hence the noise is also white. The effects of gaussian white noise on probability of error in data transmission over telephone circuits has already been presented in Chapter 6, which described digital modulation techniques used in data modems.

Impulse noise is characterized by large peaks that are much higher than the normal peaks of channel (thermal) noise. Thus with the occurrence of impulse noise, the probability of data error is even higher than that caused by gaussian noise. Measurement is made by counting the number of impulses above a selected threshold in a certain test period. Impulse noise can be caused by natural sources, such as lightning, but most impulse noise can be traced to such sources as switching transients and normal maintenance, repair, and installation activities in the telephone plant. Because of the nature of impulse noise, mathematical characterization is not practical. Instead, direct measurement of a particular channel is used. The AT&T specification for voice channels limits impulse noise to 15 counts in 15 min, using a threshold of −22 dBm [3].

Examples of single-frequency interference are crosstalk and intermodulation. Crosstalk occurs when an unwanted signal path is coupled with the desired signal path. Intermodulation noise results from the presence of nonlinearities that cause intermodulation products, some of which appear in the passband. For data transmission, these single-frequency interferers have greatest effect on multiplexers that transmit several low-speed sources, such as teletype signals, over a single voice channel. Whereas gaussian noise is uniformly distributed over the voice band, a single-frequency interferer may present a higher noise level at a particular slot in the voice channel. The AT&T standard voice channel specifies single-frequency interference to be at least 3 dB below the thermal noise level [3].

12.3.4   Phase Jitter
Phase jitter arises from incidental frequency or phase modulation about the desired transition times of the data signal. Phase jitter contributes to data errors by reducing the receiver margin to other impairments. Sources of phase jitter include instabilities in power supplies and in oscillators used to generate carrier frequencies in FDM equipment. The most common frequency components of jitter are found at power frequencies (say 50 or 60 Hz), ringing currents (say 20 Hz), and at harmonics of these frequencies. The AT&T limit on jitter in the voice channel is 10° peak-to-peak for the frequency band of 20 to 300 Hz and 15° peak-to-peak for the band 4 to 300 Hz [5].

12.3.5   Level Variation
Standard practice in FDM systems is to set transmission levels to control crosstalk and intermodulation. Allowed level variations are also prescribed. For example, the channel loss for a 1004-Hz reference frequency is specified as 16 dB ± 1 dB by AT&T [3]. However, variations in the prescribed level occur due to short-term and long-term effects. Hence a data modem receiver must incorporate automatic gain control.

Short-term level variations occur typically over a period of a few seconds and may be due to maintenance actions such as patching, automatic switching to standby equipment, or fading in microwave radio systems. Longer-term level variations are caused primarily by temperature changes, component aging, and amplifier drift. AT&T limitations on level variations are ±3 dB for short-term variations and ±4 dB for long term, relative to the nominal 16-dB channel loss specification [3].

12.4 Conditioning

Of the transmission parameters described in Section 12.3, two critical parameters, attenuation distortion and envelope delay distortion, can be corrected by equalization. The remaining characteristics described in Section 12.3 can be corrected only by careful selection of the circuit configuration, which involves routing the circuit to avoid certain types of media or equipment. Equalization may be done by separate amplitude and envelope delay compensation networks or simultaneously in a single network. In either case, the objective is to flatten the amplitude and envelope delay response to prescribed levels.

An equalizer thus performs two functions. Using the amplitude of the reference frequency, the equalizer adds attenuation or gain to other frequencies until the entire amplitude response is flat. The second function of the equalizer is to add delay to center frequencies in order to match the delay found at the band edge. The overall result is to flatten the amplitude response and envelope delay characteristic to within a specified tolerance relative to values at the reference frequency. In practice, it is not possible to equalize amplitude and envelope delay completely over the entire channel bandwidth.

Equalization can be performed by the user's equipment, by the telephone company's equipment, or by a combination of the two. A user-provided modem is likely to include a form of equalization, which may be fixed or adjustable. Fixed or "compromise" equalizers are based on the average characteristics of a large number of telephone circuits. Adjustable equalizers are of two types: manual and automatic. For high-speed data transmission at rates above 4800 b/s, automatic adaptive equalization is now a standard feature in modems. Thus newer modems are designed to operate over basic (unconditioned) lines.

Equalization provided by the telephone company is known as **conditioning** and is available in different grades for voice-band channels. These grades correspond to different sets of allowed attenuation and envelope delay distortion. The required grade of conditioning depends on the bandwidth of the data signal; larger bandwidths require a greater degree of equalization. Table 12.2 shows the grades of circuit conditioning available on AT&T facilities. The various levels of conditioning are specified as C1 through C5. The attenuation and envelope delay characteristics of the basic voice channel are shown for comparison with conditioned circuits. Conditioned circuits are also prescribed by the CCITT for international leased circuits according to the following recommendations [6]:

- Rec. M.1020 (equivalent to AT&T C2): conditioning for use with modems that do not contain equalizers
- Rec. M.1025: conditioning for use with modems that contain equalizers

- Rec. M.1040: ordinary quality for applications that do not require conditioning

Circuit conditioning may be performed at various points in the transmission system. Most commonly, circuits are equalized at the receiver, which is termed **postequalization.** If the line characteristics are known, **preequalization** may be employed in which the transmitted signal is predistorted. If the predistortion amounts to the inverse characteristic of the line, then the overall amplitude and envelope delay response will be flat at the receiver. For long-haul international circuits, additional equalization may be required at the gateway nodes. This form of equalization, which combines the effects of postequalization and preequalization, is known as **midpoint equalization.**

Another type of conditioning, high-performance data conditioning or type D, specifies performance of two parameters: signal-to-noise ratio and harmonic distortion. Type D conditioning is independent of C conditioning and is achieved by careful selection of transmission facilities. The need for D conditioning arises with high-speed modems (≥ 9.6 kb/s) whose performance is limited by noise and distortion.

12.5 Voice-Band Modems

Digital transmission over existing FDM telephone facilities was initially limited to modems operating over single voice-band channels. The earliest forms of voice-band modems used binary FSK. Rates up to 1800 b/s can be conveniently transmitted in a nominal 4-kHz channel using FSK. For binary FSK, two frequencies are required within the voice band, one used for a mark (1) and the other for a space (0). For example, CCITT Rec. V.23 specifies binary FSK for a 600/1200-b/s modem operating in the general switched telephone network, as indicated here:

	Mark Frequency	Space Frequency
Mode 1: up to 600 b/s	1300 Hz	1700 Hz
Mode 2: up to 1200 b/s	1300 Hz	2100 Hz

For slower rates, the voice-channel bandwidth can be divided into separate frequency subbands to accommodate several data signals. This form of FDM is known as voice frequency telegraph (VFTG) or voice frequency carrier telegraph (VFCT). Each data channel is allocated a certain frequency subband. FSK modulation is most commonly used, where the space and mark frequencies are contained within a specific subband. Typical applications range from multiplexing two 600-b/s channels into a composite 1200 b/s to multiplexing 24 75-b/s channels into a composite 1800 b/s. Standards for VFCT modulation format have been developed for several applications. In these modulation plans, frequencies are usually spaced uniformly across the voice-channel bandwidth. For 75-b/s telegraph channels, 12 channels can be handled with 120-Hz spacing, 16 channels with 85-Hz spacing, and 24 channels with 60-Hz spacing.

Differential phase-shift keying (DPSK) has been commonly used to extend modem transmission rates beyond the limitations of FSK modems. Today, four-phase DPSK is a universal standard for 2400 b/s, as described by CCITT Rec. V.26; similarly, eight-phase DPSK has become a universal standard for 4800 b/s, as described by CCITT Rec. V.27. Table 12.3 and Figure 12.6 present characteristics of these two and other standard high-speed voiceband modems. DPSK is generally limited to rates up to 4800 b/s because of the susceptibility of PSK systems to phase jitter found on voice channels.

Further increases in bandwidth efficiency were introduced with the use of single sideband (SSB) and vestigial sideband (VSB) modulation. Modems designed with SSB or VSB proved to be complex and vulnerable to channel perturbations, however, especially amplitude and envelope delay distortion. The use of these linear modulation techniques introduced the need for automatic equalization. Both preset and adaptive modes of equalization have been used. (See Chapter 5 for a detailed description of equalization techniques.) Preset equalization uses a training sequence to fix the equalizer settings; adaptive equalization provides continuous updating. The adaptive equalizer is preferable and has become a universal feature in high-speed modems. To ensure convergence of the equalizer, data scrambling is required and this feature too has become standard in high-speed modems.

Advanced modulation and spectrum shaping techniques are today used to achieve up to 19.2-kb/s transmission rates. Filtering schemes such as partial response and raised cosine, described in Chapter 5, are used to control or eliminate intersymbol interference. Such spectrum shaping is often used in combination with the modulation technique in high-speed modems. For rates at 9600 b/s and higher, quadrature amplitude modulation (QAM) and various forms of amplitude/phase modulation (AM/PM) have been used. At 9600 b/s, AM/PM with a 16-state constellation is part of a universal standard, CCITT Rec. V.29, as indicated in Table 12.3 and Figure 12.6c. A 64-state constellation that has been implemented with a 16-kb/s modem also uses AM/PM as illustrated in Figure 12.6d [8]. 64-QAM has been used for both 14.4-kb/s and 16-kb/s modems. Attendant with these advances in modulation techniques, sophisticated equalization and carrier recovery have

TABLE 12.3 Transmission Characteristics of Standard High-Speed Voice-Band Modems [7]

Characteristic	CCITT Rec. V.26 ter	CCITT Rec. V.27 ter	CCITT Rec. V.29	CCITT Rec. V.32	CCITT Rec. V.33
Data rate	2400 b/s ± 0.01%	4800 b/s ± 0.01%	9600 b/s ± 0.01%	9600 b/s ± 0.01%	14400 b/s ± 0.01%
Mode of operation	Two-wire, full-duplex	Two-wire, full-duplex	Four-wire, full-duplex	Two-wire, full-duplex	Four-wire, full-duplex
Modulation rate	1200 symbols/s ± 0.01%	1600 symbols/s ± 0.01%	2400 symbols/s ± 0.01%	2400 symbols/s ± 0.01%	2400 symbols/s ± 0.01%
Carrier frequency	1800 ± 1 Hz	1800 ± 1 Hz	1700 ± 1 Hz	1800 ± 1 Hz	1800 ± 1 Hz
Spectrum shaping	100% raised cosine	50% raised cosine	N/A	N/A	N/A
Modulation type[a]	4-DPSK	8-DPSK	16-state AM/PM	Trellis coding with 32-signal space	Trellis coding with 128-signal space
Equalization	Fixed compromise or automatic adaptive	Automatic adaptive	Automatic adaptive	Automatic adaptive	Automatic adaptive
Scrambler	Self-synchronizing with length $2^{23}-1$	Self-synchronizing with length 2^7-1	Self-synchronizing with length $2^{23}-1$	Self-synchronizing with length $2^{23}-1$	Self-synchronizing with length $2^{23}-1$

[a]For signal constellations see Figures 12.6 and 6.40

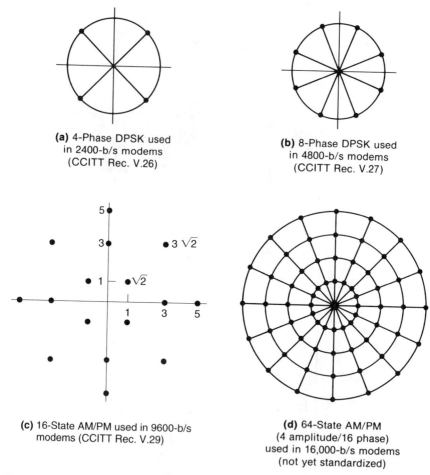

(a) 4-Phase DPSK used
in 2400-b/s modems
(CCITT Rec. V.26)

(b) 8-Phase DPSK used
in 4800-b/s modems
(CCITT Rec. V.27)

(c) 16-State AM/PM used in 9600-b/s
modems (CCITT Rec. V.29)

(d) 64-State AM/PM
(4 amplitude/16 phase)
used in 16,000-b/s modems
(not yet standardized)

FIGURE 12.6   Standard Signal Constellations for Voice-Band Modems

been incorporated into modems to the point where conditioned lines may
not be required.

Trellis coding, error control, data compression, and echo control have
been added to second generation high-speed modems to achieve greater
throughput and improved performance. The CCITT V.32 modem employs
8-state trellis coding with a nominal 4-dB coding gain, as described earlier
in Chapter 6, to provide a 9.6-kb/s data rate. Current generation 14.4-kb/s
modems—based on CCITT Rec. V.33 (four-wire), V.32*bis* (two-wire), and
V.17 (Group Three facsimile)—have a 128-QAM signal constellation used
in conjunction with 8-state trellis coding. The nonlinear convolutional

encoder used with these 14.4-kb/s modems is identical to that used with the V.32 modem (see Chapter 6). CCITT Study Group XVII is considering the use of multidimensional trellis coding for 19.2/24-kb/s modems. All these CCITT recommendations include fallback rates and standards to allow operation at lower rates over marginal circuits [7].

12.5.1   Error Control

To provide better performance in the presense of transmission errors, modem manufacturers began to add error correction. To ensure compatibility, the CCITT has published Rec. V.42, which specifies a new high-level data link control-based protocol called Link Access Procedure for Modems (LAPM). LAPM specifies that data be transmitted in frames whose structure is illustrated in Figure 12.7. Each frame has the following fields [7]:

Flag Fields
All frames are delimited by the 8-bit pattern 01111110, known as the flag. The flag found before the address field is known as the opening flag, while the flag located after the frame check sequence is referred to as the closing flag. A single flag may be used as the closing flag for one frame and the opening flag for the next.

Address Field
The address field is used to identify the originator that transmitted the frame or destination that is to receive the frame. For point-to-point circuits, this field is not needed but is included for sake of uniformity. An address is normally 8 bits but may be extended to 16 bits.

Control Field
The control field identifies the type of frame, which will be a command or response. Three types of control field formats are specified: numbered information transfer (I format), supervisory functions (S format), and unnumbered information transfers and control functions (U format).

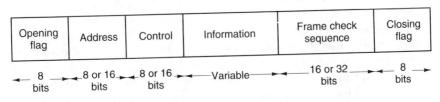

FIGURE 12.7   LAPM Frame Structure

Information Field
The information field is present only in I frames and some U frames. The maximum number of octets that can be carried by the information field is normally 128 but different values may be used for the two directions of data transmission.

Frame Check Sequence Field
The frame check sequence is a cyclic redundancy check (CRC) code that is applied to the remaining bits of the frame, excluding the flags. CCITT Rec. V.42 provides error-correcting procedures for voice-band modems using the following versions of $P(x)$:

$$CRC-16 = x^{16} + x^{12} + x^5 + 1$$

$$CRC-32 = x^{32} + x^{26} + x^{23} + x^{22} + x^{16} + x^{12} + x^{11}$$
$$+ x^{10} + x^8 + x^7 + x^5 + x^4 + x^2 + x + 1$$

The CRC is normally set at 16 bits but the 32-bit code may be used if the frame length or line error rate is excessive. See Chapter 11 or reference [9] for a further discussion of CRC codes.

12.5.2 Data Compression

The idea behind data compression is the removal of as much of the redundancy in the raw data input as possible. There are several techniques that are possible in voice-band modem applications. One of these techniques, known as **run-length** coding, removes a run of identical characters and replaces it with a much shorter set of characters. Another commonly used scheme is **Huffman** coding, in which code lengths are related to their probability of occurrence. The Huffman code increases the information rate by decreasing the average size of transmitted data characters. Further improvement in compression is possible if start and stop bits are removed from asynchronous characters before they are transmitted.

As a follow-up to V.42, the CCITT has approved V.42bis, which introduces a data compression standard for voice-band modems. V.42bis is based on a version of the Lempel-Ziv data compression algorithm that is similar to techniques used in computer file compression [10]. The encoder specified in V.42 is capable of operation in two modes, a transparent mode and a compressed mode. The choice of modes is determined by a test of data compressibility, in which the efficiency of the data compression algorithm is estimated. When the compression mode is activated, the algorithm converts a string of characters from the DTE into a fixed-length code word. The string–code word relationship is stored in two dictionaries, one used at the encoder and the other used in conjunction with error control at the decoder. Character strings are matched to dictionary code words by starting with a single character and then adding additional characters so long as there is a code word that corresponds to the newly created string. Procedures are specified for updating, in which strings can be added to the dictionary, and for deletion, in which infrequently used strings are deleted to allow reuse of code words.

12.5.3 Echo Control

First-generation high-speed modems such as V.27, V.29, and V.33 were designed to operate over four-wire, full-duplex leased lines. The main drawback to these modems is their inability to operate over standard dial-up lines. Whereas leased lines employ a four-wire circuit for the entire communications path, a dial-up line uses a two-wire local loop to connect each modem to a central office, as shown in Figure 12.8. Four-wire circuits are then used to interconnect central offices. Hybrid circuits are used to connect the two-wire circuit to the four-wire circuit and to isolate the modem's transmitter from its receiver. Because impedances of the two-wire and four-wire circuits are not perfectly matched by the hybrid circuits, reflections of the transmitted signal occur both at the near end and far end, causing interference at the modem's receiver. The solution to this echo problem is the use of an echo canceller placed between the modem's transmitter and receiver, as shown in Figure 12.9. As a result, in full-duplex operation, both modems can utilize the entire bandwidth of the telephone channel.

The echo canceller generates a replica of the real line echo which is then subtracted from the incoming signal. A perfectly replicated echo will cancel the true echo reflected back from the hybrid circuits so that the modem receiver input consists only of the desired signal. In voice-band data applications (as opposed to voice applications), the echo canceller gets its input from the data symbols at the input to the transmitter. The fact that there are two types of echo, near end and far end, each with different characteristics,

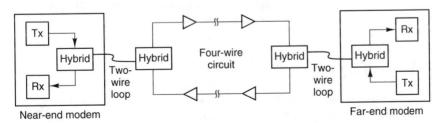

FIGURE 12.8   A Typical Dial-Up Telephone Line Connection

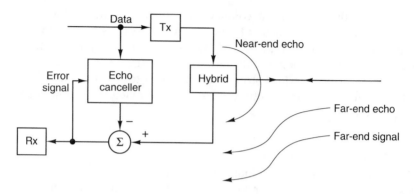

FIGURE 12.9   Data-Driven Echo Canceller Used in Modem

suggests that the echo canceller must have two different sections, a "near canceller" and a "far canceller." The requirements for these two cancellers are quite different. Near-end echo is characterized by a short time delay, usually less than 20 ms, and by a large amplitude, which can be up to 10 dB below the transmitted signal. Far-end echo has a much lower level, but a longer delay, which varies depending on the transmission distance from a few ms to 600 ms for a satellite link. The standard design approach calls for the use of transversal filters with variable tap coefficients. A minimum mean square error criterion is commonly used to adapt the tap coefficients. The number of taps in the delay line is governed by the expected range of the echo time delay, but the far canceller requires a larger number of taps than the near canceller due to the delay characteristics of far-end echo. Tap coefficients are updated under the control of the error signal as shown in Figure 12.9 [11].

CCITT Rec. V.26*ter*, V.27*ter*, and V.32 prescribe standards for two-wire, full-duplex modems with echo cancellation operating at 2.4, 4.8, and 9.6 kb/s, respectively. Although the echo canceller design and operation need not be specified, the operations needed to train the echo canceller are given in these recommendations. The calling and answering modems exchange known bit sequences until a sufficient degree of echo cancellation is established. These sequences may also be used to train the adaptive equalizer in the receiver.

12.6   Wideband Modems

For data rates above 24 kb/s, higher levels in the FDM hierarchy are required for modem operation. The FDM group with 48 kHz bandwidth can be accessed with a group modem to provide wideband data transmission. Earliest applications of group modems allowed data rates of 19.2 or 38.4 kb/s using simple ASK, FSK, or PSK modulation. These choices of data rate were extensions of the teletype hierarchy given by 75×2^n b/s, with n equal to an integer. Standards for group modems are now based on an $8n$-kb/s standard, which has evolved from the 8-kHz sampling rate for digitized voice. CCITT Recs. V.35, V.36, and V.37 for group modems are based on this $8n$-kb/s standard, as shown in Table 12.4. For data rates above 72 kb/s, automatic adaptive equalization techniques have been applied to group modem design, resulting in rates of up to 168 kb/s, as specified by CCITT Rec. V.37.

The next level in the FDM hierarchy is the 240-kHz bandwidth supergroup. Data rates on the order of 250 kb/s have been transmitted via a supergroup modem, such as the Western Electric 303 data set. Using efficient modulation types and automatic adaptive equalization, supergroup modems have been demonstrated for rates up to 1544 kb/s, although there is insufficient demand for the introduction of such rates.

An increasing demand for wideband digital transmission over existing microwave systems led to the development of wideband modems that interface directly with an analog radio. Frequency modulation was once a worldwide standard in microwave radio systems in conjunction with standard FDM equipment for multichannel voice transmission. With increased use of PCM repeatered lines, however, modems were developed to allow direct interface of PCM and TDM equipment with analog FM radios, thereby eliminating the need for converters between FDM and TDM systems.

The basic functions of a digital FM microwave radio system are shown in Figure 12.10. The transmitting TDM combines digital channels into a single serial bit stream. The multiplexer output is encoded into an M-ary baseband signal, which is then shaped by the transmit filter. After transmission over the FM radio link, the demodulated signal is passed through the receive filter and decoder to recover the TDM signal. Typical choices of filter characteristic are raised cosine or partial response, split equally between the

TABLE 12.4   Transmission Characteristics of CCITT-Recommended Group Modems [7]

Characteristic	CCITT Rec. V.35	CCITT Rec. V.36	CCITT Rec. V.37
Data rates	48 kb/s	48, 56, 64, and 72 kb/s	96, 112, 128, 144, and 168 kb/s
Modulation type	AM, suppressed carrier	3-level partial response	7-level partial response
Equalization	N/A	N/A	Automatic adaptive
Scrambler	Self-synchronizing with length $2^{20}-1$	Self-synchronizing with length $2^{20}-1$	Self-synchronizing with length $2^{20}-1$

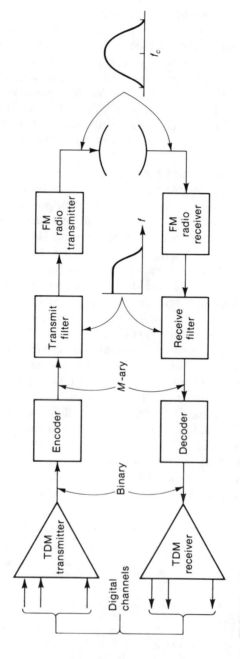

FIGURE 12.10 Digital Transmission Over FM Microwave

transmitter and receiver as shown in Figure 12.10. Implementation alternatives are to build the encoder and filter into the radio or multiplexer [12] or as separate equipment [13].

Digital FM has been used to provide radio transmission of multiple T1 and T2 digital channels. For example, the TD-2 FM radio used by AT&T has been configured for digital transmission of three T2 signals, at a total bit rate of 20.2 Mb/s, using four-level encoding with 50 percent raised-cosine filtering [14]. Partial response filtering is also commonly used in digital FM implementations in both three-level [15] and seven-level [16] versions. Figure 12.11 plots the transmit and combined filter characteristic, normalized to a single T1 (1.544-Mb/s) channel, for one implementation of three-level partial response.* The RF spectrum depends on the baseband spectrum and the frequency deviation of the radio. For example, using the partial response filter of Figure 12.11, Figure 12.12 indicates the RF spectrum occupied by an eight-T1-channel multiplexer (12.6 Mb/s) with 4-MHz FM deviation of the radio. For this application, 99 percent of the power is contained within a 14-MHz bandwidth. This configuration of an eight-T1-channel TDM with built-in partial response filtering operating over FM radio links with a 14-MHz bandwidth has been commonly used in U.S. military communication systems [17].

12.7 Transmultiplexers

The interface between FDM and TDM systems can be accomplished by use of back-to-back analog and digital channel banks with individual voice channels connected in between. An alternative to this conventional approach is the **transmultiplexer,** which directly translates FDM into PCM signals and vice versa. The advantage of transmultiplexers stems from the use of a single piece of equipment versus two channel banks. These advantages are realized while meeting or exceeding the performance for a tandem connection of PCM and FDM channel banks. The primary applications of transmultiplexers are the interface of digital switches with analog transmission facilities and the interface of analog with digital transmission facilities.

In the design and application of transmultiplexers, three standard configurations have been adopted:

1. Translation between the 60-channel supergroup and two 30-channel, European standard, PCM multiplexers

---

*In Figure 12.11 the high end of the frequency scale is normalized. To obtain the actual filter response, multiply the frequency scale by the number of T1 channels to be serviced by the multiplexer. For example, a four-channel multiplexer requires a combined filter response that is down 3 dB at four times 400 kHz or 1.6 MHz.

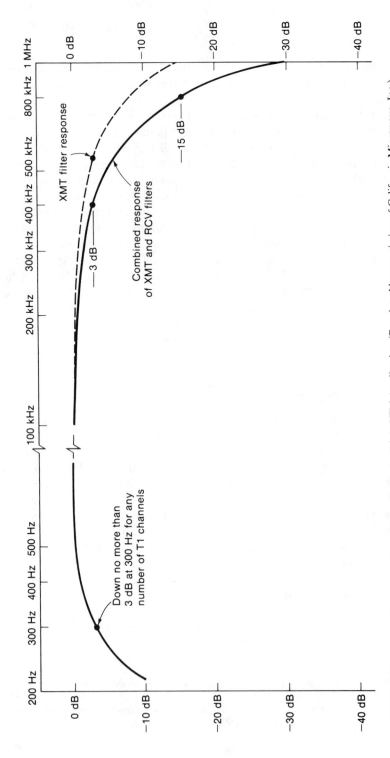

FIGURE 12.11 Partial Response Filter Characteristic for Digital FM Application (Reprinted by permission of California Microwave, Inc.)

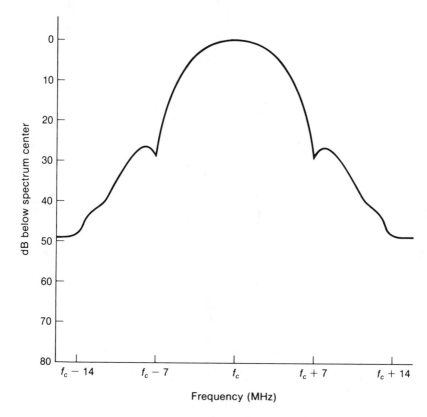

FIGURE 12.12   RF Spectrum of 12.6-Mb/s Digital FM Radio Using Partial Response

2. Translation between two 12-channel groups and the 24-channel, North American standard, PCM multiplexer
3. Translation between two 60-channel supergroups and five 24-channel, North American standard, PCM multiplexers

The first two configurations have been recognized by the CCITT in Recs. G.793 and G.794, respectively [18]. Salient characteristics described in these CCITT recommendations are digital and analog interfaces, correspondence between analog (3 kHz) and digital (64 kb/s) channels, synchronization of the transmultiplexer with PCM and FDM equipment, and operation with different types of signaling.

The algorithms used in FDM-TDM translation in general are based on the use of digital signal processing as illustrated in Figure 12.13. The FDM signal is digitized, and processing such as filtering, modulation,

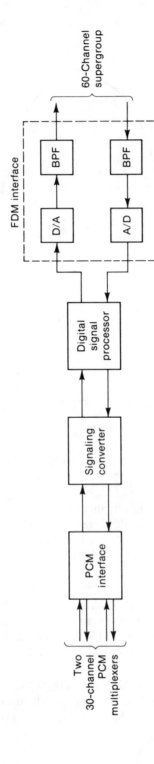

FIGURE 12.13 Block Diagram of 60-Channel Transmultiplexer

and amplification is performed on the digital representation of the FDM signal to produce a PCM signal. In the reverse direction, the PCM signal is digitally processed to produce a digital version of a FDM signal. A digital-to-analog converter is then used to produce the conventional FDM signal. Many design approaches have been used [19], and no single technique is considered standard practice in the telephone industry today.

12.8   Hybrid Transmission Systems

With the growing need for digital transmission of nontelephone signals, such as data, facsimile, and visual telephony, hybrid transmission systems have been developed that carry FDM voice and digital services on the same transmission media. Existing FDM transmission systems can be adapted to simultaneously carry a digital signal within the baseband of a cable or microwave system. Three basic methods have been used. Their names indicate where the digital spectrum is located with respect to the FDM spectrum: **data under voice** (DUV), **data in voice** (DIV), and **data above voice** (DAV), which is sometimes called **data over voice** (DOV).

As part of its Digital Data Service, AT&T used its existing microwave radio systems to carry 1.544-Mb/s data via the DUV technique [20]. The AT&T U600 mastergroup occupies the band from 564 to 3084 kHz, so that the lower 564 kHz of baseband spectrum is available for data transmission. Use of the AT&T L600 mastergroup requires removal of the lower two supergroups (60 to 564 kHz) to provide the same 564-kHz spectrum for data. Figure 12.14 indicates the scheme used to transmit 1.544-Mb/s data within the lower 564-kHz frequency band of the baseband spectrum. A clock recovery circuit extracts a 1.544-MHz clock from the 1.544-Mb/s bipolar input and distributes it with a 772-kHz half-rate clock to the appropriate transmitter circuits. A scrambler converts the bipolar signal to a unipolar format and then scrambles the signal to prevent discrete spectral components that could interfere with the FDM channels. The encoder converts the binary input to a four-level rate of 772 kilosymbols per second. The class 4 partial response filter shapes the digital signal, resulting in a seven-level signal with a spectral null at 386 kHz. The DUV receiver reconstructs the 1.544-Mb/s bipolar signal from the seven-level partial response signal by the inverse functions of the transmitter.

The use of data above voice (DAV) requires that the data spectrum be translated above the FDM spectrum. FSK, PSK, and digital AM can be used to place the data in the desired frequency spectrum. An advantage of

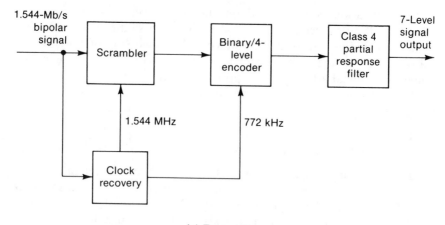

(a) Transmitter

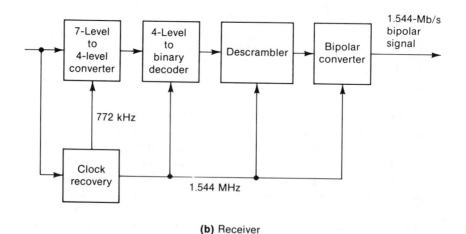

(b) Receiver

FIGURE 12.14   Data Under Voice (DUV) System Used for Transmission of 1.544 Mb/s

the DAV technique over DUV is that FDM signals and any service channels, which normally may be located at the lower end of the spectrum, do
not require translation. DAV has been adopted by Canadian National and
Canadian Pacific for transmission of 1.544 Mb/s over a 6-GHz FDM
microwave system [21]. A DAV system for 2.048-Mb/s transmission over
FDM systems has been recommended by CCITT Rec. G.941 [22] and has
been developed in hardware [23].

The data in voice (DIV) technique has not been as commonly used as DUV or DAV. Nevertheless, a provision for 6.312-Mb/s DIV transmission in place of a 1232-kHz mastergroup is given in CCITT Rec. G.941 [18].

The advantage of hybrid digital/analog transmission is the ability to carry digital and analog signals over a single microwave or cable system when there is not sufficient analog or digital traffic to justify separate systems. Hybrid transmission has greater equivalent digital capacity and performance than an analog system using modems to carry data within FDM channelization. The AT&T version of DUV described here provides approximately 4-bps/Hz spectral efficiency, for example. Hybrid techniques also allow regeneration of the digital signal with every radio or cable repeater. Finally, hybrid transmission is more efficient than using PCM coding for voice channels. Using standard 64-kb/s PCM, 600 voice channels would require 40 Mb/s compared to the 2.5-MHz bandwidth required for a 600-channel FDM mastergroup.

12.9 Summary

With the predominance of the telephone network in today's communication systems, the application of digital transmission must often coexist with analog transmission facilities. Frequency division multiplex (FDM), in which telephone channels are allocated different portions of the spectrum, is still found in many telephone networks. Each of the different multiplex levels in the FDM hierarchy is used to combine lower-level FDM signals, but digital signals can also be accommodated via modems, devices that convert a digital signal into a format compatible with FDM transmission. The North American FDM hierarchy is composed of the 12-channel group that occupies a 48 kHz bandwidth, the 60-channel supergroup that occupies a 240-kHz bandwidth, the 600-channel mastergroup that occupies approximately a 2.5-MHz bandwidth, the 3600-channel jumbogroup that occupies a 17 MHz bandwidth, and the 10,800-channel jumbogroup multiplex that occupies a 60-MHz bandwidth. The CCITT and North American hierarchies differ above the supergroup level—the CCITT hierarchy has a 300-channel mastergroup and a 900-channel supermastergroup (see Table 12.1).

Data transmission over 4-kHz telephone channels is affected by attenuation distortion, envelope delay distortion, noise, phase jitter, and level variation. For most voice-band modems, attenuation and group delay distortion are the dominant sources of degradation. Attenuation distortion is specified by a limit placed on the loss at any frequency relative to the loss at a reference frequency. Envelope delay distortion is specified by the maximum variation in envelope delay. Both attenuation and envelope delay distortion can be corrected by use of equalization, which adds attenuation or gain to flatten the amplitude response and adds delay to flatten the group delay. Equalization may be included in the modem or supplied by the telephone company. Equalizers found in modems may be fixed or adaptive, although automatic adaptive equalization is standard at rates above 4800 b/s. Equalization provided by the telephone company is termed conditioning and is available in different grades (see Table 12.2).

Voice-band modems provide transmission of data rates up to 24 kb/s over 4-kHz telephone channels. Early modem applications, however, were limited to about 1800 b/s and used frequency-shift keying (FSK). Rates up to 4.8 kb/s are provided by use of differential phase-shift keying (DPSK). Higher data rates have been achieved through use of combined amplitude/phase modulation (AM/PM), QAM, and trellis coding. Modem characteristics for data rates to 14.4 kb/s have now been standardized by the CCITT (see Table 12.3). Rates above 24 kb/s require the use of higher levels in the FDM hierarchy. Group modems, for example, provide transmission of rates up to 168 kb/s in a 48-kHz group bandwidth. Demands for even higher data rates have led to the replacement of the entire FDM equipment with a modem that interfaces directly with analog radio or cable systems.

Alternative approaches to conventional modems have been developed to provide greater efficiency in combining analog and digital transmission. The transmultiplexer, for example, directly translates FDM into PCM signals and vice versa, thereby avoiding the use of back-to-back PCM and FDM channel banks. Existing FDM systems have been adapted to carry data by insertion of the data signal under, in the middle, or above the FDM spectrum. These hybrid transmission systems are known respectively as data under voice (DUV), data in voice (DIV), and data above voice (DAV). When both digital and analog signals are to be carried over a single microwave or cable system, hybrid transmission systems provide a more bandwidth-efficient approach than the use of individual modems for data signals or PCM for voice channels.

References

1. Members of the Technical Staff, Bell Telephone Laboratories, *Transmission Systems for Communications* (Winston-Salem: Western Electric Company, 1971).
2. CCITT Blue Book, vol. III.2, *International Analogue Carrier Systems.* (Geneva: ITU, 1989).
3. *Data Communications Using Voiceband Private Line Channels,* Bell System Tech. Ref. Pub. 41004 (New York: AT&T, 1973).

4. *Transmission Parameters Affecting Voiceband Data Transmission—Description of Parameters,* Bell System Tech. Ref. Pub. 41008 (New York: AT&T, 1974).
5. *Notes on the Network* (New York: AT&T, 1980).
6. CCITT Blue Book, vol. IV.2, *Maintenance; International Voice-Frequency Telegraphy and Facsimile, Internationally Leased Circuits* (Geneva: ITU, 1989).
7. CCITT Blue Book, vol. VIII.1, *Data Communication Over the Telephone Network* (Geneva: ITU, 1989).
8. RADC TR-76-311, "16 kb/s Data Modem Techniques," Rome Air Development Center, Rome, N.Y., October 1976.
9. W. Peterson and D. Brown, "Cyclic Codes for Error Detection," *Proceedings of the IRE,* January 1961.
10. A. Lempel and J. Ziv, "A Universal Algorithm for Sequential Data Compression," *IEEE Trans. on Information Theory,* vol. IT-23, no. 3, May 1977, pp. 337–343.
11. K. Murano, S. Unagami, and F. Amano, "Echo Cancellation and Applications," *IEEE Comm. Mag.,* January 1990, pp. 49–55.
12. W. E. Fleig, "A Stuffing TDM for Independent T1 Bit Streams," *Telecommunications* 6(July 1972):23–32.
13. J. L. Osterholz and M. K. Klukis, "Spectrally Efficient Digital Transmission Using Analog FM Radios," *IEEE Trans. on Comm.,* vol. COM-27, no. 12, December 1979, pp. 1837–1841.
14. C. W. Broderick and R. W. Gutshall, "A 20 Mbps Digital Terminal for TD-2 Radio," *Conference Record ICC 1969,* June 1969, pp. 27-21–27-26.
15. T. L. Swartz, "Performance Analysis of a Three-Level Modified Duobinary Digital FM Microwave Radio System," *Conference Record ICC 1974,* June 1974, pp. 5D-1–5D-4.
16. A. Lender, "Seven Level Correlative Digital Transmission Over Radio," *Conference Record ICC 1976,* June 1976, pp. 18-22–18-26.
17. *PCM/TDM System Design Verification Test Program,* U.S. Defense Communications Agency, Reston, VA, February 1972.
18. CCITT Blue Book, vol. III.4, *General Aspects of Digital Transmission Systems; Terminal Equipments* (Geneva: ITU, 1989).
19. S. L. Freeny, "TDM/FDM Translation as an Application of Digital Signal Processing," *IEEE Comm. Mag.,* 18(January 1980), pp. 5–15.
20. K. L. Seastrand and L. L. Sheets, "Digital Transmission Over Analog Microwave Radio Systems," *Conference Record ICC 1972,* June 1972, pp. 29-1–29-5.
21. K. Feher, R. Goulet, and S. Moris, "1.544 Mbit/s Data Above FDM Voice and Data Under FDM Voice Microwave Transmission," *IEEE Trans. on Comm.,* vol. COM-23, no. 11, November 1975, pp. 1321–1327.
22. CCITT Blue Book, vol. III.5, *Digital Networks, Digital Sections and Digital Line Systems* (Geneva: ITU, 1989).
23. H. Panschar and O. Ringelhaan, "Data Above Baseband Modem for Analog Radio Relay Systems," Siemens Telecom Report 2, Special Issue, "Digital Transmission," 1979, pp. 142–143.

Problems

12.1 CCITT Standard V.22 for voiceband data transmission via modems uses the following signal constellation, in which only the bold points are transmitted when channel conditions are known to be poor, and the full constellation when channel conditions are good.

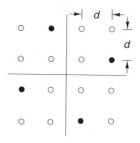

(a) Identify the two modulation techniques used for the good versus poor channel.

(b) Assuming the symbol rate remains constant, how different is the bit rate for the good versus poor channel?

(c) Determine the average transmit power for each modulation technique as a function of d, and compare the two answers.

12.2 Assuming asynchronous transmission of ASCII characters with one start bit, two stop bits, and one parity bit, what is the information rate for a 300-baud modem that uses 4-PSK?

12.3 For the following forms of differentially coherent PSK, show a table that assigns bit(s) to phase change. Then show the phases of the transmitted signal corresponding to the binary sequence 110011110100: (a) 2-DCPSK; (b) 4-DCPSK; and (c) 8-DCPSK.

12.4 A 9.6-kb/s modem is to be designed in which the designer has to choose between 16-PSK and 16-QAM.

(a) Draw the signal constellation for 16-PSK and 16-QAM with both having the same peak amplitude A.

(b) Determine the bandwidth required (for either modulation technique) if raised-cosine filtering is used with $\alpha = 0.2$.

(c) Determine the average transmit power for each modulation technique as a function of A, the peak amplitude.

(d) For 16-QAM, using Gray coding, assign binary values to the 4-bit words represented by the 16 points.

(e) For 16-QAM, determine the I and Q signal values to generate each of the 16 signals, assuming that the 4 allowed signal levels of the I and Q components are +3, +1, −1, and −3.

12.5 You are given a voice-band modem that uses 8-PSK with 50 percent raised-cosine filtering for 4800-bps data transmission.
(a) Draw the signal constellation and show the assignment of bits to symbols using Gray coding.
(b) Find the transmission bandwidth and spectral efficiency.
(c) What is the E_b/N_0 in dB of this modem for a $S/N = 10$ dB?

12.6 For the CCITT Rec. V.33 14.4-kb/s modem described in this chapter:
(a) Draw a figure, like that given in Figure 6.39, to show the general form of the trellis coder.
(b) Show the signal constellation.

12.7 CRC-12 is a cyclic redundancy check with polynomial

$$P(x) = x^{12} + x^{11} + x^3 + x^2 + x + 1$$

(a) Describe the error-correcting properties of this CRC.
(b) Show a circuit consisting of a shift register with feedback paths to EX-OR gates required to implement CRC-12.
(c) For a message signal 110100100111, generate the 12-bit CRC.
(d) Show the receiver circuit for CRC-12 and demonstrate that the remainder equals 0 for the message of part (c).
(e) If the transmitted message is the 6-bit stream 111111, and if the received stream is 111100, show that the CRC will detect the errors.

12.8 For the CRC-16 polynomial given in this chapter:
(a) Show the circuit necessary to divide by this polynomial.
(b) Find the FCS for the message consisting of 16 consecutive 1's.
(c) By displaying the register contents step-by-step, show that the shift register at the receiver contains all 0's after the message and FCS have been entered.

12.9 Repeat above for the CRC-32 polynomial given in this chapter, assuming that the message length is 32 bits consisting of all 1's.

12.10 Consider an FDM group with bandwidth of 48 kHz that is to carry data over a channel corrupted with additive white gaussian noise.
(a) What is the capacity in kb/s if the signal-to-noise ratio is 25 dB?
(b) What is the spectral efficiency for part (a)?
(c) What is the minimum SNR required for a data rate of 168 kb/s?

(d) Assuming use of 7-level partial response, what is the probability of error for part (c)?

12.11 Consider the DUV technique used by AT&T to transport 1.544 Mb/s in their FDM systems.

(a) What theoretical bandwidth is required for the choice of 7-level partial response?

(b) Sketch the partial response filter characteristic.

(c) Calculate the probability of error for an S/N of 25 dB.

13

Digital Transmission Networks

OBJECTIVES

- Describes the development of private digital transmission networks that provide greater intelligence and control at the customer's premise
- Discusses the application of digital transmission to new, emerging personal communication networks, that provide mobile, wireless connections to the individual
- Covers both narrowband and broadband integrated services digital network (ISDN)

13.1 Introduction

With the invention of the transistor, the world entered a new Information Age that will mature in the next two decades. The global telecommunication network today forms the infrastructure for this information-based society. Not surprisingly, telecommunications is the world's fastest-growing industry, owing to the demands of society, the rapid growth of technology, and the declining cost of telecommunication services.

The telephone network is now and will continue to be the cornerstone of the telecommunication industry. In 1992, the U.S. telephone network encompassed 220 million telephones and handled over one billion telephone calls a day. However, the demand for data, video, and other telecommunication services will cause major changes in today's telephone networks. The development of microelectronics has resulted in low-cost memory, greater computing power, the microprocessor, and other technological advances that have created a requirement for efficient, low-cost communications. Simultaneously, the computer and microelectronics industries have penetrated the telecommunication industry in the form of digital communication systems. In the future, the distinction between the telephone and computer, and other home or office terminals, will blur. Tomorrow's telephones will be "smart terminals" that integrate features now unavailable or provided by separate terminals. This integration of services is made possible by the pervasive use of "1's and 0's" in telecommunication networks. Digital technology will be extended all the way to the customer's premises. This anticipated availability of a completely digital network has led to the development of a concept for an **integrated services digital network (ISDN)**, now standardized by the CCITT and being implemented worldwide.

Apart from the traditional services of voice, data, and video, new services are emerging from the fields of data processing, broadcast video, and interactive video. The marriage of telecommunications with data processing has resulted in "telematic" services such as facsimile, teletex, videotext, and electronic mail, which are most conveniently carried with digital transmission. Video services such as teleconferencing, interactive CATV, and direct broadcast by satellite are also entering the market, although with the exception of teleconferencing these services have been provided via analog transmission. Introduction of these new services has led to the concept of "automated offices" for improved office efficiency and productivity and "wired households" for improved home environment [1].

Digital transmission will continue to play an important role in the evolution of telecommunication networks. Historically, the transmission plant of telecommunication networks has been first to be digitized, while switches and terminals have remained analog. Breakthroughs in digital transmission technology and declining transmission costs allow the creation of new networks and services. Already the introduction of digital switches has led to integration of switches and transmission systems, which has been termed the **integrated digital network (IDN)** by the CCITT. New technologies such as local area networks, private networks, and cellular telephone networks blend transmission with switching and networking. In the future, we can expect transmission systems to merge even further with switches, terminals, and networks, so that transmission systems will be inseparable from telecommunication systems.

13.2   Private Networks

Advances in customer premise equipment and the proliferation of digital transmission has led to the growing use of **private networks** by large corporations and organizations. Advantages of private networks include lower cost compared to public networks, user control of the network, and use of technology tailored to the specific requirements of the network. Users of such a network have control over their network through use of intelligence in the customer premise equipment (CPE). The role of the carrier may be reduced to the provision of leased lines, typically T1 or E1, for transmission interconnectivity of the nodes in the network. Network management by the user can include equipment configuration, traffic analysis, bandwidth allocation, circuit layout, network reconfiguration, and performance monitoring. The definition of private networks can also be expanded to include interconnection of local area networks (LANs) and the use of public network offerings to augment services or flexibility of the user network. But the basic element common to virtually every private network is the **networking multiplexer,** which provides the intelligence and control at the customer's premise.

13.2.1 *Networking Multiplexers*

Digital multiplexers had their beginnings in the digitization and multiplexing of voice channels. The digital multiplex hierarchies that developed provided fixed time slot assignment to each channel, which allowed digital switches and cross-connects to access individual DS-0 channels. Data services were accommodated by using separate multiplexers dedicated to data channels or by using subrate ports within digital channel banks. Voice or data users were given full period connectivity without regard to their activity factors, resulting in unused bandwidth during channel idle time. Multiplexers were also connected on a point-to-point basis, which required $N(N-1)/2$ transmission links to fully interconnect N terminating locations. In response to the growth in data requirements, the resulting pressure to integrate voice and data services, and the availability of private line networks, CPE multiplexer vendors have developed networking multiplexers that dramatically improve the efficiency and flexibility of digital transmission. These multiplexers introduce switching to allow networks to be defined and controlled by the intelligence contained in the multiplexer. The first such equipment was designed for T1 or E1 networks, but T3 multiplexers are now available with many of these same networking features.

A number of properties are common to most networking multiplexers, while other properties differ from one vendor's product to another. The following list provides salient properties:

- *Multiplexing orientation* indicates whether the multiplexer uses circuit, packet, or hybrid multiplexing and switching. Circuit switched multiplexers use a bit or byte orientation in which the circuit is established as a time slot within a frame. Packet techniques use fixed-length or variable-length packets consisting of overhead for addressing, error correction, and an information field usually from only one source. The terms *fast packet* [2] and *wideband packet* [3] have been used to distinguish these forms of packet switching from X.25 packet switching. An example of a fixed-length, 24-byte packet is shown in Figure 13.1 that matches the 193-bit frame length of conventional T1 multiplexers; another choice of packet length is based on cell relay standards that prescribe a 53-byte packet. Hybrid schemes are a combination of circuit and packet switching whereby bursty traffic is packetized and full-period traffic is circuit switched.

- *User services* include standard offerings such as voice, data, and video, but may include interfaces to data communication networks such as LANs and public network services
- *Number of aggregates* is the largest number of aggregates available at a single node. A typical number of aggregates is 16 T1s, indicating that up to 16 different nodes can be directly accessed from a single node.
- *Capacity* is the total switching throughput of the multiplexer, including connections made port to port, port to aggregate, and aggregate to aggregate. Since each type of traffic being serviced (voice, data, and so on) will have different overhead, the throughput will vary with the number and type of connections.
- *Bandwidth* on demand can be made on a dynamic basis, by assigning bandwidth only when a user needs it. A voice user would be given bandwidth only when going "off-hook" and a data user only when indicating "ready to send." Bandwidth can also be reserved in advance for special applications that are required by time of day.
- *Bandwidth compression* can provide greater bandwidth efficiency by use of statistical multiplexing for data and speech interpolation and low-bit-rate coding for voice. A common choice for voice channels has been variable-rate ADPCM (as described in Chapter 4), with rates of 48, 32, 24, and 16 kb/s, depending on traffic congestion and the type of voice-band signal.

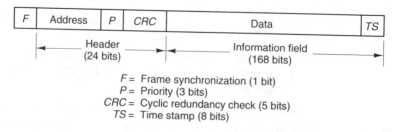

F = Frame synchronization (1 bit)
P = Priority (3 bits)
CRC = Cyclic redundancy check (5 bits)
TS = Time stamp (8 bits)

FIGURE 13.1   Typical Structure of Packet as Used in Networking Multiplexer

- *Bandwidth contention* results when the multiplexer capacity is exceeded by the demand from the user connections, as will happen in statistical multiplexing and speech interpolation schemes. With all bandwidth in use, any other connected users will see a busy signal until bandwidth is freed. However, before reaching a blocking condition, certain user traffic such as voice can have its bandwidth effectively reduced by dropping packets or samples.

- *Bandwidth efficiency* is a function of the overhead required for each type of port and aggregate signal. Overhead will vary for each type of port, with asynchronous signals requiring more overhead than synchronous signals. Aggregate overhead provides the intermachine signaling required to convey multiplexer configuration, status information, and control signals.

- *Network routing* determines the path used to connect one user to another. The multiplexer intelligence required to perform routing may be *distributed*, with all nodes involved in routing decisions, or *centralized*, where a single node has routing responsibility. The routing method may be *table-driven*, which means that a predefined list determines the path, or *algorithm-driven*, which is based on various parameters analyzed by an algorithm. Parameters used to determine routing may include cost, delay, distance, security requirements, and choice of media.

- *Automatic rerouting* provides restoration of a connection without manual intervention after a node or transmission line failure has occurred. The time to reroute includes the time to select a new path and to send the appropriate circuit set-up messages, and will depend on the number of nodes traversed and the length of the transmission links involved.

- The *maximum number of nodes* that can be addressed in a single network determines the maximum possible size of the network. A typical value is 250 nodes.

- *Nodal delay* is the processing delay encountered at a node by a circuit routed through that node. Typical delays are a few milliseconds, but delay may vary with the data rate of the circuit.

- *Maximum number of hops* is the maximum number of internodal links allowed to complete a connection. This limit may be imposed by the vendor or by the user to limit delay accumulation.

- *Traffic balancing* is the ability to balance the load among the nodes of a network. This feature allows a more evenly distributed spare capacity and limits the number of circuits affected by a node failure.

- *Priority bandwidth assignment* allows the multiplexer to resolve contention among user requests for bandwidth by assigning a priority to each user. Higher priority circuits can preempt lower priority circuits or be restored first in the event a failure occurs.

- *Inverse multiplexing* allows a user high-speed circuit to be divided into and carried by multiple low-speed circuits. The high-speed circuit need not be constrained to standard rates, while the multiple low-speed rates are selected from standard rates (for example, DS-0s or DS-1s).
- *Partitioned subnetworks* split the aggregate signal(s) among several user communities who each have control over a part of the aggregate signal(s). These logical partitions can create several virtual networks from a single physical network. Users outside a particular subnetwork community then do not have access to that part of the network.
- *Network management* provides several functions, including fault diagnosis, configuration management, performance assessment, and traffic statistics, often using graphics-oriented computers.

13.2.2   Local, Metropolitan, and Wide Area Networks

Local area networks (LAN) are used to interconnect computers, terminals, word processors, facsimile, and other office machines within a building, campus, or small geographic area. A single definition of a local area network is difficult to achieve because of the divergent applications and design alternatives, but generally the characteristics include:

- Geographically local with distances limited to a few miles
- Multiple services often possible, including voice, data, and video
- High-speed transmission, with typical data rates in the range of 50 kb/s to 150 Mb/s
- Predominance of cable transmission media, including twisted pair, coaxial cable, and fiber optics, although wireless LANs are also used
- Some type of switching technology, typically packet switching
- Some form of network topology and network control
- Owned by a single company

Thus various technologies are necessary in local area networks, including transmission, switching, and networking. Here we will examine the alternative choices for transmission media.

Twisted-pair cable although limited in bandwidth and distance is nevertheless still a popular choice. Both shielded and unshielded twisted pair is used, with rates up to 16 Mb/s commonly used. For example, shielded twisted cable at rates of 1, 4, and 16 Mb/s is specified for Token Ring LANs. Coaxial cable is presently the most commonly used medium for LANs, although use of optical fiber is expected to grow rapidly. LAN transmission on coaxial cable falls into two categories: baseband and broadband. In the baseband mode, data rates to 50 Mb/s have been transmitted by using baseband coding schemes such as Manchester coding. The most prominent of the baseband

LANs is Ethernet, which transmits data at 10 Mb/s. The broadband mode of transmission uses a RF modem, which results in greater noise immunity and the ability to provide multiple FDM channels for simultaneous voice, data, and video services. Optical fiber is most commonly used with the fiber distributed data interface (FDDI), which operates at 100 Mb/s, but twisted pair has been used as well.

LANs are typically limited to transmission distances of a few miles. To extend the distance spanned by a LAN, **metropolitan area networks** (MANs) and **wide area networks** (WANs) have been developed, all based on some version of packet switching. Several standards have been developed to support these networks, such as **frame relay** and **switched multimegabit data service (SMDS)**. Frame relay services are based on variable-length packets transmitted at higher speed and without error control as compared to conventional X.25 packets. Standards for frame relay are based on CCITT Rec. I.122 [4], CCITT Rec. Q.921 [5], and ANSI T1.606 [6]. SMDS is a public packet-switched service that uses fixed-length packets and high-speed transmission (T1 and T3). Characteristics of SMDS are based on the IEEE 802.6 Metropolitan Area Network standard [7]. Frame relay and other such packetized services are commonly found as a standard feature in networking multiplexers, and in some cases a packetized architecture has driven the multiplexer design [2]. Frame relay and SMDS, however, are both considered precursors of broadband ISDN, which uses a form of packet switching similar to SMDS, as will be described in Section 13.4. As these forms of metropolitan and wide area networks evolve toward ISDN, networking multiplexers can also be expected to evolve to ISDN standards.

13.3   Personal Communication Networks

Mobile, wireless communications already exist for the individual in the form of cordless and cellular telephones. These technologies are the basis for the ongoing development of personal communication networks (PCN) and services. Today, three basic elements are necessary to connect the mobile user to the public network: the user terminal, a base station, and a switch. A mobile user communicates with the base station via a mobile radio. The base station is connected to a switch via fixed radio links or cable. These elements may all be terrestrial, or some may be satellite based. Each base station serves many wireless users, while each switch connects many base stations to the public switched network. Many of today's cellular telephone systems are already saturated, and new technology will be needed to satisfy the growing demands for personal, mobile communication.

The idea behind personal communication is to extend the ability of cellular and cordless telephones so that a user can communicate person-to-person regardless of physical location. Ideally, PCN provides each user a personal phone number that follows that person no matter where he or she may be located, in the office, at home, or traveling. Additional features could include the ability to communicate different types of information signals, in addition to voice, such as voice mail or electronic messages.

The principal transmission issues facing the design of PCN, as discussed here, are spectrum allocation, voice coding, modulation, and access technique.

13.3.1   Terrestrial PCN

In a terrestrial mobile radio network, an area is served by a single repeater located at a point designed to maximize coverage through proper line-of-sight clearance. To serve a large number of users, however, a service area must be split into smaller subareas called cells. These cells allow the reuse of frequencies to increase the number of possible users. To accomodate an even larger number of users, the approach taken in personal communication networks is to use smaller cells, called microcells, together with advanced digital technology. Smaller cells allows a greater degree of frequency reuse and therefore greater capacity per unit area. The objective is to increase the availability and reduce congestion, enabling more users to be accommodated. As the cell size goes down, however, the complexity of the network increases, and the cost per subscriber goes up. Cellular systems and PCN as well are therefore initially limited to those geographic areas with a population density large enough to generate sufficient revenues.

Numerous technologies have been used for cellular communications, and there are at least six different, incompatible standards worldwide [8]. First-generation cellular systems were based on analog technology, with a combination of frequency modulation and frequency-division multiple access. The Advanced Mobile Phone System (AMPS) standard used in the United States is an example of this first-generation technology. Each voice channel is carried in a 30-kHz bandwidth, up to 416 duplex channels are carried in a 25-MHz spectrum allocation, and the two cellular service providers (called channel A and channel B in the United States) operate in either the 824-849 MHz band or the 869-894 MHz band. Second-generation technology is based on digital technology, including digital voice coding and time-division multiple access. Interim Standard 54, approved by the TIA for use in the United States, calls for a dual-mode cellular radio that retains the AMPS technology but adds a digital mode based on QPSK modulation and CELP voice coding. The frequency spacing of IS-54 remains at 30 kHz; the voice coding rate is about 8 kb/s and the transmission rate (with overhead) is 13 kb/s. Three voice channels can be time-division multiplexed within each 30-kHz bandwidth [9]. Another example of this second-generation technology is the European Groupe

Special Mobile (GSM), which combines frequency division of carriers spaced 200 kHz apart with time division of eight voice channels per carrier. GSM modulation is MSK, and voice coding is a form of LPC with regular pulse excitation. The voice coding rate is 13 kb/s, and the transmission rate (with overhead) is 22.8 kb/s [8].

Third-generation cellular systems and first-generation PCNs will also be based on digital technology, where choice of the access scheme is the most contentious technical issue. TDMA and code-division multiple access (CDMA) are the leading contenders as the access system. A comparison of these competing access techniques was given earlier, in Chapter 5, with the conclusion that CDMA offers several advantages over TDMA [10,11]. The 100 to 200 MHz of bandwidth needed to support a new cellular or PCN system is also a contentious issue because of frequency congestion and lack of separate frequency allocations, at least in the United States. Two approaches are possible, the establishment of a discrete band and the sharing of an existing band. A discrete band would require the relocation of existing users and the approval of regulatory bodies such as the FCC in the United States. The frequency band of 1700-2300 MHz is preferred internationally, where licenses for personal communication services have been granted in Europe. In the United States, these frequencies are currently used by point-to-point microwave. CDMA has been proposed as a means of sharing this portion of the spectrum with PCN applications, and field trials have already been conducted [12]. Other technical issues to be resolved are the choices of cell size, voice coding, and handset. Cell sizes on the order of 1200 ft by 1200 ft have been proposed that would greatly increase the user density that could be served. Voice coding at a rate between 4 and 8 kb/s is being proposed for PCN [13]. Low power levels made possible by use of microcells and adaptive power control have reduced the weight and cost of the handset required for PCN.

13.3.2 Satellite PCN
Satellite PCN systems employ satellites as base stations and repeaters. Two approaches have been taken: satellites in geosynchronous orbit and satellites in a low-earth orbit. Coverage provided by the satellite depends on the type of antenna. Single-beam antennas provide global coverage; multibeam antennas allow coverage of multiple, smaller areas and also reuse of frequencies, similar to terrestrial cellular systems.

INMARSAT is an example of a geostationary satellite system that provides connectivity of mobile users, typically ship-to-shore. Several services are available, including Standard A (voice-band channel for telephony, facsimile, and telex), Standard B (digital voice, data, or telex at 9.6 kb/s), Standard B1 (voice at 16 kb/s), Standard M (voice at 4.2 kb/s), or Standard

C (600-b/s data). Next generation (INMARSAT-III) satellites will use spot beams to provide localized coverage and better usage of frequency; power and bandwidth will be dynamically allocated to global and spot beams to respond to changing traffic demands [14]. Typical design objectives for next generation geosynchronous satellites are based on multibeam antennas to provide 10 to 20 beams (cells) with a radius of 800 to 1600 km [15]. Advanced satellites will also provide on-board processing and support intersatellite links, to avoid double-satellite hops [16].

Low-orbit satellites offer several advantages over geosynchronous satellites for PCN, including higher signal levels, shorter delays, and compatibility with terrestrial cellular systems. The concept for Iridium, for example, provides 77 satellites, each with 37 beams covering a radius of 690 km. The Iridium constellation of satellites consists of 11 satellites in 7 planes, each plane equally spaced in longitude. Intersatellite links would provide handover of users as satellites pass from view. Connectivity to the ground is via gateways that provide access to public networks. Voice coding is based on 4.8-kb/s vector sum excited linear prediction (VSELP), modulation is QPSK, and access is a combination of TDM and FDM. Other planned services include radiodetermination, two-way messaging, paging, facsimile, and data. These services are intended to be worldwide and full period [17].

13.4   Integrated Services Digital Network

The natural evolution of the public switched telephone network and the emergence of digital communication technology has led to the development of a concept known as the integrated services digital network (ISDN). The principle behind the ISDN concept is end-to-end digital connectivity that will support a wide range of voice and nonvoice services. The ISDN will be initially based on current digital networks that use 64-kb/s switched digital connections. Initial services under consideration include voice, data, facsimile, teletex, and videotext; full-motion video is likely to be one of the last services implemented. The CCITT has been the international focal point in defining and standardizing the ISDN concept. The original concept of ISDN within the CCITT was based on the use of existing T1 and E1 rates; this original version is now termed **narrowband ISDN.** The latest concept of ISDN is based on higher data rates, hence the term **broadband ISDN.** The CCITT I series of recommendations lay the groundwork for both narrowband and broadband ISDN.

13.4.1   Narrowband ISDN

Narrowband ISDN provides two basic types of interface, the **basic rate interface (BRI)** and the **primary rate interface (PRI)**. As shown in Figure

13.2, the basic rate interface provides two types of channels, the *bearer* (B) channel at 64 kb/s and the *delta* (D) channel at 16 kb/s. The B channel can carry digital voice or data, but can also be subdivided via lower-level multiplexing as prescribed by CCITT Rec. I.460 [18]. As specified in CCITT Rec. Q.931 [19], the D channel's primary function is to provide out-of-band signaling for B channel services, but it may also be accessed for telemetry and packet-switched data. The basic rate interface is defined as $2B + D$, for a total of 144 kb/s to the subscriber. Subscribers requiring higher rates can use the primary rate interface, either 1.544 Mb/s (DS-1) or 2.048 Mb/s (E-1). The 1.544-Mb/s version of PRI contains 23 B channels and 1 D channel, as specified by CCITT Rec. I.431 [18]. A multiple-PRI configuration may be arranged with a single D channel common to all the B channels, allowing all but 1 DS-1 to contain 24 B channels. In the E-1 format, there are 30 B channels, 1 D channel, and 1 channel for overhead including frame synchronization. Two other rates have been defined for narrowband ISDN, a 384-kb/s H0 channel and a 1536-kb/s H11 channel, both of which can be carried within a PRI; another rate that has been proposed is the 1472-kb/s H10 channel. By adhering to the CCITT PRI standards, a user with a primary rate interface to an ISDN network can have the B channels routed to different destinations within the network or reconfigured to accommodate changing channel (bit rate) requirements.

The ability of the transmission plant to support these narrowband ISDN services is most limited in the local loop, which consists primarily of metallic two-wire loops. Various transmission techniques have been studied and proposed for full-duplex digital transmission over two-wire loops [20]. The two most popular choices have been time-compressed multiplexing and echo cancellation. Time-compressed multiplexing (TCM) shares the loop between the two directions of transmission by transmitting bursts of data in each direction at a rate slightly higher than twice the nominal rate [21]. In the United States, a standard has been adopted for the basic rate interface based on an echo canceller and a 4-level line code called *2B1Q* (2 binary, 1 quaternary) [22]. Overhead and synchronization bits operating at a 16 kb/s rate are added to the $2B + D$ user channels, for a total transmission rate of 160 kb/s between the subscriber and the network side of the loop.

13.4.2   Broadband ISDN
The CCITT defines broadband ISDN as any service requiring transmission channels greater than that of a primary rate. The intent of the CCITT, however, is to create a comprehensive ISDN that encompasses both narrowband and broadband services. Because conventional digital switches are not suitable for broadband services, another form of switching has emerged

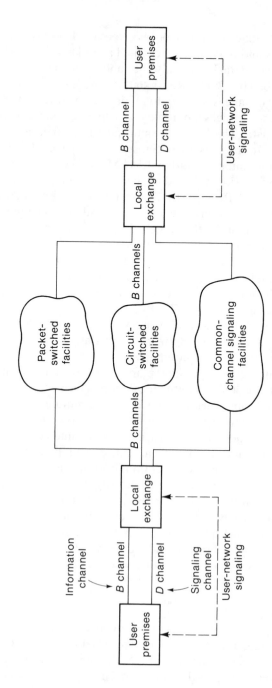

FIGURE 13.2 Functional Model of Narrowband ISDN

upon which broadband (and narrowband) ISDN services would be carried. **Asynchronous transfer mode (ATM)**, a form of packet or cell switching, is identified by the CCITT as the target solution. Copper lines also are unsuitable for broadband transmission, a problem solved with the introduction of optical fiber in the subscriber line network.

Broadband services defined by the CCITT are divided into two types, interactive and distribution, as shown in Figure 13.3. Each service type is further subdivided by the type of information and degree of user control. *Conversational services* provide two-way exchange between users of video, data, or text; examples include video telephony, videoconferencing, video surveillance, and transfer of files, documents and images. *Messaging services* offer user-to-user communications via store and forward; examples include video or document mail, analogous to today's voice mail and electronic mail. *Retrieval services* allow the user to retrieve stored information available to the public, such as videotext, video, images, documents, and data. *Distribution services without user presentation control* are broadcast services that a user may access but not control, like today's broadcast and cable television, although other information could be included such as text, graphics, and images. *Distribution services with user*

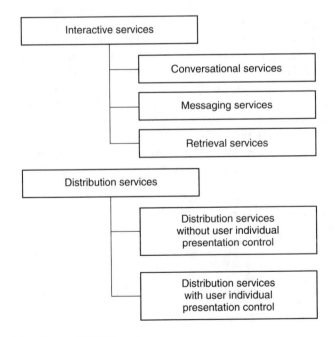

FIGURE 13.3   Types of BISDN service

presentation control are also broadcast services but ones in which a user may control the start and order of presentation; examples include text, graphics, sound, and images.

Although variable-bit-rate services are inherent in BISDN, certain standard user or *bearer* rates have been selected by the CCITT to help bridge the gap between existing digital hierarchies and new BISDN services. These broadband rates along with previously selected narrowband rates are shown in Table 13.1. All rates are prescribed to be multiples of 64 kb/s. The H4 rate is also specified to be at least four times the H21 rate and three times the H22 rate. These rates are considered suitable for video telephony, videoconference, conventional television distribution, and high-definition television distribution. Other rates are expected, especially those required for BISDN to compete with the high-speed data services now provided by local, metropolitan, and wide area networks.

Asynchronous Transfer Mode

The choice of asynchronous transfer mode (ATM) for BISDN marks a shift from the principles of a synchronous digital hierarchy (SDH) and its corresponding synchronous transfer mode (STM). The drawback to an STM approach in accommodating flexible and changing user data rates is its rigid structure and dependency on 64-kb/s modularity. STM is best suited to digital telephony and fixed-rate services, and was the basis for the choice of narrowband ISDN user rates. ATM removes these limitations of STM by choice of a switching fabric that can support bursty as well as fixed-rate services. The term asynchronous here is not meant to imply asynchronous multiplexing as described in Chapter 4, but rather a form of packet multiplexing and switching suitable for all types of traffic. Different from STM, ATM does not assign specific, periodic time slots to channels but instead segments usable bandwidth into fixed-length packets called *cells*. Small cells are used along with statistical multiplexing to provide time transparancy to delay sensitive services like voice.

TABLE 13.1   ISDN Channel Rates

Narrowband		Broadband	
D	16 or 64 kb/s	H21	32.768 Mb/s
B	64 kb/s	H22	43 to 45 Mb/s
H0	384 kb/s	H4	132 to 138.24 Mb/s
H11	1.536 Mb/s		
H12	1.92 Mb/s		

As prescribed by CCITT Recs. I.150 [4] and I.361 [18], ATM cells consist of a 5-byte header followed by 48 bytes of information, for a total of 53 bytes per cell. The number of cells transmitted per unit of time is dependent on the bit rate of the service. The capacity of the transmission facility can be shared among a large number of signals of differing bit rates to serve various requirements. These cells are multiplexed at the user-network interface using a label rather than a position (as done with time slot interchange) to identify each cell and to allow switching through the ATM nodes in the network. The header is used to identify the payload and to provide routing information. As shown in Figure 13.4, there are two cell header structures corresponding to the user-network interface (UNI) and the network-node interface (NNI). The functions performed by the header subfields are:

- *Generic Flow Control (GRC)* controls the flow of information from the customer's premise for different types of services
- *Virtual Path Identifier (VPI)* establishes a path for network routing
- *Virtual Channel Identifier (VCI)* establishes connections through the use of nodal tables
- *Payload Type (PT)* indicates whether the cell contains user or network information
- *Cell Loss Priority (CLP)* identifies cells of lower priority that can be discarded depending on network conditions
- *Header Error Control (HEC)* provides a CRC calculation over the header field for error detection and correction, which reduces cell loss and misrouting due to transmission errors

Virtual channels allow various services, both fixed and variable bit rate, to be offered over a single interface without the need for separate channel interfaces. The virtual path (VP) is a group of virtual channels (VC) that share the same VPI value, are carried on the same transmission facility, and are switched as a unit. The different VPs carried within the same transmission facility are distinguishable by their VPI. VC routing is performed at VC connecting points (a VC switch), where the VCI values of incoming VCs are translated to the VCI values of the outgoing VCs. Likewise, VP routing takes place at VP connecting points (a VP cross-connect), where VPI values of incoming VPs are translated to the VPI values of the outgoing VPs. VCs are typically controlled on the basis of customer demand, while VPs are typically controlled by the network operation system [23].

BISDN Transmission
Figure 13.5 is the BISDN protocol reference model specified by CCITT Rec. I.121, which shows the ATM layer just described between the physical layer

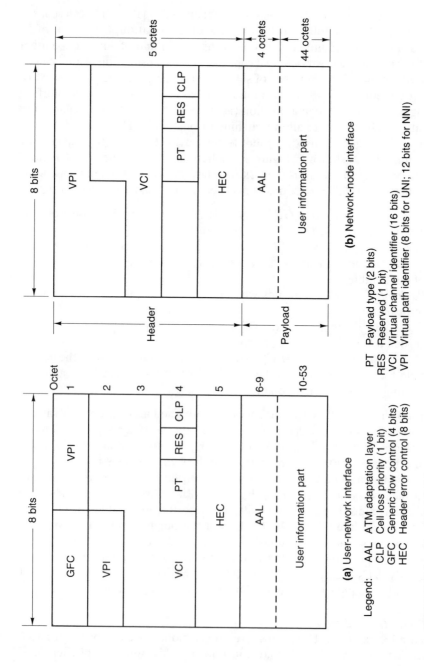

(a) User-network interface

(b) Network-node interface

Legend:
AAL ATM adaptation layer
CLP Cell loss priority (1 bit)
GFC Generic flow control (4 bits)
HEC Header error control (8 bits)

PT Payload type (2 bits)
RES Reserved (1 bit)
VCI Virtual channel identifier (16 bits)
VPI Virtual path identifier (8 bits for UNI; 12 bits for NNI)

FIGURE 13.4 The ATM Cell Structure

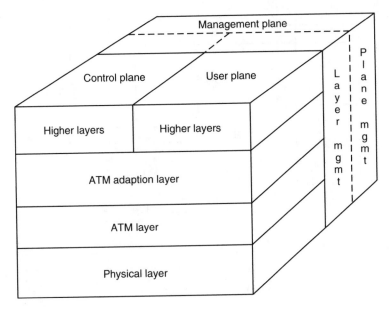

FIGURE 13.5   The BISDN ATM Protocol Reference Model

and the ATM adaptation layer [4]. The ATM adaptation layer (AAL) maps services into the information fields of generic ATM cells. The information field of the cell may then contain further network overhead to support the functions of the adaptation layer. These AAL functions, which are dependent on the type of service, provide end-to-end network features that are not included in the cell header functions. Higher layers provide support to certain services as well as signaling protocols. The physical layer provides the transmission infrastructure that carries the ATM cell stream, by mapping the cells into the transmission payload.

Two data rates have been standardized, at 155 Mb/s and 622 Mb/s, for the physical layer. Two options are identified by the CCITT for the physical interface, one based on the existing CCITT Synchronous Digital Hierarchy (SDH) and the North American Synchronous Optical Network (SONET), and the other on an ATM cell-based scheme. The only option that has been standardized is the SDH and SONET interface, as defined by the CCITT [24] and ANSI [25]. In the SDH-based approach, the structure is based on the STM-1 signal format, as shown in Figure 13.6. In this structure, the cell stream is mapped into the Container Type 4 (C-4), which occupies a 9-row-by-260-column space within the STM-1. Since each time slot is an 8-bit byte and the frame repetition rate is 8000 frames per second, the cell transfer rate is

$$9 \text{ rows} \times 260 \text{ columns} \times 8 \text{ bits/byte} \times 8000 \text{ frames/second} = 149.76 \text{ Mb/s}$$

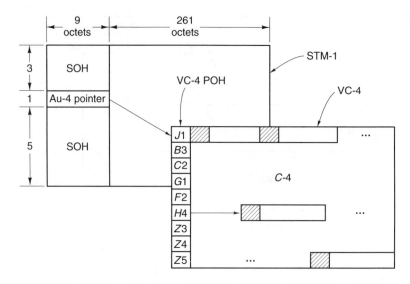

FIGURE 13.6   SDH-Based Structure of BISDN at 155 Mb/s (Adapted from [26])

The C-4 plus the path overhead is then mapped into a Virtual Container Type 4 (VC-4). In turn, the VC-4 plus a pointer are mapped into the STM-1. The only difference between this version of the STM-1 and that described in Chapter 8 is that here the C-4 contains ATM cells. To allow faster cell delineation, a pointer mechanism has been added that allows cell boundaries to be recovered on a frame by frame basis. This pointer is located in the H4 byte of the VC-4 path overhead, as indicated in Figure 13.6. In all other respects, the same SDH/SONET transmission standards described in Chapter 8 apply here. The CCITT recommends the use of STM-1 at both the network-node interfaces and the user-network interfaces. To achieve the higher bit rate of 622 Mb/s, four STM-1 signals can be byte interleaved to form an STM-4 frame. Alternatively, a concatenated STM-4c structure can be used that provides slightly higher payload capacity [26,27].

Example 13.1_____

Determine the effective payload capacity of the 622-Mb/s NNI rate for two cases, one based on the use of STM-4 and the other based on STM-4c.

Solution   In the case of the STM-4, the first nine columns of the STM-1 signal contain the section overhead (SOH) and the administration unit

(AU-4) pointer, while the tenth column contains the path overhead (POH). The payload capacity for the STM-1 is therefore

260 columns $\times$ 9 rows $\times$ 8 bits/byte $\times$ 8000 frames/sec = 149.760 Mb/s

The STM-4 payload is then four times that of the STM-1, or 599.04 Mb/s. For the case of an STM-4c, we know that the pointer of the first STM-1 in a STM-4c applies to all STM-1s in the concatenation. Thus the payload capacity of the STM-4c occupies 260 columns of the first STM-1 and 261 columns of the last three STM-1s, or

(260 columns $\times$ 9 rows $\times$ 8 bits/byte $\times$ 8000 frames/sec) + 3(261 columns $\times$ 9 rows $\times$ 8 bits/byte $\times$ 8000 frames/sec) = 600.768 Mb/s

---

13.4.3   ISDN Performance

The use of asynchronous transfer mode in ISDN introduces a number of performance issues that are different from those associated with conventional synchronous multiplexing. With ATM, user signals may be fixed or variable rate; in fact, the source signals can change bit rates in a variety of ways, for example, by simply being turned off and on, by varying in steps, or by varying continuously. With such bit rate variation allowed, the flow of information as seen by the ATM multiplexer will have peaks that could lead to excessive delays, variation in delays, and even lost information. Congestion control in ATM is provided through the use of buffering and by imposing constraints on the user signals. Buffer sizes are selected by trading off low probability of buffer overflow (and resulting cell loss) against acceptable delay. Violation of constraints placed on user data flow can result in penalties such as the dropping of cells, either automatically or in the event of congestion. Cells can also be lost or misdelivered due to transmission errors that corrupt the cell header. Transmission errors detected but uncorrected by the header result in discarding of that cell to protect against misdelivery. A misdelivered cell can occur, however, if a VCI is corrupted by transmission errors not detected by the HEC. Transmission errors undetected by the header can thus lead to the insertion of extraneous cells due to misdelivery. Variable delays in cell transmission lead to a form of jitter that requires some type of compensation at the receiver. As a result of these various ATM effects, new performance parameters are needed to account for the effects of cell loss, cell delay, and cell delay jitter on quality of service.

Today's packet-switched networks for data communications provide a means of detecting lost or corrupted packets and requesting retransmission of a particular packet. This procedure is effective for non-real-time services but impractical for real-time services such as telephony and television. With real-time services, lost or corrupted cells cannot be retransmitted so some means of mitigating their effect is needed. To compensate for cell loss due to congestion, the source information can be divided into high priority and low priority fields. High priority fields are identified as nondiscardable and low priority fields are marked as discardable. During periods of network congestion, discardable fields are subject to cell loss. The source coding and decoding algorithms must be able to operate effectively with or without the discardable fields. Lost or misdelivered cells caused by transmission errors can be minimized by use of error correction coding for the cell information field, beyond the error protection already inherent in the cell header. The dropping or insertion of cells can be detected by numbering the cells at the transmitter and monitoring the sequence of cells at the receiver. Cell delay jitter can be compensated at the receiver by use of a buffer, whose size is determined by the maximum jitter expected in the network. The functions implied by these various methods to protect against the effects of ATM are contained in the ATM adaptation layer.

Speech signals can be accommodated by ATM using either fixed bit rate algorithms, such as 64-kb/s PCM or 32-kb/s ADPCM, or variable rate techniques such as embedded ADPCM and speech interpolation. Whatever the coding scheme, the speech samples are converted to cells, where up to 4 bytes (the ATM adaptation layer) of the 48-byte payload can be used to specify coding type, sequence number, and the like, while the remaining 44 bytes are reserved for speech samples. To best match the bursty nature of speech with ATM, variable rate coding with speech interpolation has been the most common choice [28]. The most significant effect on speech from ATM processing is the loss of cells from transmission errors or congestion control. Several techniques have been proposed to recover from lost cells, including the use of substitution patterns, interpolation between samples, and cell discarding [29], with the last technique favored for low-bit-rate speech. The loss of cells due to congestion control can be compensated by discarding only lower priority cells; for example, pairs of cells can be formed in which the first cell contains the more significant bits and the second contains the less significant bits [30]. This scheme is similar to that of the voice packetization protocol given in CCITT Rec. G.764 and described in Chapter 4, but in ATM the packets are organized in cells rather than blocks. Using a coding scheme such as embedded ADPCM allows the rate to be varied and cells to be dropped, anywhere in the network, without coordination with the receiver. A cell loss rate of up to 5 percent in either embedded ADPCM or 64-kb/s PCM can be tolerated with no noticeable degradation in voice quality [31]. Delay effects on speech are also of concern in ATM networks, where studies have shown that constant bit rate services such as 64-kb/s PCM will experience mean delays on the order of 50 μs per node and variable bit rate services such as embedded ADPCM will experience delays of less than one ms per node [30].

Video coding over ATM networks presents some of the same problems and solutions just described for speech signals. Both constant and variable bit rate coders are possible with ATM, but most video coders use a compression algorithm that leads to a variable coding rate suitable for ATM. As suggested by CCIR Rep. 1240 [32], one possible approach is to provide a minimum bit rate guaranteed even during network congestion but also allow a maximum bit rate for critical picture sequences. The detection of lost and misdelivered cells, compensation of delay variation, and detection and correction of errors are performed by the ATM adaptation layer, using techniques similar to those used with speech.

Example 13.2

A video signal coded at a T3 rate (44.736 Mb/s) is transmitted via ATM using 48 bytes of information per cell. If the probability of cell loss is 10^{-8}, what is the mean time between cell losses?

Solution   The cell transmission rate for a T3 signal is

$$\text{cell rate} = \frac{44.736 \times 10^6 \text{ bps}}{(48 \text{ bytes/cell})(8 \text{ bits/byte})} = 116{,}500 \text{ cells/s}$$

The mean time between cell loss is then

$$\text{mean time} = \frac{1}{(\text{Prob. of cell loss})(\text{cell rate})} = 14.3 \text{ min}$$

13.5   Summary

Digital transmission now dominates long-haul, multi-channel telecommunication systems worldwide. Other applications have more recently emerged that indicate digital transmission as the choice for extension of various services to the individual user.

Used at the customer premise, networking multiplexers have facilitated the introduction of private digital transmission networks. The user maintains control over the network via intelligence inherent in the multiplexers. These networking multiplexers can be applied to local, metropolitan, and wide area networks. Standardized services based on packet switching have also emerged that support high-speed data communications in both private and public digital networks.

Personal communication networks are an extension of mobile, wireless communications that provide the individual access to a variety of telecommunication services from any location. Digital transmission in the form of low-bit-rate voice, spectrally efficient digital modulation, and time-division or code-division multiple access form the basis for emerging PCN. Both terrestrial and satellite based versions of PCN are possible. Terrestrial systems are based on a derivative of cellular radio systems, but with improved spectral efficiency through advanced digital transmission technology. Satellite systems employ satellites, either in geosynchronous or low-earth orbit, to provide connectivity to the mobile user.

The proliferation of digital transmission and switching has led to the development and implementation of integrated services digital networks. Narrowband ISDN provides a number of B channels at 64 kb/s plus a D channel at 16 kb/s directly to the user. Broadband ISDN introduces higher-bandwidth services suitable for video and many other wideband signals. To support these broadband as well as narrowband services, a new switching technology has been adopted, known as asynchronous transfer mode, based on a form of high-speed packet or cell switching. ATM can be carried via SONET or SDH transmission, as already prescribed by CCITT standards.

References

1. E. B. Carne, "New Dimensions in Telecommunications," *IEEE Comm. Mag.* 20(January 1982), pp. 17–25.
2. J. P. Cavanagh, "Applying the Frame Relay Interface to Private Networks," *IEEE Comm. Mag.,* March 1992, pp. 48–64.
3. W. Giguere, "New Applications of Wideband Technology," *1990 International Conference on Communications,* pp. 324.1.1–324.1.3.
4. CCITT Blue Book, vol. III.7, *Integrated Services Digital Network (ISDN) General Structure and Service Capabilities* (Geneva: ITU, 1989).
5. CCITT Blue Book, vol. VI.10, *Digital Subscriber Signalling System No. 1 (DSS 1), Data Link Layer* (Geneva: ITU, 1989).
6. ANSI T1.606-1990, *Integrated Services Digital Network (ISDN)—Architectural Framework and Service Description for Frame-Relaying Bearer Service,* 1990.
7. W. R. Byrne and others, "Evolution of Metropolitan Area Networks to Broadband ISDN," *IEEE Comm. Mag.,* January 1991, pp. 69–82.
8. D. J. Goodman, "Trends in Cellular and Cordless Communications," *IEEE Comm. Mag.,* June 1991, pp. 31–40.
9. EIA/TIA Interim Standard IS-54, "Cellular System Dual Mode Mobile Station-Base Station Compatibility Specification," May 1990.

10. D. L. Schilling and others, "Spread Spectrum for Commercial Communications," *IEEE Comm. Mag.*, April 1991, pp. 66–79.
11. A. J. Viterbi, "Wireless Digital Communication: A View Based on Three Lessons Learned," *IEEE Comm. Mag.*, September 1991, pp. 33–36.
12. D. L. Schilling and others, "Broadband CDMA for Personal Communications Systems," *IEEE Comm. Mag.*, November 1991, pp. 86–93.
13. N. J. Jayant, "High-Quality Coding of Telephone Speech and Wideband Audio," *IEEE Comm. Mag.*, January 1990, 10–20.
14. P. Wood, "Mobile Satellite Services for Travellers," *IEEE Comm. Mag.*, November 1991, pp. 32–35.
15. A. D. Kucar, "Mobile Radio: An Overview," *IEEE Comm. Mag.*, November 1991, pp. 72–85.
16. J. H. Lodge, "Mobile Satellite Communications Systems: Toward Global Personal Communications," *IEEE Comm. Mag.*, November 1991, pp. 24–30.
17. J. L. Grubb, "The Traveller's Dream Come True," *IEEE Comm. Mag.*, November 1991, pp. 48–51.
18. CCITT Blue Book, vol. III.8, *Integrated Services Digital Network (ISDN)— Overall Network Aspects and Functions, ISDN User-Network Interfaces* (Geneva: ITU, 1989).
19. CCITT Blue Book, vol. VI.11, *Digital Subscriber Signalling System No. 1 (DSS 1), Network Layer, User-Network Management* (Geneva: ITU, 1989).
20. D. T. Huang and C. F. Valenti, "Digital Subscriber Lines: Network Considerations for ISDN Basic Access Standard," *Proc. IEEE* (February 1991): 125–144.
21. B. Bosik and S. Kartalopoulos, "A Time Compression Multiplex System for a Circuit Switched Digital Capability," *IEEE Trans. on Comm.*, vol. COM-30, no. 9, September 1982, pp. 2046–2052.
22. ANSI T1.601-1991, *Integrated Services Digital Network (ISDN)—Basic Access Interface for Use on Metallic Loops for Application on the Network Side of the NT Layer (Layer 1 Specification)*, January 1991.
23. M. Kawarasaki and B. Jabbari, "BISDN Architecture and Protocol," *IEEE Journal on Sel. Areas in Comm.* (December 1991): 1405–1415.
24. CCITT Blue Book, vol. III.4, *General Aspects of Digital Transmission Systems; Terminal Equipment* (Geneva: ITU, 1989).
25. ANSI T1.105-1990, *Digital Hierarchy—Optical Interface Rates and Formats Specification*, 1990.
26. J. Anderson, "Progress on Broadband ISDN User-Network Interface Standards," *1990 Global Telecommunications Conference*, pp. 900.4.1–900.4.7.
27. H. Bauch, "Transmission Systems for the BISDN," *IEEE Mag. of Lightwave Telecommunication Systems*, August 1991, pp. 31–36.
28. N. Kitawaki and others, "Speech Coding Technology for ATM Networks," *IEEE Comm. Mag.*, January 1990, pp. 21–27.
29. J. Suzuki and M. Taka, "Missing Packet Recovery Techniques for Low-Bit-Rate Coded Speech," *IEEE Journal on Sel. Areas in Comm.* (June 1989): 707–717.

30. K. Sriram, R. S. McKinney, and M. H. Sherif, "Voice Packetization and Compression in Broadband ATM Networks," *IEEE Journal on Sel. Areas in Comm.* (April 1991): pp. 294–304.
31. K. Kondo and M. Ohno, "Variable Rate Embedded ADPCM Coding Scheme for Packet Speech on ATM Networks," *1990 Global Telecommunications Conference,* pp. 405.3.1–405.3.5.
32. Reports of the CCIR, 1990, Annex to Vol. XII, *Television and Sound Transmission* (Geneva: ITU, 1990).

Index